Springer Tracts in Mechanical Engineering

About this Series

Springer Tracts in Mechanical Engineering (STME) publishes the latest developments in Mechanical Engineering - quickly, informally and with high quality. The intent is to cover all the main branches of mechanical engineering, both theoretical and applied, including:

- Engineering Design
- Machinery and Machine Elements
- Mechanical structures and stress analysis
- Automotive Engineering
- Engine Technology
- Aerospace Technology and Astronautics
- Nanotechnology and Microengineering
- Control, Robotics, Mechatronics
- MEMS
- Theoretical and Applied Mechanics
- Dynamical Systems, Control
- Fluids mechanics
- Engineering Thermodynamics, Heat and Mass Transfer
- Manufacturing
- Precision engineering, Instrumentation, Measurement
- Materials Engineering
- Tribology and surface technology

Within the scopes of the series are monographs, professional books or graduate textbooks, edited volumes as well as outstanding Ph.D. theses and books purposely devoted to support education in mechanical engineering at graduate and post-graduate levels.

More information about this series at http://www.springer.com/series/11693

Jerzy A. Sładek

Coordinate Metrology

Accuracy of Systems and Measurements

Springer

Prof. Jerzy A. Sładek
Laboratory of Coordinate Metrology
Cracow University of Technology
Krakow
Poland

ISSN 2195-9862 ISSN 2195-9870 (electronic)
Springer Tracts in Mechanical Engineering
ISBN 978-3-662-56924-5 ISBN 978-3-662-48465-4 (eBook)
DOI 10.1007/978-3-662-48465-4

Springer Heidelberg New York Dordrecht London

Softcover reprint of the hardcover 1st edition 2016

Printed on acid-free paper

Springer-Verlag GmbH Berlin Heidelberg is part of Springer Science+Business Media
(www.springer.com)

Contents

Chapter 1
Introduction

Abstract This chapter describes coordinate metrology as an important scientific research area originating from old metrology of geometric quantities. It defines the field and describes the history of the first studies; moreover, it presents applications in production engineering, quality systems, and research works. The first constructions of coordinate measuring machines and their development up to the modern state are discussed. Innovative solutions of contact and noncontact systems are included along with multisensor systems and redundant systems such as measuring arms and laser trackers. Coordinate metrology is also based on imaging and measuring systems using structured light, photogrammetry, laser triangulation, systems based on the measurement of beam returning time (TOF, time of flight), and also computed tomography (CT) and magnetic resonance imaging (MRI). The problem of the coordinate systems' accuracy and monitoring in accordance with the current standards and recommendations is touched upon, including a new trend in this field, the matrix method (MM) developed by the author. The broad and difficult issue of the measurement accuracy and methods of its assessment including simulation methods (so-called virtual CMMs) is also touched upon. The role of the Laboratory of Coordinate Metrology at Cracow University of Technology (LCM CUT) created by Professor Jerzy A. Sładek in studies on the development of coordinate metrology is presented at the end of chapter.

Manufacturing of better and better products and expanding the boundaries of knowledge are connected to the development of the science of measurements, coordinate metrology. Coordinate metrology provides a scientific basis for carrying out measurements and three-dimensional geometric object imaging with the use of coordinate measuring systems. The ranges are from micro- to nanodimensions. The accuracy and realization time are matched to the manufacturing rhythm. Because of new technologies and the development of science, the scope of applications of these systems is constantly growing and the measurement accuracy is increasing.

Origins of coordinate metrology may be found in works of French mathematicians, working independently of each other, such as Pierre de Fermat (1601–1665) and Rene Descartes (1596–1650), to whom we owe the idea of using the coordinate system for description of the material object—reference system relation.

J.A. Sładek, *Coordinate Metrology*, Springer Tracts in Mechanical Engineering,
DOI 10.1007/978-3-662-48465-4_1

Measurement of three-dimensional objects marks the section of the space with the geometry of the measured object given as a point coordinate in a reference coordinate system. This discovery resulted in the description of dimensions and their relations, and on the other hand, it joined geometry with algebra. Together with the development of analytic geometry, it turned out that the coordinate system allows us to study geometric properties in the field of vector volume. It was discovered in the nineteenth century by German mathematician Hermann Gunther Grassmann (1809–1877). This resulted in introducing the axiomatic, and creating thus the scientific, basis of modern coordinate metrology [82]. The basis developed by Grassmann could not directly lead to methods and measuring tool construction, as the level of technical development limited them. Although the first coordinate measuring systems allowed the determination of point coordinates, in practice it came down to the realization of length measurements in the direction of the three coordinates. The first coordinate measuring machines (CMMs) were developed by the Scottish Ferranti (1956) and the American Moore Tool Company (1957). Also the constructions carried out by Swiss SIP (1961) and by engineer Franco Sartorio, cofounder of Italian DEA (1962), shouldn't be forgotten. However, only the development of computer technology and computer science, which occurred in the late 1960s, made it possible to use fully the basis of vector geometry and Gaussian expansion calculation. This, combined with the technique of identification of points on the surface of the measured object (which was related to the design of contact probes), led to construction of the first numerically controlled coordinate machine. It was done by Zeiss, which developed the UMM 550 measuring machine and introduced it to the market in 1973. It can be concluded that Zeiss began modern coordinate technology and its industrial application.

Coordinate metrology also covers imaging technologies based on scanning surfaces and whole objects with the use of optical technique and computed tomography. Without a coordinate measuring technique it is currently impossible to imagine production engineering, including rapid prototyping, reverse engineering, and industrial design or biomedical systems technique. Moreover, this technique is widely used in inventory of works of art and in monument protection (virtual museums), architecture, and development of building construction. The innovations in the field of electronics, mechanics, mechatronics, optics, and computer science are widely used here. The innovations could not be used without the scientific development in the field of coordinate metrology [20, 79, 80] which took over and expanded the traditional metrology of geometric quantities, becoming a platform for most modern technological solutions. Coordinate metrology is constantly developing and exceeding boundaries of knowledge through new calibration technologies and methods of assessment of measurement accuracy.

The basis of coordinate measurement techniques are:

Coordinate measuring machines

- Contact operating in 3D system
- Optical operating in 2D system
- Multisensor 2D/3D

Coordinate redundant systems, which include:

- Articulated arm coordinate measuring machine (AACMMs)
- Laser tracker systems (LTSs)

Noncontact coordinate measurement and imaging systems, which include:

- Systems using structured light
- Photogrammetric systems
- Systems performing laser triangulation
- Systems based on the measurement of beam returning time (TOF)
- Computed tomography (CT) systems
- Magnetic resonance (MRI) systems

Contact and noncontact systems of microgeometry and surface topography measurement.

Figure 1.1 presents the coordinate measuring systems. This figure shows not only technical solutions for various application fields and different sizes of measured objects but also their common characteristic, which is nowadays a multi-sensor character of these systems.

Coordinate measuring systems are today's most technologically advanced solutions used in production engineering. Practically every new installation

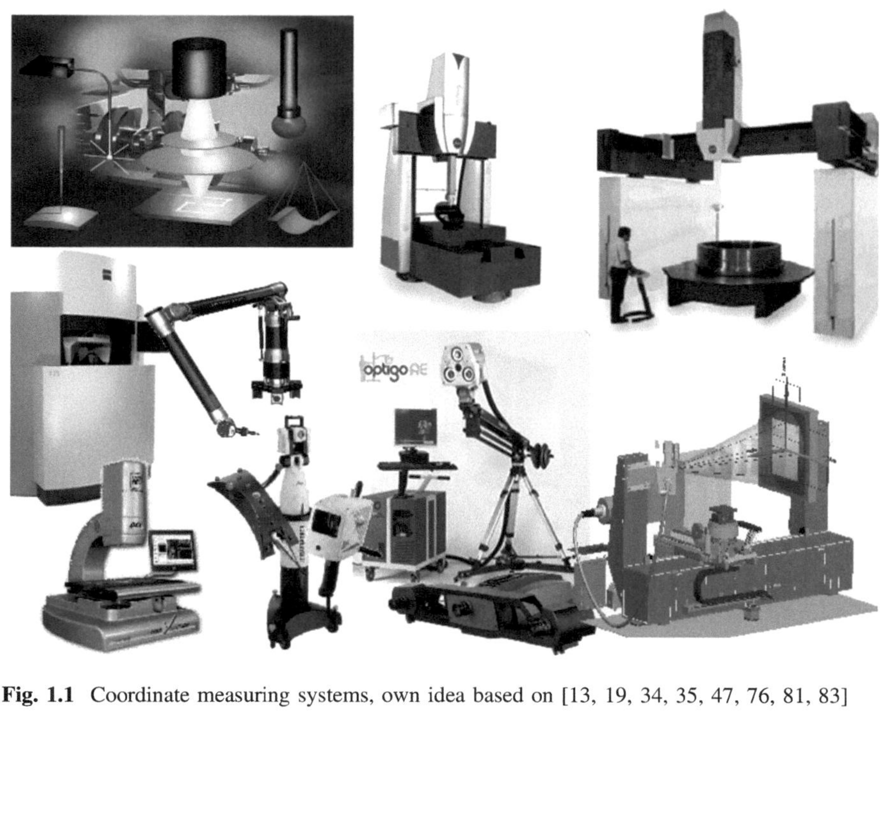

Fig. 1.1 Coordinate measuring systems, own idea based on [13, 19, 34, 35, 47, 76, 81, 83]

determines a high level of development and innovativeness of an industrial plant or research laboratory. It is assumed to consider the number of coordinate systems used as a measure of the level of technological development not only of a plant but also of a whole region or country [67]. Only the most advanced companies are able to adapt this technology. In our country, despite continuous development, the level of saturation with coordinate measurement technology is still low, and therefore investments supported by European Union resources are indispensable. At the same time the demand for specialists in the field of coordinate metrology grows, along with the importance of calibration laboratories specialized in this technique.

At the beginning of 2014, as maximum permissible errors of the coordinate system indicated in three-dimensional volume $E = 0.1 + L/1000$ µm should be taken [35, 42, 62] (where L is the measured length given in millimeters). Accuracy limits are determined here by the technology of both manufacturing and verification of standards, which are necessary to link the measuring system with the measurement unit. Figure 1.2 presents one of the most accurate coordinate machines in the world, Dutch IBS's Isara 400, with potential measurement accuracy better than 100 nm in the total measuring volume (400 × 400 × 100) for which the calibration standard and technology are being searched [8, 35, 42].

From the manufacturing engineering point of view the key role is measurement accuracy, which should be matched to manufacturing tolerance. This problem was already noted by Professor Georg Berndt (1880–1972) in the first half of the twentieth century. He formulated "the golden rule of measurement technique" (*Die Mefiungsicherheit beim Nachprufen einer Toleranz—T nur ein Zehntel bis ein Funftel von—T betragen soll*: measurement uncertainty with specified tolerance T should be from one tenth to one fifth of this tolerance [translated by the author; 4]). This rule not only set out a new direction of the development of measuring tool construction, but also opened a wider view on the process of measurement accuracy assessment. Since the end of the 1960s measurement accuracy assessment has been considered a complex process in which the accuracy of the measuring tool (device, system) is only one component [5]. When developing this issue for coordinate

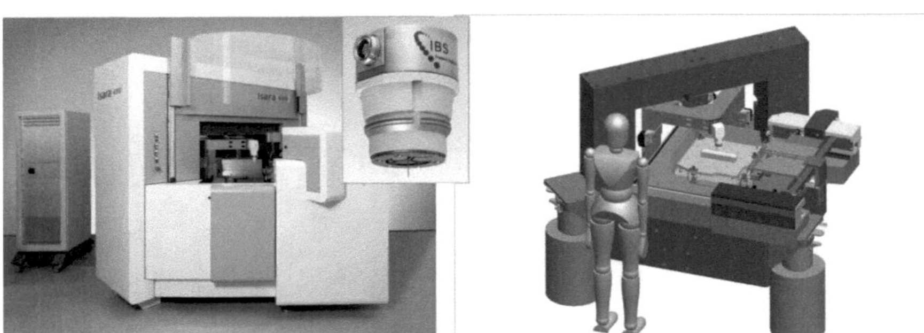

Fig. 1.2 Nanocoordinate measuring machine, the system of Dutch company IBS—ISARA 400—with contact probe Triscelion [8, 21]

measurement, the difference in relation to the classical length measurement was noticed. The measurement model is also different and because of that the description of the error field and sources of its variability are also different. Therefore two scientific problems were identified that were closely related to each other: the accuracy of coordinate measuring systems and the accuracy of coordinate measurements.

The first scientific problem was related to the calibration task (the determination of indication error by connection with length unit) and also with the task of development of the methodology that allows current assessment of the coordinate system accuracy and the monitoring of its accuracy in the total exploitation time period together with the task of its maintenance in certain limits. The accuracy assessment of the coordinate system performed on the basis of length measurement is the result of works such as [40] signed later by the working group "Me" CIRP, under the direction of Stoferle [6] and work of the Pfeifer group [45]. These teams introduced terms including measuring tool uncertainty (understood accordingly to DIN) as a one-, two- and three-coordinate, including both random errors and uncorrected systematic errors. They also proposed to express it in the form of a linear function. Initially a series of recommendations signed by Coordinate Measuring Machine Manufacturers Association (CMMA), *Verein Deutscher Ingenieure und Verband der Elektrotechnik* (VDI/VDE), and then national standards ASME, ONORM, BS, and JIS were developed [1, 3, 7, 44, 78]. They finally take the form of the ISO 10360 series standard and operate with terms of indication error, maximum errors, and maximum permissible errors (MPE) [23]. Because of this approach, we can now use a coherent system for measuring system accuracy assessment. It is possible to fulfill requirements of ISO 9001 standards [22] regarding the monitoring of the measuring equipment. The author of this book who is also the inventor of the original matrix method [52, 55], presents a different view on the issue of coordinate system accuracy assessment, considering the result of a system indication test in selected reference points as the basis of information about coordinate system accuracy.

The second scientific problem, which was formulated in the field of coordinate metrology, is the assessment of measurement accuracy. This problem has not yet found an effective solution, although it has been undertaken by many researchers, as evidenced by a large number of publications. Thus there is still no complete description of the accuracy assessment problem. There is no comprehensive collection of information of methods available for users that could help in determination of the efficient way to solve the task of coordinate measuring system accuracy assessment. All of this motivated the author to undertake this issue and to describe it based on the scientific work carried out on his own and by the team directed by him.

The problem of coordinate measurement accuracy assessment arises strongly when there is the necessity to determine the conformity of measured object dimensions with the specification, as described in ISO 14253-1. A broader view on this issue leads to determination of the measuring system's ability to realize the planned task. It should be done by assessing a relation of real measurement uncertainty to manufacturing tolerance. Deciding upon dimensional compliance with geometric product specification (GPS) is currently becoming one of the main tasks for manufacturing engineering, which with a simultaneous economic imperative, carries the risk of making the wrong decisions. Their effects can be seen in the form of numerous product claims. This is especially noticeable in the field of such industries as: motorization, aerospace, telecommunication, home appliances, or in the intensively developing medical devices industry. The important role here is that of the possibility of coordinate measurement accuracy determination. In the case of coordinate measuring techniques (CMTs) it is an especially difficult task that is not always clear. Therefore, both users and manufacturers of this technology ignore this problem, taking the accuracy of a measuring device as a measurement accuracy, determined for a simple length-measuring task. It is then wrongly calculated directly from the equation of maximum permissible error. The accuracy designated in that way is significantly different from the real accuracy of the realized task and leads directly to wrong decisions in the production field. This is confirmed by evaluations carried out by the Aukom Association realized for German industry [2, 28, 33] as well as the author's observations in the domestic market [55, 64, 67]. Therefore, the issue of the accuracy of realized measurement has a crucial importance in the case of coordinate measuring technique and the accuracy of measuring devices (given as MPE) is only one of the measurement accuracy components. It is known that the measure of measurement accuracy is its uncertainty, thus considering the problem of coordinate measurement uncertainty it was assumed that coordinate measurement is a classic indirect measurement. After analyzing it, the two trends of theoretical approach to this issue may be noted.

The first assumes that the simplest length measurement described as the difference between two measuring point coordinates is the indirect measurement, having its source in relation to traditional length and angle metrology. As an example of indirect measurement in traditional metrology the measurement of the cone convergence with the use of measurement rolls can be taken. It consists of direct length measurements and of the calculated angle of the cone [23, 25–27, 29–32]. It is similar to determining the diameter of a circle as a length measured from point coordinates. That approach leads to searching for the solution of the accuracy assessment method within the classical approach of complex standard uncertainty calculation. The uncertainty can also be estimated with use of some kind of simplification, assuming that the variation area of the given independent components is described, for example, with normal distribution. Then the estimation of the standard uncertainty can be done by calculating the product of the width of the given area corresponding to the possible indication error (obtained for a given length, e.g., from a calibration certificate) and the inverse of the assumed coverage factor. Of course, it should be related to the area of maximum errors of the coordinate system used [16–18, 46].

The second trend, as before, allows the conclusion that the coordinate measurement is an indirect measurement, assuming that the determination of measuring point coordinates is the direct measurement. This, not the length measurement, should be the basis for determination of the measuring system accuracy. The set of standard coordinates, not the standard lengths, should be the reference. This completely changes the construction of a standard understood as a set of individual equally spaced points. The consequence of this approach is the assumption that at every point of contact the error accumulation appears. Therefore, the measure of these errors is given as a vector that consists of the sum of CMM geometric errors vector, probe head errors, and errors of contact. It has its systematic and random part—described with an uncertainty ellipsoid [41, 79, 80]—and forms the basis for a new concept developed by the author, the vector model of coordinate measurement accuracy [49, 55, 56]. The problem of measurement uncertainty propagation is also connected with this. It is done by compensatory algorithms used to calculate ideal elements, shape parts measured in multiple points. Undoubtedly, contributions of Lotze, Hartmann and Teichmann [12, 14, 15, 36–39, 75] should be distinguished, especially their development of this issue for the two-dimensional case, which was subsequently developed by the author for the three-dimensional case [55, 56]. The presented second trend of measurement accuracy assessment is also represented by the matrix method developed by the author [51–53, 55, 66, 69, 70].

This method is independent of the method of measuring point identification. It can thus be applied both in contact measurement with CMM and in noncontact measurement, for example, in a computed tomography system. Standards are used in the MM method with reference shape elements in the form of balls, holes, or grids virtually created from points in the examined volume with the use of laser trackers and laser tracers.

The consequence of such an approach is modern simulation models, virtual machines, the latest tool for uncertainty estimation, developed in LCM CUT, using artificial neural networks [50, 51, 53–55, 65, 68] or the Monte Carlo method [10, 11, 62, 73]. An issue of considerable importance is the unique concept of modeling of the very recently popular redundant systems, articulated arm coordinate measuring machines, developed as well in LCM CUT under [43, 71, 74]. It shows that CMM modeling is not only limited to measuring machines, but it may be applied to other systems. This is important because it opens new research areas that may also include optical, laser, or tomography systems. Therefore, it can be concluded that the problematics of development and, above all, dissemination of methods of coordinate measurement accuracy assessment are key tasks for the development of metrology used in production engineering. This is confirmed by results of the author's long-term research [55, 57–59, 61, 64, 67, 72]. It can be concluded from them that the industry is waiting for effective and clear methods for measurement accuracy assessment and their absence is one of the barriers of high-tech industry development. This brought additional inspiration for the development of this book. Here, the methods tested in laboratory conditions are discussed and the current state of the world's research on this topic is presented.

This book is organized as follows. Chapter 2 describes the theoretical basis together with the author's concept of identification of reproducibility error of measuring points, as the basis of the matrix method of CMM error identification and the assessment of their accuracy. Chapter 3 describes classical (nonsimulation) methods of accuracy assessment including ISO 15530 [24] standard requirements and procedures used in the Laboratory of Coordinate Metrology at Cracow University of Technology, also with selected examples of their use, included in Annex A. Chapter 4 gives the methods for identification of coordinate measuring machine errors with the use of plate standards, laser interferometers, or laser tracers together with descriptions of software such as Kal-Kom or TraC-Cal, with examples given in Annexes B, C, and D. This chapter also describes the method of software correction of CMS accuracy based on determination of CAA matrix components and the basis for the construction of models and virtual coordinate systems. Chapter 5 describes their use in simulation methods for measurement uncertainty assessment. Chapter 6 expounds on the problem of monitoring of CMS as an important task required for maintaining, together with a system of accredited laboratories, measurement traceability. Chapter 7 describes the directions for further research on measurement accuracy defined by the author of this book.

The book contains not only the author's own work, but also work done under his inspiration and under his direction in the research team of the Laboratory of Coordinate Metrology at Cracow University of Technology (LCM CUT). This work was carried out in the form of scientific research and development projects (grants) and PhD theses (in which the author was the supervisor) by: PhD Marcin Krawczyk, PhD, Grzegorz Sokal, PhD Robert Kupiec, PhD Michal Muzyka-Zmudzki, PhD Magdalena Kupiec, PhD Ksenia Ostrowska, PhD Krzysztof Gawlik, and PhD Adam Gąska and also by PhD students and graduate students from the Faculty of Mechanical Engineering of Cracow University of Technology tutored by the author.

Cracow University of Technology has over 30 years of experience in the field of coordinate metrology and technology. In 1979 the first coordinate measuring machine and the first laser interferometer used in Polish universities, but above all one of the world's most accurate measuring machines, the PMM 12106 Leitz Messtechnik, were installed here. LCM CUT is constantly developing, undertaking new research tasks, and gaining new equipment, for example, the unique LaserTracer system of Etalon AG.

In scientific terms LCM CUT continues scientific metrology traditions at the Cracow University of Technology, and has important and pioneering achievements. Here the first PhD thesis and afterwards postdoctoral thesis dedicated to coordinate metrology were completed. By the end of 2014 in LCM CUT 12 PhD theses in that field were prepared. Several major scientific research and development projects were also realized, including European ones [9, 48, 60, 63, 77].

LCM CUT has the status of a laboratory accredited by the Polish Centre for Accreditation (Accreditation no. AP 132) in the field of calibration of coordinate systems and objects and elements with use of coordinate methods. Its calibration and measurement capabilities confirmed by comparative tests allow the inclusion of LCM CUT to the best coordinate measurement laboratories in Europe.

In conclusion, the author wants to thank the scientific reviewers, Professors Eugeniusz Ratajczyk and Wladyslaw Jakubiec, for their valuable comments which made this book better. He also thanks to the whole team and permanent coworkers of the Laboratory of Coordinate Metrology of the Faculty of Mechanical

Engineering at Cracow University of Technology for many years of very good and productive cooperation. This work is also one of its effects.

The author wants this book to expand and disseminate knowledge in the field of coordinate metrology. It will fulfill its role when it is used not only by undergraduate and PhD students, but also by engineers and metrologists working in industry. Because of them the examples of methods of accuracy assessment based on works carried out in LCM CUT as an accredited calibration laboratory are also placed herein. By describing solutions of virtual coordinate systems the state of worldwide research on this important issue is shown. Another objective is to inspire young scientists to take up and develop this interesting and innovative theme, and in the case of metrologists from industrial laboratories, to implement the system of measurement accuracy assessment.

References

1. American National Standard ASME B89.1.12-90—Methods for the Performance Evaluation of Coordinate Measuring Machines. ASME, New York (1991)
2. Aukom, www.aukom.com
3. British Standards Institution. Norma BS 6808: Coordinate Measuring Machines. British Standard Institute, London
4. Berndt, G.: Grudlagen und Gerate technischer Langenmessungen. Julius Springer, Berlin (1921)
5. Berndt, G., Hultzsch, E., Weinhold, H.: Funktionstoleranz und MeBunsicherheit. Wissenschaftliche Zeitschrift der Universitat Dresden **17**, 465–471 (1968)
6. Committee "Me", A proposal for defining and specifying the dimensional uncertainty of multi-axis measuring machines. Annals of the CIRP, vol. 27/2/1978
7. Coordinate Measuring Machine Manufacturers Association: Norm CMMA: accuracy specification for coordinate measuring machines. CMMA, London, 2nd edn (1989)
8. Donker, R., Widdershoven, I., Spaan, H.: ISARA 400: a large measurement volume ultra-precision CMM. In: Sładek, J., Jakubiec, W.: Advances in Coordinate Metrology, Monograph, pp. 50–56. University of Bielsko-Biala. (2010)
9. EVIGEM—European Virtual Institute for geometry measurements EVIGeM)—grouping best research facilities in the field of geometric metrology where the only one representative of Poland is Laboratory of Coordinate Metrology (2002–2007). The EVIGEM purpose is to provide the knowledge in the field of geometric measurements, calibration services and to undertake technical and scientific issues in EU countries
10. Gąska, A.: Modeling of accuracy of coordinate measurement with use of Monte Carlo Method. Ph.D. theses, Cracow University of Technology, Faculty of Mechanical Engineering, 2011 (Advisor–Prof. J.A. Sladek)
11. Gąska, A.: Different uses of Monte Carlo Method in coordinate metrology. In: Xth International Scientific Conference Automation in Production Planning and Manufacturing, Zilina, Maj 2009
12. Geise, G., Hartmann, M.: Geometrische Aspekte bei Ausgleichproblemen in der Koordinatenmesstechnik. Konferenz INFERT 82, Dresden 1–2 Sept 1982
13. GOM Optical measuring Techniques, www.gom.com
14. Hart, H., Lotze, W., Woschni, E.G.: Messgenauigkeit. Verlag Technik, Berlin (1987)
15. Hartmann, W.W., Geise, G.: Displaydarstellungen als Entscheidungshilfe fur Messauswertungsstrategien in der Koordinatenmesstechnik. Feingeratetechnik 1984/1

16. Hernla, M.: Aufgabenspezifische MeBunsicherheit bei Koordinatenmessung TM 64 7/8, pp. 286–293. Oldenburg Verlag (1997)
17. Hernla, M. Unsicherheiten einfach abschaetzen. QZ Qualitaet und Zuveraessigkeit, 45 (2000)
18. Hernla, M.: Messumsicherheit bei Koordinatenmessungen. Expert Verlag (2007)
19. Hexagon Metrology. www.cognitens.com
20. Hocken Robert, J., Pereira Paulo, H.: Coordinate Measuring Machines and Systems, 2nd edn. CRC Press, Cleveland (2011)
21. IBS Precision Engineering. www.ibspe.com
22. ISO 9000: Series of standards for quality management and quality assurance system
23. ISO 10360 1–12 Geometrical Product Specifications (GPS)—Acceptance and reverification tests for coordinate measuring machines (CMM)
24. ISO 15530 Geometrical product specifications (GPS)—Coordinate measuring machines (CMM): Technique for determining the uncertainty of measurement
25. Jakubiec, W.: Analityczne wyznaczanie niepewności pomiaru we współrzędnościowej technice pomiarowej.(Analilitical evaluation of the measurement uncertainty in coordinate measurement technique). University of Bielsko-Biała, Bielsko-Biała. Scientific Books 22 (2008)
26. Jakubiec, W., Starczak, M.: EMU—friendly software for estimation of measurements uncertainty for CMM. In: 8th International Symposium on Measurement and Quality Control in Production, Erlangen 2004. VDI Verlag GmbH, Dusseldorf (2004)
27. Jakubiec, W., Starczak, M.: Metodyka wyznaczania niepewności pomiarów współrzędnościowych (Methodology of evaluation of the coordinate measurements uncertainty). In: VIth Internationa Scientific Conference Coordinate Measuring Technique
28. Jakubiec, W., Weckenmann, A.: Metro-e-learn—European project for manufacturing metrology. In: Xth National, 1st International Scientific-Technical Conference Metrology in Production Engineering. Cracow University of Technology, Cracow (2003)
29. Jakubiec, W., Plowucha, W., Starczak, M.: The estimation of coordinate measurements uncertainty in industrial conditions. In: VIIth International Scientific Conference Coordinate Measuring Technique. University of Bielsko-Biała, Bielsko-Biała. Scientific Books 22 (2006)
30. Jakubiec, W., Plowucha, W., Starczak, M.: EMU—oprogramowanie do analitycznego wyznaczania niepewności pomiarów współrzędnościowych (Software for analytical evaluation of the coordinate measurements uncertainty). In: Sładek, J., Jakubiec, W. (eds.) Advances in Coordinate Metrology, pp. 83–89, Monograph. University of Bielsko-Biała, Bielsko-Biała (2010)
31. Jakubiec, W., Plowucha, W., Starczak, M., Wizner, M.: Wdrażanie oprogramowania do analitycznego wyznaczania niepewności pomiarów współrzędnościowych (Implementation of the software for analytical evaluation of the coordinate measurements uncertainty). In: Sładek, J., Jakubiec, W. (eds.) Advances in Coordinate Metrology, Monograph, pp. 90–97. University of Bielsko-Biała, Bielsko-Biała (2010)
32. Jakubiec, W., Plowucha, W., Starczak, M.: Analytical evaluation of the coordinate measurements uncertainty. In: Sładek, J., Jakubiec, W. (eds.) Advances in Coordinate Metrology, Monograph, pp. 169–176. University of Bielsko-Biala (2010)
33. Keferstein, C., Marxer, M., Jakubiec, W., Weckenmann, A., Beetz S.: EUKOM—European training for coordinate metrology. In: 8th International Symposium on Measurement and Quality Control in Production, Erlangen. VDI Verlag GmbH, Duesseldorf (2004)
34. Leica Geosystems AG. www.metrology.leica-geosystems.com
35. Leitz. www.leitz-metrology.com
36. Lotze, W.: Prufkorper fur Koordinatenmessgeraten. Feingeratetechnik 30 (1981), und VDI-Z 123, pp. 154–155 (1981)
37. Lotze, W.: Rechnergestutzte Koordinatenmesstechnik. Werkstatt und Betrieb 113 1980/6, pp. 391–395
38. Lotze, W., Teichmann, U.: Einfluss von Gestalt—und Lageabweichungen auf die Unsicherheit der Rechnergestutzehn Koordinatenmessung. Feingeratetechnik **8**, 139–343 (1976)

39. Lotze, W., Teichmann, U.: Genauigkeit und Prufung von Koordinatenmessgeraten. Feingeratetechnik **35**/86, 339–342
40. Matthias, E., Schultschik, R.: Definition und Beschreibbarkeit der dreidimensionalen Messunsicherheit. Bericht von ETH Zurich IWF (1976)
41. Nawara, L., Kowalski, M.: Analiza błędów w pomiarach przestrzennych (Analysis of errors in spatial measurements). Mechanik **12**/80, 674–679
42. Neugebauer, M.: Precision size and form measurements with a micro CMM F25. In: Sładek, J., Jakubiec, W. (eds.) Advances in Coordinate Metrology, Monograph, pp. 44–49. University of Bielsko-Biala (2010)
43. Ostrowska, K.: Accuracy assessment method for measurements done with the use of articulated arm coordinate measuring machines. Ph.D. theses, Cracow University of Technology, Faculty of Mechanical Engineering, 2010 (Advisor–Prof. J.A. Sladek)
44. ONORM M 1380–1386—Koordinatenmesstechnik Vornormen
45. Pfeifer, T., Bambach, M.: Definition und Prufung von Kriterien zur Bestimmung systematischer und zufalliger Fehler von Dreikoordinatenmessgeraten. Forschungsberichte des Landes Nordhein-Westfalen No. 2856
46. Pressel, H.G.: Genau messen mit Koordinatenmessgeraeten. Expert Verlag, Renningen-Malmsheim (2003)
47. Romer. www.romer.eu
48. Sitnik, R., Sładek, J., et al.: Opto-numeryczny system do pomiaru elementów geometrycznych zintegrowany z Współrzędnościową Maszyną Pomiarową (Opto-numeric system integrated with Coordinate Measuring Machine for measuring geometrical elements). Research grant no. 3 T10C 010 29
49. Sładek, J.: Assessment of the accuracy of contact probe heads used in coordinate measuring machines. Ph.D. theses, Cracow University of Technology, Faculty of Mechanical Engineering (1990)
50. Sładek, J.: Strategy of applying coordinate metrology in quality assurances systems with regards to a measuring accuracy. In: Ford, D.G., Postlethwaite, S.R. (eds.) Laser Metrology and Machine Performance, III, pp. 497–508. Computational Mechanic Publications, Southampton–Boston (1997)
51. Sładek, J.: Modelowanie współrzędnościowych maszyn pomiarowych – macierzowa metoda identyfikacji błędów (Modeling of coordinate measuring machines—matrix method for errors identification). VIII Konferencja Naukowo-Techniczna pt. Metrologia w technikach wytwarzania maszyn, pp. 437–444. Szczecin University of Technology, Szczecin (1999)
52. Sładek, J.: Errors identification and measurement accuracy assessment of coordinate measuring machines (CMM). Advances in Technology of Machines and Equipment, vol. 23(4), pp. 113–136. Postępy Technologii Maszyn i Urządzeń, kwartalnik PAN (1999)
53. Sładek, J.: Model wirtualny WMP z zastosowaniem do oceny dokładności realizowanych pomiarów w warunkach czasu quasi-rzeczywistego (CMM virtual model with the application to accuracy assessment of measurements realized in equations of quasi- real time). In: IVth International Scientific Conference—Coordinate Measuring Technique, pp. 193–208. Lodz University of Technology Scientific Books, branch in Bielsko-Biała, no. 53, Bielsko-Biała (2000)
54. Sładek, J.: A neural network model of CMM applied for measurement accuracy assessment. In: Proceedings IMECO World Congress, Vienna, 23–26 Sept 2000, vol. 11, pp. 281–286
55. Sładek, J.: Modelowanie i ocena dokładności maszyn oraz pomiarów współrzędnościowych (Modeling and accuracy assessment of coordinate machines and measurements). Cracow University of Technology Scientific Books, Mechanika no. 87, Cracow (2001)
56. Sładek, J.: Conception of description of coordinate measurement accuracy—on the basis of determination 3D—point uncertainty error. J. Mech. Eng. **53**(6) (2002)
57. Sładek, J.: The relationship between measuring machines performance specifications and their real capability to solve given measurement tasks, METROMEET 2005 International Conference on Industrial Dimensional Metrology Bilbao (Spain), 7–8 April 2005, Euskalduna Jauregia Invited Lecture

58. Sładek, J.: Opracowanie systemu oceny dokładności pomiarów oraz metody identyfikacji własnego pola błędów współrzędnościowych maszyn pomiarowych (Developing of a system for assessing the accuracy of measurements and of a method for identification coordinate measuring machines machine's errors field). Final Report of the Research Project funded by the KBN under contract No. PB-0921/T07/99/1 1999–2001
59. Sładek, J.: Metoda oceny dokładności pomiarów realizowanych redundantnymi systemami współrzędnościowymi (RSW) [Method for accuracy assessment of measurements done with the use of redundant coordinate systems (RCS)]. Research Project No. N505 255935, Ministry of Science and Higher Education (2008–2010)
60. Sładek, J.: Badania rozwojowe nad wzorcowaniem (z wykorzystaniem Laser Tracera) i wyznaczanie niepewności pomiarów dla systemów współrzędnościowych (Development researches on calibration (with the use of Laser Tracer) and determination of measurements uncertainty for coordinate systems). Grant no. 0869/R/T022010/10, National Centre of Research and Development (2010–2013)
61. Sładek, J., Gawlik, K.: Looking for uncertainty of measurement—virtual machines based on the matrix method using artificial neural networks. In: IV International Congress on Precision Machining 2007, Sandomierz-Kielce (2007)
62. Sładek, J., Gąska, A.: Modelling of CMM probe head errors. In: Proceedings of 10th ISMQC—TC14 IMEKO, 5–9 Sept 2010, Osaka Japan
63. Sładek, J., Kowalski, M.: Opracowanie wirtualnej wielowspółrzędnościowej maszyny pomiarowej z zastosowaniem do badań i korekcji błędów obiektów rzeczywistych i optymalizacji pomiarów (Development of virtual multi-coordinate measuring machine with the application for researches and correction of errors of real objects and for measurements optimization). PB 1367/T0795/08 KBN, Research Grant (1995–1998)
64. Sładek, J., Krawczyk, M.: Metody oceny dokładności pomiarów współrzędnościowych (Methods for coordinate measurements accuracy assessment). Pomiary Automatyka Kontrola (Measurement Automation and Monitoring) No. 9, vol. 53, pp. 478–481 (2007)
65. Sładek, J., Rakoczy, R.: CMM virtual modelling applied for measuring accuracy assessment. In: Proceedings of 1st International Workshop on CMM Calibration, 1–2 June 1999. Published by The Czech Metrological Institute, Prague Czech Republic
66. Sładek, J., Rakoczy, R.: Wykorzystanie koncepcji sztucznych sieci neuronowych do modelowania współrzędnościowej maszyny pomiarowej (WMP) przy zastosowaniu wzorca przestrzennego (The use of the concept of artificial neural networks for modeling the coordinate measuring machine (CMM) with the use of spatial standard). Works of Institute of Machine Technology and Production Automation, Polish Academy of Sciences, The Committee on Machine Building, Technology Bases Section, Scientific Book no. 63, pp. 193–206. Cracow University of Technology (1999)
67. Sładek, J., Szewczyk, D.: Wykorzystanie Simulatora I++ w nauczaniu metrologii współrzędnościowej i pracach badawczych (The use of Simulator I++ in coordinate metrology teaching and in research works). In: Proceedings of XIVth National, Vth International Scientific-Technical Conference Metrology in Production Engineering, Warsaw University of Technology, Sept 2011
68. Sładek, J., Rakoczy, R., Szwajkowski, A.: Application of neural networks for modelling coordinate measuring machines. In: Kunzmann, H., Waeldele, F., Wilkening, G., Corbet, J., MacKeow, P., Weck, M., Huemmler, J.: Progress in Precision Engineering and Nanotechnology, vol. 1, pp. 323–326, PTB-Braunschweig u. Berlin P.u.Oe-Verlag (1997)
69. Sładek, J., Kowalski, M., Ryniewicz, A., Juras, B., Muzyka-Zmudzki, M.: The method of the identification of the Orientation of control standard plater for the assessment of large coordinate measuring machine's (LCMM's) accuracy. In: Measurement 2005, 5th International Conference, Smolice Slovakia (2005)
70. Sładek, J., Kowalski, M., et al.: Identification method of the CMM measurement subspace localization based on location errors. In: VII International Scientific Conference Coordinate Measuring Technique, Bielsko-Biala, vol. 22(7), pp. 291–301 (2006)

71. Sładek, J., Ostrowska, K., Gacek, K.: Kinematic metrological model of the coordinate measuring arm (MCMA). In: XIX IMEKO World Congress Fundamental and Applied Metrology, Lisbon, Portugal, 6–11 sept 2009
72. Sładek, J., et al.: System metrologicznego nadzoru nad dokładnością maszyn pomiarowych robotów i obrabiarek z wykorzystaniem wzorców interferometrycznych laserowych systemów śledzących jako podst. wzrostu jakości produkcji przemysłu maszynowego (The metrological supervision system on accuracy of measuring machines and robots and machine tools with the use of interferometric standards of laser tracking systems as base for the increase in the quality of production in engineering industry). Research and Development Project no. (R03 029 01): implemented in 2006–2009
73. Sładek, J., Juras, B., Krawczyk, M., Gąska, A.: Modelling of the probing system errors by use of Monte Carlo Method. In: Sładek, J., Jakubiec, W. (eds.) Advances in Coordinate Metrology, Monograph. University of Bielsko-Biała (2010)
74. Sładek, J., Ostrowska, K., Gąska, A.: Wirtualne współrzędnościowe ramię pomiarowe (WWRP) (Virtual articulated arm coordinate measuring machine (VAACMM). Pomiary Automatyka Kontrola (Measurement Automation and Monitoring), 01/2010
75. Teichmann U.: Metrologische Probleme bei der Pruefung von koordinatenmessgeraeten. Feingeraetentechnik **3** (1978)
76. Tesa technology. www.tesabs.ch
77. TRACES—Transnational Calibration Expert System (2005–2007) C517456 TRACES Project—European Research and Development Project implemented under e-Ten to develop new type of service targeted to the industry using the most advanced measurement technologies in the production of machines and tools. Its participants beside Laboratory of Coordinate Metrology were: leading German scientific institute Physikalisch Technischen Bundesanstalt in Braunscheig—(German Office of Measures) as the project coordinator, Laboratory of Coordinate Metrology in Cracow University of Technology—Poland, Danish Technical University in Copenhagen, CMI—Czech Metrology Institute, Spanish research center UNIMETRIK, Italian CERMET, German calibration center DKD-1—FEINMESS, Czech VUOS and ETALON AG Germany, 2005–2007
78. VDI/VDE 2617 Genauigkeit von Koordinatenmessgeraten Kenngrossen und deren Pruefung
79. Weckenmann, A.: Koordinatenmesstechnik—Flexible Strategien für Funktion und fertigungsgerechtes Prüfen Carl Hanser Verlag München, Wien 2012
80. Weckenmann, A.: Ubersicht uber Abnahme und Uberwachungsverfahren fur Mehrkoordinaten-Messgerate. VDI-B 378, pp. 1–10 (1980)
81. Werth Messtechnik GmbH. www.werth.de
82. Wirtz, A.: Sind Geometriemerkmale nach DIN und ISO mit der Drei-Koordinaten-Messtech-nik erfassbar, VDI-B 529 (1984)
83. Zeiss. www.zeiss.pl/imt

Chapter 2
Measurement Uncertainty and Requirements of Production System. Selected Issues of Measurement Uncertainty Theory

Abstract This chapter describes the theoretical bases together with the author's concept of identification of reproducibility error of the measuring point (REMP), as the basis of the matrix method (MM) used for CMM error identification and the assessment of their accuracy. Also the importance of coordinate metrology for quality management and the evaluation of measurement uncertainty as a key task in deciding product geometry compliance with its specifications (GPS, geometrical product specification) are pointed out. There is also presented one of the most accurate coordinate measuring machines in the laboratory, with the air-conditioning system ensuring thermal stability in the range of ±0.05 °C. This chapter presents the theory of measurement uncertainty, the vector concept of describing coordinate measurement accuracy, and the REMP. It discusses the results of the author's original work concerning the determination of coordinate system accuracy at the measuring point and presents it as the basis for the new concept of CMS accuracy assessment. It also examines the issue of standards construction and the methodology of their application, which includes the possibility of identifying the accuracy at the measuring point. The discussion on coordinate measurement uncertainty in accordance with international standards and the author's research results, including the method using a calibrated object or standard, the concept of a multiposition method with the use of a noncalibrated object or simulative, analytical and expert methods is also presented.

2.1 Coordinate Measurement During Production Process

The acceptance of system solutions in the field of quality management and their later certification for compliance with the ISO 9000 series of standards means that many companies have to face concepts such as uncertainty, calibration, and metrological traceability. The implementation of these concepts into metrological practice has to be provided by a huge number of metrological standards and recommendations. The most important issue is the evaluation of measurement

J.A. Sładek, *Coordinate Metrology*, Springer Tracts in Mechanical Engineering,
DOI 10.1007/978-3-662-48465-4_2

uncertainty. It should be treated as a key task in deciding about the compliance of the product geometry with its specifications (GPS, geometrical product specification) [36]. We get used to the fact that in advanced technology industries, especially in the aerospace, space, and arms industries, the emphasis is put on the evaluation of measurement uncertainties, but recently the problem is also becoming crucial for other industries. It is particularly important for the automotive industry and its suppliers, producers of household appliances, medical and telecommunication devices, or mechatronic systems. In the case of small and medium-sized enterprises (SMEs) it can be noted that the use of accuracy assessment not only for customers, but also for internal processes, is not only useful, but necessary. It is mainly related to narrow tolerance ranges of feedback on tools and measuring systems development (particularly the coordinate ones), which in turn is reflected in product reliability and safety. A similar problem arises in the research and calibration laboratories, certified for compliance with PN-EN ISO/IEC 17025 [88]. But of course in this case the requirements of measurement accuracy are usually determined by higher precision requirements. By performing dimensionally shaped evaluations for each produced object, its shape is compared with the designed one. In traditional technical documentation the geometric object is presented as a view or cross-section.

Figure 2.1 includes formal information such as size, shape, location, and condition of the surface, together with the relevant tolerances. Tolerances describe the geometrically limited space, which sets the limits for the real geometric shape changes. The measurement instruments used in the metrology of geometric quantities were constructed to perform direct measurement of one characteristic. Other necessary dimensions were calculated by simple operations based on the results of direct measurements. It did not allow carrying out comprehensive measurement of spatial objects produced for the rapidly developing industry of the early 1970s. Common use of numerically controlled machine tools brings in new design possibilities, but also new measurement problems, especially in the case of outlines with variable curvature (gears, cams, etc.). In addition there was a whole group of indirect measurements (e.g., distance between axes of holes) for dimensionally expanded objects such as a corpus. The consequence of this was the introduction of a large number of specialized measurement instruments for the separate measurement of each dimension. It was an obstacle for the processes of control automation and thus for effectiveness growth. The control took too much time in comparison to the processing time and generated difficulties in the integration of the material and information flow.

The need to ensure production quality necessitated the search for new efficient measurement methods. The problem was solved by the coordinate measuring technique (CMT), originally based on measuring machines (CMM) that made representations of measured objects. Figure 2.1 presents one of the most accurate coordinate measuring machines in a modernized laboratory with an air-conditioning system maintaining temperature stability within ±0.05 °C.

The basic concepts used here are position and direction vector. Position vectors represent the location of measuring points, and direction vectors the angular

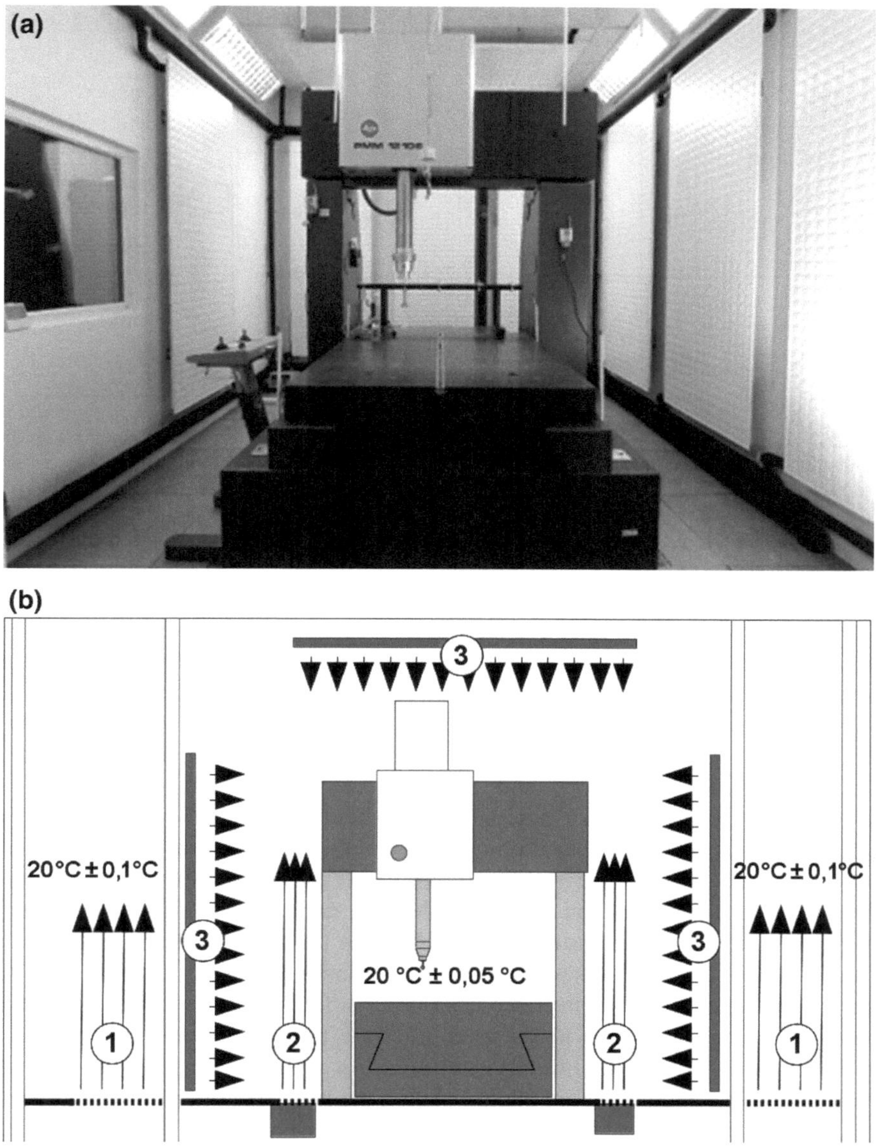

Fig. 2.1 **a** Multicoordinate measuring machine—PMM12106 Leitz—Laboratory of Coordinate Metrology at Cracow University of Technology; **b** air-conditioning system with thermal stability 0.05 °C. *1* isolation zone, *2* air curtain, *3* water panels

location in the object's space created from these points, such as straight lines, planes, and the like. Dimensional characteristics are represented by scalar quantity (length, radius of the circle or cylinder, cone angle). The representation of the

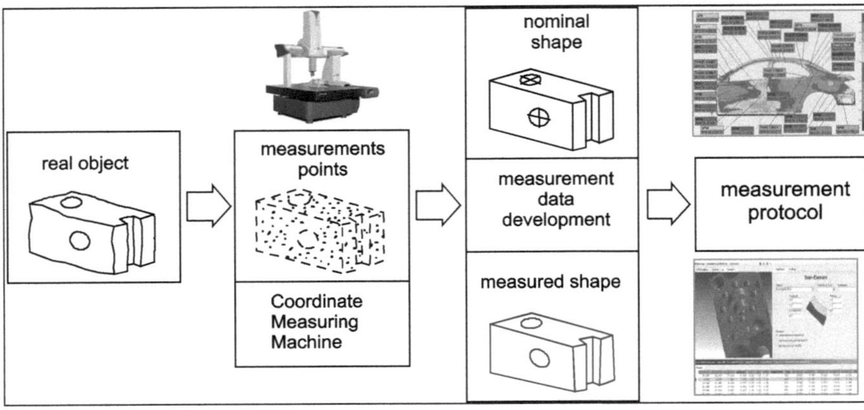

Fig. 2.2 Use of the coordinate measuring technique to geometric quantity measurements with the schematically indicated current method of measurement protocol preparation [10, 106]

measured object, however, involves the problem of determining its reference surface, having the shape of the assumed nominal surface.

The application of coordinate technique in the measurement process is shown in Fig. 2.2. As shown in the figure, in the coordinate measurement technique identified points are situated on a shell, and their coordinates are used exactly to calculate the ideal surface. Its location depends on the coordinates of individual points, and for its determination it requires, from the geometric point of view, a specified number of the points, whereas the measured surface (the real one) has the shape that results from production technology, and differs from the assumed nominal shape. Therefore to describe it more precisely, there should be identified as many points that represent it as possible, an almost infinite number. However, based on technical and economic criteria, their number is optimized to an indispensable minimum. This number is still substantially higher than the number that results from a geometric description of the assumed shape, because the ideal profile is usually calculated by the Gauss least squares method or minimum zone by Chebyshev. The difference in the location of the reference (ideal) surfaces creating an “image” of the measured object in relation to the location of the real surface determines a measurement error. Therefore in the analysis of errors, attention should be paid to the basic problem, which is the difference between the location of the measuring point, indicated in the coordinate system and used later to calculate the locked reference area, and its real location. The difference appears due to a whole complex of factors that can be broadly defined as the sources and causes of errors that occur in measurements with the use of coordinate machines. The CMT opened up new opportunities for automated quality control, finding usage in research laboratories, measurement chambers, and above all, in modern production engineering, where characteristics such as flexibility in making the measuring tasks, speed, and accuracy are of essential importance. Because of that CMT becomes an essential part of the whole quality assurance system. According to the experience of the last several years it can be said that the organization of today's dimensionally shaped control systems in the machine construction field should definitely be based on this technique. It also comes from the need to control more and more parts designed and produced using the computer techniques (CAx) that enable the development of information linkages within the production system.

2.2 Measurement Uncertainty

Measurement of any quantity is correlated with the need for its evaluation. The measurement is always an inaccurate operation, due to physical conditions, and it means that the value obtained as a result of the measurement (determined in the process of measurement) differs from its true value. The measure of this difference is measurement accuracy. Therefore its determination is one of the basic tasks that should be performed in the measurement process. The result of measurement in the form of a standalone single measured value has little practical significance; only after measurement accuracy is added to it, can the comprehensive meaning of the measurement be obtained. The generally accepted interpretation of the accuracy is the range on the value axis, set around the true value x_t as

$$[x_t - \Delta_{\min}x, x_t + \Delta_{\max}x] \tag{2.1}$$

where:

$$\Delta_{\min}x, \Delta_{\max}x > 0$$

In this range the previously obtained individual measurement results are kept. Those values of differences between these results, established during the process of measurement, and the true value, are known as the *error of measurement*. In coordinate metrology we use the term *deviation*, understood as the difference between the indicated and the reference value. The obtained limit value of deviation from the true value (or the one accepted instead) is known as the *maximum measurement error* [20, 92], or (taking the specified range of variability of obtained values, in accordance with [21]), as the measurement uncertainty range. In the case of coordinate measurements, the axial-range interpretation of measurement accuracy is extended to a spatial vector, maintaining the directionally axial interpretation (coordinates) [106]. According to [92] *measurement accuracy* is defined as closeness of agreement between a measured quantity value and a true quantity value.

Concluding the reflections on the accuracy of determining the measurement result, it can only be said about the true value x_t of the measured quantity that it lies within a given designable area. From the theory of measurement point of view, its mathematical description bears significant importance, that is, measurement accuracy modeling, because it determines whether the resulting data—results of measurements—can be used for the identification of the measured object. Determination of the measurement error can be understood as a mathematical model of measurement accuracy and in accordance with [21] it can be divided into two types: deterministic and random models.

The *deterministic model* [21] assumes that the unknown true value of the measured quantity x_t lies within an area of uncertainty. The deterministic model is used to describe the measurement accuracy instruments. For the given instrument (or instrument type) the broad enough uncertainty range is determined (constant or as a value function of the measured quantity) which for certain ranges of variation of influential quantities (e.g., temperature) contains all errors of measurements performed with the given instrument in the entire measurement range. Maximum error describing the uncertainty range in a deterministic model is a component of the metrological characteristics of the given measuring instrument. Determination of the maximum error of measurement is associated with the concept of measurement traceability, which requires parameters characterizing the inaccuracy of measurement and instrument having a reference to generally accepted primary standards, implemented as national or international standards.

The *random model* [21] assumes that in addition to the uncertainty range the level of confidence is also determined. It is the probability p that the unknown true value of the measured quantity x_t lies within the uncertainty range, and resolves the error of a single measurement into two components: *systematic error* understood as the difference between the average of an infinite number of measurement results of the same measured quantity carried out under repeatability conditions, and the true value of the measured quantity; and *random error* defined as the difference between the measurement result and the average of an infinite number of measurement results of the same measured quantity carried out under repeatability conditions. Systematic error is the same for all observations or changes according to a certain rule together with the change of conditions. *Random error*, equal to the difference between the total error and systematic error is generally different for individual measurements and takes positive and negative values, therefore it is modeled with a random variable.

In industrial practice, the measurement with CMM is generally performed only once. Therefore the use of the random model requires acceptance of a hypothetical method of measurement repetition, which can be used in the case of simulation of real measurements (while maintaining in this simulation the conditions of repeatability and reproducibility). In the random model of measurement accuracy the following cases may be specified:

(a) *Direct multiple measurement with random error only*. This is typical for repetitive measurements, which are dedicated to set a coordinate of a single measuring point, assuming the elimination of systematic errors, for example, for the study of a random part of a contact probe error.
(b) *Direct multiple measurement with random and systematic error*. It is possible only to determine a single measuring point and is used practically in the research environment.

(c) *Indirect measurement with the use of directly measured quantities, the accuracy of which is described by a deterministic or random model.* This applies to measurements carried out on CMM in industrial conditions, in which a single measurement dominates; in this case determination of parameters of measured shape elements is based on directly determined coordinates of measuring points.

A random model including both random and systematic error is a model already adopted in the standard PN–71/N-02050. However, the classical metrology, in the opinion of [21] does not provide a clear recipe enabling the user to carry out calculations of maximum error of measurements affected by errors with a systematic and random component. In fact, the classical approach requires identification and removal of the systematic error in the form of appropriate corrections and if it is not possible to determine the systematic error, the value of which is "sufficiently small" in comparison with measurement accuracy, systematic errors in the calculation of uncertainty should be considered as random. In coordinate technology the adoption of a random model with the possibility of eliminating systematic error, typical of the work of Lotze and Hartmann [19, 26, 27, 61–65, 125], again becomes important because of the intensive development of CMM error correction techniques by the CAA method [2, 5, 7, 8, 16, 22, 72, 79, 81, 96, 99, 106, 137, 138]. Nowadays, because of the development of correction technology based on CAA, a significant reduction of the impact of a systematic error without its technical elimination is possible [6, 29, 128]. Some arguments call into question the usefulness of systematic error elimination [29, 48, 51, 79, 131] instead of performing the software correction, because in consequence the residual systematic errors are treated as random ones during measurement uncertainty calculation.

An already classical theory of measurement accuracy (measurement error) has been replaced by the so-called *uncertainty theory*, the formally recognized and universal method for determining measurement accuracies, adopted by all international metrological organizations.

In general, *measurement uncertainty* is a range around the measurement result, which with some given probability contains the true value of the measured quantity. Therefore the measurement uncertainty consists of two values, with the real value located somewhere in between.

The basic set of definitions commonly used in Poland is "International vocabulary of metrology—Basic and general concepts and associated terms," (VIM) the PKN ISO/IEC Guide 99 revised in 2010 [92], in terms of measurement uncertainties determination, supplemented by the ISO *Guide, Guide to the Expression of Uncertainty in Measurement (GUM) ISO 1993/1995,* and published in 1999 by the Central Office of Measures in Polish version—*Wyrażanie niepewności pomiaru przewodnik,* Główny Urząd Miar, Warszawa 1999 z dodatkiem do wydania polskiego, J.M. Jaworskiego [21]. This guide contains a description of the measurement uncertainty model and guidelines of methods for its determination, the usage of which provides complete comparability of measurement results.

Modern understanding of the measurement uncertainty concept appeared in the German metrological literature, and its continuation is the definition of this term by the German standard DIN 1319 [11] "Fundamentals of Metrology" developed in 1997 (already based on GUM), but referring to the earlier definition. Although this standard has taken from GUM concepts and mathematical models for determining measurement uncertainty, the concept of uncertainty itself is slightly different from the definition adopted there. In the DIN 1319 standard uncertainty is defined as "The parameter obtained during the measurement along with the measurement result and is used to define the variation of true value of the measured quantity." This definition coincides with the conventionally true value of the measured quantity, whereas GUM defines uncertainty as "The parameter related to the result, which characterizes the spread of values that can be substantially attributed to the measured quantity." The definition describing the term "uncertainty" in the DIN 1319 standard states that uncertainty is not only a statistical dispersion of measurement results, which can be reduced by performing a large number of measurements and calculating the average value, but it contains a systematic part (known and unknown), associated with, for example, a reference standard. It means the confirmation of a connection between the measured value and national/international standards, an important component in industrial practice. According to the ISO 10012 standard [84], uncertainty is defined as the result of measurements made to estimate the range that contains a true value of the measured quantity, usually with a given reliability. In Poland, the uncertainty has already been defined by PN–71/N-02050 [91] as the dispersion of measurement results determined by maximum errors (Fig. 2.3).

Finally, according to the VIM, measurement uncertainty is defined as a non-negative parameter characterizing the dispersion of the quantity values being attributed to a (*measurand*) quantity true value/quantity intended to be measured based on the information used. This parameter could be, for example, the standard deviation, the standard uncertainty (or its multiple), as well as half of the range that has a certain expanded probability. In accordance with the above-mentioned

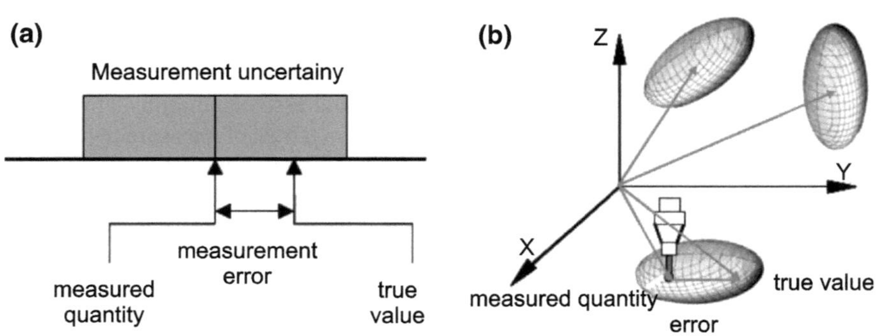

Fig. 2.3 Graphical representation of the meaning of error and measurement uncertainty: the axial system (scalar) (**a**), spatial (vector) (**b**)

information of measurement uncertainty cited after DIN 1319, VIM accepts that uncertainty contains components from systematic influences, such as components associated with corrections and values assigned to reference standards and definitional uncertainty. Sometimes estimated systematic effects are not corrected, and instead the measurement uncertainty components corresponding to them are introduced. It is therefore assumed that measurement accuracy is described by its uncertainty. The practical development of this definition in the form of guidelines for evaluation of measurement uncertainty is given in the GUM Guide [21]. It introduces the concept of the *standard uncertainty u,* understood as the uncertainty of the measurement result expressed as a standard deviation and *expanded uncertainty U*, given as a quantity determining the range around the measurement result, which covers a large part of the value distribution that reasonably can be attributed to the measured quantity (resulting from the assumed confidence level).

$$U = k \cdot u \tag{2.2}$$

where:

U expanded uncertainty

u standard uncertainty

k coverage factor that depends on the assumed confidence level p (determined from the t distribution or normal distribution)

It is generally assumed that $k = 3$ for $p \approx 99.73$ % or $k = 2$ for $p \approx 95.45$ %).

In practical conditions of measurements $t = 2$ is assumed, because the confidence level $p = 95$ % is usually assigned.

When the measurement result is obtained from values of other quantities, then the standard uncertainty is called *combined standard uncertainty* u_c: it is an estimate of variance and assuming n independent components of standard uncertainty u_i it is determined in accordance with the uncertainty propagation rules stated as Eq. (2.3):

$$u_c^2 = k_1 u_1^2 + k_2 u_2^2 + \cdots k_i u_i^2 + \cdots k_n u_n^2 \tag{2.3}$$

where:

u_i standard uncertainty components

k_i weight factors for ith uncertainty factor

Standard uncertainty or combined standard uncertainty components may be determined, as suggested by the definition of uncertainty [21, 92], by two methods:

(1) Based on statistical analysis of a series of measurement results
(2) Based on assumptions about the possible variation of a given uncertainty component (type of distribution, the variation range), allowing the estimation of standard deviation

In accordance with the recommendations of [21], uncertainty components are divided into two categories, depending on their calculation method:

(A) Uncertainties that have been calculated by statistical methods
(B) Uncertainties that have been estimated by other methods

This division aims to indicate two different ways of calculating the uncertainty components and facilitating discussion; the classification does not aim to indicate the differences in the nature of the components calculated by different methods. Both methods of calculation are based on probability distribution, and uncertainty components calculated by both the first and the second method are defined in terms of the variance or standard deviations.

Therefore, the standard uncertainty that is, the measurement uncertainty expressed as a standard deviation or as an estimate of this deviation, determined as *standard uncertainty type A*, is calculated from the probability density function obtained from the observed frequency distribution, whereas the *standard uncertainty type B* is calculated on the basis of the assumed probability density function, based on the confidence degree for probability of the given event appearance.

Therefore the determination of the type A uncertainty is of experimental character. Based on repeating n measurements of a given quantity w, the individual results of which are marked as w_i, their arithmetic mean w, as the best estimation of the expected value μ_w should be determined. As a measure of the variability, the estimate of the variance of the random variable is taken. In the case of coordinate measurements, if particular coordinates of the measuring points were treated as independent random variables, the variance of the random variable w for each independent coordinate of the measuring point (x, y, z) of the probability function $p(w)$ would be determined by the expression:

$$\sigma^2(w) = \int (w - \mu_w)^2 p(w) dw \tag{2.4}$$

where w takes the appropriate values for particular coordinates: x, y, z, and μ_w is the expected value of w and takes appropriate values μ_x, μ_y, μ_z.

As the estimator of the variance $\sigma^2(w)$ the experimental standard deviation $s^2(w_i)$ is accepted. It characterizes, as has already been mentioned, the variability of observed values w_i, or more precisely, their dispersion around the average w:

$$s^2(w_i) = \frac{1}{n-1} \sum_{i=1}^{n} (w_i - \bar{w})^2 \tag{2.5}$$

where w_i means the ith from n independent observations of a random variable w. As the estimator of expected value μ_w, the arithmetic mean w from n observations is taken:

$$\bar{w} = \frac{1}{n}\sum_{i=1}^{n} w_i \tag{2.6}$$

If as a result of the measurement the average of the number of repetitions is taken, then the correct measure of the uncertainty of the measurement result is the variance of the arithmetic mean. The variance of the arithmetic mean of a series of n independent observations w_i of a random variable w is determined by the expression:

$$\sigma^2(\bar{w}) = \frac{\sigma^2(\bar{w}_i)}{n}$$

and its estimate is given by the experimental variance of the mean:

$$s^2(\bar{w}) = \frac{s^2(w_i)}{n} = \frac{1}{n(n-1)}\sum_{i=1}^{n}(w_i - \bar{w})^2 \tag{2.7}$$

Experimental variance of an average $s^2(\bar{w})$ and experimental standard deviation of an average $s^2(\bar{w})$ determine numerically how correctly the expected value μ_w of variable w is estimated. Each of them can be used as a measure of the standard uncertainty $\bar{w}$. According to [21] (Sect. 4.2.3), the expressions $u(w) = s(w_i)$ are called the *standard uncertainty type A*.

In contrast to the type A method, which is based on frequency distributions, the *type B method* for calculation of uncertainty components is based on the a priori data distributions, which means that for the estimate w_i of the input quantity w (not determined from repeated observations), the estimate of its variance $u^2(w_i)$ or the standard uncertainty $u(w_i)$ is determined by scientific analysis based on all available information about the possible variation of the input quantity w.

In relation to the coordinate measurement uncertainty determination done using the B method, the set of this information may include:

- Previous measurement data
- Measuring experience based on the knowledge of phenomena and characteristics of the standards or instruments used
- Manufacturers' specifications
- Data obtained during calibration

However, it should be noted that the determination of standard uncertainty by the type B method can be as reliable as the calculation carried out by the type A method. That situation may occur especially when calculation by the type A method is based on a relatively small amount of independent information [82].

The type *B* method of determining the uncertainty uses an error propagation process (random variables). Each input quantity is described by its best estimate and standard deviation as the standard uncertainty associated with the estimate, and then for these data the probability distribution propagation is used. In this way, the estimate of output quantity is obtained and associated with its standard uncertainty. This model, however, has one major defect. Namely, the output quantity is usually described by a random variable that has a Gaussian or *t* distribution. As is known, these distributions are symmetric, and their extreme is achieved for the expected value. In practice, however, it often happens that an output quantity does not have symmetrical distribution. In this case much better results may be had by the use of the Monte Carlo (MC) method [70]. The MC method is a numerical method developed by Stanislaw Ulam, implemented for mathematical modeling of complex processes where the value is difficult to determine by an analytical method. It is suitable for uncertainty estimation, which confirms the GUM supplement describing its application [124]. The use of the MC method for probability propagation is a random sampling from the distribution function. The propagation process is also worth mentioning (adding random variables). The method is based on determination of the probability distribution function for the output quantity by analyzing the function of distributions assigned to the input quantities, which have a certain influence on output quantity.

The use of the MC method can be presented as shown in Fig. 2.4 [124]. There are three main stages highlighted: formulation of the problem, the propagation, and the summary stage.

1. Formulation of the problem:
   (a) Determination of the output quantity
   (b) Determination of the input quantities, those on which the output quantity is dependent
   (c) Construction of a model (mathematical), linking the input quantities with the output quantity
   (d) Assigning appropriate probability density functions (PDF) to input quantities, based on the knowledge and known information
2. Propagation:
   PDF sampling from input quantities and use of mathematic measurement model for obtaining the PDF for the output quantity
3. Summary:
   PDF application for the output quantity in order to determine:
   (a) The expected value of this quantity being its estimate
   (b) The standard deviation of this quantity being its standard uncertainty
   (c) The coverage interval containing the values of input quantity with appropriate probability *p*

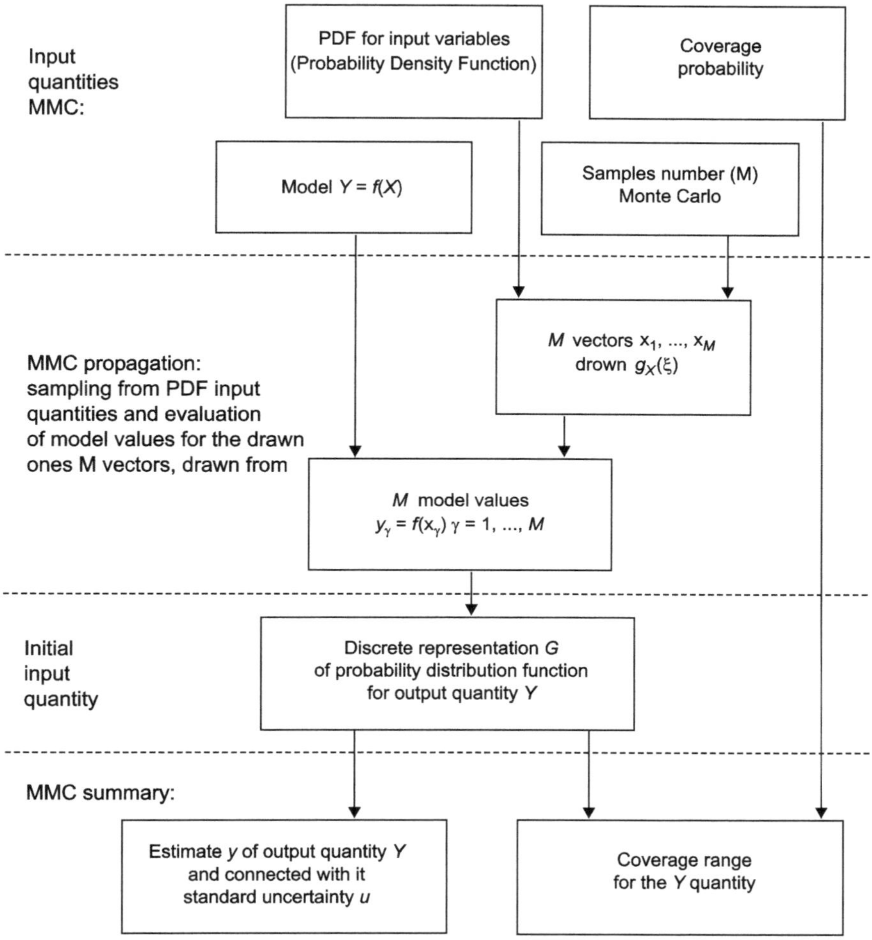

Fig. 2.4 MC method usage stages [116]

The main advantages of the application of the MC method in the uncertainty estimation are:

- Efficiency of calculations for complex or nonlinear uncertainty models
- Increase of accuracy of the Y estimate determination for nonlinear models
- Improvement of accuracy of standard uncertainty estimation for Y estimate in nonlinear models, especially when PDFs without a Gaussian distribution or t distribution are assigned to input quantities
- Accurate determination of the coverage interval in the case when the output quantity is not described by a Gaussian or t distribution

2.3 Vector Concept of Describing Coordinate Measurement Accuracy: Measuring Point Reproducibility Error

Coordinate measurement is essentially the determination of the basic measurement information quantum in the form of measuring point coordinates. This task is carried out as a direct measurement, and the other tasks of this metrological process are indirect measurements, and that is why the construction of the coordinate measuring system accuracy model should be based on it. However, the process of determination of any point coordinate and the analysis of the errors should be described first. In the case of systems of the contact measuring point identification as a basis for model construction, the reproducibility error of the measuring point (REMP) formulated in [99] was adopted. It is characterized by a vector $\bar{P}_a a$, which represents a difference between the position vector of the probe tip contact point $\bar{P}_{rz}$ with the measured surface, and a position vector of the point indicated by the machine $\bar{P}_m$, treating $\bar{P}_{rz}$ as a position vector of the real measuring point. From the metrological point of view vector $\bar{P}_a$ is a vector that characterizes CMM accuracy in a given point of its measuring volume, as shown in Fig. 2.5.

$$\bar{P}_a = \bar{P}_{rz} - \bar{P}_m \tag{2.8}$$

However, to make the model complete, the concept of position error given in papers [51, 74, 76, 99] and probe error has to be recalled. The position error $\bar{P}_p$[1] in accordance with [51] and [106] is the difference between a position vector of a contact tip (being at the limit of sending the impulse of contact) and measurement surface $\bar{P}_{km}$ common point, and a vector $\bar{P}_m$ of a point indicated by the machine. Therefore the $\bar{P}_p$ vector cumulates inside the errors associated with the kinematics (21 error components) and the errors of displacement measuring system:

$$\bar{P}_p = \bar{P}_{km} - \bar{P}_m \tag{2.9}$$

According to this definition, the shape deviation of the contact tip sphere, the impact of the measured object, and most of all the variable errors of the probe, the values of which are described by the FBG (α β, BG) function (Sect. 4.3) are not included here. Therefore the model for a single measuring point, with probe errors included and reported here as vector coordinates $\bar{P}_g(x, y, z)$, appears as

$$\bar{P}_a = \bar{P}_p + \bar{P}_g = \bar{P}_{rz} - \bar{P}_m \tag{2.10}$$

In the case of multistylus probes, indexing probes, and redundant systems (measuring arms and laser tracer) styli, this record should be supplemented with an

[1]Note: Do not confuse with the positioning error defined in Sect. 4.1.

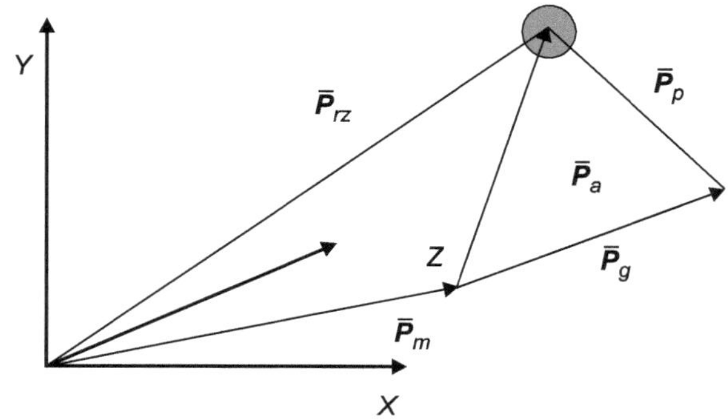

Fig. 2.5 Accuracy error and components of the reproducibility error of the measuring point (REMP) [105]

orientation vector P_o, which together with the vector of probe errors defined before creates a vector P_{go}. It is defined as their sum (2.11), as shown in Fig. 2.6:

$$\bar{P}_{go} = \bar{P}_g + \bar{P}_o \tag{2.11}$$

However, the total (defined for the whole measuring space) CMM accuracy model can be adopted using the differentiator given above in the form of a vector $\bar{P}_a$, applying it to the whole volume V of dimensions R^3. If we assume that to each point $P(x, y, z) \in V \subset R^3$ a vector $\bar{P}_a$ is assigned, a vector of the CMM error field will be characterized, with its components in the form of:

$$\bar{P}_a(p) = \bar{i}Pa_x(p) + \bar{j}Pa_y(p) + \bar{k}Pa_z(p) \tag{2.12}$$

where:

$\bar{P}_a(p)$	vector field of measuring point reproducibility errors at the p point in measuring volume V of given CMM
Pa_x, Pa_y, Pa_z	components of error field in Cartesian CMM coordinate system
$\bar{i}, \bar{j}, \bar{k}$	versors of coordinate system axes

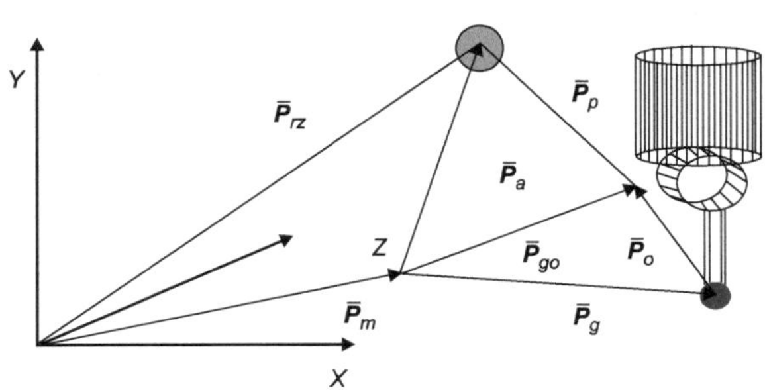

Fig. 2.6 Accuracy error and reproducibility error components including probe error and orientation of the measuring stylus for probes with indexing mechanisms and redundant systems, for example, measuring arms

The creation of the error model is therefore based on the determination of relation (2.12). However, it should be considered that in real terms, apart from systematic factors, measuring point reproducibility errors are influenced by random factors. It is therefore necessary to characterize the random error field, which for the CMM measuring volume can be considered to be continuous with a discrete parameter, according to [51, 106, 133] (See Fig. 2.7). If the coordinate system U is specified in a three-dimensional volume, then a random vector field $\bar{U}(P_a)$ can be written as

$$\bar{U}(P_a) = \bar{i}U_x(P_a) + \bar{j}U_y(P_a) + \bar{k}U_z(P_a) \tag{2.13}$$

where

U_x, U_y, U_z = component uncertainties in x, y, z system

On the basis of the findings described previously [75, 77, 99, 106] we can formulate the following statement: the part of error that comes from the contact probe head dominates in random component error.

Analyzing the issue of random error participation, the work [26, 63, 65] of Lotze and Hartmann should be adduced. They deal with the determination of the uncertainty area for two-dimensional issues, only in [106] the uncertainty area for spatial (three-dimensional) measurements was determined, assuming that the random error for coordinate measurement has vector character [13, 99, 106, 107] and is usually described by a matrix of variances and covariances [3, 106]:

$$\mathbf{S} = \begin{bmatrix} \sigma_x^2 & \rho_{xy}\sigma_x\sigma_y & \rho_{xz}\sigma_x\sigma_z \\ \rho_{xy}\sigma_x\sigma_y & \sigma_y^2 & \rho_{yz}\sigma_y\sigma_z \\ \rho_{xz}\sigma_x\sigma_z & \rho_{yz}\sigma_y\sigma_z & \sigma_z^2 \end{bmatrix} \tag{2.14}$$

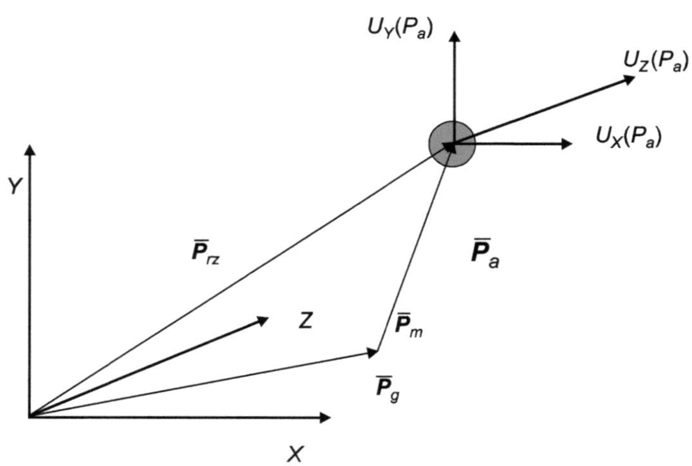

Fig. 2.7 Random components $U_x(P_a)$, $U (P_a)$, $U_z(P_a)$ of the measuring point reproducibility error $\bar{P}_a$

where:
σ_x standard deviation of error
ρ_{xz} covariance between coordinates

In practice the described matrix has the form of estimates of variances and covariances:

$$\mathbf{S} = \begin{bmatrix} s_x^2 & r_{xy}s_xs_y & r_{xy}s_xs_z \\ r_{xy}s_xs_y & s_y^2 & r_{yz}s_ys_z \\ r_{xz}s_xs_z & r_{yz}s_ys_z & s_z^2 \end{bmatrix} \tag{2.15}$$

Random errors during multirepetition of measurements, under set conditions, change in unpredictable ways both in the sign and in absolute value. Considering the performance of a typical contact probe head and other components of CMM error it can be concluded that random errors are the sum of many influences, and have not been included either because of the small effect on the behavior of the machine, or because it has not been possible to measure them. Their impact appears during the contact process and affects the determined measuring point coordinate. In the case of random errors the following assumptions can be adopted:

- Positive and negative errors are equally probable (symmetry).
- Possibility of making the big mistake is less probable than the small one (concentration).

Because of these two reasons, it can be concluded that random errors are of normal distribution. The equation of the distribution (probability density function) for the general n-dimensional case by [3, 106, 107] takes the form:

$$\begin{gathered} (x) = \frac{1}{\sqrt{\det(S)\cdot(2\pi)^n}}\exp\left(-\frac{Q}{2}\right) \\ Q(x) = (x-\mu)^T \cdot S^{-1}(x-\mu) \end{gathered} \tag{2.16}$$

for which:

$$x = \begin{bmatrix} x_1 \\ \vdots \\ x_n \end{bmatrix} \quad \mu = \begin{bmatrix} \mu_1 \\ \vdots \\ \mu_n \end{bmatrix}$$

where:
x random vector
μ expected value of random error

If the case of measurement in the flat system (two-dimensional) is taken as in [3, 106, 107], then relations (4.7) and (4.8) take the form:

$$\mathbf{S} = \begin{bmatrix} s_x^2 & r_{xy}s_xs_y \\ r_{xy}s_xs_y & s_y^2 \end{bmatrix}$$
$$\varphi(x,y) = \frac{1}{2\pi s_xs_y\sqrt{1-r_{xy}^2}}\exp\left(-\frac{Q}{2}\right) \tag{2.17}$$
$$Q(x,y) = \frac{1}{1-r_{xy}}\left[\frac{(x-\mu_x)^2}{s_x^2} - 2r_{xy}\frac{(x-\mu_x)(y-\mu_y)}{s_xs_y} + \frac{(y-\mu_y)^2}{s_y^2}\right]$$

If the coordinate system is moved to the point (μ_x, μ_y) and rotated so that the covariance r equals zero, Eqs. (2.16) turn into:

$$\mathbf{S} = \begin{bmatrix} s_{\mathrm{I}}^2 & 0 \\ 0 & s_{\mathrm{II}}^2 \end{bmatrix}$$
$$\varphi(x,y) = \frac{1}{2\pi s_{\mathrm{I}}s_{\mathrm{II}}}\exp\left(-\frac{Q}{2}\right) \tag{2.18}$$
$$Q(x,y) = \frac{x^2}{s_{\mathrm{I}}^2} + \frac{y^2}{s_{\mathrm{II}}^2}$$

When determining values of the function φ (x, y) for points (s_{I}, 0), ($-s_{\mathrm{I}}$, 0), (0, s_{II}), (0, $-s_{\mathrm{II}}$) we obtain:

$$\varphi(-s_{\mathrm{I}},0) = \frac{1}{2\pi s_{\mathrm{I}}s_{\mathrm{II}}}\exp\left(-\frac{\left(\frac{-s_{\mathrm{I}}}{s_{\mathrm{I}}}\right)^2}{2}\right) = \frac{1}{2\pi s_{\mathrm{I}}s_{\mathrm{II}}}\exp\left(-\frac{1}{2}\right)$$
$$\varphi(s_{\mathrm{I}},0) = \frac{1}{2\pi s_{\mathrm{I}}s_{\mathrm{II}}}\exp\left(-\frac{\left(\frac{s_{\mathrm{I}}}{s_{\mathrm{I}}}\right)^2}{2}\right) = \frac{1}{2\pi s_{\mathrm{I}}s_{\mathrm{II}}}\exp\left(-\frac{1}{2}\right) \tag{2.19}$$

As can be seen, the value of the function at these points is the same. Now by cutting function ^(x, y) with plane parallel to the XY plane at the height of this point, we obtain:

$$\begin{aligned}\varphi(x,y) &= \frac{1}{2\pi s_{\mathrm{I}}s_{\mathrm{II}}}\exp\left(-\frac{\frac{x^2}{s_{\mathrm{I}}^2}+\frac{y^2}{s_{\mathrm{II}}^2}}{2}\right)\\ &= \frac{1}{2\pi s_{\mathrm{I}}s_{\mathrm{II}}}\exp\left(-\frac{1}{2}\right)\exp\left(-\frac{\frac{x^2}{s_{\mathrm{I}}^2}+\frac{y^2}{s_{\mathrm{II}}^2}}{2}\right)\\ &= \exp\left(-\frac{1}{2}\right) - \frac{\frac{x^2}{s_{\mathrm{I}}^2}+\frac{y^2}{s_{\mathrm{II}}^2}}{2}\\ &= -\frac{1}{2}\frac{x^2}{s_{\mathrm{I}}^2} + \frac{y^2}{s_{\mathrm{II}}^2} = 1\end{aligned} \tag{2.20}$$

It is an ellipse with axes that describe main standard deviations. In the case of a three-dimensional Eq. (2.16), after moving the coordinate system origin to the point μ, they take the form:

$$\begin{aligned} \varphi(x,y,z) &= \frac{1}{2\sqrt{2\pi^3}\cdot s_x s_y s_z\sqrt{1-r_{xy}^2-r_{yz}^2-r_{xz}^2+2r_{xy}r_{yz}r_{xz}}}\exp\left(-\frac{Q}{2}\right) \\ Q(x,y,z) &= \frac{1}{1-r_{xy}^2-r_{yz}^2-r_{xz}^2+2r_{xy}r_{yz}r_{xz}} \\ &\left(\begin{array}{c} \frac{(1-r_{yz})x^2}{s_x^2}+\frac{(1-r_{xz})y^2}{s_y^2}+\frac{(1-r_{xy})z^2}{s_z^2} \\ -2\frac{(r_{xy}-r_{yz}r_{xz})xy}{s_x s_y}-2\frac{(r_{yz}-r_{xy}r_{xz})yz}{s_y s_z}-2\frac{(r_{xz}-r_{xy}r_{yz})xz}{s_x s_z} \end{array}\right) \end{aligned} \tag{2.21}$$

After determining the values and eigenvectors **S**, and then transforming the system in a system in which versors are eigenvectors of a matrix of variances and covariances, Eqs. (2.19) take the form:

$$\begin{aligned} \mathbf{S} &= \begin{bmatrix} s_{\mathrm{I}}^2 & 0 & 0 \\ 0 & s_{\mathrm{II}}^2 & 0 \\ 0 & 0 & s_{\mathrm{III}}^2 \end{bmatrix} \\ \varphi(x,y,z) &= \frac{1}{2\pi\sqrt{2\pi}s_{\mathrm{I}}s_{\mathrm{II}}s_{\mathrm{III}}}\exp\left(-\frac{Q}{2}\right) \\ Q(x,y,z) &= \frac{x^2}{s_{\mathrm{I}}^2}+\frac{y^2}{s_{\mathrm{II}}^2}+\frac{z^2}{s_{\mathrm{III}}^2} \end{aligned} \tag{2.22}$$

When values of the function $\varphi(x, y, z)$ are determined for points on the axes having values corresponding to standard deviations, the formula describing the ellipsoid with axes s_{I}, s_{II}, s_{III} is obtained (2.22).

If the coordinate system is not rotated by matrix **S** eigenvectors, the equation of an ellipsoid of axes rotated relative to coordinate system axes is obtained. It has been proved that for the case of a three-dimensional measurement, the obtained shape is the ellipsoid with axes s_{I}, s_{II}, s_{III}. It can be concluded that the random field area, corresponding to the uncertainty area, takes the shape of an ellipsoid (Fig. 2.8). This is compatible with suggestions about the shape of this area contained in the literature [50, 74, 99, 106, 107, 136].

It can be said that the ellipsoid represents the area in which the probability of finding the true value of the measured point is *pp*. Therefore inasmuch as the standard deviation values constitute a reference point for the uncertainty area, the ellipsoid illustrates this state in three-dimensional volume.

In order to verify the developed equations, the studies based on gaining knowledge of the impact of random errors have to be carried out. They are based on checking the real performance of the measuring machine probe head and redundant

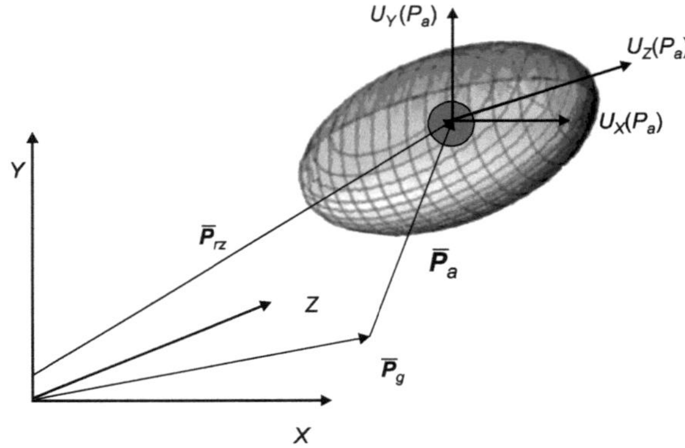

Fig. 2.8 The shape of the uncertainty area around the measuring point defined in the form of an ellipsoid

system stylus (articulated arm coordinate measuring machine, AACMM) for different approach angles.

In the case of AACMM the sphere was measured at 15 points, and the measurements were repeated 32 times. The sphere was placed in the center of the AACMM measuring volume to minimize the effect of systematic components of measurement error.

The variance, standard deviation, and covariance of coordinates of the points determining the center of the sphere were calculated:

$$\begin{aligned} \mathrm{Var}(X) = 2.961 \times 10^{-3} \mathrm{Stadev}(X) &= 0.054 \\ \mathrm{Cvar}(X, \mathrm{Y}) &= -4.809 \times 10^{-4} \\ \mathrm{Var}(Y) = 1.698 \times 10^{-3} \mathrm{Stadev}(Y) &= 0.041 \\ \mathrm{Cvar}(X, \mathrm{Z}) &= 2.591 \times 10^{-4} \\ \mathrm{Var}(Z) = 2.548 \times 10^{-3} \mathrm{Stadev}(Z) &= 0.050 \\ \mathrm{Cvar}(Y, Z) &= -5.585 \times 10^{-5} \end{aligned}$$

where:

Var variance
Stdev standard deviation
Cvar covariance

A matrix of random errors for the AACMM has been designated:

$$S = \begin{pmatrix} 2.961 \times 10^{-3} & -1.078 \times 10^{-6} & 7.119 \times 10^{-7} \\ -1.078 \times 10^{-6} & 1.698 \times 10^{-3} & -1.162 \times 10^{-7} \\ 7.119 \times 10^{-7} & -1.162 \times 10^{-7} & 2.548 \times 10^{-3} \end{pmatrix}$$

Eigenvalues of the **S** matrix, which are ellipsoid axes:

$$w = \begin{pmatrix} 1.698 \times 10^{-3} \\ 2.961 \times 10^{-3} \\ 2.549 \times 10^{-3} \end{pmatrix}$$

Eigenvectors of the **S** matrix, which are the versors of the axes system:

$$z = \begin{pmatrix} 8.536 \times 10^{-4} & 1 & -1.723 \times 10^{-3} \\ 1 & -8.538 \times 10^{-4} & -1.354 \times 10^{-4} \\ 1.359 \times 10^{-4} & 1.723 \times 10^{-3} & 1 \end{pmatrix}$$

Therefore the ellipsoid of the uncertainty equation characterizing the AACMM random errors field has the form [69] shown in Fig. 2.9.

Analyzing random errors of AACMM [80, 109], the following regularities are observed.

- The value of the z variable is on the same level as x and y, which may mean that the friction and adhesion forces do not affect the measurement.
- The ellipsoid is rotated in relation to the reference system; it is associated with an orientation error of the stylus.
- The axis parallel to the direction of the approach in AACMM is comparable to y, conversely to the case of the conventional coordinate measuring machine (CMM), where the axis parallel to the direction of approximation is shorter by an order of magnitude.
- The area of random errors confirms that the device may be used interchangeably with conventional CMM.

As for AACMM in the case of the measuring machine, the sphere was measured on PMM12106 at 15 points and 32 times. After calculation in accordance with the equations given above, graphs shown in Figs. 2.10, 2.11 and 2.12 were drawn. They show projections of points and the determined uncertainty area. The uncertainty ellipsoid is shown by projections on the planes *XY*, *YZ*, and *XZ*, respectively.

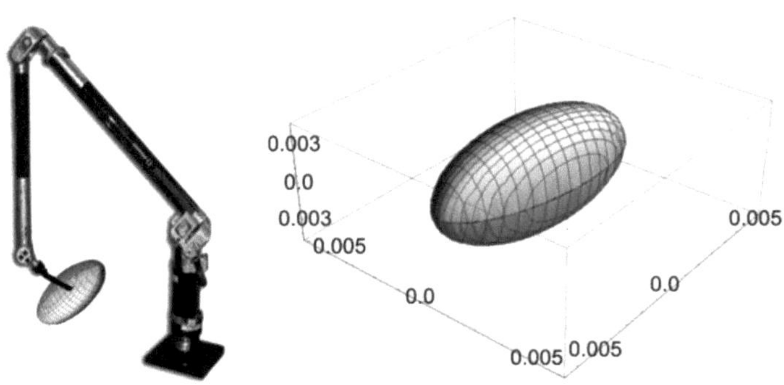

Fig. 2.9 AACMM uncertainty ellipsoid

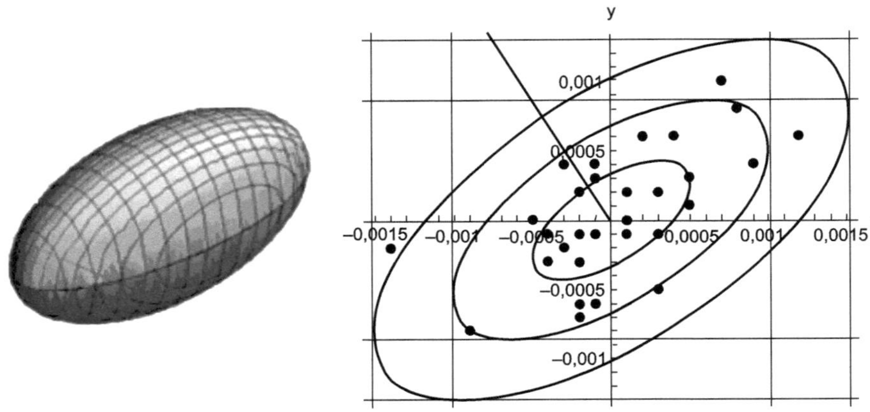

Fig. 2.10 Projection of ellipsoid of uncertainty on the *XY* plane [106]

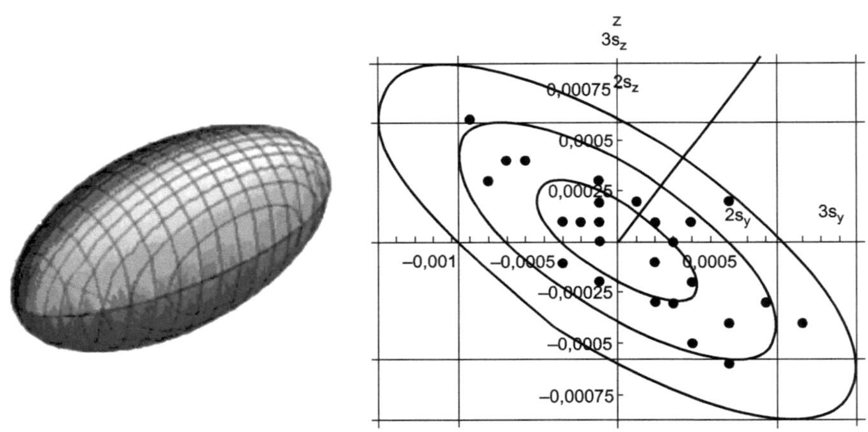

Fig. 2.11 Projection of ellipsoid of uncertainty on the *YZ* plane [106]

The analysis of several simulations of the implementation of the probe performance during contact with the curved surface (the case of simultaneous impact of sources of random errors in all axes on the random error field formation), taking into account the change in the angle between the measurement direction and the surface normal, allows us to draw these conclusions:

- The uncertainty ellipsoid axes take locations in accordance with the direction of measurement movement.
- The axis parallel to the direction of approximation is the shortest one.
- The standard deviations (uncertainty ellipsoid axes) of measuring point coordinates in the orthogonal directions to the measurement direction are several times, even several dozen times, greater than in the parallel axis, and the largest ones are perpendicular to the measurement plane.

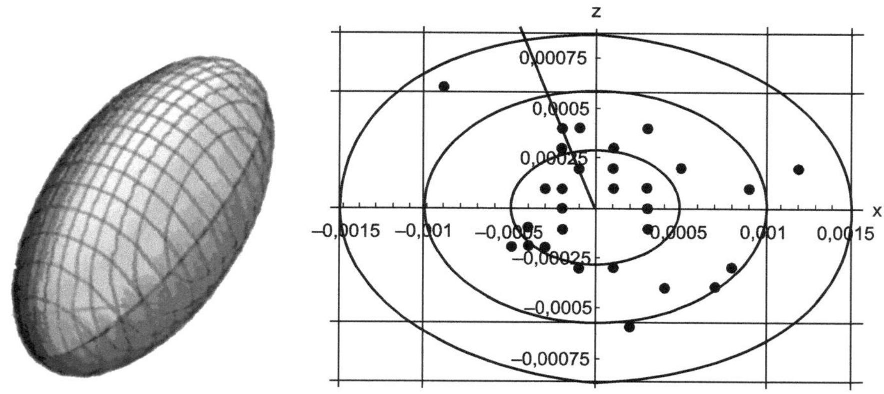

Fig. 2.12 Graph of uncertainty ellipsoid in the *XZ* plane [106]

For these reasons, the process of probe head model development can be really focused on modeling interactions in the contact direction (tip approach). This confirms the adoption of the PEF identification concept [99, 106] (in accordance with Sect. 3.5) and the assumption of the reasonable simplification in the form of transferring the matter of CMM random error identification to the group of errors independent of the location.

However, from the concept and the model of REMP analysis and in accordance with the works [106, 107] it can be stated that the basic difficulty, from a technical point of view, is a determination of the value and direction of the measuring point reproducibility error for any point of the so-defined error field, and of the dimensions of the uncertainty field for its determination. It is also clear that the REMP determination for all points from the measuring range of the machine is impossible, because each point can be achieved from any number of directions and reaching it is limited by the location and shape of the measuring tips. It is therefore necessary to define some, possibly small, number of points, for which it is necessary to determinate error vectors that will most precisely characterize a given coordinate system in metrological terms. This is possible if the method of CMM accuracy identification were based on the determination of REMP vectors for a selected number of measuring points.

Accepting the multisensor direction of the development of coordinate systems (with contact and noncontact determination of the coordinate of a single point or its cloud), it can be clearly said that the essence from the measurement theory point of view is the determination of the measuring point as a direct measurement. Actually this should be the basis for the measuring system accuracy determination, more than the already known length measurement. This idea was formulated for the first time by the author in [99] and developed in [106], and now with the methods and measuring systems development the technical capabilities of its implementation have been established. The most general conception of description and assessment of the coordinate system accuracy, including all contact and noncontact systems, such as computed tomography (CT) and optical scanners, was described for the first time in [106] in the form of the MM. This issue is discussed in Sect. 4.4 of this book. The MM is based on the determination of the set (grid) of reference points, and its aim is to define its own coordinate measuring system error field. The reference points are obtained as points on the material standard surface or as the centers of constant curvature elements, for which an explicitly repeatable measurement strategy can be defined. A better explanation of this concept gives the idea based on using a standard cube for the CT system calibration presented in Fig. 2.13. It reveals the relationship of the REMP described by Eq. (2.8) with its random components. The uncertainty area is defined in the form of an ellipsoid created around the measuring point as shown in Fig. 2.8. This idea is so universal that it corresponds well with the concept of metrological use of CT and the single voxel as a measure of its resolution.

It can also be successfully used for optical scanner error identification. Such research was carried out in the framework of [98] and published in [120, 121]. However, the modern concept of assessing the accuracy of CMS in the reference points keeps waiting for comprehensive implementation and it should become a subject of detailed study in scientific work.

From a theoretical point of view, to determine the value and the direction of a so-defined error, the most appropriate measuring system is an independent system, "external" in relation to the checked device, and with higher base accuracy. In technical terms, the implementation of such a system is still difficult. Such a system, analogous to the (GPS) global positioning system, also known as internal GPS, requires reference points and the possibility to measure the distance using the above-mentioned multilateration method [2, 13, 14, 17, 18, 33, 72, 73, 78, 97, 126, 127, 131, 139, 140]. The biggest hopes are put on laser tracking systems LTS [14, 32, 33, 49, 59, 60, 73, 97], the accuracy of which is currently sufficient, for example, for large coordinate measurement machines. In the case of more accurate systems, the

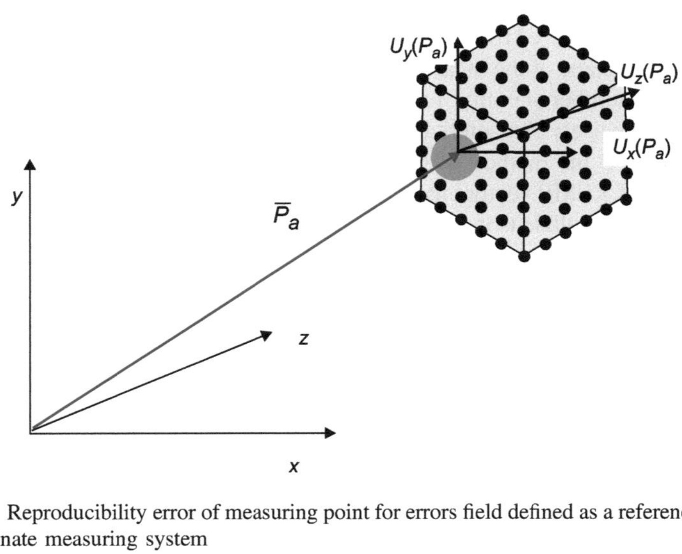

Fig. 2.13 Reproducibility error of measuring point for errors field defined as a reference point grid for coordinate measuring system

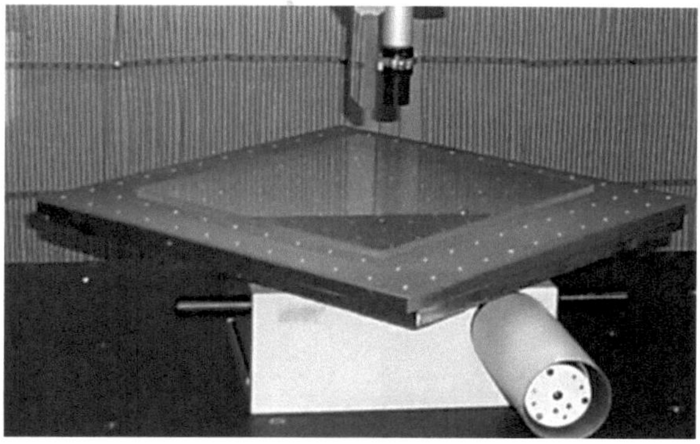

Fig. 2.14 PTB plate standard made of a thermostable material: Zerodur [83]

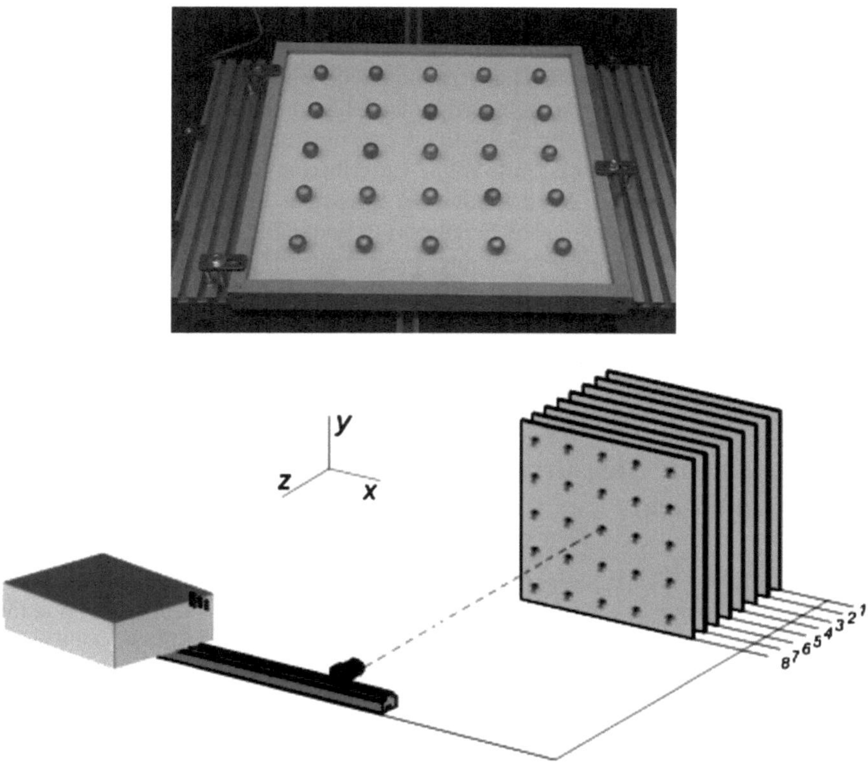

Fig. 2.15 Plate standard for structured-light optical scanners and the construction method of measuring point reference grid for system accuracy assessment [98]

Fig. 2.16 Standard for computed tomography, which allows the construction of the reference grid for accuracy assessment of these systems [15]

LaserTracer with accuracy comparable to the most accurate step gauges can be applied. For optical systems flat and spatial standards are sufficient. In two-dimensional systems a standard used by the PTB could be applied. The example of such a standard is shown in Fig. 2.14. In the case of structured-light optical scanner sphere plates, such as proposed in [58, 98], described in [120, 121], and shown in Fig. 2.15, could be applied. Another solution, consistent with the concept shown in Fig. 2.12, is a standard developed by Feinmess [15] in the form of a hole cube made of material with specific X-ray permeability, presented in Fig. 2.16. Such a standard enables us to carry out the accuracy identification of a coordinate measuring system at the reference points for the CT system. Of course all these systems implement the MM idea [106].

2.4 Coordinate Measurement Uncertainty and Regulatory Requirements

In addition to the GUM guide, among descriptions having great practical importance for measurement assessment, both in the area of quality control and also in the accuracy assessment of measuring instruments, the ISO 14253 standard—Inspection by measurement of workpieces and measuring equipment [85]—should be mentioned. It consists of three parts:

- ISO 14253-1: Decision rules for proving conformity or nonconformity with specifications
- ISO 14253-2: Guidance for the estimation of uncertainty in GPS measurement, in calibration of measuring equipment and in product verification
- ISO 14253-3: Guidelines for achieving agreements on measurement uncertainty statements

This standard is dedicated to geometric quantity measurements and is based on the indications contained in [21].

The ISO 14253-1 standard introduces the concept of specification that here means the tolerance of an object property (dimension). According to GPSs (ISO/TR 14638:1995, *GPSs—Masterplan*) [36, 86] it contains the decision rules that require narrowing the tolerance by the measurement uncertainty to prove conformity with specifications, and extending the tolerance by the measurement uncertainty to prove nonconformity with the specifications. In technical drawings tolerance limits are usually clearly defined. However, when we intend to determine that the real object is set inside or outside the tolerance range, and additionally the measurement is loaded with an uncertainty, then this definition may not be clear (Figs. 2.12 and 2.13).

The purpose of the ISO 14253–1 standard is to regulate these concerns, which may contribute to costly disputes between customers and buyers. Therefore, the standard defines three zones, which are shown in Figs. 2.17 and 2.18.

The compliance zone with the tolerance is limited by the measurement uncertainty at the area borders. Noncompliance zones are set outside the measurement uncertainty ranges. Uncertainty ranges are areas where the compliance or noncompliance cannot be determined.

Measurement uncertainty always acts against the one who performs a measurement, and therefore presents the judgment of compliance or noncompliance. In the ISO 14253-2 [87] there is a suggestion of an algorithm that enables the determination of geometric quantity measurement uncertainty based on the acceptance of an iteration-determined uncertainty budget (called PUMA, procedure of uncertainty management), Fig. 2.19.

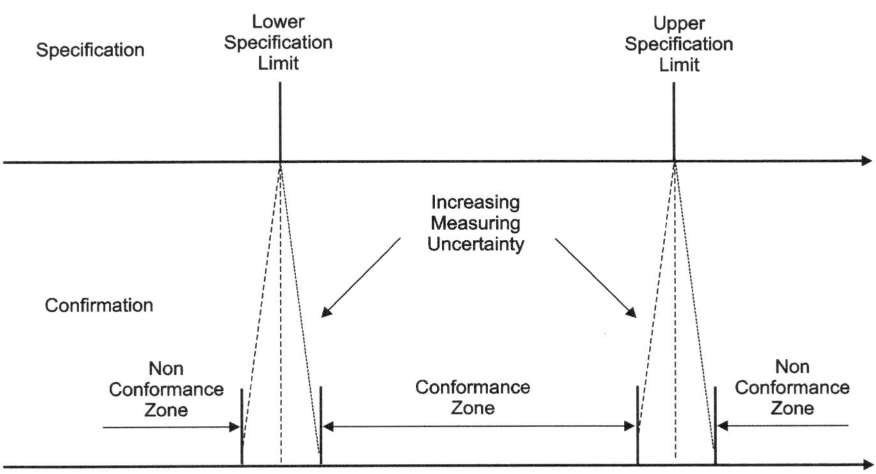

Fig. 2.17 Specification with two clear limits down and top and the method of its conformation [86]

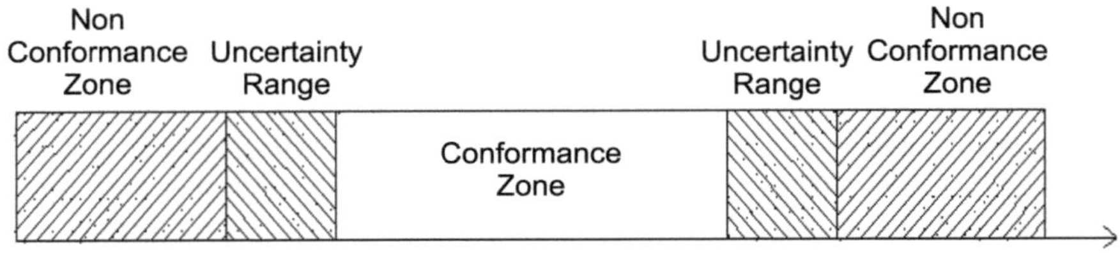

Fig. 2.18 Conformance and nonconformance zones with the specification by ISO 14253–1 [86]

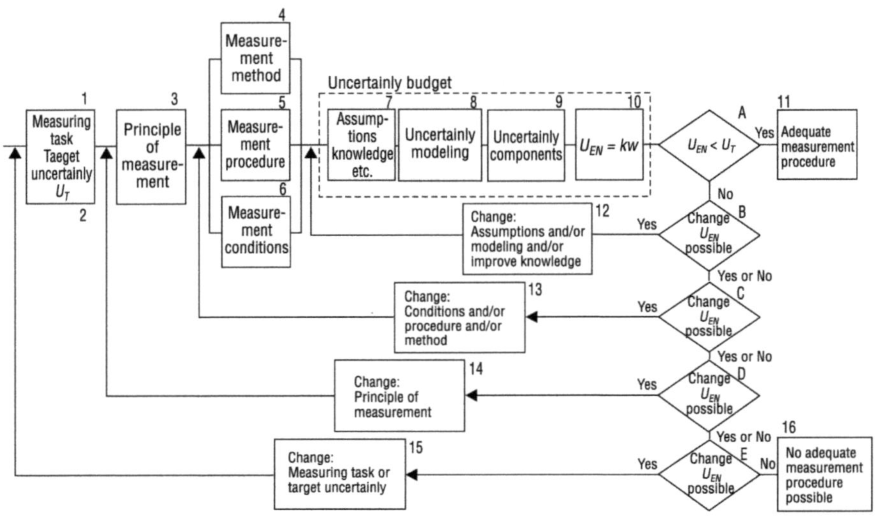

Fig. 2.19 Iterative method of measurement uncertainty evaluation, PUMA [87]

The iterative method for determining uncertainty consists of four steps visualized in Fig. 2.18. After the development of the uncertainty budget, according to requirements set by the quality system, they constitute these variables:

- Assumptions and/or model and/or increase knowledge
- Conditions and/or procedure and/or method
- Measurement principle
- Measuring task or the aim of uncertainty determination

The ISO 9000 series of standards [34] distinguishes in many places the problem of measurements and monitoring product quality and, related to it, inspection of measuring equipment and its calibration. These standards have contributed to the implementation of the concept of uncertainty in industry as a measure of the accuracy of production, the basic concept of determination of product compliance with requirements.

PN/EN ISO 9001:2009 [89] standard in Sect. 7.6—Control of monitoring and measuring equipment states:

> … The organization shall determine the monitoring and measurement to be undertaken and the monitoring and measuring equipment needed to provide evidence of conformity of product to determined requirements. The organization shall establish processes to ensure that monitoring and measurement can be carried out and are carried out in a manner that is consistent with the monitoring and measurement requirements. Where necessary to ensure valid results, measuring equipment shall:
>
> - be calibrated or verified, or both, at specified intervals, or prior to use, against measurement standards traceable to international or national measurement standards. Where there are no such standards, the basis used for calibration or verification shall be recorded,
> - be adjusted or re-adjusted as necessary;
> - have identification in order to determine its calibration status;
> - be safeguarded from adjustments that would invalidate the measurement result;
> - be protected from damage and deterioration during handling, maintenance and storage.
>
> In addition, the organization shall assess and record the validity of the previous measuring results when the equipment is found not to conform to requirements. The organization shall take appropriate action on the equipment and any product affected.
>
> Records of the results of calibration and verification shall be maintained …

In addition, the ISO 10012–1 standard contains the requirements to carry out and control calibration on a regular basis, increasing its importance in ensuring product quality.

Furthermore, the standard contains the requirement to carry out documentation of the uncertainty determination method for the measurements. When determining the uncertainty the ISO 10012-1 [84] standard does not permit any derogations and at the same time refers to ISO 9004-1 [90] Chap. 13. There are also recommendations related to measuring device quality assurance and their connections with units by a system of calibrations. Among these recommendations an undisturbed chain of links can be created through a system of independent accredited institutions (calibration laboratories of measuring equipment; manufacturers do not provide this independence, which is a common reason for failures to meet system requirements). In conclusion, it should be noted that the system requirements are fully compliant with the essential requirements contained in the GUM guide [21]. Chapter 6 describes the ISO 10360 standard for controlling coordinate system accuracy. A set of German recommendations VDI/VDE 2617 commenting on the standard determinations and often outpacing its determinations is also of significant importance.

In industrial plants, particularly in the automotive industry, different internal directives are used for measurement uncertainty management in the quality management system, consisting of examination of the measurement capability. For example, companies that apply such directives are Bosch, Daimler Benz, and Ford [119]. The foundation for the process of measuring resource ability examination is the development of proprietary recommendations. In the case of the Bosch and Daimler Benz companies it is the ratio of the measurement uncertainty to the

Table 2.1 Confrontation of measurement capability calculation procedures on example of the three companies

Przedsiębiorstwo	$\mathbf{c_g}$	$\mathbf{c_{gk}}$	Wymaganie
Bosch	$\frac{0.2 \cdot T}{6 \cdot s_M}$	$\frac{0.1 \cdot T - a_s}{3 \cdot s_M}$	>1.33
Ford	$\frac{0.15 \cdot 6\sigma_p}{3 \cdot s_M}$	$\frac{0.075 \cdot 6\sigma_p - a_s}{3 \cdot s_M}$	>1.00
Daimler Benz	$\frac{0.133 \cdot T}{2 \cdot s_M}$	$\frac{0.133 \cdot T}{a_s + 2 \cdot s_M}$	>1.33

Symbols
T Tolerance of checked dimension
σ_p Production process standard deviation
s_M Measurement results standard deviation for repeated measurements
a_s Sum of measurement result systematic errors derived from the standard calibration

tolerance, and in Ford the dispersion of the production process. Although their definitions in the recommendations are identical, the rules of their calculation are different. The principle of the capacity calculation is shown for the example of three producers in Table 2.1.

Although the index $\mathbf{c_g}$ covers only the dispersion of the measured value, the component $\mathbf{c_{gk}}$ takes into account the deviation of the average value from the standard calibration value, and thus takes into account the systematic deviations of the measurement result.

Essentially in all discussions on measurement capability it is required that the result be corrected of systematic errors. Systematic errors of measuring resources are included in the form of measurement capability component c and thus added to the dispersion of the measurement result. This shows that in industry there is a need to close systematic errors in measurement uncertainty [40, 111]. A request formulated in the GUM [21] to correct systematic errors often turns out to be impractical from the measuring technique point of view. For many of the measuring resources such an eventuality is impossible to achieve, or uneconomical, however, confrontations of various detection methods are carried out and calculations are made in order to assess the measuring equipment.

To confirm the capability of the measurement process, tasks connected with measurement uncertainty must be known and remain within an acceptable relation to corresponding tolerances of an object. Measurement uncertainty depends not only on the measurement process itself, but also on dispersion of measured object material properties and on production conditions [135]. The measurement process capability is marked by g_{pp}. It is the ratio of the expanded uncertainty U $(k = 2)$ to the tolerance T of measured value:

$$g_{pp} = 2 \times \frac{U}{T} \tag{2.23}$$

The g_{pp} parameter permits us to state that measurement-expanded uncertainty is consistent with the specification. This uncertainty is specific for a given measuring task, so that the tolerance T associated with the product specification applies only to the given task.

Assessment of the measurement capacity can be defined as the sum of knowledge of the error sources in the measurement process, which is an important requirement for the assessment of the task of uncertainty determination. In coordinate measurements error sources are specified, and the examples are presented in Fig. 3.1. The measurement process capability is confirmed when the ratio of the expanded uncertainty U and tolerance T does not exceed the G_{pp} limit value [135]:

$$g_{pp} = 2 \times \frac{U}{T} \leq G_{pp} \tag{2.24}$$

Generally, the limit value of G_{pp} is between 0.2 and 0.4. However, higher values can be adopted due to the specificity of tools and measuring systems or because of difficulties in measurement realization. Also in accordance with the historical Berndt principle G_{pp} can be taken at 0.5. However, the primary principle is here: the smaller the limit value is, the greater the cost of the measurement. The final choice of the upper limit values should be made with full awareness of the fact that the regulation of the production process may be insufficient. Therefore assessment of the real measurement accuracy gains key importance. In the case of coordinate measurements the determination of their accuracy is a difficult task, but the developed methods are:

- Method using calibrated artifact or standard (ISO/TS 15530–3) [37]
- Multiposition method with the use of noncalibrated artifact, based on specification ISO CD TS 15530–2 [35]
- Error budget: uncertainty estimation based on the identification of the sources of their individual components
- Analytical method [42–47]
- Expert method based on relations from the model of maximum geometric errors for the measuring task [12, 23–25, 31, 38, 93]
- Simulation methods, [1, 4, 9, 12, 28, 30, 39, 41, 48, 52–57, 66–68, 71, 94, 95, 100–106, 108–110, 112–115, 117, 118, 122, 123, 128–132, 134]

On the basis of the implementation method and availability, we can distinguish classical methods available for each user, and simulation methods, based on the use of a virtual measuring system, the design and installation of which are described in Chap. 5.

References

1. Abacerli, A.J., Pereira, P.H., Calonego, N.: A case study on testing CMM uncertainty simulation software (VCMM). J. Braz. Soc. Mech. Sci. Eng. **XXXII**(1/9) (2010)
2. Balsamo, A., Pedone, P., Ricci, E., Verdi, M.: Low-cost interferometric compensation of geometrical errors; CIRP Ann. Manuf. Technol. **58**(1), 459–462 (2009)
3. Baran, W.: Teoretyczne podstawy opracowania wyników pomiarów geodezyjnych (Teoretical base for development of geodetic measurements results) PWN, Warsaw (1983)
4. Beaman, J., Morse, E.: Experimental evaluation of software estimates of task specific measurement uncertainty for CMMs. Precis. Eng. **34**(1), 28–33 (2010)
5. Belforte, G., Bona et al.: Coordinate measuring and machine tools selfcalibration and error correction. Ann. CIRP **36**(1):359–364 (1987)
6. Bosch, J.A.: Coordinate Measuring Machines and Systems. Marcel Dekker Inc., New York (1995)
7. Busch, K., Kunzmann, H., Waldele, F.: Numerical error correction of CMM. In: Proceedings of the International Symposium on Metrology for Quality Control in Production, Tokyo, pp. 270–282 (1984)
8. Curran, E., Phelan, P., Quick check error verification of coordinate measuring machines. J. Mater. Process. Technol. pp. 155–156, pp. 1207–1213 (2004)
9. van Dorp, B., Haitjema, H., Delbressine, F., Bergmans, R., Schellekens, P.: Virtual CMM using Monte Carlo methods based on frequency content of the error signal. In: Decker, J.E., Brown, N. (Eds.) Proceedings of the SPIE 4401 Recent Developments in Traceable Dimensional Measurements', SPIE (2001)
10. Dutschke, W.: Fertigungsmesstechnik. Taubner Verlag, Stuttgart (1993)
11. Deutsches Institut fur Normung. (1996) DIN 1319-1: Fundamentals of Metrology. Part 3: Evaluation of Measurements of a Single Measurand, Measurement Uncertainty
12. EASYTRAC Project: "Easier and cheaper traceability in industry by up-to-date methods of calibration, "EU Contract No. G6RD-CT-2000-00188 Work package 6.4 "Reversal technique to calibrate gear and thread standards"—Final Report from DTU—Calibration of Thread Gauges on CMMs
13. Estler, W.T., Edmundson, K.L., Peggs, G.N., Parker, D.H.: Large-scale metrology—an update. CIRP Ann. Manufac. Technol. **51**(2), 587–609 (2002)
14. ETALON AG, www.etalon.de
15. FEINMESS GmbH & Co. KG, www.feinmess.com
16. Flack, D.: CMM Verification. Measurement Practice Guide No. 42. New Delhi, National Phisical Laboratory (2001)
17. Fletcher, S., Longsta, A., Myers, A.: Investigation into the accuracy of a proposed laser diode based multilateration machine tool calibration system. J. Phys.: Conf. Ser. **13**, 398–401 (2005)
18. Franke, M., Hartig, F., Wendt, K.: Measuring large 3D structures using a portable 4-arm laser interferometer. In: Sładek, J., Jakubiec, W. (eds.) Advances in Coordinate Metrology, Monograph, pp. 35–42. University of Bielsko-Biała, Bielsko-Biała. ISBN 978–83–62292–52–3 2010 (2010)
19. Geise, G., Hartmann, M.: Geometrische Aspekte bei Ausgleichproblemen in der Koordinatenmesstechnik. Konferenz INFERT 82, Dresden 1–2 Sept 1982
20. Główny Urząd Miar—Central Office of Measures (Warsaw 1996)—International vocabulary of basic and general terms in metrology (VIM)—second edition 1993 ISO
21. Główny Urząd Miar—Central Office of Measures (Warsaw 1999) Guide to the Expression of Uncertainty in Measurement (GUM) ISO 1993/1995 with the addition to polish edition of J. M. Jaworski
22. Halupczok, E.: Definicja dokładności współrzędnościowej techniki pomiarowej na przykładzie maszyny UPMC CARAT (Definition of accuracy of coordinate measurement

technique on the example of UPMC CARAT machine) Carl Zeiss Industrielle MeBtechnik GmbH, Esslingen
23. Hansen, H.N.: A database system for uncertainty estimation in coordinate metrology. Metrology for Quality Control in Production. In: Proceedings of 6th ISMQC IMEKO
24. Hansen, H.N., Trapet, E.: An approach to uncertainty estimation in coordinate metrology. In: Kunzmann, H., Waldele, F., Wilkening, G., Corbet, J., MacKeow, P.P., Weck, M., Huemmler, J. (eds.) Progress in Precision Engineering and Nanotechnology, vol. 1, pp. 323–326. PTB—Braunschweig u. Berlin P.u.Oe. (1997)
25. Hansen, H.N., De Chiffre, L., Savio, E.: Traceability in coordinate metrology. In: PRIME 2001 International Conference, pp. 363–368 (2001)
26. Hart, H., Lotze, W., Woschni, E.G.: Messgenauigkeit. Verlag Technik, Berlin (1987)
27. Hartmann, W.W., Geise G.: Displaydarstellungen als Entscheidungshilfe fur Messauswertungsstrategien in der Koordinatenmesstechnik. Feingeratetechnik 1984/1
28. Haitjema, H., van Dorp, B., Morel, M., Schellekens, P.H.J.: Uncertainty estimation by the concept of virtual instruments. In: Decker, J.E., Brown, N. (eds.) Proceedings of the SPIE 4401 Recent Developments in Traceable Dimensional Measurements', SPIE (2001)
29. Hartig, F.: How to work with the on-line virtual CMM. In: Proceedings of Workshop on Trace- bylity of CMM—PTB, Braunschweig October 9–10 (1997)
30. Herbst, C., Tutsch, R.: Simulation of micro coordinate measuring machine for parallel measurement of microstructures. Sładek, J., Jakubiec, W.(eds.) Advances in Coordinate Metrology, pp. 15–24. University of Bielsko-Biala, Bielsko-Biala (2010)
31. Hernla, M.: Messumsicherheit bei Koordinatenmessungen. Expert Verlag (2007)
32. Hocken, R.: Tree-dimensional metrology. Ann. CIRP **26**(2), 378–382 (1977)
33. Hughes, E.B., Wilson, A., Peggs, G.N.: Design of a high-accuracy CMM based on multilateration techniques. CIRP Ann. Manuf. Technol. **49**(1), 391–394 (2000)
34. ISO 9000: Series of standards for quality management and quality assurance system
35. ISO/CD TS 15530–2: Use of multiple measurement strategie
36. ISO/TR 14638:1995, Geometrical Product Specifications (GPS): Masterplan
37. ISO/TS 15530–3:2004: Use of calibrated workpieces or standards
38. ISO/TS 15530–5 (Draft) GPS: Use of expert judgement
39. ISO/TS 15530–4:2008: Evaluating tasc-specific measurement uncertainy using simulation
40. Jaeger, G.: Limitations of precision length measurements based on interferometers. Measurement **43**(5), 652–658 (2010)
41. Jakubiec, W., Starczak, M.: Identification of parameters of virtual coordinate measuring machines model. In: IVth International Scientific Conference—Coordinate Measuring Technique, Lodz University of Technology Scientific Books, branch in Bielsko-Biała, no. 53, Bielsko-Biała 2000, pp. 115–118
42. Jakubiec, W., Starczak, M.: EMU—Friendly software for estimation of measurements uncertainty for CMM. In: 8th International Symposium on Measurement and Quality Control in Production. Erlangen 2004. VDI Verlag GmbH, Dusseldorf (2004)
43. Jakubiec, W., Starczak, M.: Metodyka wyznaczania niepewności pomiarów współrzędnościowych (Methodology of evaluation of the coordinate measurements uncertainty) In: VIth International Scientific Conference—Coordinate Measuring Technique, Bielsko- Biała. University of Bielsko-Biała, Scientific Books 10 (2004)
44. Jakubiec, W.: Analityczne wyznaczanie niepewności pomiaru we współrzędnościowej technice pomiarowej (Analilitical evaluation of the measurement uncertainty in coordinate measurement technique) University of Bielsko-Biała. Scientific Books 22, Bielsko-Biała (2008)
45. Jakubiec, W., Płowucha, W., Starczak, M.: EMU—oprogramowanie do analitycznego wyznaczania niepewności pomiarów współrzędnościowych (Software for analilitical evaluation of the coordinate measurements uncertainty). In: Sładek, J., Jakubiec, W. (eds.) Advances in Coordinate Metrology, pp. 83–89. University of Bielsko-Biała, Bielsko-Biała (2010)

46. Jakubiec, W., Płowucha, W., Starczak, M., Wizner, M.: Wdrażanie oprogramowania do analitycznego wyznaczania niepewności pomiarów współrzędnościowych (Implementation of the software for analilitical evaluation of the coordinate measurements uncertainty). In: Sładek J., Jakubiec, W. (eds.) Advances in Coordinate Metrology, pp. 90–97. University of Bielsko-Biała, Bielsko-Biała (2010)
47. Jakubiec, W., Płowucha, W., Starczak, M.: Analytical evaluation of the coordinate measurements uncertainy. In: Sładek, J., Jakubiec, W. (eds.) Advances in Coordinate Metrology, pp. 169–176. University of Bielsko-Biała, Bielsko-Biała (2010)
48. Joskowicz, L., Ostrovsky-Berman, Y., Myers, Y.: Efficient representation and computation of geometric uncertainty: The linear parametric model. Precis. Eng. **34**(1), 2–6 (2010)
49. Koch, K.P., Peter, R., Weisig, S., Koordinatenmessung mit einem Lasertiangulationstaster. Feinwerktechnik & Messtechnik F&M 95, Heft 6 (1988)
50. Kotulski, Z., Szczepiński, W.: Rachunek błędów dla inżynierów (Calculus of errors for engineers). WNT Warsaw (2004)
51. Kowalski, M.: Analiza pomiarów na współrzędnościowych maszynach pomiarowych obarczonych błędami kształtu (Analisys of measurements done with the use of coordinate measuring machines affected by form errors). Doctoral Dissertation, Cracow University of Technology (1993)
52. Kowalski, M.: Model wirtualnej wielowspółrzędnościowej maszyny pomiarowej i jej zastosowanie (Model of virtual multi-coordinate measuring machine and its application) In: Proceedings of VIth Scientific-Technical Conference Metrology in Production Engineering, pp. 218–223. Rzeszow University of Technology (1995)
53. Kowalski, M.: Wybrane aspekty realizacji modelu wirtualnej współrzędnościowej maszyny pomiarowej (Chosen aspects of realization of the virtual coordinate measuring machine model), pp. 117–126. Polish Academy of Sciences—KBM, Scientific Book no. 63, Works of Institute of Machine Technology and Production Automation, Cracow University of Technology (1999)
54. Kowalski, M.: Zastosowanie wirtualnej WMP do wyznaczania i minimalizacji błędów pomiarowych (The use of virtual CMM for determintaion and minimization of measurement errors). In: Proceedings of VIIIth scientific-technical conference Metrologia w technikach wytwarzania maszyn (Metrology in Machine Production), vol. II, pp. 401–408. Szczecin–Międzyzdroje 13–15 IX (1999)
55. Kowalski, M., Sładek, J.: Opracowanie wirtualnej wielowspółrzędnościowej maszyny pomiarowej z zastosowaniem do badań i korekcji błędów obiektów rzeczywistych i optymalizacji pomiarów (Development of virtual multi- coordinate measuring machine with the application for researches and correction of errors of real objects and for measurements optimization) PB 1367/T0795/08 KBN Research Grant (1995–1998)
56. Kowalski, M., Sładek, J., et al.: Opracowanie wirtualnej wielowspółrzędnościowej maszyny pomiarowej z zastosowaniem do badań i korekcji błędów obiektów rzeczywistych i optymalizacji pomiarów pomiarów (Development of virtual multi- coordinate measuring machine with the application for researches and correction of errors of real objects and for measurements optimization). Final Report of Research Project PB 1367/T07/95/08, Cracow University of Technology, not published (1998)
57. Kowalski, M., Sładek, J., Rakoczy, R.: Opis wirtualnej współrzędnościowej maszyny pomiarowej (Decription of virtual coordinate measuring machine). In: IIIrd International Scientific Conference—Coordinate Measuring Technique, pp. 71–80. Lodz University of Technology Scientific Books, branch in Bielsko-Biała, No. 44, Bielsko-Biała (1998)
58. Kupiec, M.: Optyczno-stykowa metoda pomiarów współrzędnościowych (Optic-contact method for coordinate measurements). Doctoral Dissertation—Cracow University of Technology, Faculty of Mechanical Engineering 2008 work realized under research grant with Department of Photonics Engineering of Institute of Micromechanics and Photonics—Faculty of Mechatronics at Warsaw University of Technology no. 3 T10C 010 29 nt: Optonumeryczny system do pomiaru elementów geometrycznych zintegrowany z

Współrzędnościową Maszyną Pomiarową (Opto-numerical system for measuring the geometrical elements integrated with Coordinate Measuring Machine) (2005–2008)
59. Lau, K.C., Hocken, R.J.: Tree and five axis laser tracking systems. US Patent No. 4714339 (1987)
60. Leica Geosystems, A.G.: www.metrology.leica-geosystems.com
61. Lotze, W.: Rechnergestutzte Koordinatenmesstechnik. Werkstatt und Betrieb **113**(6), 391–395 (1980)
62. Lotze, W.: Prufkorper fur Koordinatenmessgeraten. Feingeratetechnik 30, und VDI- Z 123 1981, pp. 154–155 (1981)
63. Lotze, W., Teichmann, U.: Einfluss von Gestalt – und Lageabweichungen auf die Unsicherheit der Rechnergestutzehn Koordinatenmessung. Feingeratetechnik **8**, 139–343 (1976)
64. Lotze, W., Teichmann, U.: Genauigkeit und Prufung von Koordinatenmessgeraten. Feingeratetechnik **35**, 339–342 (1986)
65. Lotze, W., Hartmann, M.W., Bressel, E.: Leistungsstand der rechnergestutzten Koordinatenmesstechnik an der TU Dresden. Feingeratetechnik **9**, 387–390 (1982)
66. Mailhe, J., Linares, J.M., Sprauel, J.M., Bourdet, P.: Geometrical checking by virtual gauge, including measurement uncertainties. CIRP Ann. Manuf. Technol. **57**, 513–516 (2008)
67. Maihle, J., Linares, J.M., Sprauel, J.M.: The statistical gauge in geometrical verification: Part I. Field of probability of the presence of matter. Precis. Eng. **33**(4), 333–341 (2009)
68. Maihle, J., Linares, J.M., Sprauel, J.M.: The statistical gauge in geometrical verification. Part II. The virtual gauge and verification process. Precis. Eng. **33**(4), 342–352 (2009)
69. Metoda Macierzowa (Matrix Method—MM)—Microsoft Excel Application—Cracow University of Technology Laboratory of Coordinate Metrology (2000)
70. Metropolis, N., Ulam, S.: The monte carlo metod. J. Am. Stat. Assoc. **44**(247), 335–341 (1949)
71. Metrosage: www.metrosage.com
72. Moustafa, S., Gerwien, N., Haertig, F., Wendt K.: Comparison of error mapping techniques for coordinate measuring machines using the plate method and Laser Tracer. In: XIX IMEKO World Congress Fundamental and Applied Metrology, Lisbon, Portugal, 6–11 Sept 2009
73. Nakamura, O., Goto, M., Toyoda, K., Tanimura, I., Kurosawa, A.: Development of coordinate measuring systems with tracking laser interferometers. Ann. CIRP **40**(1), 523–530 (1991)
74. Nawara, L., Kowalski, M.: Analiza błędów w pomiarach przestrzennych (Analysis of errors in spatial measurements). Mechanik **12**, 674–679 (1980)
75. Nawara, L., Sładek, J.: Investigations of measuring heads errors influence on the measuring accuracy of multi-coordinate machine. In: 7th International Conference on Automated Inspection and Product-Control, pp. 305–312. Birmingham England (1985)
76. Nawara, L., Kowalski, M., Sładek, J.: Badanie maszyn pomiarowych 3-koordynatowych ze względu na dokładność. PTMiU (Examination of 3-coordinate measuring machines due to accuracy PTMiU). Quarterly PAN (Polish Academy of Sciences). **2**, 141–150 (1983)
77. Nawara, L., Kowalski, M., Sładek, J.: The Influence of kinematics errors on the profile shapes by means of CMM. Ann. CIRP **37**(1), 398–401 (1989)
78. Nawara, L., Kowalski, M., Sładek, J.: Pewna metoda oceny błędów maszyn pomiarowych (Sure method for assessment of measuring machines errors). Mechanik **1**, 14–19 (1987)
79. Neumann, H.J i innii Praezisionmesstechnik in der Fertigungs mit Koordinatenmessgeraeten. Expert Verlag (2005)
80. Ostrowska, K.: Accuracy assessment method for measurements done with the use of articulated arm coordinate measuring machines. Doctoral Dissertation, Cracow University of Technology, Faculty of Mechanical Engineering (2010)
81. Ozono, S., Takamasu, K.: Data processing and calibration of coordinate measuring machine. Int. J. Jpn. Soc. Prec. Eng. **29**(2), 105–108 (1995)

82. Pfeifer, T., Bambach, M.: Definition und Prufung von Kriterien zur Bestimmung systematischer und zufalliger Fehler von Dreikoordinatenmessgeraten. Forschungsberichte des Landes Nordhein-Westfalen No. 2856
83. Physikalisch-Technische Bundesanstalt: www.ptb.de
84. PN-EN ISO 10012:2003- Systemy zarządzania pomiarami—Wymagania dotyczące procesów pomiarowych i wyposażenia pomiarowego (Measurement management systems —Requirements for measurement processes and measuring equipment)
85. PN/EN/ISO 14253: Kontrola wyrobów i sprzętu za pomocą pomiarów (Inspection by measurement of workpieces and measuring equipment)
86. PN/EN/ISO 14253-1: Reguły decyzji przy orzekaniu zgodności lub niezgodności ze specyfikacją (Decision rules for proving conformance or non-conformance with specifications)
87. PN/EN/ISO 14253-2: Przewodnik do oceny niepewności pomiarów podczas wzorcowania sprzętu pomiarowego i kontroli wyrobów (Guidance for the estimation of uncertainty in GPS measurement, in calibration of measuring equipment and in product verification)
88. PN-EN ISO/IEC 17025:2005: Ogólne wymagania dotyczące kompetencji laboratoriów badawczych i wzorcujących (General requirements for the competence of testing and calibration laboratories)
89. PN-EN ISO 9001:2009: Systemy zarządzania jakością. Wymagania (Quality management systems—Requirements)
90. PN-EN ISO 9004:2001: Systemy zarządzania jakością—Wytyczne doskonalenia funkcjonowania (Quality management systems—Guidelines for performance improvements)
91. PN–71/N-02050: Metrologia nazwy i określenia (Metrology names and terms)
92. Polski Komitet Normalizacji: Międzynarodowy słownik metrologii—Pojęcia podstawowe i ogólne oraz terminy z nimi związane (VIM). PKN-ISO/IEC Guide 99:2010 ("International vocabulary of metrology—Basic and general concepts and associated terms (VIM)") (2010)
93. Pressel, H.G.: Genau messen mit Koordinatenmessgeraeten. Expert Verlag, Rennigen-Malmsheim (2003)
94. Schwenke, H.: Abschatzung von Messunsicherheiten durch Simulation an Beispielen aus der Fertigungsmesstechnik. Dissertation der Technischen Universitat Chemnitz—PTB-F 36 Braunschweig Juli (1999)
95. Schwenke, H., Trapet, E., Waldele, F.: Calibration of coordinate measuring machines to improve and to know the measurement uncertainty. In: IIIth International Scientific Conference—Coordinate Measuring Technique, Lodz University of Technology Scientific Books, branch in Bielsko-Biała, No. 44, Bielsko-Biała, pp. 213–232, 1998
96. Schwenke, H., Franke, M., Hannaford, J., Kunzmann, H.: Error mapping of CMMs and machine tools by a single tracking interferometer. CIRP Ann. Manuf. Technol. **54**(1), 475–478 (2005)
97. Schwenke, H., Knapp, W., Haitjema, H., Weckenmann, A., Schmitt, R., Delbressine, F.: Geometric error measurement and compensation of machines—an update. CIRP Ann. Band **57**(2), 660–675 (2008)
98. Sitnik, R., Sładek, J., et al.: Opto-numeryczny system do pomiaru elementów geometrycznych zintegrowany z Współrzędnościową Maszyną Pomiarową (Opto-numeric system integrated with Coordinate Measuring Machine for measuring geometrical elements) Reseach grant no. 3 T10C 010 29
99. Sładek, J.: Assessment of the accuracy of contact probe heads used in coordinate measuring machines. Ph.D. Thesies—Cracow University of Technology, Faculty of Mechanical Engineering (1990)
100. Sładek, J.: Metody oceny dokładności WMP a prognozowanie dokładności pomiarów (Methods for accuracy assessment of CMM vs forecasting of measurement accuracy). In: IInd National Scientific Conference (with international participation)—Coordinate Measuring Technique, pp. 211–221. Lodz University of Technology Scientific Books, branch in Bielsko-Biała, Bielsko-Biała (1996)

101. Sładek, J.: Metrological assessment of CMMs and the possibilities of forecasting a measurement accuracy. In: 13 Conference, BIAM'96, Zagreb, Croatia, pp. D15–D183 June 18–21 1996
102. Sładek, J.: Strategy of applying coordinate metrology in quality assurances systems with regards to a measuring accuracy. In: Ford, D.G., Postlethwaite, S.R. (eds.) Laser Metrology and Machine Performance, pp. 497–508. Computational Mechanic Publications, Southampton-Boston (1997)
103. Sładek, J.: Errors identification and measurement accuracy assessment of coordinate measuring machines (CMM). Adv. Technol. Mach. Equip. Postępy Technologii Maszyn i Urządzeń, Quarterly PAN (Polish Academy of Sciences). **23**(4), 113–136 (1999)
104. Sładek, J., Model wirtualny WMP z zastosowaniem do oceny dokładności realizowanych pomiarów w warunkach czasu quasi-rzeczywistego (CMM virtual model with the application to accuracy assessment of measurements realized in equations of quasi- real time). In: IVth International Scientific Conference Coordinate Measuring Technique. pp. 193–208. Lodz University of Technology Scientific Books, branch in Bielsko-Biała, No. 53, Bielsko-Biała (2000)
105. Sładek, J.: A neural network model of CMM applied for measurement accuracy assessment. In: Proceedings IMECO World Congress, Vienna, vol. 11, 23–26 Sept 2000, pp. 281–286
106. Sładek, J.: Modelowanie i ocena dokładności maszyn oraz pomiarów współrzędnościowych (Modeling and accuracy assessment of coordinate machines and measurements), No. 87. Cracow University of Technology Scientific Books—Mechanika, Cracow (2001)
107. Sładek, J: Conception of description of coordinate measurement accuracy—on the basis of determination 3D—point uncertainty error. J. Mech. Eng. **53**(6) (2002)
108. Sładek, J.: Opracowanie systemu oceny dokładności pomiarów oraz metody identyfikacji własnego pola błędów współrzędnościowych maszyn pomiarowych (Developing of a system for assessing the accuracy of measurements and of a method for identification coordinate measuring machine's errors field). Final Report of the Research Project funded by the KBN under contract No. PB-0921/T07/99/1
109. Sładek, J.: Metoda oceny dokładności pomiarów realizowanych redundantnymi systemami współrzędnościowymi (RSW) (Method for accuracy assessment of measurements done with the use of redundant coordinate systems (RCS)) Research Project No. N505 255935 (2008–2010)
110. Sładek, J., Kowalski, M.: Anwendung des Virtuellen normales zur Schätzung der Genauigkeit bei Mehrkoordinaten-Messungen. 6 Internationales Symposium von Donau Adria Assoziation für Automation und Messtechnik, DAAAM, Automation und Metrologie: Mensch, Technik, Umwelt: TU Krakow, Fakultät Maschinenbau, Cracow 24–26 Okt. 1995, pp. 187–188
111. Sładek, J., Krawczyk, M.: Rozwój metrologii współrzędnościowej w systemach zapewnienia jakości. Zagadnienia jakości i środowiska eksploatacji maszyn (Development of Coordinate Metrology in quality assurances systems. The quality and machine operating environment issues), Wrocław, 2004. Inżynieria Maszyn, Agenda Wydawnicza Wrocławskiej Rady FSN NOT
112. Sładek, J., Rakoczy, R.: Wykorzystanie koncepcji sztucznych sieci neuronowych do modelowania współrzędnościowej maszyny pomiarowej (WMP) przy zastosowaniu wzorca przestrzennego (The use of the concept of artificial neural networks for modeling the coordinate measuring machine (CMM) with the use of spatial standard) Works of Institute of Machine Technology and Production Automation, Polish Academy of Sciences, Committee on Machine Building, Technology Bases Section, Scientific Book no. 63, Cracow University of Technology, 1999, pp. 193–206
113. Sładek, J., Rakoczy, R., Szwajkowski, A.: Application of neural networks for modelling coordinate measuring machines. In: Kunzmann, H., Waeldele, F., Wilkening, G., Corbet, J., MacKeow, PP., Weck, M., Huemmler, J. (eds.) Progress in Precision Engineering and Nanotechnology, vol. 1, pp. 323–326. PTB-Braunschweig u. Berlin P.u.OeVerlag (1997)

114. Sładek, J., Rakoczy, R., Majdosz. R.: Realizacja i badanie powiązań informatycznych CAD WMP CAD z uwzględnieniem możliwości prognozowania dokładności pomiarów (Realisation and testing of the CAD CMM CAD IT couplings, considering the aspect of possibilities of measurement accuracy prognostication). In: IVth International Scientific Conference—Coordinate Measuring Technique, pp. 209–221. Lodz University of Technology Scientific Books, branch in Bielsko-Biała, no. 53, Bielsko-Biała (2000)
115. Sładek, J., Rakoczy, R., Majdosz, R.: Research and realisation of the CAD-CMM-CAD couplings, considering the aspect of possibilities of measurement accuracy prognostication. In: Proceedings of the 2nd International Conference on Surface Machining and Measurements of Sculptured Surfaces, Cracow 20–23 Sept 2000, pp. 71–84
116. Sładek, J., Krawczyk, M., Ostrowska, K., Gąska, A.: Zastosowanie metody Monte Carlo do wyrażania niepewności pomiarów współrzędnościowych na przykładzie metody wielopozycyjnej (The use of the Monte Carlo method for estimation of coordinate measurements uncertainty on the example of multi-position method). University of Bielsko-Biała Scientific Books no. 81, Bielsko-Biała (2008)
117. Sładek, J., Ostrowska, K., Gacek, K.: Kinematic metrological model of the coordinate measuring arm (MCMA). In: XIX IMEKO World Congress Fundamental and Applied Metrology, Lisbon, Portugal
118. Sładek, J., Ostrowska, K., Gąska, A.: Wirtualne współrzędnościowe ramię pomiarowe (WWRP) (Virtual articulated arm coordinate measuring machine (VAACMM), Pomiary Automatyka Kontrola (Measurement Automation and Monitoring), 01/2010
119. Sładek, J., Ostrowska, K., Gacek, K., Bryndza, M.: Designation of operator impact on errors of measurements realized by coordinate measuring arm. In: Sładek, J., Jakubiec, W. (eds.) Advances in Coordinate Metrology. University of Bielsko-Biała, Bielsko-Biała (2010)
120. Sładek, J., Sitnik, R., Kupiec, M., Blaszczyk, P: The hibrid coordinate measurement system as a response to industrial requirements, Metrol. Meas. Syst. **XVII**(1), 109–118` (2010)
121. Sładek, J., Blaszczyk, P.M., Kupiec, M., Sitnik, R.: The hybrid contact-optical coordinate measuring system. Measurement **44**, 503–510 (2011)
122. Sommer, K.D., Siebert, B.R.L.: Systematic approach to the modelling of measurements for uncertainty evaluation. Metrologia **43**, 200–210 (2006)
123. Summerhays, K.D., Baldwin, J.M., Campbell, D.A., Henke, R.P.: Application of simulation software to coordinate measurement uncertainty evaluation. In: Proceedings of Aspe "Uncertainty Analysis in Measurement and Design", June–July 2004
124. Suplement 1 to the "Guide to the expression of uncertainty measurement". Propagation of distribution using Monte Carlo method, JCGM (2006)
125. Teichmann, U.: Metrologische Probleme bei der Pruefung von koordinatenmessgeraeten. Feingeraetentechnik, 3 (1978)
126. Teoh, P.L., Shirinzadeh, B., Foong, C.W., Alici, G.: The measurement uncertainties in the laser interferometry-based sensing and tracking technique. Measurement **32**(2), 135–150 (2002)
127. Takatsuji, T., Goto, M., Kirita, A., Kurosawa, T., Tanimura, Y.: The relationship between the measurement error and the arrangement of laser trackers in laser trilateration. Measur. Sci. Technol. **11**(5), 477 (2000)
128. Trapet, E.: Introduction to traceability with the virtual CMM. In: Proceedings of Workshop on Tracebylity of CMM, PTB-Braunschweig 9–10 Oct 1997
129. Trapet, E., Waeldele, F.: The virtual CMM concept. In: Ciarlini, PP., Cox, M.G., Pavese, F., Richter, D. (eds.) Advanced Mathematical Tools, II World Conference Scientific, Singapore 1996, pp. 239–247
130. Trapet, E., et al.: Tracebility of coordinate measurements according to the method of the virtual measuring machine. European Commission SMT- EDG XII/C/5,Project No. Mat.1 CT94-0076 final report
131. Trapet, E., Franke, M., Hartig, H., Schwenke, H., Waldele, F., Cox, M., Forbes, A., Delbressine, F., Schellekens, P., Trenk, M., Meyer, H., Moritz, G., Guth, Th., Wanner, E.:

Tracebility of coordinate measurements according to method of virtual measuring machine. PTB F–35, Braunschweig (1999)
132. Trenk, M., Franke, M., Schwenke, H.I.: The "Virtual CMM", a software tool for uncertainty evaluation—practical application in an accredited calibration lab. In: ASPE Proceedings: Uncertainty Analysis in Measurement and Design, July 2004
133. Treib T.: Error budgeting—applied to calculation and optimisation of the volumetric error field of multiaxis systems. Ann. CIRP **36**(1), 365–368 (1987)
134. VDI/VDE 2617 Genauigkeit von Koordinatenmessgeraten KenngroBen und deren Prufung. Blatt7.Ermittlung der Unsicherheit von Messungen auf Koordinatenmessgeraten durch Simulation VDI/VDE- Richlinien ver. 2008
135. VDI/VDE 2617 Genauigkeit von Koordinatenmessgeraten KenngroBen und deren Prufung. Blatt 11 Messaufgabenbezogene Messunsicherheit VDI/VDE- Richlinien entwurf 2010
136. Weckenmann, A.: Ubersicht uber Abnahme und Uberwachungsverfahren fur Mehrkoordinaten-Messgerate. VDI-B 378, 1–10 (1980)
137. Wendt, K., Schwenke, H., Waldele, F., Krawczyk, M., Kniel, K.: Error mapping of large CMM's by sequential multi-Iateration using a lasertracker. In: Materials of conference. EUSPEN, 2001
138. Zhang, G., Veale, R., Charlton, T., Borchardt, B., Hocken, R.: Error compensation of coordinate measuring machines. Ann. CIRP **34**(1), 445–447 (1985)
139. Zhang, G.X., LI, X.H., Lin, Y.B., Liu, S.G., Liu, X.L., LI, X.F., Guo, J.B., Qiu, Z.R., Zhao, S.Z., Jiang, C.Z., Fan, Y.M.: A study on the optimal design of laser-based multi-lateration systems. CIRP Ann. Manuf. Technol. **52**(1), 427–430 (2003)
140. Zhuang, H., Motaghedi, S.H., Roth, Z.S., Bai, Y.: Calibration of multi-beam laser tracking systems. Robot. Comput. Integr. Manuf. **19**(4), 301–314 (2003)

Chapter 3
Classic (Nonsimulative) Methods of Measurement Accuracy Assessment

Abstract In this chapter, sources and causes of errors of coordinate measuring systems (CMSs) are described and defined. Their impact on measurement uncertainty is also presented. The classic (nonsimulative) methods of coordinate measurement accuracy assessment, including ISO 15530 standard requirements and also procedures used also in the Laboratory of Coordinate Metrology at Cracow University of Technology (LCM CUT), as well as selected examples of their use are also described. A detailed description of both the method using a calibrated object or standard (accordingly to ISO 15530 Part 3) and the method for a noncalibrated object (multiposition method) is presented in this chapter, and practical applications are described in Annex A. Furthermore, new findings on these methods in the form of Monte Carlo method application to estimate uncertainty in the multiposition and substitution method, as well as the issue of estimation of freeform profile measurement uncertainty, are reported here. Findings on the difficult problem of estimating measurement uncertainty for calibrations based on the identification of error sources and on creation of an error budget are also presented. This chapter also describes methods based on dependencies resulting from the model of maximum permissible errors (MPEs) of a CMS and analytical method. It ends with the conclusion that it is necessary to create a method that will be more universal and easier in direct realization. It is also important to make it possible to determine the measurement uncertainty for every single point in the CMM measuring volume, and consequently, to determine measurement uncertainties for all measuring tasks that were performed. The research on simulative methods and the concept of so-called virtual CMM described in Chap. 5 are also introduced here.

Chapter 2 shows that the assessment of measurement accuracy is closely related to the issue of assessment of the accuracy of the measuring device. These issues can hardly be separated, nevertheless the estimation of measurement accuracy is a wider issue, including within its scope the measuring device error with factors affecting it, shown in Fig. 3.1.

Based on a metrological practice it can be said that quite often, for the purpose of single measurement accuracy analysis, the error mainly considered is the error coming from the machine itself. Also quite often, due to lack of other options, the

J.A. Sładek, *Coordinate Metrology*, Springer Tracts in Mechanical Engineering,
DOI 10.1007/978-3-662-48465-4_3

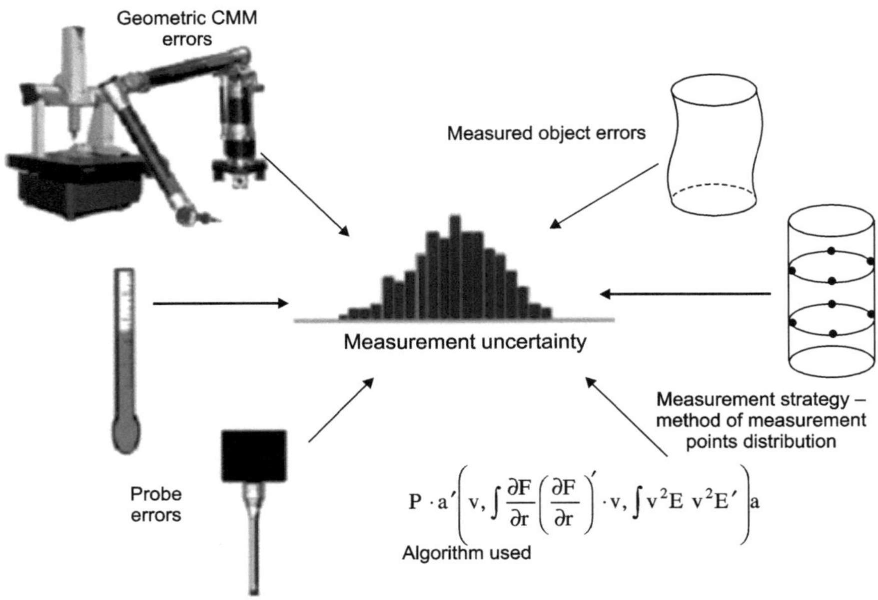

Fig. 3.1 Sources and causes of errors of coordinate measuring systems and their impact on measurement uncertainty [24]

CMM user who assesses measurement accuracy of even a complex object, is forced to refer to an equation of the CMS maximum permissible errors (MPEs) and to use the length measurement, the range of which (dimension) corresponds to the task carried out. This often leads to wrong decisions and also indicates a lack of the necessary knowledge of accuracy assessment methods. Furthermore, these methods are still being discussed and have not been clearly established. Most coordinate measurement laboratories try to develop methodologies often based on preliminary projects of standards or recommendations. A good example is the history of agreements of the ISO 15530 standard series. This issue, as has been distinguished many times before, is of crucial importance for coordinate metrology. The first work on this issue came from West European Calibrate Committee (WECC), where, following the requirements of the GUM guide [5], two methods shown in Fig. 3.2 were defined. These are: a method of direct comparison, realizing in general assumption of the type *A* identification of measurement uncertainty and a method involving determination of the so-called error budget, based on a system of reasoning, leading to uncertainty determination by the type *B* method.

Overall accuracy assessment methods included in the group of classical, analytical, and expert ones, according to the classification presented in the chapter are mainly developed under laboratory conditions and their effective adaptations are more and more often taken up in industry. In developing the first implementations of methods for assessing measurement accuracy, special attention should be paid to

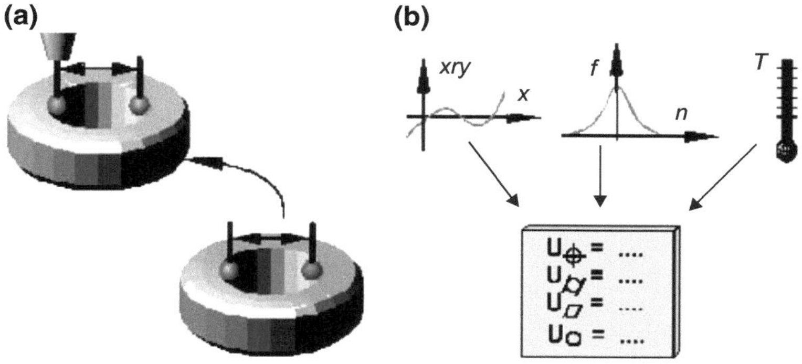

Fig. 3.2 Methods of measurement error evaluation: the substitution method(s), the analytical method (**a**), error budget (**b**) [54]

the method using a calibrated object or standard, derived from the classical comparative method, and to the method of a noncalibrated object with multiple measurement strategy. Both can be implemented in conditions of industrial laboratories. Therefore, examples of implementation of the noncalibrated object method as the more universal one and more often used in practice in the LCM CUT are posted in Annex A.1. This should also facilitate development of procedures complying with the requirements of these methods.

3.1 Method Using the Calibrated Object or the Standard

Experimental methods for measurement accuracy assessment are based, as defined in Chap. 2 and in accordance with method *A*, on the rule of the comparison, and have been used for about a century to measure standards such as gauge blocks or gauges. In turn, the experimental methods of coordinate measurement uncertainty estimation are based on the statistical evaluation of measurement errors observed in relation to the calibration value of a reference object. The user must perform a suitable number (>20) of measurements under various conditions, and during measurement real objects can be used. Such an approach seems to be very simple and clear from the user's point of view. Its consequence is the development of ISO technical specifications presenting the method using a calibrated object or standard included in the ISO/TS 15530-3:2004 [9], which defines the conditions and the method for coordinate measurement uncertainty estimation. This method makes possible experimental determination of realized measurement accuracy using the calibration of a reference object or standard of size and shape similar to the measured object. It is also the basis for elaboration of the calibration procedures in laboratories [2, 31, 37, 38, 41, 46, 48, 52, 53, 58]. It has been derived from the comparative method. The sources of errors are not important for the measurement results performed with it, but

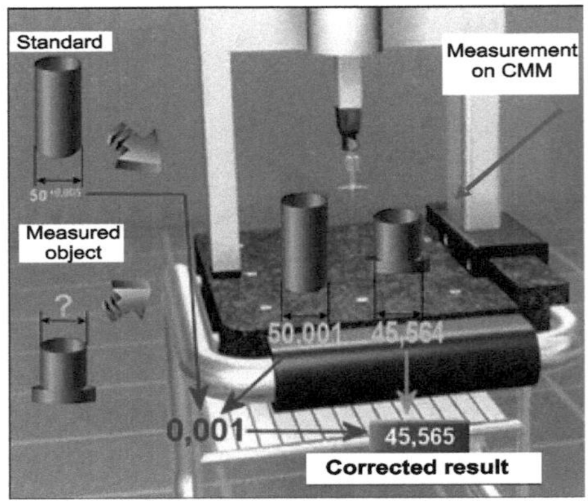

Fig. 3.3 Idea of substitution method

the assessment of their effects in the form of deviations from the nominal dimension is. The principle of this method is shown in Fig. 3.3. The CMS is used as a comparator between the standard and the measured object.

The measurement procedure is carried out in the following order. At the beginning the measurement of the standard is performed. The result obtained in this way is interpreted as the CMM error for this measuring task. Then, the measurement of an object is performed and the error value is added to the obtained result. The quantity obtained in this way is assumed to be the true value of the measured geometric feature.

This method is based on a comparison of given geometric feature measurement results with the values of corresponding geometric features of the standard. It does not identify clearly the causes of errors, but carries out the assessment of their effects in the form of determination of deviations from the reference dimension. To carry out a measurement it is necessary to use a material standard (e.g., in the form of a gauge block or a standard ring depending on the measuring task) having dimensional similarity to the measured dimension.

To properly carry out an experiment, the following conditions should be kept:

1. Measurement strategy:
   (a) Location and orientation of the measuring object and the standard in the same place of the workspace; that is, the object and the standard must be situated on the same axis, plane, and at the same height.
   (b) The strategy for collecting measuring points must be the same, that is, the same number and method for collecting measuring points.
   (c) The same type and kind of a probe head and a configuration of styli.
2. Ambient condition similarity regarding:
   (a) Temperature
   (b) Humidity
   (c) Time

Note: At this stage, it is recommended to dismount the standard and to locate it again in another part of the workspace, so that measurement conditions are as close as possible to industrial conditions.

Based on the information given above, the principle of the method is to measure on CMS the features of an object and of a standard at the same location in the machine measuring volume. The standard nominal dimension is compared with the result of the standard measurement obtained from CMS; this is the coordinate system calibration stage. In the second stage the deviation obtained during the calibration stage is included in the presented result of the measured quantity. For example, in order to determine the uncertainty of a cylindrical hole diameter, the analysis should be done with a reference ring or other cylindrical standard with diameter close to the measured object (Table 3.1).

The main factor affecting the accuracy of the method is the acceptable differences between the measured dimension and the standard representing this dimension. Permissible tolerances are presented in Table 3.1 [9].

The biggest advantage of this method is the simplicity of its implementation. The knowledge of CMM component errors is not required here, but the equal conditions of measurement realization must be kept. Also, the machine does not have (theoretically) to have a very high accuracy, but it should have good temporary repeatability, which is easier to achieve than high (long-term) accuracy. In practice, however, this method requires a very sophisticated system to ensure stable conditions for measurement realizations and the use of a high accuracy machine, because the use of CMM that should be recalibrated frequently is not cost effective. Therefore the substitution method is used in calibration laboratories, and aside from that, mainly in mass control of tests carried out in special measurement centers (Fig. 3.4, Table 3.2).

The substitution method can be used in mass production, where from an economic point of view, it becomes profitable. However, it is then required to ensure stable conditions for measurement realization (compensation of the effects of thermal conditions for both CMM and the measured object should be done).

Table 3.1 Permissible deviations between the measured object and the standard used in the comparative method [9]

Examined parameter	Measurement realization conditions
Geometric dimensions	±10 % of length when dimension is bigger than 250 mm or ±25 mm when dimension is smaller than 250 mm ±5° angular orientation
Material—thermal expansion, hardness, etc.	Identical
Measurement strategy	Identical
Probe configuration	Identical
Measuring force at the contact point	±20 %

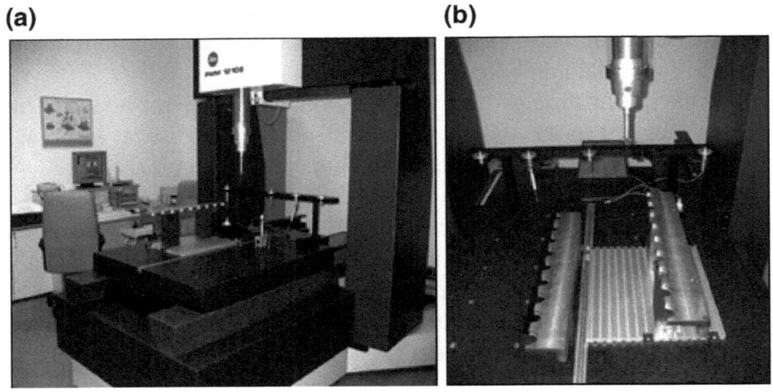

Fig. 3.4 Standard calibration on CMM, comparative measurement

Table 3.2 Example statement of results of measurement procedure

	Object 1	Object 2	Object 3
Measurement cycle 1	Distribution 1 $^{11}\mathbf{y}$	Distribution 5 $^{12}\mathbf{y}$	Distribution 9 $^{13}\mathbf{y}$
Measurement cycle 2	Distribution 2 $^{21}\mathbf{y}$	Distribution 6 $^{22}\mathbf{y}$	Distribution 10 $^{23}\mathbf{y}$
Measurement cycle 3	Distribution 3 $^{31}\mathbf{y}$	Distribution 7 $^{32}\mathbf{y}$	Distribution 11 $^{33}\mathbf{y}$
Measurement cycle 4	Distribution 4 $^{41}\mathbf{y}$	Distribution 8 $^{42}\mathbf{y}$	Distribution 12 $^{43}\mathbf{y}$
Standard deviation	$^{1}S_w$	$^{2}S_w$	$^{3}S_w$

Distributions of measurement points for three measured objects, ^{ij}y—measurement result of *ith* measurement cycle, of *jth* measured object

Figure 3.5 shows the concept of industrial use of the substitution method. In this way a given percentage of the production is tested, which ensures stability of accuracy, and simultaneously does not interfere with the production process. However, the biggest defect of this method is its extremely low versatility. In addition there should be expected high costs associated with the calibration and periodic standard checking to enable monitoring of the whole production.

The broadly accepted substitution method is defined according to [9] as a calibrated object method. It is shown in Fig. 3.6 and is a specific sequence of measurements of measuring objects carried out in the same way and under the same conditions as the standard measurements on CMS. The measurements have to be made for one or more calibrated objects, for which the measurement results are already known. The difference between the measurement result and the known measurement value of the calibrated object is used to calculate the measurement uncertainty for a given CMM.

The value of uncertainty obtained by this method is affected by the impact of uncertainty components caused by:

- Measuring procedure
- Calibrated object

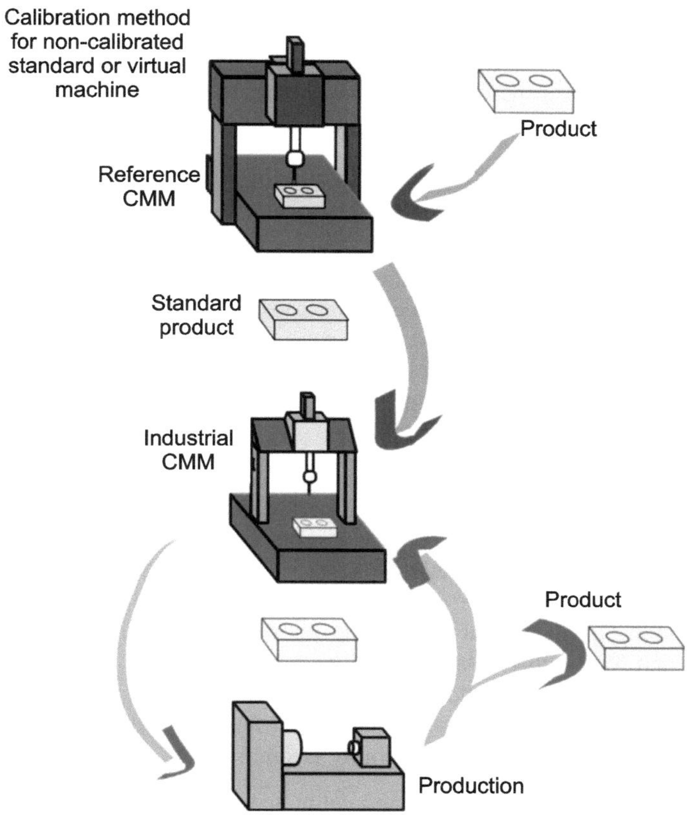

Fig. 3.5 Diagram of substitution method use for monitoring production quality [38]

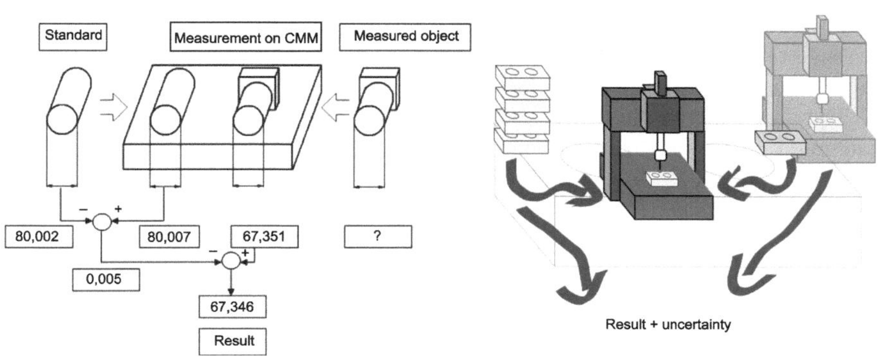

Fig. 3.6 Calibrated object method scheme [38]

- Properties of measured objects, such as surface roughness, coefficient of thermal expansion, other mechanical properties, and so on

According to [9], to perform the estimation of measurement uncertainty properly, any changes occurring in the measurement environment should be taken into account.

Therefore, it can be treated as an extension of the classical comparative method described in earlier versions of the ISO 15530-3 standard. However, there are some differences that can be noted even in the strategy of measurement itself. To perform the analysis properly, a measurement of five randomly selected details in at least ten different distributions of points should be carried out. The calibrated object or standard method also has its requirements for a measuring machine (generally for a coordinate system), its environment, measured object, and also for the measurement procedure, which states [9] that the measuring machine must meet these requirements:

- After the standard procedure of switching the machine on, preparation for the measurement of an object should be performed, that is, configuration of the probe tips, zeroing of the CMM scales and the probe tip calibration with an appropriate tool (e.g., calibration sphere).
- The machine used for the experiment should use parameters consistent with assumptions of its producer, and the temperature correction procedures for CMM and for an object itself should be used.
- An important factor in addition to temperature compensation is its stability; recommendation [9] emphasizes that during measurement, the CMM environment should have temperature stability ensured.

When it comes to the measured and calibrated object, the similarity conditions are identical as in the case of measurements performed with the substitution method (Table 3.1). A measured object should represent the same technological model, specified permissible deviations, and roughness (waviness) of surface, so that the probe has physical contact exactly in the same way as during the measurements of the calibrated object. The calibrated object should be calibrated with known and high enough accuracy, meeting the requirements of the measuring task.

The procedure employed for the measurement uncertainty estimation of a noncalibrated object should meet the requirements:

- Environmental condition stability
- Use of at least one calibrated object
- Measurement conditions for a noncalibrated object should be the same (or very similar) as for measurement of a calibrated object and taking into account:
  - Measuring tip configuration adequate for the given measuring task.
  - Mounting and manipulating of an object.
  - Measurement time and periods between the consecutive measuring points collection.

Measurement time and the CMM speed of movements associated with it. The measurement of a noncalibrated object can be carried out using one of the two methods:

1. *Nonsubstitution method*: Only a noncalibrated object is used here and the CMM uncorrected indication is considered to be the measurement result (Fig. 3.7).
2. *Substitution method*: In this case a calibrated object and noncalibrated object are measured in order to obtain additional systematic error correction on CMM. In Fig. 3.8 a scheme of such an approach is shown.

In both methods a single measurement cycle consists of the measured object mounting and of single or multiple measurement. As distinguished from, for example, the multiposition method (described in Sect. 3.2), position and orientation of the measurement object are indefinite in the whole range of the CMM measuring volume. The uncertainty estimation should be done as follows.

The calibrated object should be measured first. In the next turn the measured object is measured, so that the uncertainty estimation can be carried out. The calibrated and measured objects should fulfill the conditions listed in Table 3.1.

For example, if only one calibrated object is measured, the number of results needed to assess the uncertainty totals up to at least 10 measurement cycles and the performance of at least 20 measurements on the calibrated object. In total, the minimum number of measurement cycles to be performed sums up to 20 full cycles.

During uncertainty evaluation the position and orientation of the calibrated object should change, having in mind the limitations of the measurement procedure, described in [9]. To maintain identical measurement conditions from the beginning to the end of the experiment, CMM should move on the same positions. That also

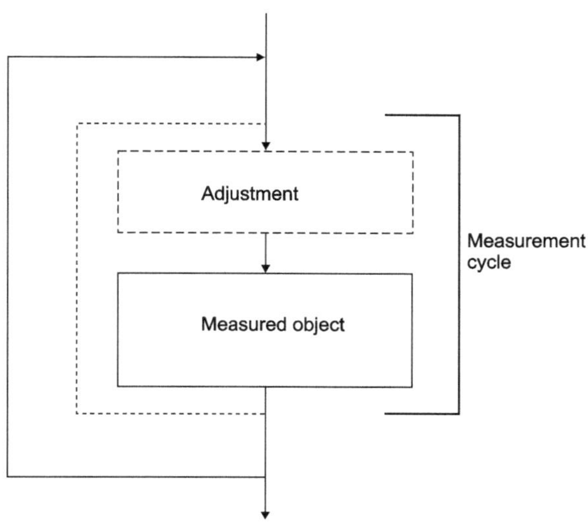

Fig. 3.7 Nonsubstitution method on CMM [9]

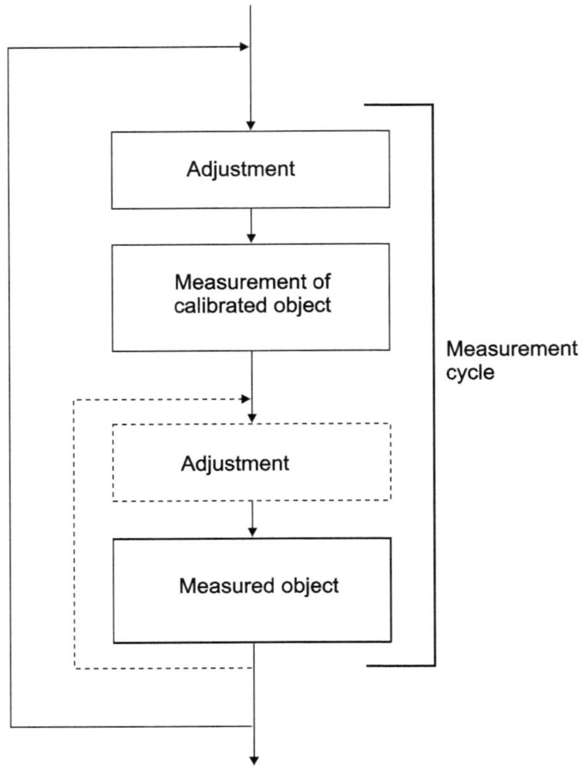

Fig. 3.8 Substitution method [9]

gives the possibility of minimizing the effect of systematic error on the measurement result. Calculation of measurement uncertainty is based on the expression:

$$U = k \cdot \sqrt{u_c^2 + u_p^2 + u_w^2 + u_b^2} \tag{3.1}$$

While performing measurements, four main uncertainty components should be taken into account:

- u_c: Standard uncertainty of the calibration of material standard, established during calibration.
- u_p: Standard uncertainty of the measurement procedure depending on the measuring task.
- u_w: Standard uncertainty of determination of the measured object form error resulting from random changes in the material and influences during production, it is also dependent on the measurement strategy.

- u_b: Standard uncertainty associated with the determination of the material standard measurements systematic error, which for a given measuring task is:

$$b = \bar{y} - x_c \tag{3.2}$$

where
$\bar{y}$ arithmetic mean of indications y of CMM in given measurement cycle
x_c true value obtained from the calibration of the calibrated object

It should be noted that in the case of fulfillment of demands set out in Table 3.1 the systematic error should be corrected.

The k coverage factor is chosen according to the recommendations [9] as $k = 2$, to obtain results at confidence level of $p = 0.95$.

The component of the uncertainty associated with the uncertainty of standard dimension $\boldsymbol{u_c}$ is determined by the expression:

$$u_c = \frac{U_c}{k} \tag{3.3}$$

where
U_c expanded standard uncertainty of the standard
k coverage factor

The two values are taken from the calibration certificate of the standard, and therefore are used in the calculations as the known quantities.

The uncertainty component associated with the uncertainty of the measurement procedure u_p carried out is determined using the expression:

$$u_p = \sqrt{\frac{1}{n-1}\sum_{i=1}^{n}(y_i - \bar{y})^2} \tag{3.4}$$

where
n number of measurements
y arithmetic mean of CMM indications y in the ith measurement cycle

The component associated with the determination of the measured object u_w form error is an element of uncertainty, which without a doubt is the most difficult to estimate. It is mostly dependent on the changes taking place during the production process of the measured object and indirectly also on the measuring strategy of object measurements. The range of this uncertainty covers changes in material and surfaces of produced objects. If all measured objects used in the experiment have been chosen including the similarity conditions contained in Table 3.1, then the effect of u_w may be insignificant or even negligible.

However, if the person performing the measurement has doubts regarding the compliance of objects with the similarity conditions, this component must be taken into account, by using the measurement of at least a few pieces from different production parts. Also, when there are suspicions about not fulfilling the required production process capability or machine capability (the capability understood as repeatability, described by parameters such as C_p, C_{pk} and C_m, C_{mk}), the specific measurement method described below should be taken into account [50]. The measurement strategy in this case, requires that:

- Five different details must be measured in turn. All should be chosen from the production process in the most random way, with the assumption that objects are produced in series or in mass.
- All parts should be measured independently, the probe tips should be calibrated each time before the measurement, and the CMM coordinate system should be transferred to the object coordinate system.
- At least three different distributions of points should fall on one noncalibrated object, and none of these distributions should be repeated for any given detail.
- At least 10 measurements should be performed, changing the distribution of measuring points, no matter how many measured objects were chosen.
- The postulate of using at least 10 different distributions of points should be fulfilled on one whole experiment.

Accordingly, when the person performing the experiment can use only one object and still wants to estimate the measurement uncertainty correctly (and thus the measuring machine), 10 different distributions of points should be used. When there are more objects, for example, three, each of the two details should be measured at least at three different sc. distributions. The recommended number of distributions is four.

But returning to the method of calculating the u_w uncertainty component, its total value is calculated from Eq. (3.5):

$$u_w = \sqrt{\frac{1}{n_w} \cdot \sum_{j=1}^{n_w} \left({}^{j}S_w\right)^2} \tag{3.5}$$

where

n_w number of details (samples) used in the measurement

${}^{j}S_{w}$ standard deviation of the sample calculated from the deviations from the nominal value of the calibration of the calibrated object. This value is calculated by using the equation:

$$ {}^{j}S_{w} = \sqrt{\frac{1}{(n_{w}-1)}\sum_{i=1}^{n_{w}}\left({}^{ij}y - \overline{{}^{j}y}\right)^{2}} \tag{3.6} $$

where $\overline{{}^{j}y}$ is the arithmetic mean of the measurement results obtained in the jth distribution of the object. For a given distribution of points and the number of objects that have to be measured, this value is given by the equation:

$$ \overline{{}^{j}y} = \frac{1}{n_{w}} \cdot \sum_{i=1}^{n_{w}} {}^{ij}y \tag{3.7} $$

In addition to the estimation of the variability of calibrated object proposed here, the estimation of the impact of environmental conditions is also important. In the case of geometric quantities, the factor that most influences the measured dimension is temperature change. According to the ISO standard, the reference temperature for geometric measurements is a temperature of 20 °C. Any deviation from this temperature should result in its correction according to the equation:

$$ l_{c} = l \cdot (1 - (20 - t) \cdot \alpha) $$

where

l determined length

t temperature measured during measurements

α coefficient of thermal expansion taken for given material of calibrated element

For determination of uncertainty of temperature-corrected measurement, the issue related to the accuracy of determination of the thermal expansion parameter should be taken into account. In the case of standards it is possible to find the proper data in the calibration certificate; in the case of an object the issues of thermal expansion are not always identified completely. The uncertainty associated with the thermal expansion may be described by the equation:

$$ u_{wt} = l \cdot (20 - t) \cdot u_{a} $$

where u_{a} is a parameter associated with the standard uncertainty of determining the coefficient of thermal expansion.

As a result of this method of implementation the quantity of the measured object feature is obtained, together with the uncertainty determined in relation not to the specification of the measuring machine producer (i.e., MPE), but to the uncertainty of the calibrated standard. The obtained uncertainty result is therefore dependent only on the particular measuring task and includes all possible environmental conditions that can affect the final measurement result.

Without any doubt the measurement itself is an easy to do operation, nevertheless its noise, seemingly easy to specify, may become very difficult for identification and interpretation. As an example, the impact of random factors in the production process on the value of the uncertainty component u_w can be taken.

The concluding mechanism of the calibrated object method described in [9] in addition to essential benefits, obtained by experimental approach to the determination of coordinate measurement uncertainty (simplicity of implementation, realistic results of uncertainty estimation), carries some limitations. It is connected with calibration objects that have to be sufficiently stable and accurate and with method ability to perform calibration only for sufficiently low uncertainties of the measured object.

The calibrated object method can also be used periodically as a tool to check CMM accuracy. Figure 3.9 presents a calibration object developed by the Eumetron Company [3, 4]. It is used to control the accuracies of measuring machines in accordance with the requirements of the method based on experimental assessment of measurement accuracy described in [10, 57]. The method of carrying out the test should be based on the principle of comparative measurement. The difference between the reference and actual values should in this case be smaller than the specified expanded uncertainty U. Otherwise, if the reason for this inconsistency cannot be found, it is necessary to verify again the obtained result and if the uncertainty determination was carried out in accordance with [9], it can be called calibration.

In Table 3.3 some common sources of uncertainty in combination with methods of their determination are shown. The above-mentioned list of uncertainty components may be incomplete because it is impossible to identify all components, which may affect the final measurement result.

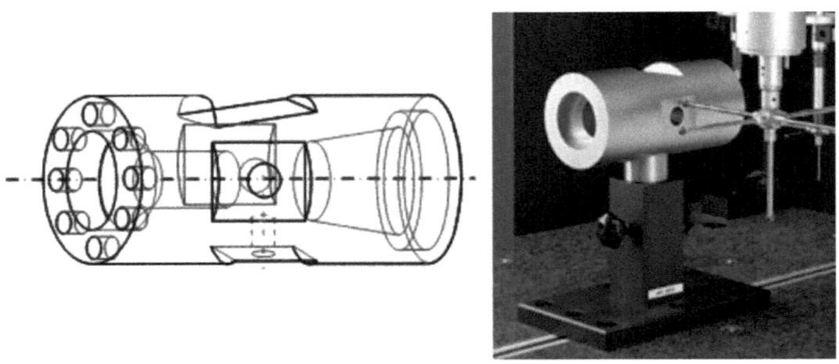

Fig. 3.9 Reference object [3, 4] for monitoring the accuracy of the measuring machines

Table 3.3 Uncertainty sources and their inclusion in uncertainty estimation with calibrated object or standard method

Uncertainty component	Method of uncertainty determination (according to GUM)	Symbols in calibrated object method
Geometric CMM errors		
Effect of CMM temperature		
Slip of CMM slides		
Systematic errors of measuring probe system		
CMM recurrence		
Resolution of CMM ruler		Estimated as sum of uncertainty
CMM temperature gradients	*A*	u_p
Random errors (statistical) of measuring probe system	*A*	u_p
Uncertainty of tip change	*A*	u_p
Errors caused by the procedure (fastening, manipulating, etc.)	*A*	u_p
Errors caused by contaminations	*A*	u_p
Errors caused by measurement strategy	*A*	u_p
Calibrations of calibrated object	*B*	u_c
Differences between measured and calibrated objects (see Table 3.1)	*B*	u_c
Roughness	*B*	u_c
Shape errors	*A* or *B*	u_w
Coverage factor	*A* or *B*	u_w
Flexibility	*A* or *B*	u_w

3.2 Noncalibrated Object Method (Multiposition Method)

The method that uses a noncalibrated measurement object with a multirepeatable measurement strategy (multiposition method) was developed during the work of TC13/WG10/ISO and described in [8]. Its specificity is generally similar to the substitution method, but the multiposition method is simpler to implement, and its use leads to determining real measurement uncertainty, independent of the limit Eq. (1.2) [31, 32, 38, 41, 55]. The procedure during realization of the method is based on one that allows some modifications to achieve the best fit with conditions of real measurement [2]. It describes the manner of object measurement, using for this purpose different distributions of measuring points, as well as different orientations of the object in the machine measuring volume. As a result, it is possible to calibrate an object, which in series or high-volume series production conditions can be used as a production standard (Annex A.1). The method requires that the calibration and uncertainty estimation be individually determined for each geometric characteristic of the measured object.

The basic principle of this method is to conduct several measurements of a given object characteristic with changing the distribution of measuring positions and orientations of the object in CMM volume. Analysis of the results is made on the basis of the analysis of standard deviations of the measured characteristic, separating two components of uncertainty:

- The impact of CMM repeatability for a particular task combined with the distribution of measuring points and form deviations, waviness, and roughness of the measured object surface u_{rep}
- The effect of CMM geometric errors in connection with the influence of the measuring tip qualification process u_{geo}

In those cases where the measured quantity is the distance or position, three other components of uncertainty are additionally determined:

- The length measurement average error is determined while measuring the length standard. This component is calculated by taking into account the uncertainty of calibration of the length standard and repeatability of measurements performed with this method. In addition, all length measurement results should be corrected using this value. If the correction is not performed, the distance average error is taken as a systematic component of calculation of measurement uncertainty of measured object characteristic E_L.
- The average error of probe tip diameter and standard deviation related to that error. It is included as a systematic component of measurement uncertainty of measured object characteristic E_D.
- The uncertainty of temperature correction of CMM, the measured object and length standard is estimated in accordance with knowledge about how the machine and measured object react to temperature changes and which temperature change has the greatest effect on the measurement result u_{temp}.

For the calibration result of a given characteristic the average result from all measurements corrected additionally by the average measurement error is taken.

The procedure is based on the assumption that for any geometric characteristic (e.g., length, angle, etc.) such orientations of the measured object are sufficient, for which the sources of most errors of the measuring machine remain independent (e.g., three various position errors in the direction of three axes, three various squareness errors in three main planes of the coordinate system, etc.). Independence of each uncertainty component must be estimated at least in four measurement cycles, presented schematically in Figs. 3.10 and 3.11, showing the changeable positions of the object in relation to the machine coordinate system. Also the uncertainty components are given.

During the measurements, stable conditions must be kept. Especially, the corrections done by the coordinate machine software should be turned on. The procedure for uncertainty estimation with the multiposition method includes the requirements:

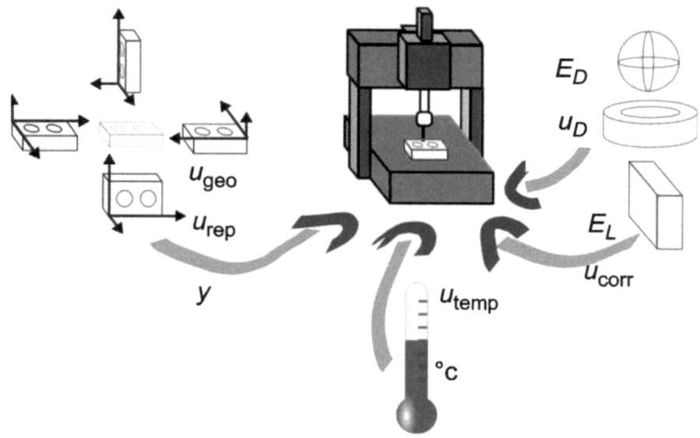

Fig. 3.10 Concept of the multiposition method [38]

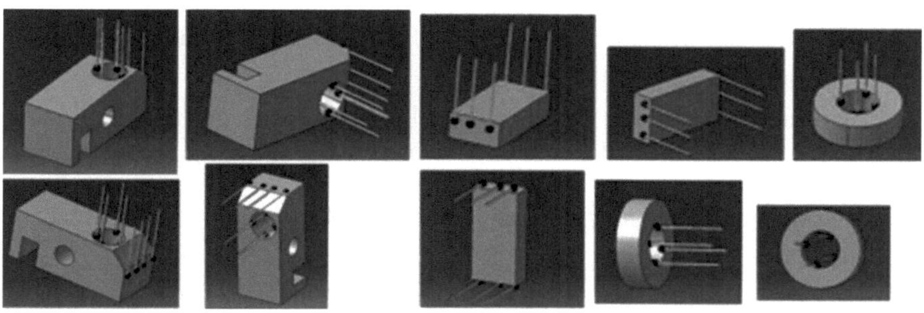

Fig. 3.11 Measurement of an object and reference elements in the following locations [41]

- The characteristics of the measured object must be determined in such a way that it is possible to measure them at any specified orientation.
- The material length standard, for example, step gauge, should meet similarity requirements to the measured length of a given geometric characteristic. Length standard is usually represented by the distance between two centers of reference elements (e.g., ball-bar standard) or the distance between two planes where normal vectors have the same direction (e.g., step gauge).
- For the inside diameter standard (e.g., the ring standard), and/or external diameter standard such as a reference sphere, diameters should be less than 20 mm or 10 % of the length of the shortest CMM axis.
- Configurations of measuring styli, appropriate for the measuring task, should allow us to make the measurements in at least four orientations.
- The measuring machine should be equipped with the necessary equipment for fastening objects.
- It is necessary to equip the measured object and standards with calibrated thermometers for environment temperature measurements.

3.2.1 Measurement of an Object Characteristic

Measurements of the characteristic of an object should be performed normally in at least five different (randomly changeable) distributions of measuring points (sampling strategies). All strategies should be equivalent in terms of the analyzed surface area and the number of points. All measurement repetitions should be performed in at least four different orientations of the measured object: one primary and three others using, if necessary, different measuring styli. Every orientation must be different; no one can be repeated. They are obtained by rotating the object relative to each machine axis, comparing it with the base orientation. They should be selected in a way that allows obtaining good and constant measurement conditions, ensuring the independence of measurement uncertainty components for each axis. Additionally, all measuring tips used for each orientation must be independently corrected: that is, if the object changes its orientation, then the qualification of the tip should be performed using the calibration ball. In the case of a machine in the Cartesian coordinate system, an object is implicitly measured in these orientations:

(1) The base orientation
(2) The orientation obtained by rotating the object 90° around the first machine axis
(3) The orientation obtained by rotating the object 90° around the second machine axis
(4) The orientation obtained by rotating the object 90° around the third machine axis

In each orientation at least $n_1 = 5$ repetitions should be done (one more than the number of orientations), totaling up to 20 measurements.

The procedure allows us to perform whole measurement cycles on more than one machine of similar characteristics. All measurements should be performed on each machine. In this case, the length measurement average error is defined as the average for all machines (from nine measurements on each machine).

It should also be noted that the four orientations cannot be determined for all measured objects. An example of such an object can be a two-dimensional flat element (e.g., thin-walled detail in the form of a flat sheet, ring), for which a maximum three different orientations (basic, rotation of 90° around one axis of machine coordinate system, then around the second one) can be obtained. For objects of this type, however, these orientations are sufficient to estimate the uncertainty components.

3.2.2 Measurements of Length Standards

The measurement of material length standards is performed similarly to measurement of their corresponding geometric characteristic. The set of orientations and

Table 3.4 Exemplary representation of measurement results of object features

	Basic orientation 1	Orientation 2 *x*-axis rotation of 90°	Orientation 3 *y*-axis rotation of 90°	Orientation 4 *z*-axis rotation of 90°
Cycle 1	Y^{11}	Y^{12}	Y^{13}	Y^{14}
Cycle 2	Y^{21}	Y^{22}	Y^{23}	Y^{24}
Cycle 3	Y^{31}	Y^{32}	Y^{33}	Y^{34}
Cycle 4	Y^{41}	Y^{42}	Y^{43}	Y^{44}
Cycle 5	Y^{51}	Y^{52}	Y^{53}	Y^{54}

number of repetitions is also chosen under similar rules. Measurements should be performed in the same temperature conditions and with the same "type" of temperature correction (e.g., "without correction" or "with correction"), as during object measurement. In each orientation, the standard measurement is performed at least three times (depending on the standard type), using the same configurations of styli, that are used to measure the object characteristic. The standard dimensions and object characteristic should be dimensionally similar, such as included in Table 3.4. Length standards do not need to be measured every time a new object is measured. The measurement result of this previously obtained standard can be used for the current calculation. However, all previous CMM parameters must be maintained for the new element measurement. The number of measuring points also has significant importance. The more numerous they are, the lower an uncertainty of the length standard correction is. The above considerations also apply to other types of standards, such as diameter standards.

3.2.3 Measurement of Diameter Standards

The ISO CD/TS 15530-2 specification [8] implies that at least one internal diameter standard and one external diameter standard should be measured. In practice, however, it is not always appropriate to use measurements described in this chapter. For example, during measurement of the distance between two parallel planes, it is sufficient to measure the length standard (e.g., the gauge block). If the distance from the axis of the hole to the plane is to be determined, it is then necessary to perform the measurement of the ring standard, the dimension of which must conform to the dimensions shown in Table 3.1.

The diameter standard must be measured at the same location in the measuring volume as the measured object was measured. Measurement of the standard is reduced to carry our measurements in three mutually orthogonal orientations (the fourth perpendicular orientation is identical to one of the previous ones). The rules for orientation selection are constant for both the measured object and the standards. The number of measuring points is set implicitly to 25 points, and the number of repetitions is at least three for each orientation, giving a total of at least 3 × 3 = 9 complete measurements.

3.2.4 Calculation of the Value of Measured Object Characteristic

As a real value of the measured object characteristic[1] the procedure considers the average from all measurements of the given characteristic, reduced by the length measurement average error (E_L) obtained from standard measurements, and the value of the correction of the standard diameter (E_D).

The corrected value y_{corr} is calculated with the equation:

$$y_{\text{corr}} = y - E_L - E_D \tag{3.8}$$

where

- y average value obtained from all measurements of given geometric characteristic of an object
- E_D correction of measured internal and/or external standard diameter
- E_L length measurement average error related to the length standard (gauge block) calculated with the equation:

$$E_L = E_{\text{Lprop}} \cdot L \tag{3.9}$$

where

- L nominal distance, position, linear dimension
- E_{Lprop} distance measurement average error. In the case of angle measurement this component is equal to zero. In other cases, it is expressed by the formula:

$$E_{\text{Lprop}} = \frac{1}{n_3} \cdot \sum_{i=1}^{n_3} \frac{{}^{i}L_{\text{measstd}} - L_{\text{calstd}}}{L_{\text{calstd}}} \tag{3.10}$$

where

- n_3 total number of measurements of diameter standard
- L_{calstd} calibration value of the material length standard used in the measurements to determine the length average error
- L_{measstd} average value of the measured length standard

[1]In this case, this feature is a distance.

3.2.5 Calculation of Measurement Uncertainty

Various mathematical definitions of the uncertainty equation can be found in the literature. It is assumed that the expanded uncertainty U is set at the 95 % confidence level. For the measurement of distance, position, or any other linear dimension, uncertainty (after [8]) is expressed as

$$U = k \cdot \sqrt{u_{\mathrm{rep}}^2 + u_{\mathrm{geo}}^2 + u_{\mathrm{corrL}}^2} \tag{3.11}$$

where

U expanded measurement uncertainty

u_{rep} standard uncertainty component derived from CMM repeatability

u_{geo} standard uncertainty component derived from CMM geometric errors

u_{corrL} uncertainty of correction determined during the distance measurement of standard

k coverage factor

However, this relationship is too general, because it does not include the component of uncertainty associated with the temperature correction.

A more complicated, but also more complete formula on the expanded uncertainty is suggested by [2] (Fig. 3.12):

$$U = |E_D| + |E_L| + k \cdot \sqrt{u_{\mathrm{rep}}^2 + u_{\mathrm{geo}}^2 + u_{\mathrm{corrL}}^2 + u_D^2 + u_{\mathrm{temp}}^2} \tag{3.12}$$

where

u_D standard uncertainty of the error difference of the measurement of internal and external diameter standard

u_{temp} standard uncertainty of the measured object, standard and environment temperature measurement

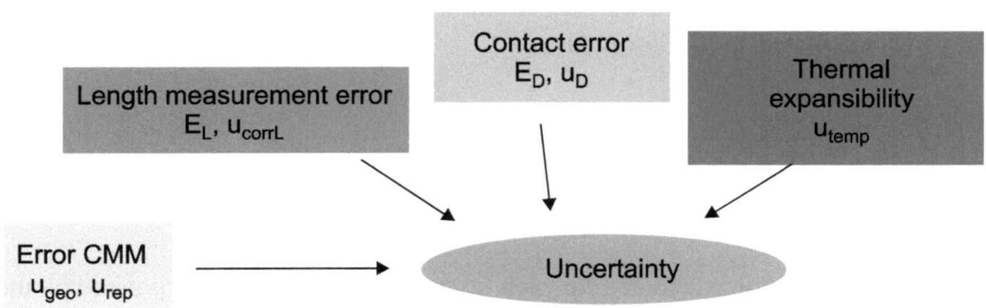

Fig. 3.12 Uncertainty components [37]

Table 3.5 Principle of selection of the coverage factor [2]

v_{eff}	1	2	3	4	5	6	7	8	9	10	11
k	13.97	4.53	3.31	2.87	2.65	2.52	2.43	2.37	2.32	2.28	2.25
v_{eff}	12	13	14	15	16	17	18	19	20	25	∞
k	2.23	2.21	2.2	2.18	2.17	2.16	2.15	2.14	2.13	2.11	2.0

k coverage factor, selected on the basis of effective degrees of freedom V_{eff} from Table 3.5

The number of effective degrees of freedom is determined using:

$$v_{\text{eff}} = \frac{u^4}{\sum_{e=1}^{m} \frac{u_e^4}{v_e}} \tag{3.13}$$

where

- V_e number of degrees of freedom associated with the eth uncertainty
- m number of uncertainty components
- u_e eth uncertainty component
- u complex uncertainty

However, for the measurement of angles, uncertainty is expressed by the formula:

$$U = k \cdot \sqrt{u_{\text{rep}}^2 + u_{\text{geo}}^2} \tag{3.14}$$

3.2.6 *Calculation of the Uncertainty Component u_{rep}*

The uncertainty component coming from CMM repeatability, u_{rep}, defined as above, is calculated by the formula:

$$u_{\text{rep}} = \sqrt{\frac{1}{n_2} \cdot \sum_{j=1}^{n_2} \left({}^{j}S\right)^2} \tag{3.15}$$

where

${}^{j}S$ standard deviation calculated for each orientation of the object, expressed by the formula:

$$ {}^{j}s = \sqrt{\frac{1}{(n_1 - 1)} \sum_{i=1}^{n_2} ({}^{ij}y - {}^{j}y)^2} \tag{3.16} $$

where

${}^{ij}y$ measurement result in the jth orientation and ith repetition (including the various points distribution)

${}^{j}y$ arithmetic mean of the measurement results obtained in the jth orientation of an object

3.2.7 Calculation of Uncertainty Component $u_{\mathbf{geo}}$

u_{geo} is the uncertainty derived from the machine geometric errors. It is calculated by:

$$ u_{geo} = \frac{1}{\sqrt{n_2}} \cdot \sqrt{\frac{1}{(n_2 - 1)} \cdot \sum_{j=1}^{n_2} ({}^{j}y - y)^2} \tag{3.17} $$

where

y the average value from all measurements which may be calculated using the equation:

$$ y = \frac{1}{n_1 \cdot n_2} \cdot \sum_{j=1}^{n_2} \sum_{i=1}^{n_1} {}^{ij}y \tag{3.18} $$

where

$n_1 \cdot n_2$ the total number of measurements of the characteristic for which the uncertainty is estimated (the number of repetitions multiplied by number of orientations).

3.2.8 Calculation of Uncertainty Component $u_{\mathbf{corrL}}$

This uncertainty is determined as

$$u_{\mathrm{corr}} = \frac{L}{L_{\mathrm{calstd}}} \sqrt{\left(\frac{U_{\mathrm{calstd}}}{2}\right)^2 + \frac{u_{\mathrm{measstd}}^2}{n_3}} \tag{3.19}$$

where

L measured length (the one being calibrated)

n_3 total number of measurements of standard

L_{calstd} correct value of the given standard characteristic, which is used in measurements to determine the average error (of length)

U_{calstd} calibration uncertainty of the used standard object

u_{measstd} component of standard uncertainty resulting from repeatability during measuring the standard diameter, expressed by the formula:

$$u_{\mathrm{measstd}} = \sqrt{\frac{1}{n_3 - 1} \cdot \sum_{i=1}^{n_3} \left({}^{i}L_{\mathrm{measstd}} - L_{\mathrm{measstd}}\right)^2} \tag{3.20}$$

where

${}^{i}L_{\mathrm{measstd}}$ measured standard value used in the ith measurement, in a series of measurements to determine the systematic error

L_{measstd} average value of standard dimension determined from all n_3 measurements

This value is calculated by the formula:

$$L_{\mathrm{measstd}} = \frac{1}{n_3} \cdot \sum_{i=1}^{n_3} \left({}^{i}L_{\mathrm{measstd}}\right) \tag{3.21}$$

3.2.9 Calculation of Uncertainty Component of Length Change Derived from Thermal Influences

Thermal influences on the measurement result are expressed as a component of complex standard uncertainty, u_{temp}, calculated using the formula:

$$\begin{aligned} u_{\mathrm{tempN}} &= L \cdot \sqrt{\begin{aligned} &\left(u_{\mathrm{TMtran}}^2 + u_{\mathrm{TMcal}}^2\right)\alpha_M^2 + \left(u_{\mathrm{TNtrans}}^2 + u_{\mathrm{TNcal}}^2\right)\alpha_N^2 \\ &+ \left(u_{\alpha\mathrm{M}}(T_M - 20\,^{\circ}\mathrm{C})\right)^2 + \left(u_{\alpha\mathrm{N}}(T_N - 20\,^{\circ}\mathrm{C})\right)^2 \end{aligned}} \\ u_{\mathrm{temp}} &= \sqrt{u_{\mathrm{tempN}}^2 + u_{\mathrm{tempW}}^2} \end{aligned} \tag{3.22}$$

where

T_W, T_M, T_N	the average temperature (of object, machine scales, length standard) in measurement time
$u_{\alpha W}$, $u_{\alpha M}$, $u_{\alpha N}$	standard uncertainty of linear thermal expansion (of object, machine scales, length standard)
α_W, α_M, α_N	coefficient of linear thermal expansion (of object, machine scales, length standard)
u_{TWtran}, u_{TMtran}, u_{TNtran}	standard uncertainty of the heat transfer for used thermometers (estimated experimentally)
u_{TWcal}, u_{TMcal}, u_{TNcal}	standard uncertainty of used thermometer indications (of object, machine, length standard)
u_{tempW}, u_{tempN}	standard uncertainty of temperature measurement (of object, machine, length standard) determined numerically

The noncalibrated object method involves many components building an adequate uncertainty model. In Table 3.3 the components that are characteristic for this method and were identified and included in the uncertainty model are shown. A more pictorial overlook at this method is presented in Fig. 3.13.

The X symbols in Table 3.6 indicate the inclusion of given uncertainty sources by its particular components in the method.

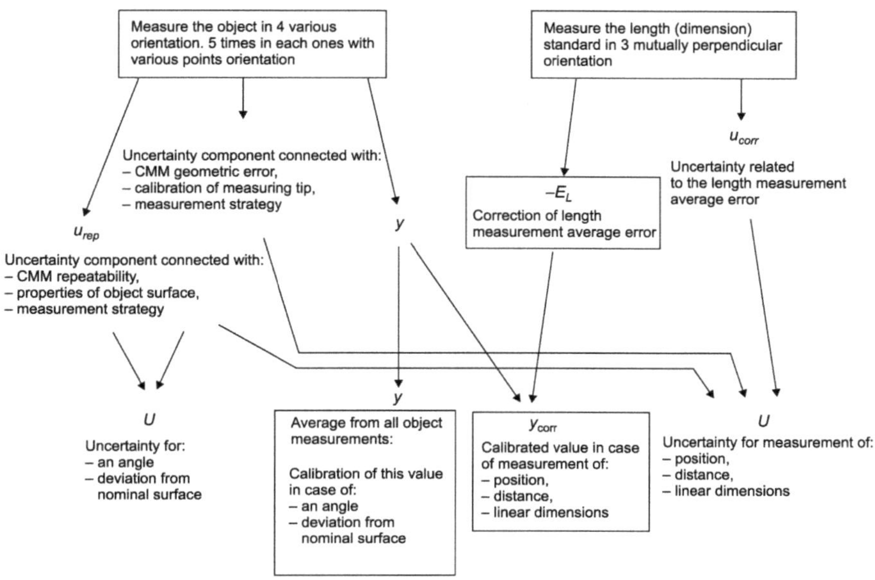

Fig. 3.13 Scheme of multiposition method

Table 3.6 Sources of uncertainty included in calibrated object method

Uncertainty caused by	Reference to		
	u_{rep}	u_{geo}	E_L, u_{corr}
	Recurrence and points collection	Geometry	Length measurement average error
Measurement repeatability point-to-point type	X		
Resolution	X		
Pollution of surface and roughness	X		
Geometric CMM errors		X	
Directional characteristic of measuring tip		X	
Uncertainty of measuring tip diameter		X	
Read too small number of points from analyzed surface	X		
Fit the coordinate system to object	X	X	
Length average error			X
Temperature difference, drift	X	X	X
Changes of temperature gradients	X	X	X
Mounting, manipulating		X	
Correction of CMM temperature		X	X
Correction of measured object temperature		X	X
Dynamic errors arising during scanning	X		

3.3 Monte Carlo Method for Uncertainty Determination in Multiposition and Substitution Method

The multiposition and substitution methods described assume stable probability distributions for uncertainty components in conditions for their use. They create proper conditions for the application of the Monte Carlo method [45, 49] described in Chap. 2. According to assumptions of the multiposition method, by adopting formula (3.1) from Sect. 3.2 as a model of measurement and by inserting in Eq. (3.1) appropriate expressions for E_L and E_D the measurement model takes the form [44]:

$$y_{\text{corr}} = y \cdot (1 + \alpha_{\text{wN}}\theta_{\text{wN}}) - L\frac{1}{3 \cdot n_3} \cdot \sum_{j}^{3} \sum_{i=1}^{n_3} \frac{{}^{ij}L_{\text{measstd}} * (1 + \alpha_{\text{wW}}\theta_{\text{wW}}) - L_{\text{calstd}}}{L_{\text{calstd}}} - \frac{1}{n_4} \cdot \sum_{i=1}^{n_4} \left({}^{i}D_{\text{measint}} - D_{\text{calint}}\right) \tag{3.23}$$

where

y_{corr}	corrected diameter, mm
y	average from all measurements, mm
α_{wN}, α_{wW}	coefficient of thermal expansion of an object and standard, °C^{-1}
θ_{wN}, θ_{wW}	temperature deviation from the reference temperature 20 °C of an object and standard, °C
L	nominal value of measured distance, mm
$^{iJ}L_{measstd}$	results of length standard measurement, mm
L_{calstd}	nominal length of measured standard, mm
$^{i}D_{measint}$	results of internal diameter standard measurements, mm
D_{calint}	nominal diameter of diameter standard, mm

The Monte Carlo method does not include only one value as the average from the measurement series, but examines the system behavior for different values taken from the probability distribution for input quantities. Of course, these values oscillate somehow around the average value. Therefore the equation may be written as

$$y_{corr} = y \cdot (1 + \alpha_{wN}\theta_{wN}) - L \cdot \frac{L_{measstd} \cdot (1 + \alpha_{wW}\theta_{wW}) - L_{calstd}}{L_{calstd}} - (D_{measint} - D_{calint}) \tag{3.23a}$$

Already knowing the mathematical model of the measurement, the appropriate probability density functions should be assigned to each input quantity. Depending on the available information on each input quantity, the best probability distribution describing them should be assigned to each. This issue has been fully described in the supplement to GUM [19]. Let's use an example. If for the certain input quantity *Xi* an average value, a standard uncertainty associated with its determination and a number of degrees of freedom of this process are already known, then the PDF with *t* distribution is assigned to it. But if we know only that a certain input quantity assumes values limited from the bottom with number "a" and from above with number "b", then the PDF with rectangular distribution is assigned to it. Probability distributions for particular input quantities are shown in Table 3.7.

In short, for the described model this process can be expressed as generating random draws from particular distributions for input quantities. The sampling is based on a generator of rectangular distribution. Having drawn a number (or a few numbers) from the rectangular distribution R (0,1) it is possible, applying appropriate transformations, to receive the randomly chosen number from any other distribution. The procedure for performing the transformations is exactly described in the Supplement [19]. A generator used here for rectangular distribution is a Wichmann–Hill generator, recommended by the Supplement [19], Annex C.1,

Table 3.7 Probability distributions assigned to particular input quantities [350]

Quantity	Probability distribution	Distribution Parameters				
		μ (mm)	σ (mm)	v	a	b
y	t_v (μ, $σ^2$)	133.95721	0.00016	16		
$α_{wN}$	R (a, b)				11×10^{-6} °C^{-1}	12×10^{-6} °C^{-1}
$α_{wW}$	R (a, b)				11×10^{-6} °C^{-1}	12×10^{-6} °C^{-1}
$θ_{wN}$	R (a, b)				0.6 °C	0.8 °C
$θ_{wW}$	R (a, b)				0.76 °C	0.96 °C
$L_{measstd}$	t_v (μ, $σ^2$)	150.00064	0.00041	8		
L_{calstd}	N (μ, $σ^2$)	149.99996	0.00025			
$D_{measint}$	t_v (μ, $σ^2$)	27.99876	0.00072	8		
D_{calint}	N (μ, $σ^2$)	27.99800	0.00015			

which passed correctness tests, and is used for validating the action of generators of pseudorandom numbers. As the number of trials of the Monte Carlo method (M) sufficient for the correct determination of the distribution of output quantity, for the probability $p = 0.95$, $M = 200{,}000$ was established. After drawing the M numbers from all distributions, the numbers are put into the model. In this way we receive the M values of the measured (output) quantity, which on the other hand, after appropriate transformations, constitutes the distribution of output quantities. However, based on this function it is possible to determine the expected value and standard deviation, as simultaneously being a standard uncertainty of measured quantity determination. The implementation of the method was carried out in the MatLab® environment and is described in [45].

In order to check the correctness of the developed model the measurements of the distance of the crank hole center from the connecting surface of the head in the cast casing of the piston pump were carried out in accordance with the multiposition method (Fig. 3.14).

According to the principles of the multiposition method the measurement uncertainty was also determined. Next, this measurement was simulated using the Monte Carlo method and its uncertainty was determined (also with the MC method). Obtained results are presented below. They are convergent among themselves, which proves that the developed model is correct.

As for the multiposition method, as well as the method using a calibrated object or standard, it is possible to develop the alternative model for determining the measurement uncertainty with the MC method application (Fig. 3.15).

Method	y_{corr}	$u(y)$
ISO/CD 15530-2	133.95580	0.00117
MC	133.95590	0.00100

Fig. 3.14 Measured object: casing of the piston pump

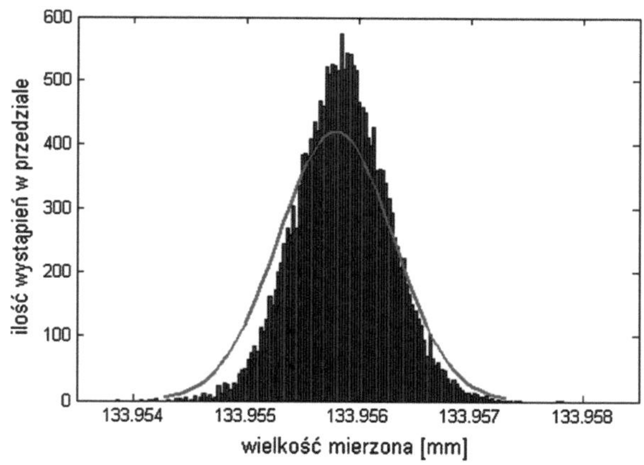

Fig. 3.15 Graphical presentation of results of multiposition method (*continuous line*) and MC method (*bars*)

According to the description and formulas from Sect. 3.2 it is possible to determine the following mathematical model, describing the measurement performed according to the substitution method.

$$y_s = L_{(p)} \cdot (1 + \alpha \cdot \Delta) - \left(y_{(w)} - x_{c(w)}\right) \tag{3.24}$$

where

y_s measured length, mm

L_p variable describing measurement results of measured object, mm

α coefficient of thermal expansion of measured object, °C^{-1}

Δ difference between actual temperature and reference temperature of measured object, °C

$y_{(w)}$ variable describing results of conducted measurements of the standard, mm
$x_{c(w)}$ variable describing standard calibration process, mm

The next stage is assigning appropriate probability distributions to input variables. Table 3.8 shows exemplary assignment of PDF to one of performed measurements.

In order to verify the correctness of the created model, measurements of a gauge block stack, multidimensional cube (Fig. 3.11, Sect. 3.1), and pump trunk (Fig. 3.14) were performed. As standards for particular measurements gauge blocks with nominal lengths selected in a way to meet the conditions of the similarity given in Table 3.1 were used. Afterwards, the uncertainties for realized measurements were determined for both the new model and methodology given by the standard specification ISO/TS 15530-3:2004. The obtained results are summarized in Table 3.9.

The main difficulties faced by metrologists using the MC method are building a measurement model adequate for the CMS being used, appropriate assignment of the distribution to particular input quantities, and performing sampling, as described in [19]. Analyzing the use of the MC method it can be concluded that it is much more practical than the traditional method of uncertainty calculation for both the substitution method and the multiposition method. Only a first application of the method is difficult, when one has to build a mathematical model and describe the process of random number generation with the chosen programming language. In

Table 3.8 Exemplary assignment of PDF to one of performed measurements

Quantity	Possibility distribution	Distribution parameters				
		μ (mm)	σ (mm)	v	a (°C)	b
L	$N(\mu, \sigma^2)$	38.9994	0.0011			
A	$R(a, b)$				11×10^{-6} °C^{-1}	12×10^{-6} °C^{-1}
Δ	$R(a, b)$				0.2 °C	0.6 °C
Y_w	$t_v(\mu, \sigma^2)$	39.9366	0.0007	23		
X_c	$N(\mu, \sigma^2)$	39.9363	0.0001			

Table 3.9 Uncertainties for realized measurements were determined for both the new model and methodology given by the standard specification ISO/TS 15530–3:2004

Measured object (mm)	ISO/TS 15530-3	Monte Carlo method
Gauge blocks	$y(s) = 38.9994$ $U = 0.0022$	$y(s) = 38.9991$ $U = 0.0020$
Cube	$y(s) = 120.0732$ $U = 0.0035$	$y(s) = 120.0730$ $U = 0.0033$
Pump casing	$y(s) = 193.8838$ $U = 0.0041$	$y(s) = 193.8829$ $U = 0.0040$

subsequent MC method applications, it is possible to use the already developed model, which after all is used in similar coordinate measurement conditions, and does not change. Therefore it is possible to use the library of previously written procedures. The only activity in this case is to provide the input values, which are appropriate distributions assigned to each input quantity. It is recommended that, while using this method, one should employ application development dedicated to a given laboratory with a library of models of carried-out measurements and the probability distributions associated with them.

3.4 Determination of Uncertainty of Freeform Profile Measurement

Measurements of freeform profiles using the coordinate technique are among the most difficult ones. This is caused by theoretical, software, and technical considerations occurring in contact measurement, where there are problems such as radius correction, or a reference to a theoretical profile formed as a set of 3-D CAD models and matching to a real profile [20, 21, 25, 28, 29, 39, 40, 42–44, 47]. A problem of noncontact measurement using laser scanners and its adequacy for object dimensions is also of importance. The general task of freeform profile measurement is shown in Fig. 3.16. The turbine blade measuring task using the contact scanning method with an indication of the number of sections—profiles and the number of points in relation to a CAD set—is also formulated.

The issue of accuracy assessment arises each time the measurement is done and in that case the scale of difficulty is also high. The methods and procedures allowing measurement uncertainty evaluation for these profiles are still to be found.

In the publications [2, 33–35] E. Savio described the concepts of usage of the group of standards with stable curvature to stage substitution of the freeform profile (Tables 3.10).

This method, using the substitution method assumptions described in Sect. 3.1, focuses on selection of such a group of standards with stable curvature making it possible to construct a modular standard from them, MFG (modular freeform gauges). It is constructed maintaining the similarity requirements formulated in Table 3.1 and allows the implementation of a comparative measurement for this object. The concepts of the method and its reference to a CAD dataset are presented in Fig. 3.17. The manner of satisfying traceability with a meter standard is shown in Fig. 3.18.

The tests included the measurement of a turbine blade profile. For the selected geometry, shown in Fig. 3.19, the estimation of profile deviation using the program based on a CAD model through contact scanning measurement for nine selected sections of a turbine blade is presented. The CAD model of the blade was taken as a nominal and the measurement was carried out with a scanning probe (Fig. 3.20). The fit of the free surface of the measured blade was carried out using three

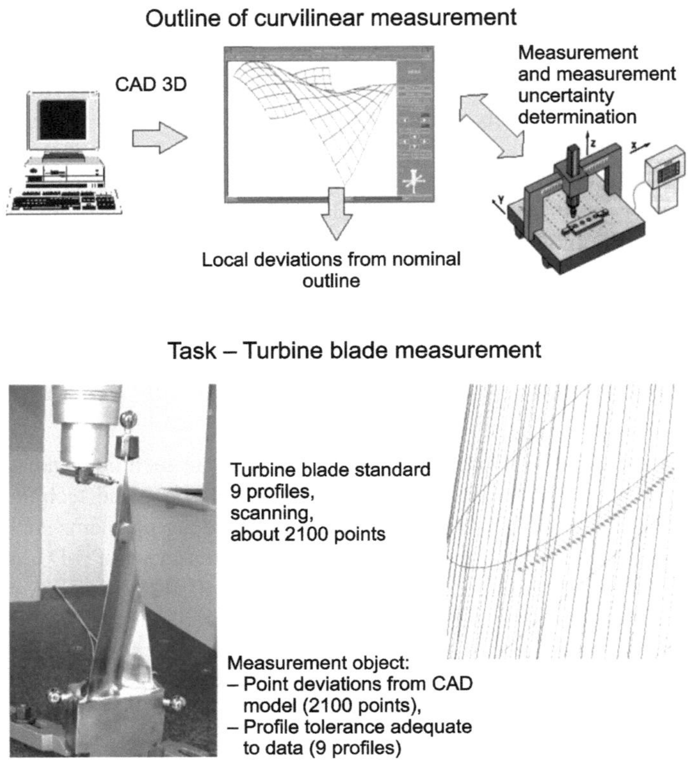

Fig. 3.16 Scheme of freeform profile measurement on CMM with reference to a 3-D CAD documentation [2]

Table 3.10 Similarity requirements formulation

d_{gi}	0.36 μm
Standard uncertainty u_{PEF}	0.14 μm
PEF_{xy} (αi)	0.64 μm
Tips system	Full “star”
d = diameter of contact ball	8 mm
l = stylus length	100 mm

reference balls and without using the best-fit method. All other measurement conditions were identical to the typical conditions of industrial measurement. In particular, the estimation of deviation of a given characteristic was carried out using a measuring point located at a distance of at least 5 mm from the beginning/end of the profile, in order to include the effect of measurement instability at profile transition points. Data analysis was performed using an external program because of the limitations of the metrological software. Estimation of checked characteristic deviation—shape of determined profile tolerance—was compared to the maximum

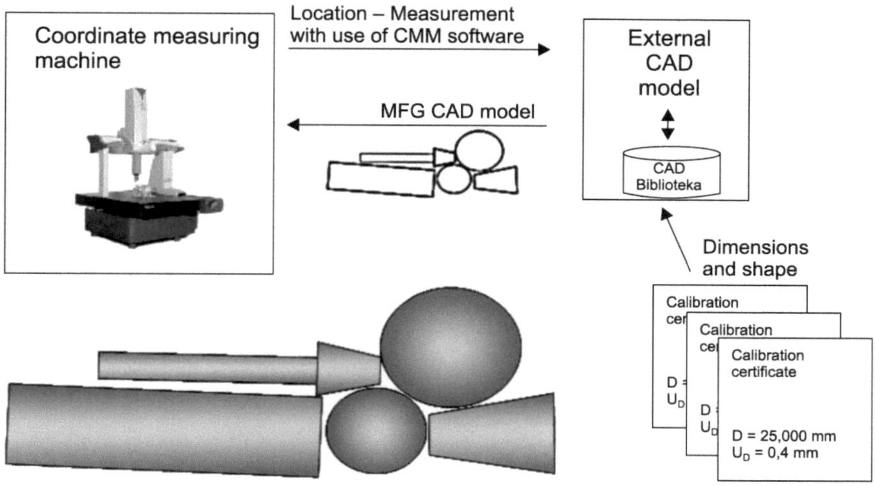

Fig. 3.17 Concepts of accuracy assessment methods for freeform profile measurement, MFG [2]

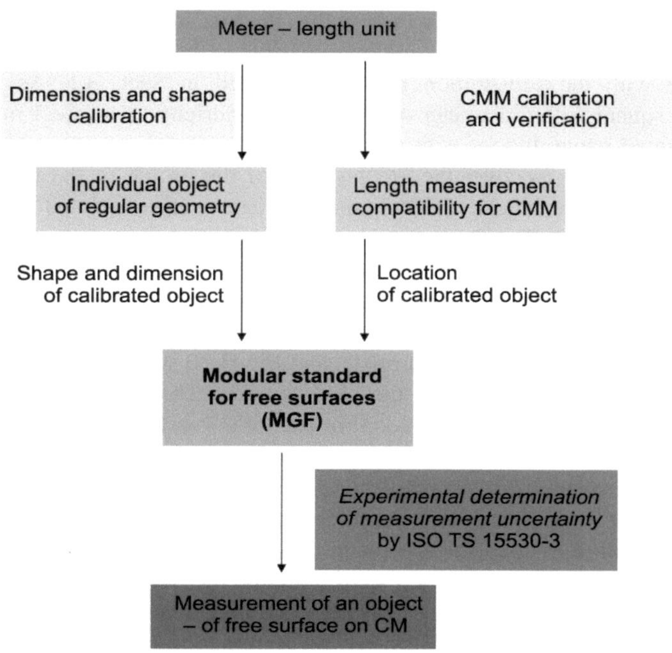

Fig. 3.18 Connection of modular method (*MFG*) with length unit (traceability) [2]

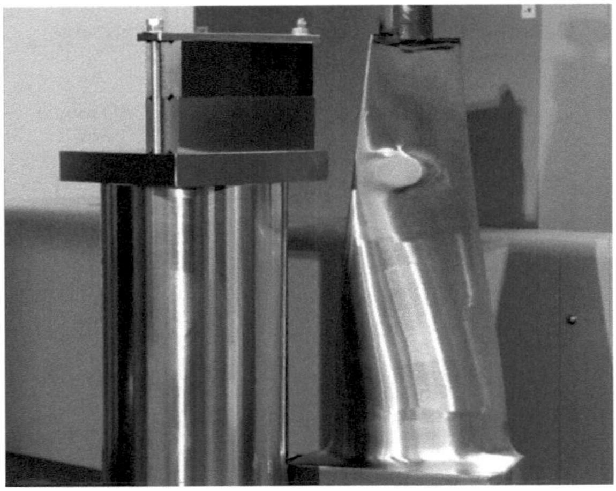

Fig. 3.19 Measured turbine blade and set of standards [2]

deviation from the nominal profile. Therefore, the measured value (mesurand) is the maximum distance from the nominal CAD model profile for all measured points of given section.

According to [34, 35] for performed studies the first step was the project and the calibration of the given MFG, as similar as possible to the assessed turbine blade, in accordance with the substitution method described in Sect. 3.1. As MFG the cylindrical square with a diameter of 100 mm (cylindricity tolerance 1 m) and two stable mounted gauge blocks with a flatness tolerance of 1 m were adopted here. They had to be able to cover the greatest part of the profile of the turbine blade geometry. The gauge blocks and cylindrical square were calibrated on CMM in very stable conditions. The CAD model related to MFG was generated considering the calibrated dimension. Calibration uncertainty was calculated using the uncertainty budget in relation to the measured value (mesurand) of shape deviation of the determined turbine blade profile. Figure 3.17 shows the projected MFG configuration appropriate for the measured blade, and Fig. 3.20 shows a CAD model with the individual nominal profiles of the blade section. The next step was to use a comparative procedure, using a standard or part of a standard according to ISO/TS 15530-3:2004 [9]. Twenty measurement cycles were performed using MFG, as required for the method for the application of a typical standard. As described in Sect. 3.1, the MFG position and orientation were analogous to the position and orientation of the measured turbine blade and the measurements were performed under the same conditions with the use of the same measuring stylus configuration and using the same metrological software. The uncertainty calculation was performed using the spreadsheet. Estimated measurement uncertainty in the area of particular profiles, according to the described procedure with implementing MFG, is shown in Fig. 3.21.

In addition, the MFG method was validated by measuring the same object in accordance with the procedure of uncertainty determination using the multiposition

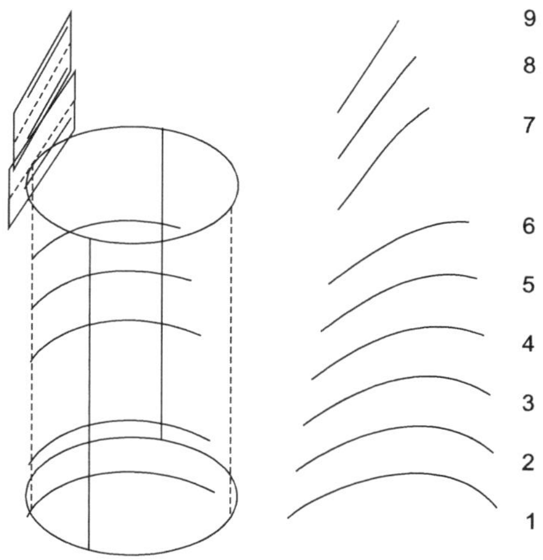

Fig. 3.20 CAD model and outline of modular standard (*MFG*), *left* side, and measured turbine profiles, *right* side [2]

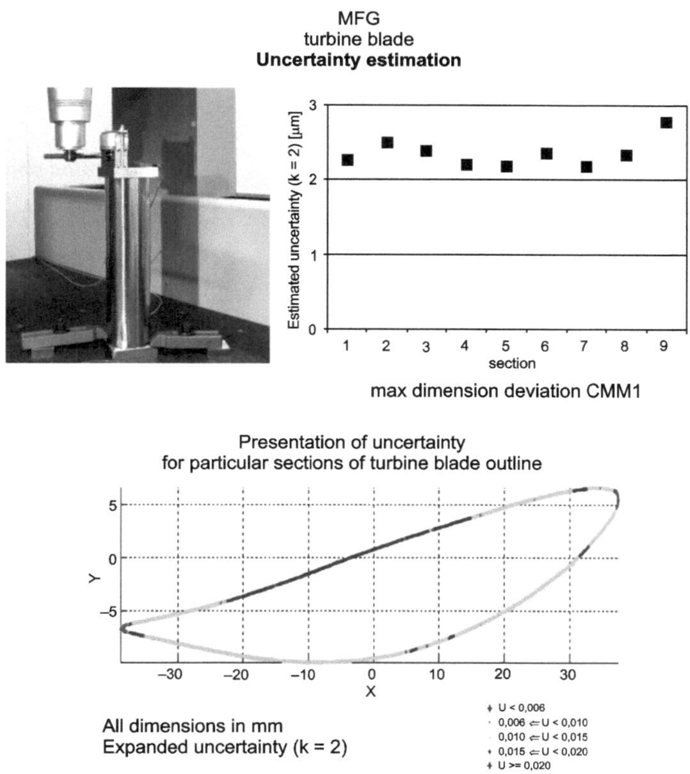

Fig. 3.21 Measurement of turbine blade and estimated measurement uncertainty for particular profiles and presentation of its variability for given profile [2]

method [34, 35]. The methodology is shown in Fig. 3.22, where the measurement principle and method of obtaining the particular positions are presented, according to requirements given in Sect. 3.2, in order to assess the reproducibility of the multiposition method and thus consider it validated.

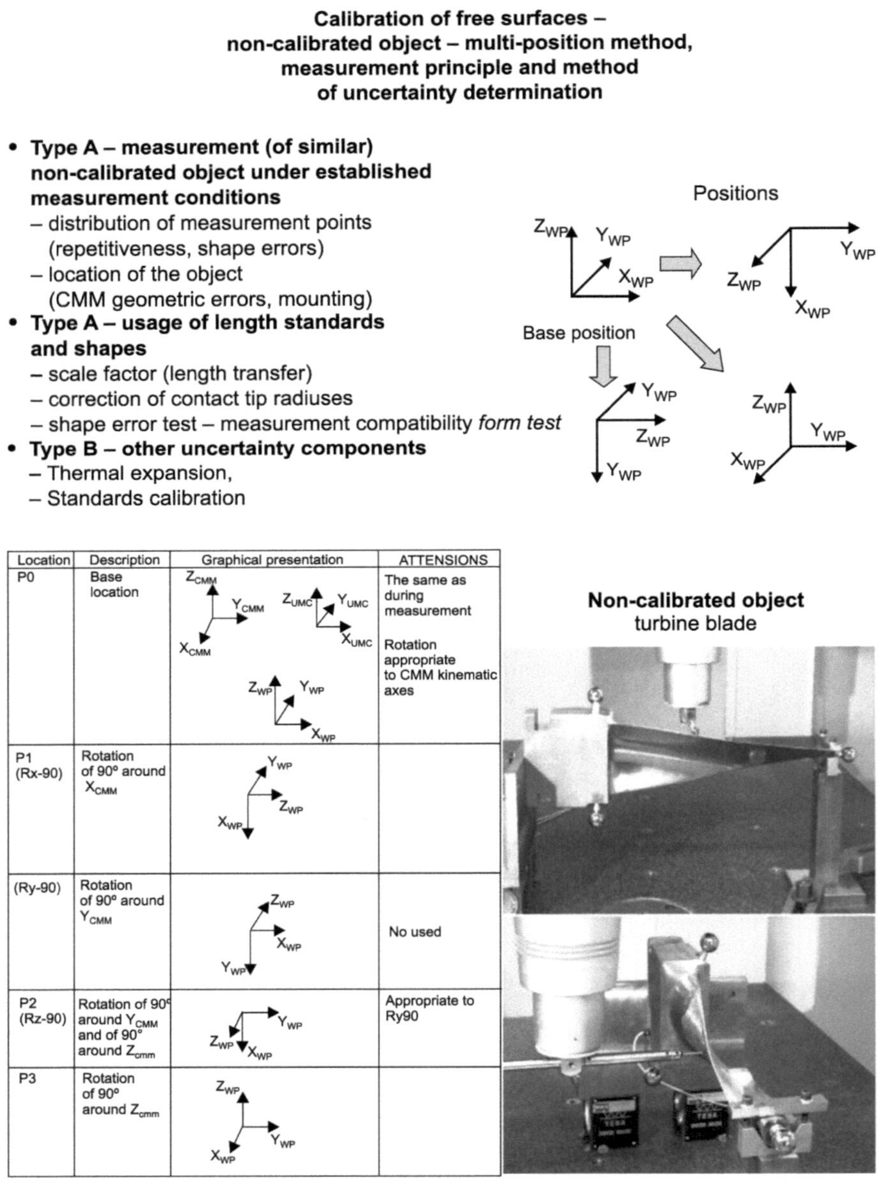

Fig. 3.22 Multiposition method of measurement uncertainty determination for turbine blade and manner of measuring object mounting [2]

In accordance with this procedure, the turbine blade was measured under different conditions on the two measuring machines. CMM1 (Zeiss UPMC 850, $U_1 = 0.4 + L/900$, μm, L, mm) CMM2 Zeiss Prismo 7, $U_1 = 2.2 + L/300$, μm, L, mm). Comparing to the typical measurement conditions there were differences in density of the points (four times greater), the measurement force (two times), and the sampling rate (ten times) [34, 35]. To assess the impact of machine geometric errors on the measurement error, the turbine blade was measured on CMM in three different positions. Each position was obtained as a result of rotation of the measured part around the machine axis with a focus on measurement feasibility. In total, 45 measurement cycles were realized. Participation of CMM errors, arising from its metrological characteristic, has been estimated for the calibration measurements using the length measurement and calibration sphere (in accordance with ISO 10360-2 [26]), under conditions as close as possible to typical operating conditions. Repeated results of the procedure, in the form of estimated uncertainties for two different machines, are shown in Fig. 3.23. It enabled determination of whether the obtained uncertainties were comparable and in consequence validate the MFG method, presented in Fig. 3.24. The use of the modular MFG standard showed the possibility to calibrate a part and to calculate measurement uncertainty without the need of possessing an appropriate reference standard (according to the requirements of ISO/TS 15530-3). The only limitation, of course, is possibilities to develop MFG for the measured parts, by possession of a set of necessary standards with steady curvature, what in fact is not difficult for a well-equipped calibration laboratory. Applying a multiposition procedure, the noncalibrated object demonstrated correctness of this procedure because of obtained compatibility for two different machines and different conditions of measurement realization. From a practical point of view the multiposition method has an advantage because there is no need to have a standard. However, difficulties may be caused by getting the necessary measurement positions and by the doubled time of the measurement. Obtained uncertainties for MFG and the accurate machine 2–3 μm (Fig. 3.21) were

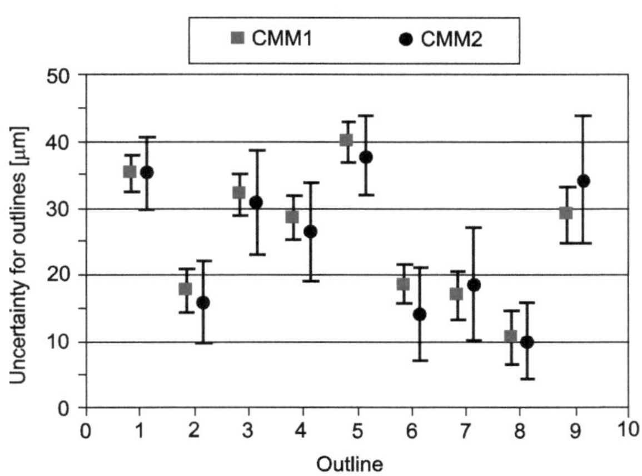

Fig. 3.23 Measurement uncertainty for turbine blade profiles: results for two machines with diversified accuracy, estimated based on the multiposition method [2]

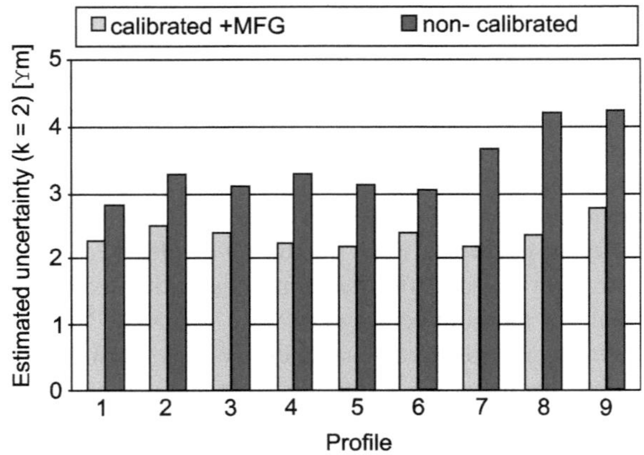

Fig. 3.24 Comparison of calibration with use of the modular standard for free surfaces (according to E. Savio) with the method of the noncalibrated object [2]

fully satisfactory particularly considering the results obtained for the multiposition method on the level of 3–4 μm. It was also satisfactory for the less accurate machine where uncertainties were on the level of 6–10 μm (Fig. 3.23). Assessing the results shown in Fig. 3.24 it should be noticed that the comparison concerns obtained uncertainties for particular profiles. From this comparison it could be concluded that uncertainties obtained for the substitution method (MFG) are lower, but it seems understandable because some uncertainty components could be included twice and the direct reference to the material standard always gives better results. However, it is more important to show the same character of the variability that confirms the correctness of the method.

The example of the freeform profile accuracy assessment presented gives a good illustration of both methods: MFG, use of the calibrated object as well as the multiposition method in the difficult application, highlighting the characteristic features that determine their application.

3.5 Measurement Uncertainty Estimation for Calibrations Based on Error Source Identification: Error Budget

The straightforward application of the *B* method leads directly to development of the measurement model and error budget. It includes the influence of the CMS and conditions of measurement realization. The influence of these factors is presented on the Ishikawa diagram shown in Fig. 3.25.

The determination of measurement accuracy based on the uncertainty budget is divided into a few stages [7, 54]. At the very beginning information necessary to estimate uncertainties of particular components should be collected. In the next stage, the model of the measurement and uncertainty components for a given source must be determined. After that, the estimate of the error in the form of standard complex uncertainty is determined. After multiplying it by the coverage factor

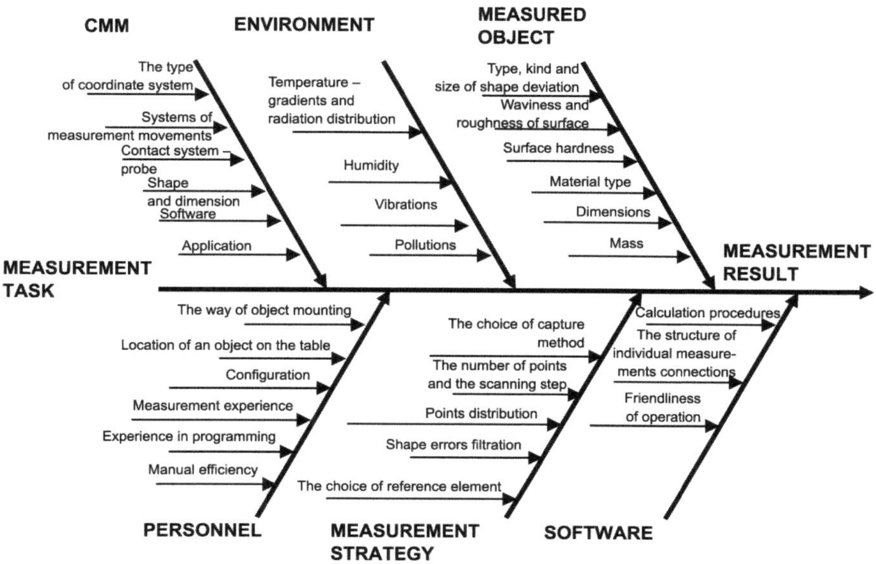

Fig. 3.25 Ishikawa diagram developed for the process of assessment of the measurement error in accordance with [37]

corresponding to the assumed probability the expanded uncertainty of the realized measurement is obtained. The uncertainty budget is created individually for each measurement, considering the conditions of its realization [37, 56]. During budget creation, the following uncertainty components should be taken into account.

- Uncertainty caused by the influence of temperature changes—U_T
- Uncertainty of contact system (probe head)—U_g
- Uncertainty of length measurement—U_k
- Uncertainty of standard calibration—U_v

The following equation shows the global uncertainty of length measurement,

$$U_K = A_K + K_K \cdot L \leq B_K \tag{3.25}$$

where

$$A_K = \sqrt{\left(A_{\mathrm{Ko}}^2 + U_g^2 + A_V^2 + A_T^2\right)} \tag{3.26}$$

$$K_K = \sqrt{(K_{\mathrm{Ko}}^2 + K_V^2 + K_T^2)} \tag{3.27}$$

$$B_K = \sqrt{\left(B_{\mathrm{Ko}}^2 + U_g^2 + A_V^2 + A_T^2 + (K_V^2 + K_T^2) - L_{\max}^2\right)} \tag{3.28}$$

A_{Ko}, A_V, A_T, K_{Ko}, K_V, K_T, B_{Ko} constants
U_{Ko}, U_v, U_g, or U_T uncertainty components

In order to calculate the total error of the length measurement the component uncertainties described in Fig. 3.26 should be calculated.

In the Laboratory of Coordinate Metrology at Cracow University of Technology the process of the estimation of measurement uncertainty based on a described method was carried out. The result of this process is given as an exemplary uncertainty budget for calibration measurements that were carried out on a PMM12106 Leitz machine.

3.5.1 Uncertainty Budget for the Calibration Procedure of the Plate Standard (Hole Plate) Calibrated on PMM12106 Leitz Machine

The uncertainty of the contact system U_g is estimated based on the work [36, 37] by analysis of standard ring measurement results. The use of different systems of probe tips for tests enables the accurate identification of probe head error and the influence of the styli changes on it.

The contact probe head error is identified in the constant area of the measuring volume in the form of a probe error function (PEF) [36, 37], defined for i = 64 measuring points spread evenly on the surface of the standard ring or reference

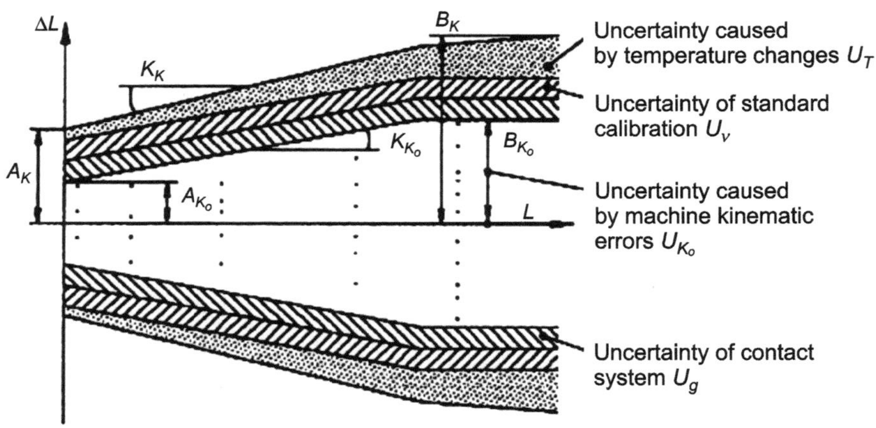

Fig. 3.26 Components of global uncertainty [7]

sphere. PEF (a.) is a function of the a. angle defined as the angle between a measurement direction and the normal to the measured surface. This function expresses the value of average radial deviation d_{gi} counted in relation to the reference circle determined using the least squares method (Fig. 3.27) and Table 3.10:

$$\mathrm{FBG_{xy}}(\alpha i) = d_{\mathrm{gi}} + 2\mathrm{u_{FBG}} \tag{3.29}$$

The other concept, applied when different systems of styli are used, allowing the accurate identification of probe head error and the influence that stylus changes have on it, is a method developed in the PTB [7, 37]. This method, also in harmony with the random model (Sect. 2.2), distinguishes the systematic and random part. The contact system uncertainty is estimated based on the measurement of the reference sphere. The measurement method is illustrated in Fig. 3.28.

The measurement is performed in two positions in measuring the area of dimensions (a, b, c), and using four different measuring styli of lengths smaller than 1/4 of the shortest area dimension, in 40 points spread evenly on the calibration ball. Measuring positions are situated in two different locations in relation to the

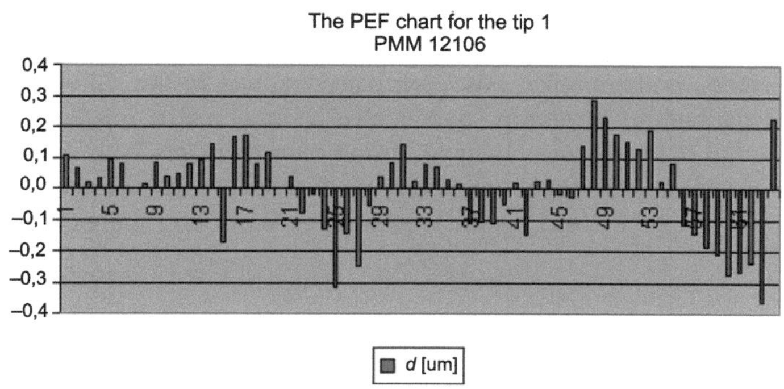

Fig. 3.27 The PEF chart for the tip 1 PMM 12106 determined by the measurement of the standard ring [37]

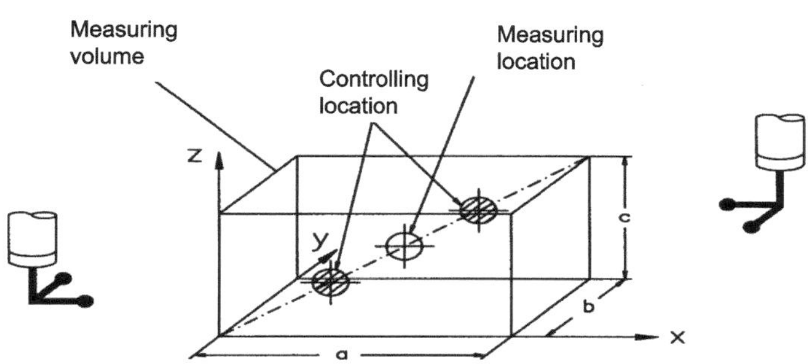

Fig. 3.28 Procedure of estimating the contact uncertainty (the PTB concept) [7]

center of the measuring area. The probe head error is identified in the form of uncertainty written with equation:

$$U_{\mathrm{gPTB}} = \Delta R + 2 \cdot s \tag{3.30}$$

where

ΔR greatest deviation from the average radius of the calibration ball determined with the least squares method (according to Gauss) including all measuring styli

s standard deviation for all measuring points

Uncertainty obtained for the probe head of PMM12106 machine is:

$$U_{\mathrm{gPTB}} = (0.36 + 2 \cdot 0.14) = 0.62\,\mu\mathrm{m} \tag{3.31}$$

This result was achieved for styli with the contact ball diameter equal to d = 8 mm and length l = 100 mm arranged in the "star" configuration.

Both expressed methods for determining probe uncertainty gave similar results for the probe head of the PMM12106 machine.

The uncertainty component of length measurement U_{Ko}, caused by the group of kinematic errors, is determined based on data obtained in the calibration process using such calibration programs as MegaKal or the matrix method. Also data provided by the producer and obtained during measurements with usage of gauge blocks are accepted. An equation describing the uncertainty of the length measurement performed on CMM, including the influence of kinematic errors, then has the form:

$$U_{\mathrm{Ko}} = A_{\mathrm{ko}} + K_{\mathrm{ko}} \cdot L \leq B_{\mathrm{Ko}} \tag{3.32}$$

where

U_{Ko} uncertainty of the length measurement performed on CMM caused by kinematic errors

A_{Ko} constant for kinematic errors

K_{Ko} dependence factor of the kinematic error group uncertainty dependent on the length

L measured distance

B_{Ko} limit value of CMM error

The uncertainty of calibration Uv depends on the length established in the process of calibration of the gauge blocks or the plate standard with balls or holes, used next to CMM calibration. Its value is affected by such partial errors as

- Irregular temperature distribution in the object
- Irregular temperature distribution on CMM
- Straightness error

- Uncertainty of the coefficient of thermal expansion
- Short-term errors of machine slides
- Contact errors

Its form conforms to the equation:

$$U_v = A_v + K_v \cdot L \tag{3.33}$$

where
A_v constant for standard calibration errors
K_v calibration uncertainty factor dependent on the length

The required data for standards are presented below.
For gauge blocks:

$$U_v = 0.06 + 1.6\,L,\ \mu\text{m}$$

For ball plate Retter no. RP 05 400–94.15:

$$U_{\text{xy}} = 0.9\,\mu\text{m} + 1.1 \cdot 10^{-6}\,L \quad U_{\text{xy}} = 0.8\,\mu\text{m} + 0.9 \cdot 10^{-6}\,L$$

For hole plate Feinmess–Robax type 415/415/83 no. 199911014:

$$U_{\text{xy}} = 0.6\,\mu\text{m} + 1.1 \cdot 10^{-6}L \quad \text{and} \quad U_{\text{l}} = 0.3\,\mu\text{m} + 0.6 \cdot 10^{-6}$$

Influence of the thermal expansion on the uncertainty of the length measurement U_T.

Uncertainty of the length measurement dependent on thermal expansion u_T is a sum of the component dependent on length $K_T \cdot L$ and the component A_T:

$$u_T = A_T + K_T \cdot L \tag{3.34}$$

where
A_T constant for errors caused by temperature changes
K_T factor dependent on uncertainty of temperature changes dependent on the length, wherein:

$$A_T = u_{T7} \tag{3.35}$$

$$K_T = \sqrt{\left(u_{T1}^2 + u_{T2}^2 + u_{T3}^2 + u_{T4}^2 + u_{T5}^2 + u_{T6}^2\right)}$$

3.5.2 Thermodynamic Model

For the PMM12106 machine in the Laboratory of Coordinate Metrology a thermodynamic model of the construction deformation was adopted. It is presented in Figs. 3.29, 3.30 and 3.31. It enables calculating individual uncertainty components associated with machine deformations caused by changes of temperature. Regular changes of temperature influence linear length changes of machine scales and of CMM mechanical parts which in turn causes deformation of slide tracks and changes of location of individual machine parts in relation to other parts. Analyzing the measuring machine structure including the spatial thermal gradient, it is possible to model thermal deformations of its basic parts, (based on [22, 23, 37, 51]), as presented in Fig. 3.29. Momentary changes of temperature for such constructions as PMM12106 have a lower significance and do not affect their deformations.

The linear and angle deformations appearing during this process are presented by relations:

$$\begin{aligned} \beta &= \operatorname{grad} T\, \alpha\, L \\ L &= \beta H \\ \frac{\Delta L}{L} &= \operatorname{grad} T\, \alpha\, H \\ \frac{\Delta D}{D} &= \frac{(\operatorname{grad} T\, \alpha\, H)}{2\sqrt{2}} \end{aligned} \tag{3.36}$$

where

β	angle of bending
L	length measured along the standard
$L + \Delta L$	length measured parallel to the standard

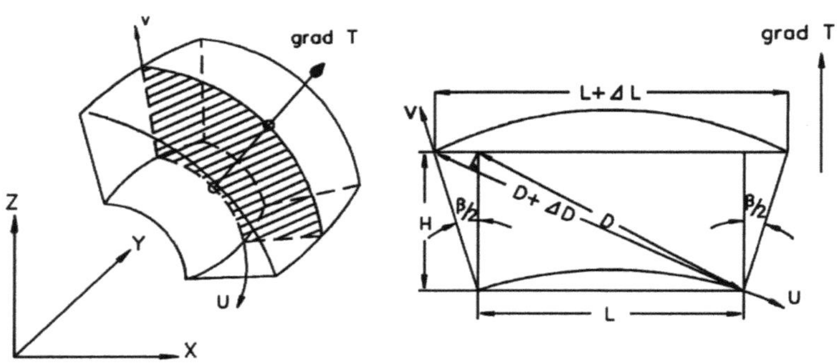

Fig. 3.29 Deformations of CMM elements caused by the thermal gradient influence [23, 37]

α	expansion coefficient of the material (from which the given CMM element is done)
H	distance from the length measurement standard
D	length of CMM element diagonal without thermal gradient participation
$\Delta D/D$	increment of CMM element diagonal length caused by thermal gradient
grad T	thermal gradient
grad T_x, grad T_y, grad T_z	gradients towards particular x, y, z axes

The spatial, normal to (area) X–Y plane (in an accepted reference system) thermal gradient essentially causes the deformation of the machine base, slides, and the table, contributing to the quill (with measuring probe head) location error marked as *xry*, according to Fig. 3.30.

The deflection of the bed and slides of the X-axis in the XY-plane is connected with the permanent thermal gradient in the direction of the Z-axis, marked in Fig. 3.30 as grad T_z. It is possible to write them as

$$xry(x) = \alpha_{\text{bed}} \cdot \text{grad}\, T_z \cdot x \tag{3.37}$$

After all, the uncertainty component of the length measurement u_{T2} (resulting from Abbe's postulate not being fulfilled in the XZ-plane), in relation to the standard placed in the X-axis, counted in the Z-axis direction (as $H_{xz\max}$) can be written:

$$u_{T2} = \frac{\Delta L_x}{L_x} = \text{grad}\, T_z \cdot \alpha_{\text{bed}} \cdot H_{xz\max} \tag{3.38}$$

where

$H_{xz\max}$ distance of cross-beam slides from the column.

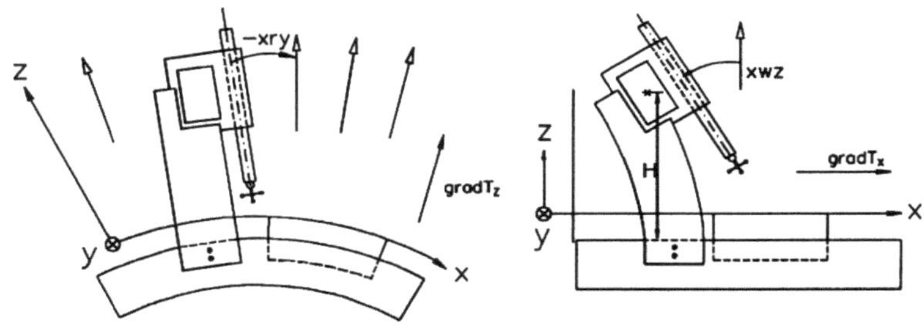

Fig. 3.30 Model of the thermal deformation of the slide base, table, and columns of the measuring machine [23, 37]

The spatial thermal gradient normal to the *YZ*-plane mainly causes deformations of measuring machine columns, change of the crossbeam location, and after all the error of the quill (with the measuring probe head) location marked as *xwz*, according to Fig. 3.30.

The spatial thermal gradients mentioned above moreover cause, in the upper part of the measuring machine, the deformation of the crossbeam of the portal, according to Fig. 3.31. Incurred errors are marked as *yrx* and *yrz*. The effect of the gradient influence in the *Y* direction causes the crossbeam deflection in the *X* direction, *xrz*. According to Fig. A.7, with the maximum distance H_{xymax} in the *Y* direction from the location of the displacement measurement standard (scale) of the *X*-axis, the component of the u_{T3} length measurement uncertainty can be expressed as

$$u_{T3} = \frac{\Delta L_x}{L_x} = \text{grad}\, T_z \cdot \alpha_{\text{bed}} \cdot H_{xymax} \tag{3.39}$$

where

H_{xymax} distance in *Y* direction from the location of the displacement measurement standard of the *X*-axis

Analogously to u_{T2} for the crossbeam, we receive the uncertainty component of length measurement u_{T4} caused by deflecting the crossbeam in the *Y* direction *yrx*(*x*) —Figure A.7—with the maximum distance H_{xymax} in the *Z* direction from the length measurement standard of the *Y*-axis, expressed as

$$u_{T4} = \frac{\Delta L_y}{L_y} = \text{grad} T_z \cdot \alpha_{\text{crossbeam}} \cdot H_{yzmax} \tag{3.40}$$

where

H_{yzmax} distance in *Z* direction from the *Y*-axis scale of crossbeam slides from the column.

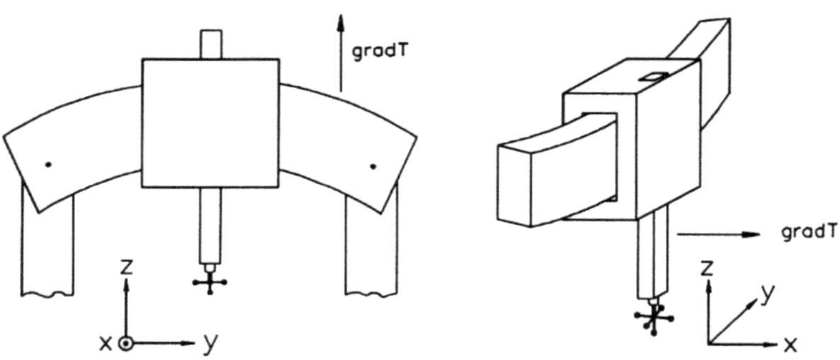

Fig. 3.31 Model of the deformation of the portal crossbeam of PMM12106 machine [23, 37]

In Fig. 3.27 there is also presented the influence of the thermal gradient in the *X* direction on the deflection of portal columns, causing nonorthogonality in the *XZ* plane, *xwz*. Nonorthogonality *xwz* influences the maximization of *D* length at a 45° angle in relation to both axes, that is:

$$\frac{\Delta D}{D} = 0.5 \cdot xwz \tag{3.41}$$

wherein *xwz* is given in the arc measure and ΔD will be a sum of maximum length measurement deviation. We receive:

$$u_{T5} = \frac{\Delta D_{\mathrm{xz}}}{D_{\mathrm{xz}}} = \mathrm{grad}\, T_x \cdot \alpha_{\mathrm{column}} \cdot \frac{H}{2} \tag{3.42}$$

where *H* will be a column height between the real place of connecting the column above the CMM bed and the real place of connecting the column with the crossbeam.

The influence of temperature change ΔT_x, ΔT_y, ΔT_z on the uncertainty of the determination of extension of optoelectronic scales of CMM is marked as u_{T1}.

The uncertainty of the thermal expansion is equivalent to a product of the temperature change and the coefficient of linear expansion of scale material. The following relations were adopted.

$$\begin{aligned} u_{T11} &= \frac{\Delta L_x}{L_x} = \Delta T_x \cdot \alpha_{\mathrm{pom}} \\ u_{T12} &= \frac{\Delta L_y}{L_y} = \Delta T_y \cdot \alpha_{\mathrm{pom}} \\ u_{T13} &= \frac{\Delta L_z}{L_z} = \Delta T_z \cdot \alpha_{\mathrm{pom}} \end{aligned} \tag{3.43}$$

The uncertainty of determining standards extension as a result of thermal changes u_{T1} is accepted as the maximum value from u_{T11}, u_{T12}, and u_{T13} for particular scales.

The uncertainty of the temperature measurement of the object was determined for the temperature change of the object ΔT_W, with the accepted expansion coefficient of the measured object α_W:

$$u_{T6} = \Delta T_w \cdot \alpha_W \tag{3.44}$$

If the correction of the object temperature is not done, we have to accept ΔT_W, determining maximum temperature deviations of the object from the reference temperature of 20 °C.

The value u_{T7} of the length measurement uncertainty component caused by zero floating because of the temperature distribution and hysteresis u_{T7} change, corresponds to the maximum stated difference (of drift) from all measurements for different plate locations both for measurements in a given direction, as well as for reverse measurements. The total uncertainty of the temperature influence u_T is counted as the sum of squares of uncertainty components from u_{T1} to u_{T7}. At the same time stable elements (only u_{T7}) and elements dependent on the length are individually calculated.

3.5.3 Description of the Hole Plate Calibration Procedure

The object for calibration is a hole plate of outside dimensions 500 × 500 mm made from aluminum alloy (AK 64) with 25 rings from high-coal chrome steel ($\sim$1.0 % C and $\sim$1.4 % Cr) situated in a neutral plane, hardened and ground with shape errors below 0.05 m. The nominal distance between ring centers is 78 mm. The hole plate was developed at Cracow University of Technology as part of research project PB 1367/T07/95/08 [22]. It is presented in Fig. 3.32. The plate has its own number and permanent denotation (PK-M 01–25–83–332–332). In the case of plate standards a coefficient of thermal expansion of the plate corpus is of significant importance. It was determined individually for the whole standard. For that purpose a special dilatometer developed in the PTB was applied (Fig. 3.33).

The temperature and relative size of changes of the analyzed object and reduced reference standard are measured during the temperature cycle (20–25–20 °C). The measurement uncertainty of the dilatometer described and used in tests is $1.5 \times 10^{-7}\ \mathrm{K}^{-1}$. As a result of the measurement cycle, the plate coefficient of thermal expansion α was received. It is equal to:

$$\alpha_{\text{plate}} = 20.5 \cdot 10^{-6}\,\mathrm{K}^{-1} \tag{3.45}$$

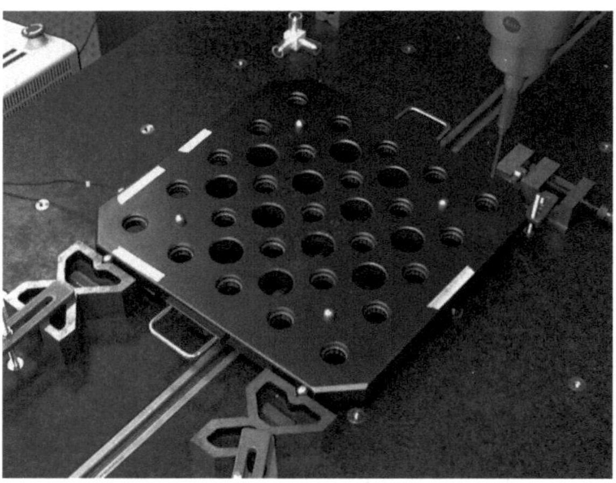

Fig. 3.32 View of the hole plate during calibration process with applied mounting system [37]

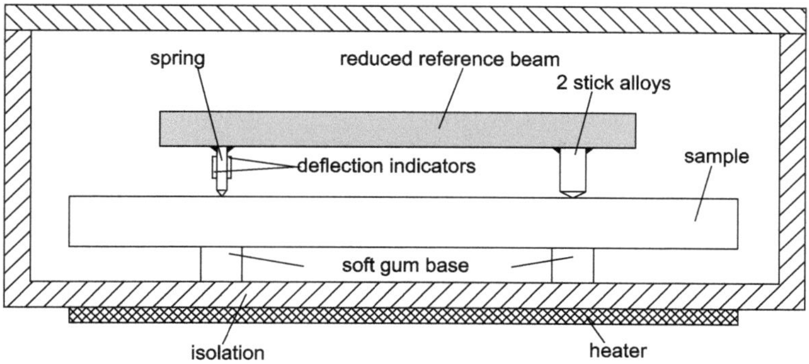

Fig. 3.33 Scheme of the dilatometer (developed in the PTB) used for setting the expansion coefficient of the plate standard

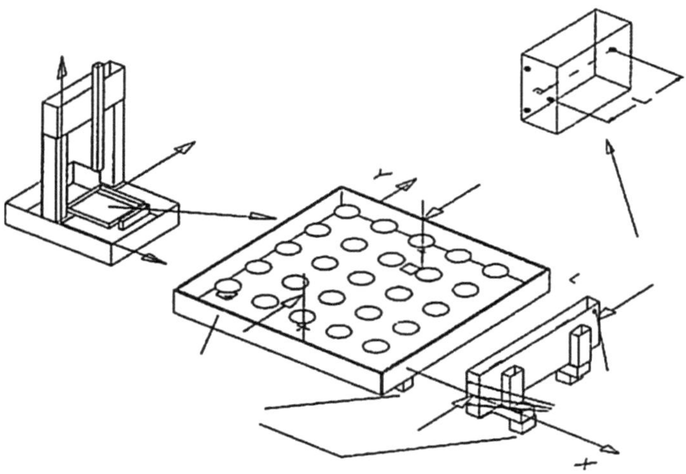

Fig. 3.34 Method of hole plate calibration measurements with use of end standards [23]

In the process of plate standard calibration the PMM12106 Leitz machine was used, with Quindos software, which fulfills requirements developed by the PTB for accredited laboratories concerning machines applied to calibration work [23, 37].

The applied reference standard is also significant. It can be the interferential standard or gauge block; in that case the machine works as the comparator presented in Fig. 3.34. In the Laboratory of Coordinate Metrology at CUT in the process of hole plate calibration the 300-mm–long gauge block was applied, at calibrated length equal to 4 × 83 mm = 332 (distance between holes of the plate), complying with the recommended similarity requirements established according to Table 3.11 [37].

Due to the measurement strategy presented in Fig. 3.35, a contact uncertainty was determined carrying out a series of 50 measurements for the same ring that allowed for determining the standard and expanded uncertainties:

Table 3.11 Uncertainty budget for hole or ball plate calibration task

Uncertainty component	Steel plate	Aluminum plate
U_g4 points/U_g5 points, ring/ball for 4/5 points	0.26 μm	0.24 μm
AV uncertainty factor of calibration standards/gauge blocks	0.06 μm	0.06 μm
KV Retter/gauge blocks	0.48 μm	0.48 μm
L (max. measured length, distance between holes/balls)	0.44, 0.332 m	0.44, 0.332 m
$U_v = AV + KV \cdot 10^{-6} L$ $U_v = 0.8 + 0.9\ L$ (Retter plate) prot. calibr. GUM) $U_v = 0.06 + 1.6\ L$ (gauge blocks) prot. calibr. GUM)	1.2 μm 0.54 μm	1.2 μm 0.54 μm
A_{Ko}	0.8 μm	0.8 μm
K_{Ko}	0.25 μm/m	0.25 μm/m
B_{Ko}	1.8 μm	1.8 μm
$U_{Ko} = A_{Ko} + K_{Ko} \leq B_{Ko}$ (wg ISO10360-2)	0.8 + 0.25 L μm	0.8 + 0.25 L μm
Max. temp. fluctuations of rulers $\Delta Tx = \Delta Ty = \Delta Tz$	0.1 K	0.1 K
Expansion coefficient of rulers α (glass)	$9 \times 10^{-6} K^{-1}$	$9 \times 10^{-6} K^{-1}$
$u_{T1-1} = \Delta L_x/L_x = \Delta T_x \cdot \Delta_{measur.}$	0.396 μm	0.396 μm
$u_{T1-2} = \Delta L_y/L_y = \Delta T_y \cdot \Delta_{measur.}$	0.396 μm	0.396 μm
$u_{T1-3} = \Delta L_z/L_z = \Delta T_z \cdot \Delta_{measur.}$	–	–
$u_{T1} = u_{T1-1} = u_{T1-2}$	0.396 μm	0.396 μm
grad T_x horizontal temp. gradient in X direction	0.3 K/m	0.3 K/m
grad T_x horizontal temp. gradient in Y direction	0.3 K/m	0.3 K/m
grad T_x vertical temp. gradient in Z direction	0.3 K/m	0.3 K/m
TREF-reference temperature	20 ± 0.2 K	20 ± 0.2 K
Expansion coefficient α bed, granite	$6 \times 10^{-6} K^{-1}$	$6 \times 10^{-6} K^{-1}$
Expansion coefficient α cross-beam, granite	$6 \times 10^{-6} K^{-1}$	$6 \times 10^{-6} K^{-1}$
Expansion coefficient α columns, steel	$8.7 \times 10^{-6} K^{-1}$	$8.7 \times 10^{-6} K^{-1}$
H_{xymax} dist. in Y dir. from displ. measur. stand. loc. X-axis	0.6 m	0.6 m
H_{xzmax} dist. of cross-beam slides form column	1 m	1 m
H_{yzmax} dist. in Z dir. from displ. measur. stand. Y-axis	0.6 m	0.6 m
$u_{T2} = \Delta L_x/L_x =$ grad $T_z \cdot \alpha_{bed} \cdot H_{xzmax}$	0.648/0.79 μm	0.648/0.79 μm
$u_{T3} = \Delta L_x/L_x =$ grad $T_z \cdot \alpha_{bed} \cdot H_{xymax}$	0.388/0.48 μm	0.388/0.48 μm
$u_{T4} = \Delta L_y/L_y =$ grad $T_z \cdot \alpha_{cross\text{-}beam.} \cdot H_{yzmax}$	0.388/0.48 μm	0.388/0.48 μm
$u_{T5} = \Delta D_{xz}/D_{xz} = \text{grad}\, T_x \cdot \alpha_{column} \cdot H/2\sqrt{2}$	–	–
Max. temperature change of calibrated standard ΔT_W	0.1 K	0.1 K
Coefficient of expansion of standard material α_W	$11.5 \times 10^{-6} K^{-1}$	$20.5 \times 10^{-6} K^{-1}$
$u_T = \Delta T_W \cdot \alpha_W$	0.51 μm	0.9 μm
u_T—drift $AT = u_T$ adopted for PMM0.5 μm/m	0.22 μm	0.22 μm
$K_T = \sqrt{\left(u_{T1}^2 + u_{T2}^2 + u_{T3}^2 + u_{T4}^2 + u_{T5}^2 + u_{T6}^2\right)}$	1.06 μm/m	1.29 μm/m
$U_T = A_T + K_T L$	0.22 + 1.06	0.22 + 1.29
$A_K = \sqrt{\left(A_{Ko}^2 + U_g^2 + A_V^2 + A_T^2\right)}$	0.34/0.88 μm	0.33/0.87 μm
$K_K = \sqrt{(K_{Ko}^2 + K_V^2 + K_T^2)}$	1.16/1.46	1.37/1.64
$B_K = \sqrt{\left(B_{Ko}^2 + U_g^2 + A_V^2 + A_T^2 + (K_V^2 + K_T^2) \cdot L_{max}^2\right)}$	0.78/1.42	0.95/1.62

(continued)

Table 3.11 (continued)

Uncertainty component	Steel plate	Aluminum plate
A_{Ko}, K_{Ko}, and B_{Ko} were omitted due to elimination of kinematic errors because of swing-round method implementation		
$U_K = A_K + K_K \cdot L \leq B_K$ calibration with gauge blocks usage	0.34 + 1.16 $L \leq 0.78$	0.33 + 1.37 $L \leq 0.95$
$U_K = A_K + K_K \cdot L \leq B_K$ calibration with ball plate (Retter) usage	0.88 + 1.46 $L \leq 1.42$	0.87 + 1.64 $L \leq 1.62$

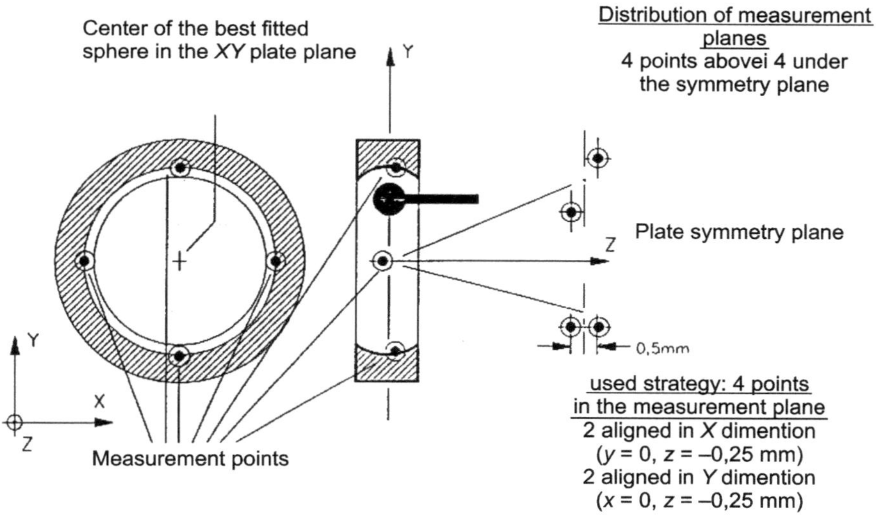

Fig. 3.35 Measurement strategy of the ring (of the hole plate) measurements [1, 37]

$$\begin{aligned} u_g 2x4pty &= 0.12\,[\mu\text{m}] \\ U_g 2x4pty &= 0.24\,[\mu\text{m}] \end{aligned} \tag{3.46}$$

Temperature monitoring is possible only with use of a high-class air-conditioning installation, which here is a special system, meeting very high requirements for air-conditioned laboratory rooms. A vertical air circulation, shown in Fig. 3.36, was applied here. Acceptable temperature fluctuations in the 24-h cycle are 20 ± 0.2 °C. In the machine measuring volume (in the place of the object measurement on CMM) they should not exceed ± 0.1. The real thermal conditions are monitored with the Ahlborn–Almemo MA5590-1V5 company system, consisting of the central unit ALMEMO 5590-1 and of 21 NTC-type sensors of resolutions 0.01 K calibrated with a mercury reference thermometer of measuring scope 18–24 K and uncertainty 0.01 K, confirmed with certificate (no. 1401E90B, given by Eichamt fur

Fig. 3.36 Scheme of air circulation in the laboratory room

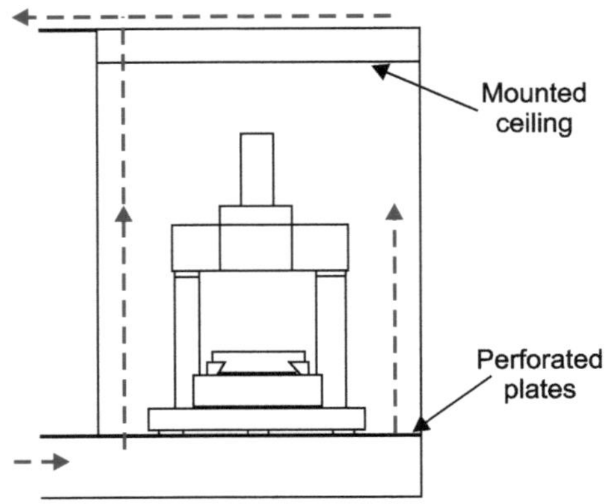

GlasmeBgerate Wertheim—DKD). The sensor distribution is presented in Fig. 3.37. Measurements were performed by readings made before and after the measurement cycle and additionally through sensors mounted directly on the CMM:

- 2 sensors on each of the scales: *X*—M00 and M01, *Y*—M26 and M27, *Z*—M28 and M29
- 4 on vertical columns: M05, M06, and M07, M08
- 2 on the portal M25 and M22
- 2 on the bed: M02 and M03
- 1 on the table: M09

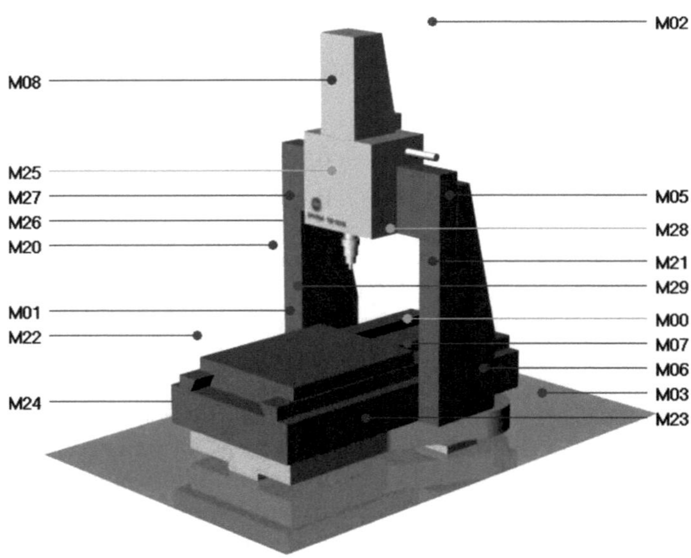

Fig. 3.37 Arrangement of temperature sensors for PMM12106 Leitz machine

Sensors of the air temperature measurement:

- 1 on the floor level M20
- 2 in the measuring volume: inside the measuring volume M23 and outside M11 and M12
- 1 on the level of the laboratory ceiling: M04

Sensors of the object temperature measurement:

- 1 sensor for the object M24

Data were recorded by the Almeno 5590–1 temperature scanner with the use of Ahlborn Control Data 4.1 software.

Deployment of this system enabled determination of deformations according to the thermodynamic machine model and determination of uncertainty components of thermal expansion of CMM construction (Eqs. 3.37, 3.38). Monitoring of the standard temperature during measurement poses a separate issue. The standard was stabilized thermally at 48 h in the vicinity of the measurement place, obtaining the maximum change of standard temperatures $\Delta_{TW} = 0.1$ K. Elimination of systematic errors, having their source in deformations of the kinematic system, is possible by applying the swing-round method in the hole plate or ball plate calibration process (Fig. 3.38). Because of that it is possible to eliminate the length measurement uncertainty component U_{Ko} caused by the group of kinematic errors, described by Eq. (3.32). This procedure allows the elimination of all systematic errors of CMM geometry in the measurement plane, except for errors of location in the CMM axis, resulting from standards errors. This value is set in the procedure presented in Fig. 3.33 and described by the uncertainty component equation of the calibration of used standards according to Eq. (3.33).

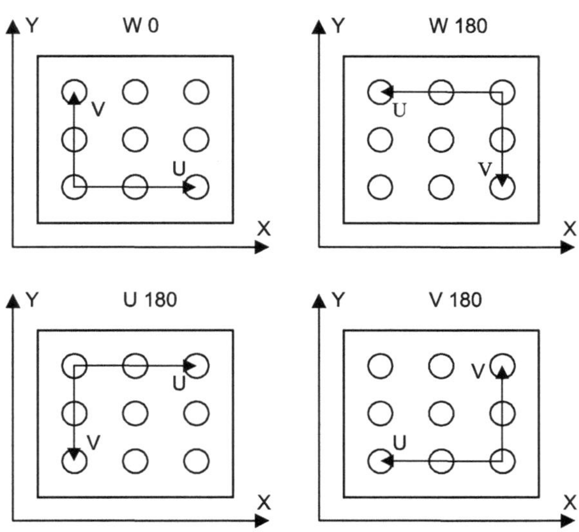

Fig. 3.38 Settings of the standard in the calibration cycle by swing-round method: *X*, *Y* axes of the machine coordinate system; *U*, *V* axes of the plate system

The value u_{T1} of the length measurement uncertainty component caused by the zero floating as a result of the temperature distribution and the hysteresis change, is limited by repeating every plate measurement cycle (“snail method”) with the upside-down sequence and gets the value of 0.5 μm/m for the PMM12106 machine (Fig. 3.39).

In the effect of plate measurement in four locations and the application of the PKal program developed in the PTB, the protocol giving coordinates of standard spheres of ball centers was obtained. The protocol also included a graphical illustration of values and directions of error vectors in particular locations as presented in Fig. 3.40. Components of the uncertainty budget for the hole plate calibration task are shown in Table 3.11.

Results of measurements carried out for two types of standards, the aluminum hole plate and the Retter ball plate, including measurement realization conditions are included in this table. Also the influence of external conditions is shown, including temperature and that of the adopted strategy, and eliminates influences of machine errors (swing-round method) and probe errors (four-point strategy in two planes). Of crucial importance is the uncertainty of the applied length standard—gauge blocks or the standard in the form of the Retter ball plate. A scope of temperature changes was set in limits of ±0.2 K.

For example, in relation to a substitution method, the uncertainty budget method presents itself as more universal and enables uncertainty estimation for any measuring task. Among the advantages is high accuracy of uncertainty estimation for a length measuring task. Also mathematical calculations do not cause many problems. One of the disadvantages of this method can be that it is realized mainly in laboratory conditions, where the appropriate technical resources conditioning the correct realization of the measurement procedure can be met. The mentioned conditions are, among others, the air conditioning adapted to measurement accuracy and the possibility of the evaluation of particular error sources.

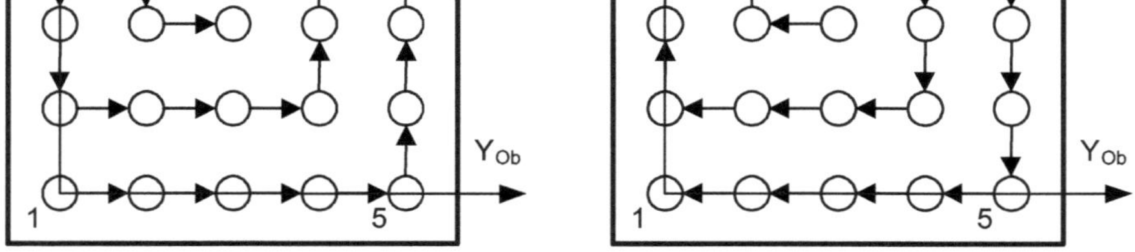

Fig. 3.39 Order of the measurement of control plate elements

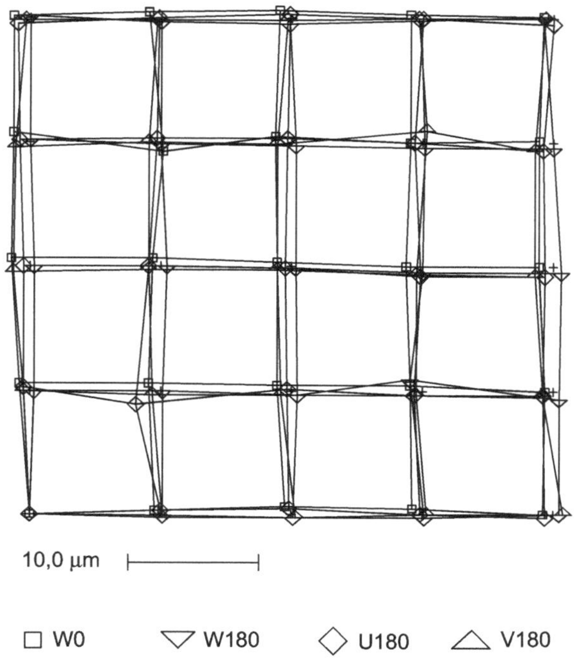

Fig. 3.40 Graphical presentation of hole plate calibration results [37]

3.6 Methods Based on Relations Resulting from the Model of Maximum Permissible Errors of Coordinate Measuring System

The uncertainty estimation based on looking for error sources and its realization in the form of an error budget gave the idea of its development and simplification, based on so-called expert knowledge or differently—metrologist's experience—that often consists of intuitive uncertainty estimation of realized measurement. The realization of this concept raises the most doubts. Admittedly it is based on the description included in the project of the ISO/TS 15530-5 standard [11], but also on the facts that an expert has to possess adequate qualifications and that this method can be applied by accredited laboratories (having management systems in accordance with ISO 17025 [27]) and national offices of measurement laboratories are stressed. It can be used then in places that fulfill the highest requirements and where every measurement procedure is documented. The method concept is closest to a proposal based on rules of estimating the uncertainty using the model of measuring machine geometric errors and contact probe head errors, and their propagations for simple measuring tasks. Its realization is based on information contained in machine indication error and MPE. Next, the maximum error of measurement is estimated, which can be interpreted as expanded uncertainty and can be identified with the application of:

- The measuring machine uncertainty in form of U_3 (according to the older version of VDI/VDE 2617–3 recommendations) for the proposal of H-G Pressel [30]
- The MPE maximum permissible error for the concept developed by M. Hernla [6]

Proposals and rules given by Pressel assume the participation of components in form of $U_3 = A + L/K$ or uncertainties of the contact probe marked as v_3, creating areas of errors around extreme points of an object's dimensions. After including values of distances between points, it enables the determination of maximum error values, under the assumption that they cover the range of error variability enabling the uncertainty estimation for simple measurement tasks.

In Hernla's concept a general assumption was that accuracy of measurement performed on a given CMM is not bigger than the indication error value specified for the maximum error equation for the assessed length. That assumption is theoretically real, when we assume that generally the values of maximum errors are bigger than the values of indication errors obtained in the process of their establishment. It is shown in Fig. 3.41. However, this assumption is not effective in practice because too many factors affect measurement error, and the area of maximum error does not determine the area of measurement error, which has already been proven in many publications and is, first of all, confirmed by results of coordinate measurement experiences [37].

The essence of the concept of uncertainty estimation given by Hernla is using the value of the maximum errors, which can be calculated using dependences resulting from the area of MPEs, characteristic for a given measuring machine. It assumes that these errors are under normal distribution, and maximum values correspond to the area of the changeability of the measurement uncertainty: expanded uncertainty with coverage factor $k = 2$. If necessary, adequate standard uncertainties should be calculated by dividing maximum values by 2. Moreover it states that for the measurement of parallelism, squareness, and slope deviations, the reference element should be longer than the tolerated element, because assuming a smaller value causes growth of the measurement error. In his concept Hernla established these rules:

- The error of dimension measurement is not bigger than that determined by dependence on the MPE length component for the dimension L/K measurement.
- The measurement error during the determination of direction or location deviation is maximally twice as big as that dependent on the length MPE component, L/K, provided that the datum is longer than the tolerated element.

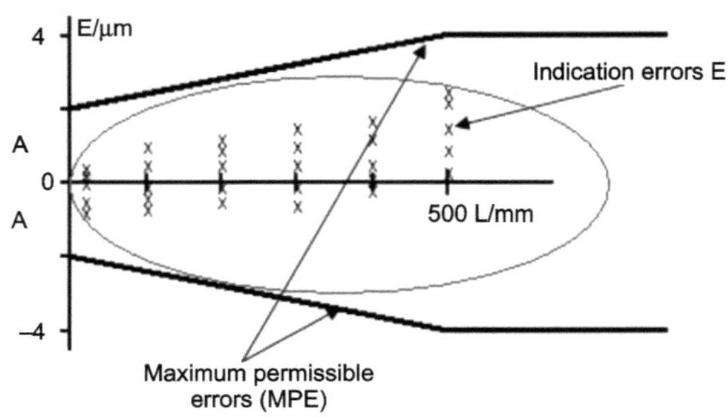

Fig. 3.41 Indication errors and maximum errors of given CMM $MPE_E = 2 + L/200$ mm

- The error of shape deviation measurement is not bigger than the quadruple value of the MPE *L*/*K* component dependent on the length (where for the *L* should be substituted the length of the biggest diagonal of measured object).
- Analyzing the influence of the temperature on the measurement result he assumes regular temperature distribution within the object and states that in such a situation the temperature does not affect the shape and direction deviations, but only dimension and distances (including position deviations).
- For simplifying, irrespective of the geometric form of the object, to estimate the temperature error for measured length *L* he inserts the biggest linear dimension of the object (length of spatial diagonal) and in that way includes the temperature influence.
- It also recognizes an influence of the sampling strategy in the case of measurement, for example, at the circle or cylinder measurement, assuming, however, regular point distribution and simultaneously implies that chosen points are burdened only with random errors. It leads to simple relations between the number of points and the uncertainty of determining the coordinates of a center. To define the total influence of particular uncertainty components it uses the model assuming the errors' additivity.

In Table 3.12 the formulae for the maximum measurement errors for dimension deviations and for form and location deviations, according to Hernla are put together.

However, the application of these relations is complicated and requires much knowledge and experience on the expert level to not overvalue or, what is even worse, undervalue the uncertainties. Hernla is conscious that used simplifications lead to such uncertainty undervaluations, however, he thinks that they are not too big; that is true, but only for simple cases. In general, this method carries the very real hazard of uncertainty undervaluation. Moreover, taking control of the proposed apparatus of standards is difficult, so that it is not suitable for industrial applications. In laboratory work, it has too low precision and cannot effectively replace the error budget.

In expert methods, as stated above, a key role is played by the MPE determined for CMM on the basis of procedures consistent with the ISO 10360: 2003 standard [26]. Recently radical new concepts of a part of this norm were introduced. Those are ISO 10360-2: 2009 and ISO 10360-5:2010 discussed in Sect. 6.1. They changed not only requirements, but above all, they started to treat the probe head system and machine kinematic system as two separate systems (e.g., the contact probe head is now checked using Part 5 of standard ISO 10360-5:2010). The problem of directly using the MPE equation for the task of measurement uncertainty estimation should be emphasized here one more time.

In order to help the reader correctly understand its meaning, the example of measurement uncertainty calculation with reference to maximum values obtained from the MPE equation of a given CMM should be noted. In Fig. 3.30 a measuring task consisting of the distance measurement between holes with a machine having the given accuracy, and the uncertainty estimation of the conducted measurement using two different methods (based on MPE and uncertainty budget) is presented. It

Table 3.12 Maximum values ΔL, ΔF, $\Delta\alpha$ of measurement errors for dimension, shape, and location deviations from the formula $MPE_E = (A + L/K)$ [µm] where l, L, D are given appropriately in mm [6]

Geometric deviation	Maximum measurement error	Comment
Length, distance, diameter, and location deviation (on plane)	$\Delta L \leq \frac{L}{K}$	L—nominal value of the measured dimension or the theoretically accurate dimension determining the position
Position deviation (in volume)	$\Delta L \leq \frac{L}{K}\sqrt{L^2 + l^2}$	L—theoretically accurate dimension determining the position l—maximum from nominal values determining the element position in perpendicular direction to the base
Coaxial and symmetry between points deviation	$\Delta L_K \leq \frac{D}{2K}$	D—maximum nominal value of the diameter
Coaxial and symmetry deviation in plane	$\Delta L_K \leq \frac{1}{K}\sqrt{\frac{D^2}{4} + L^2}$	D—maximum nominal value of the diameter or length L—nominal value of smaller length
Parallelism or slope deviation	$\Delta L_P \leq \frac{2L}{K}$	L—smaller one from nominal length values of measured elements with regard to the perpendicular distance between both elements (the longer element is a reference element)
Perpendicularity deviation	$\Delta L_R \leq \frac{2L}{K}$	L—nominal length value of the shorter side of the angle (the longer side of the angle is the base)
Slope deviation	$\Delta L_N \leq \frac{2L}{K} \sin\alpha$	L—nominal value of the triangle opposite side to the angle (the longer side of the angle is the base) α—nominal angle value
Angle deviation	$\Delta\alpha \leq \frac{2}{10^3 K} \sin^2\alpha$	α—nominal angle value
Rectilinearity deviation	$\Delta L_G \leq \frac{1}{K}$	L—nominal length value of the straight line
Flatness deviation	$\Delta F_E \leq \frac{1}{K}\sqrt{L^2 + 5l^2}$	l, L—nominal values of the shorter and longer plane sizes
Roundness deviation	$\Delta F_R \leq \frac{\sqrt{26}}{2}\frac{D}{K}$	D—nominal diameter value
Cylindricity deviation	$\Delta F_Z \leq \frac{1}{K}\sqrt{\frac{26}{4}D^2 + 10L^2}$	D—nominal diameter value L—nominal length

can be observed that the estimated uncertainty based on MPE is twice as low as reasonable due to the area of maximum errors understood as a symmetric one. It is not possible to apply the uncertainty estimation based on the MPE equation directly. Results of calculations shown in Figs. 3.43 and 3.44 are warning against it. Generally, of course, it should be recommended to use the methods described in former chapters (Fig. 3.42).

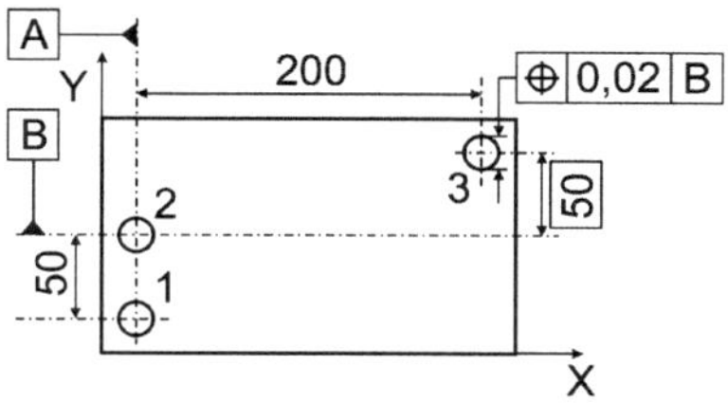

Fig. 3.42 Measured object and given measuring task performed on CMM

Uncertainty budget

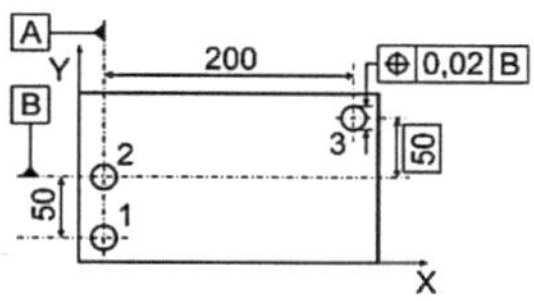

Uncertainty model $d = y_3 - y_2 + (200/50) \cdot (x_2 + x_1) + \delta_{posy} + \delta_{Sqr}$ $200 + \delta_{Strx} + \delta_{Temp}$.

Component	Value (mm)	Type/distr.	Chang. scope	Stand. uncert.	Uncert. part (μm)	Index (%)
y_3	120	*A*/normal		0.2 μm	0.2	1.4
y_2	70	*A*/normal		0.2 μm	0.2	1.4
x_2	20	*A*/normal		0.2 μm	0.8	22.6
x_1	20	*A*/normal		0.2 μm	0.8	22.6
δ_{Stry}	0	*B*/regular	0.5 μm	0.3 μm	1.2	33.7
δ_{posy}	0	*B*/regular	0.5 μm	0.3 μm	0.3	2.9
δ_{Sqr}	0	*B*/regular	5 μrad	3 μrad	0.6	11.8
δ_{Strxy}	0	*B*/regular	0.5 μm	0.3 μm	0.3	2.2
δ_{Temp}	0	Sub-budget		0.2 μm	0.2	1.4
A	50			$u_c = 1.8$ μm		
Expanded uncertainty				$U = 2\,u_c = \pm 3.6$ μm		

Fig. 3.43 Uncertainty budget for described measuring task

Uncertainty budget for length measurement task between holes

Obtained results

Length	MPE (μm)	U_{budget} (μm)
50 (2–3)	1.8	3.6
200 (2–3)	2.3	2.3
206 (2–3)	2.3	2.0

Fig. 3.44 Uncertainty budget and comparison of estimation for measuring machine MPE

- Even for the simple task we must estimate its uncertainty.
- Estimation based on MPE_E is limited and incorrect: $U = 1.8$ mm (vs. 3.6 mm).

Overall conclusion: It is not possible to treat the area of maximum MPE_E errors and measurement uncertainty U interchangeably.

Uncertainty estimation for a more complicated measuring task requires special tools and methodology such as

- Applying the ISO 15530 standard, calibrated object method
- Simulation with virtual CMM use

Task Formulation and Measurement Data:

- Accuracy of used CMM:$MPE_e = 1.6 + L/300\,\mu m$; the L is given in mm
- Accuracy of hole dimensions unspecified
- Temperature influences:

  (a) Material: steel ($a = 11 * 10^{-6}$/K)
  (b) No temperature differences for the object
  (c) The same temperature of displacement measurement standards

- Influence of torsion of the object is included in nonorthogonality of coordinate system

3.7 Analytical Method of Measurement Uncertainty Determination

The analytical method for coordinate measurement uncertainty determination and the software working on its basis were developed in the Laboratory of Metrology of Technical-Humanistic Academy in Bielsko-Biala under the direction of Professor Jakubiec [13–15]. The essence of the method [12] is based on:

- Treating the model of geometric errors containing 21 kinematic component errors as a characteristic model for given machine (similar to the simulation method described in Chap. 5).
- The assumption that in coordinate measurements the measurement uncertainty is analyzed individually for each characteristic.
- The assumption that particular geometric characteristics (dimensions, geometric deviations) are functions of differences of object points coordinates, the measurement uncertainty for a given characteristic is calculated based on this equation as the complex uncertainty (essential novelty called by Jakubiec, "The generality principle").
- Definition of functions describing the maximum differences of geometric error values; the arguments of these functions are differences of coordinates of object

points; definition of this function makes it possible to apply method B for uncertainty analyses of determination of pairs of point coordinate differences.
- The assumption that measurement models use the minimum number of object characteristic points.
- The assumption that measurement models are based on relations expressing the point–point, point–axis, point–plane distances.

3.7.1 Geometric Error Model

Geometric errors model the well-known kinematic model and are associated with the error propagation model (formula 3.47) including the construction and structural dimensions of the machine. For example, for the bridge machine with the movable portal (Fig. 3.45) the appropriate model contains (apart from 21 geometric errors, the probe head error, the temperature errors, and parameters of the measuring tip) two structural dimensions, the m and the h, and in matrix notation is presented in a following way.

$$\begin{aligned}
\begin{bmatrix} ex \\ ey \\ ez \end{bmatrix} &= \begin{bmatrix} xpx + dtx \\ xty \\ xtz \end{bmatrix} + \begin{bmatrix} ytx \\ ypy + dty \\ ytz \end{bmatrix} + \begin{bmatrix} ztx \\ zty \\ zpz + dtz \end{bmatrix} + \\
&+ \begin{bmatrix} 0 & zt & -yt \\ -zt & 0 & (x+xt) \\ yt & -(x+xt) & 0 \end{bmatrix} \begin{bmatrix} xrx \\ xry \\ xrz \end{bmatrix} + \\
&+ \begin{bmatrix} 0 & (z+zt) & (m-yt) \\ -(z+zt) & 0 & (x+xt) \\ -(m-yt) & -(m+xt) & 0 \end{bmatrix} \begin{bmatrix} yrx \\ yry \\ yrz \end{bmatrix} + \\
&+ \begin{bmatrix} 0 & -zt & (m-yt) \\ -zt & 0 & (x+xt) \\ -(m-yt) & -(m+xt) & 0 \end{bmatrix} \begin{bmatrix} zrx \\ zry \\ zrz \end{bmatrix} + \\
&+ \begin{bmatrix} 0 & zt & yt \\ -(z+zt) & 0 & -(x+xt) \\ -(m-yt) & -(x+xt) & 0 \end{bmatrix} \begin{bmatrix} ywz \\ xwz \\ xwy \end{bmatrix} + \begin{bmatrix} Px \\ Py \\ Pz \end{bmatrix}
\end{aligned} \tag{3.47}$$

3.7.2 Measurement Models

For all geometric characteristics (dimension or the geometric deviation) a measurement model is built. This model uses a minimal number of characteristic points of the object. Measurement models are built on relations expressing distances: point–point, point–axis, and point–plane. In the measurement model particular geometric characteristics are shown as functions of differences of characteristic

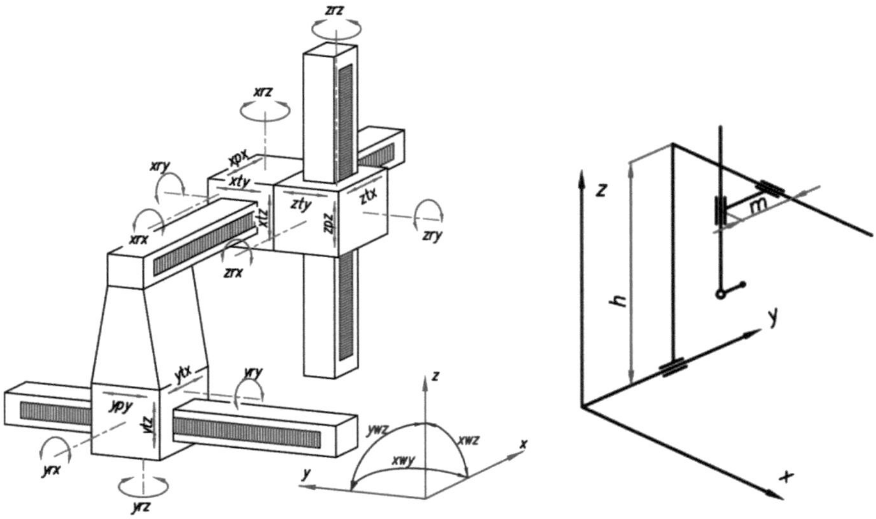

Fig. 3.45 Geometric model of machine errors: geometric errors (**a**), construction dimensions (**b**) [12]

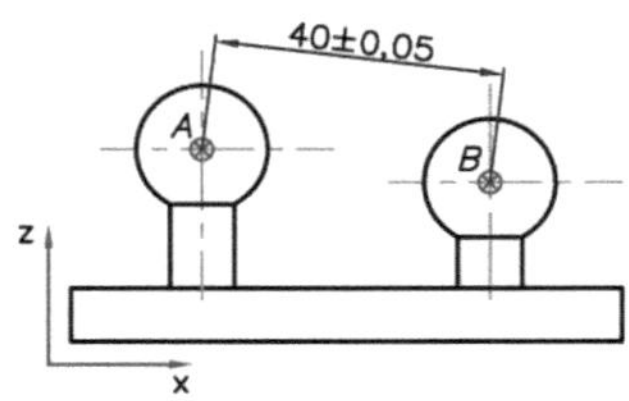

Fig. 3.46 Distance of sphere centers as the example of the task for which the uncertainty is determined using the model based on the relation expressing point–point distance [12]

object point coordinates. Particular models use from three to seven characteristic points of the measured object. These points can be points on the surface, axis, and sphere centers. Characteristic points represent the outermost points of the sampling of particular geometric elements and are distributed according to good measurement practice principles.

The model based on the point–point distance was used for determining the two sphere center distance uncertainty (Figs. 3.45 and 3.46).

This model has a form known from geometry:

$$l = \sqrt{(x_B - x_A)^2 + (y_B - y_A)^2 + (z_B - z_A)^2} \tag{3.48}$$

or in the shorter notation

$$l = \sqrt{x_{AB}^2 + y_{AB}^2 + z_{AB}^2}$$

where $x_{AB} = x_B - x_A$, $y_{AB} = y_B - y_A$ and $z_{AB} = z_B - z_A$.

The measured characteristic is expressed in the function of three coordinate differences: x_{AB}, y_{AB}, and z_{AB}.

The models based on the point–axis distance have been used for analysis of measurement uncertainty of characteristics such as straightness deviation, coaxiality deviation, axes parallelism deviation (cylindrical tolerance area), perpendicularity of axes to plane deviation, and some cases of position deviations. Examples of measurement models based on relations expressing point–axis distance are shown in Figs. 3.47, 3.48 and 3.49.

Generally the distance from *S* point to the *p*-axis, given by any *P* point belonging to this axis and by the *u* unit vector parallel to this axis, is calculated according to the equation:

$$d(S,p) = |(P - S) \times u| \tag{3.49}$$

In the first two cases the axis is given with points *A* and *B*; in the third the axis is given with the *K* point and *u* vector parallel to the *AB*-axis.

In the first two cases an equation for measured deviation (measurement model) takes the form of:

$$l = \sqrt{\frac{(y_{AS}z_{AB} - z_{AS}y_{AB})^2 + (z_{AS}x_{AB} - x_{AS}z_{AB})^2 + (x_{AS}y_{AB} - y_{AS}x_{AB})^2}{x_{AB}^2 + y_{AB}^2 + z_{AB}^2}} \tag{3.50}$$

The measured characteristic is given in the function of six different coordinates: x_{AB}, y_{AB}, z_{AB}, x_{AS}, y_{AS}, and z_{AS}.

Models based on the point–plane distance were used in uncertainty analysis of measurement of characteristics such as flatness deviation, axes parallelism deviation in the common plane and in plane normal to common, deviation of perpendicularity of axis to plane, deviation of perpendicularity of plane to axis, slope deviation, and some cases of position deviation and form deviation of determined profile and

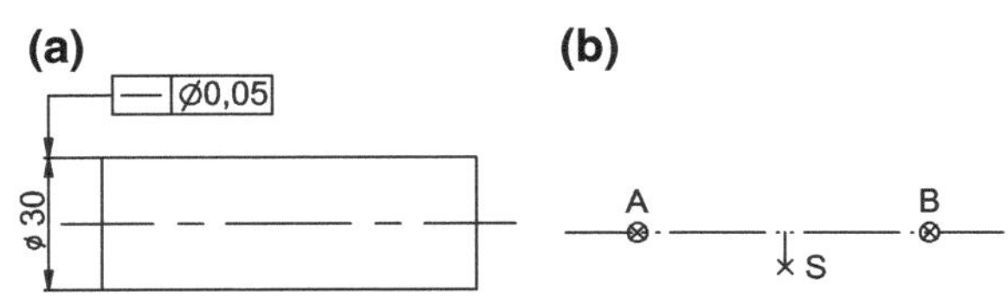

Fig. 3.47 Straightness deviation as the example of the task for which the uncertainty is determined from the model based on the relation expressing the point–point distance: technical drawing (**a**), characteristic points (**b**) [12]

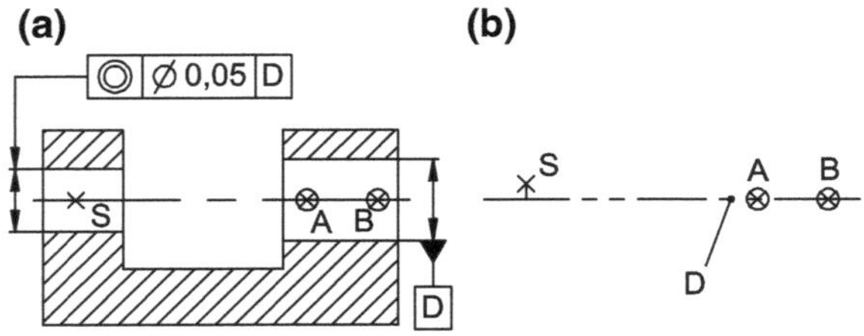

Fig. 3.48 Coaxiality deviation as an example of the task for which the uncertainty is determined from the model based on the relation expressing the point–axis distance: technical drawing with characteristic points (**a**), model (**b**) [12]

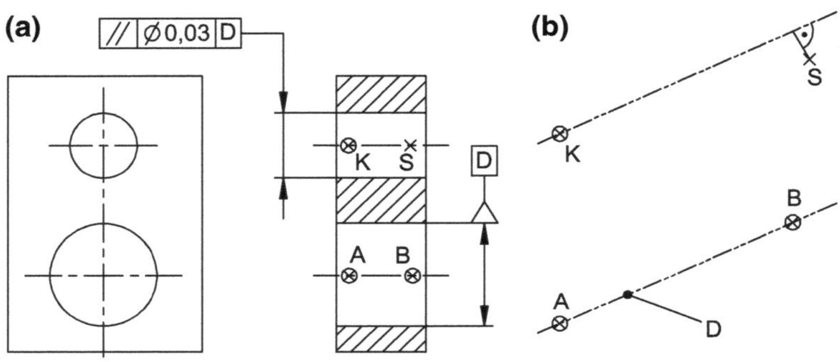

Fig. 3.49 Axes parallelism deviation (cylindrical tolerance area) as the example of the task for which the uncertainty is determined from the model based on the relation expressing the point–axis distance: technical drawing with characteristic points (**a**), model (**b**) [12]

determined surface. Examples of measurement models based on the relation expressing the point–plane distance are shown in Figs. 3.50 and 3.51.

The l distance of the S point from the p plane, given by any P point belonging to this plane and the u unit vector perpendicular to this plane is calculated according to the equation:

$$l(S,p) = |(P - S) \cdot u| \tag{3.51}$$

In the case of the flatness deviation, the p plane from the equation for a point–plane distance is given by three points A, B, and C. As a P point, in this equation one of three characteristic points (A, B, or C) may be adopted, and a normal unit vector u can be calculated according to the equation:

$$u = \frac{AB \times AC}{|AB \times AC|} \tag{3.52}$$

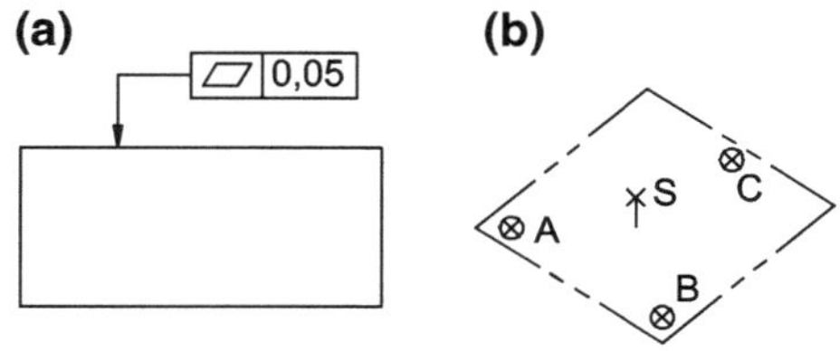

Fig. 3.50 Flatness deviation as the example of the task for which the uncertainty is determined from the model based on the relation expressing the point–plane distance: technical drawing (**a**), characteristic points (**b**) [12]

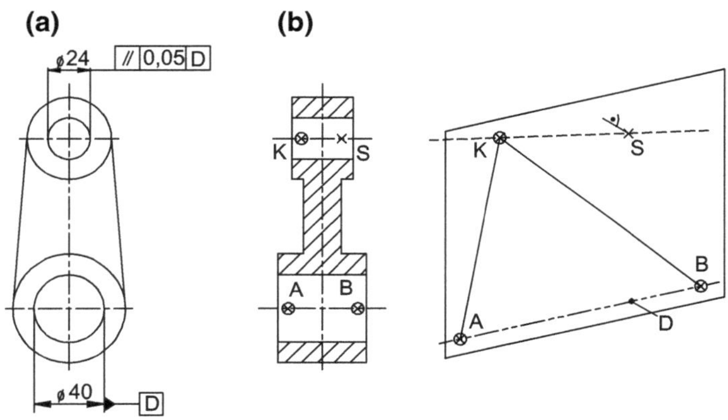

Fig. 3.51 Axes parallelism deviation (in plane normal to common) as the example of the task for which the uncertainty is determined from the model based on the relation expressing the point–plane distance: technical drawing (**a**), model (**b**) [12]

If the *A* point is adopted as the *P* point, the equation for the flatness deviation assumes the form:

$$l = \left(\frac{ax_{AS} + by_{AS} + cz_{AS}}{m} \right) \tag{3.53}$$

where

$$\begin{aligned} a &= y_{AB}z_{AC} - z_{AB}y_{AC} \\ b &= x_{AC}z_{AB} - x_{AB}z_{AA} \\ c &= x_{AB}y_{AC} - x_{AC}y_{AB} \\ m &= \sqrt{a^2 + b^2 + c^2} \end{aligned} \tag{3.54}$$

Measurement models for all kinds of dimensions and geometric deviations were appointed in a similar way. In all models the geometric characteristic is a function of coordinate differences.

3.7.3 Measurement Uncertainty as a Complex Uncertainty

Having prepared measurement models, it is possible to write an equation for complex uncertainty. For example, for the distance measurement of sphere centers this equation has the following form (in other cases many more arguments appear and formulas are much longer),

$$u_l = \sqrt{\left(\frac{\partial l}{\partial x_{\mathrm{AB}}} \cdot u_{\mathrm{xAB}}\right)^2 + \left(\frac{\partial l}{\partial y_{\mathrm{AB}}} \cdot u_{\mathrm{yAB}}\right)^2 + \left(\frac{\partial l}{\partial z_{\mathrm{AB}}} \cdot u_{\mathrm{zAB}}\right)^2} \tag{3.55}$$

in which u_{xAB}, u_{yAB}, u_{zAB} denote standard uncertainties of determining particular differences of coordinates.

In order to calculate all needed uncertainties of determining differences of coordinates the *B* type method was used and the uncertainty balance including all geometric errors and probe error was built. In the *B* type method it is necessary to know the maximum value that can be achieved by a given component error, and the corresponding probability distribution. Due to lack of other indications, the uniform distribution was adopted as the relevant distribution.

In order to determine maximum error values the calculations were carried out. The measurement error of the difference of coordinates can be calculated as the difference of errors of machine indication for two points measurement, which after resigning the matrix notation and after proper arranging gives:

$$\begin{aligned} e_{\mathrm{xAB}} = e_x(B) - e_x(A) = xpx(x_B) - xpx(x_A) + \cdots + zt_B \cdot xry(x_B) \\ - zt_A \cdot xry(x_A) + \cdots + zt_B \cdot xwz - zt_A \cdot xwz + \cdots + P_x(B) - P_x(A) \end{aligned} \tag{3.56}$$

$$\begin{aligned} e_{\mathrm{yAB}} = e_y(B) - e_y(A) = xty(x_B) - xty(x_A) + \cdots - zt_B \cdot xrx(x_B) \\ + zt_A \cdot xrx(x_A) + \cdots - (z_B + zt_B) \cdot ywz + (z_A + zt_A) \cdot ywz \\ - zt_A \cdot xwz + \cdots + P_y(B) - P_y(A) \end{aligned} \tag{3.57}$$

$$\begin{aligned} e_{\mathrm{zAB}} = e_z(B) - e_z(A) = xtz(x_B) - xtz(x_A) + \cdots + yt_B \cdot xrx(x_B) + \\ - yt_A \cdot xrx(x_A) + \cdots - (m - yt_B) \cdot xwz + (m - yt_A) \cdot xwz + \cdots \\ + P_z(B) - P_z(A) \end{aligned} \tag{3.58}$$

It can be noted that each component error is a sum of 12 errors, from which all come from different geometric errors.

$$e_{\mathrm{xAB}} = \sqrt{\sum_{i=1}^{12} e_{\mathrm{ix}}^2} \tag{3.59}$$

$$e_{\mathrm{yAB}} = \sqrt{\sum_{i=1}^{12} e_{\mathrm{iy}}^2} \tag{3.60}$$

$$e_{\mathrm{zAB}} = \sqrt{\sum_{i=1}^{12} e_{\mathrm{iz}}^2} \tag{3.61}$$

where

$$e_{1\mathrm{x}} = xpx(x_B) - xpx(x_A) \tag{3.62}$$

$$e_{4\mathrm{x}} = zt_B \cdot xry(x_B) - zt_A \cdot xry(x_A) \tag{3.63}$$

$$e_{10x} = zt_B \cdot xwz - zt_A \cdot xwz \tag{3.64}$$

$$e_{12x} = P_x(B) - P_x(A) \tag{3.65}$$

and so on.

3.7.4 Estimation of Maximum Value for the Geometric Error Difference

According to the concept applied in the type *B* method, the biggest values that each difference can take were separately estimated for all components of geometric errors. For example, in order to estimate the maximum value of $xrz_{\mathrm{MAX}}(d)$ for the measurement error of the difference of coordinates coming from $xrz(x)$ error, the knowledge of $xrz(x)$ error itself is necessary and performance of the following calculations has to be done (it is done similarly for other errors):

$$(|xrz(x) - xrz(x+d)|) \tag{3.66}$$

Geometric errors can be modeled with different functions. In the simplest case, they can be linear or square functions, but also broken lines, splines, harmonic functions, and so on. A very simple example of the $xrz_{\mathrm{MAX}}(d)$ function, for the case when the $xrz(x)$ error has a form of the linear function is shown in Fig. 3.50 [16]. This picture illustrates, among others, that two linear functions, different in the sign of the direction coefficient, have the same $xrz_{\mathrm{MAX}}(d)$ function (Fig. 3.52).

As can be seen in the presented method, in every model there appear points of all geometric elements (of both datum/data and the tolerated element) mentioned in the definition of the given characteristic. In this procedure an important difference in accordance with the simulated model can be seen. In the latter, at first geometric elements are built based on sampled point coordinates, and after that a given geometric characterization is determined from these elements.

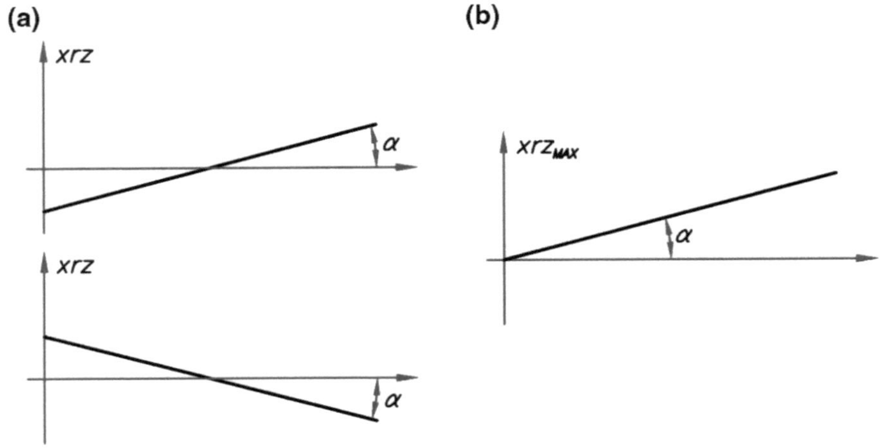

Fig. 3.52 Examples of functions: expressing measuring machine geometric errors (**a**) and, corresponding with them, functions expressing maximum value that can be taken by the error difference (**b**) [16]

3.7.5 Software

To realize the presented method special software was developed and named EMU (evaluation of measurement uncertainty for coordinate measurement). The software consists of two parts, Master and Client. In the Master part, operated only by the software supplier, construction dimensions, information of machine geometric errors (obtained on the basis of calibration results), and temperature conditions are introduced, so that the user receives the software dedicated for her machine. The Client part functioning is described in more detail below [16–18]. The characteristics operated by the EMU program were grouped together on the toolbar and shown in Fig. 3.53.

Other icons in accordance with the meanings of geometric tolerance symbols according to ISO 1101.

3.7.6 Particular Stages in the Operating Software

Stage 1. All information concerning the accuracy parameters of the measuring machine used have been entered earlier in the software by the appropriate software add-on.

Only in the case of possessing a larger number of machines, the first operation that is carried out is the choice of the machine (Fig. 3.54).

Stage 2. Operator sets the material of measured object, choosing it from base (Fig. 3.55).

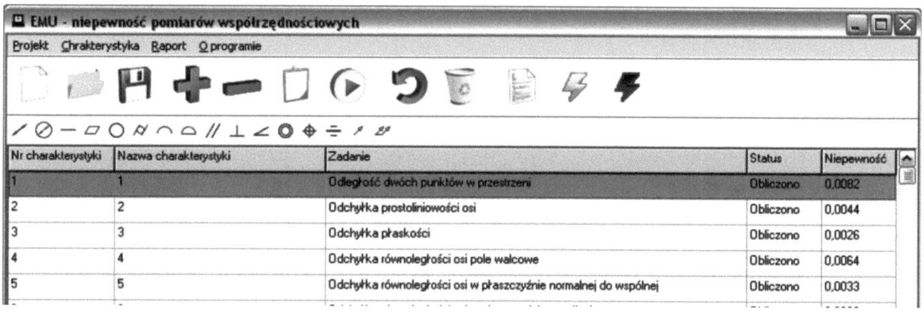

Fig. 3.53 The main EMU software window with characteristics list, for which the uncertainty determination is possible [16]

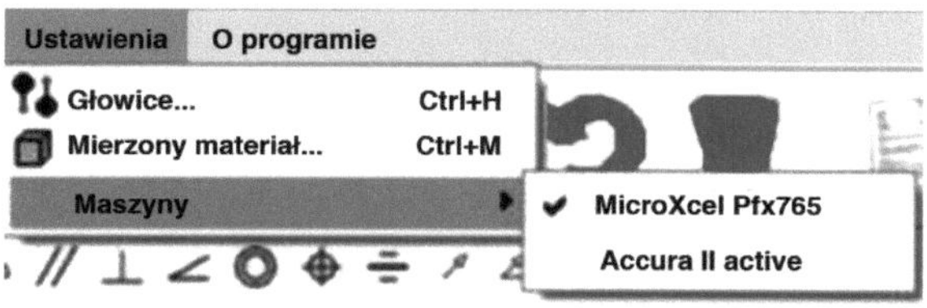

Fig. 3.54 Choice of measuring machine in the main menu: settings/machines [16]

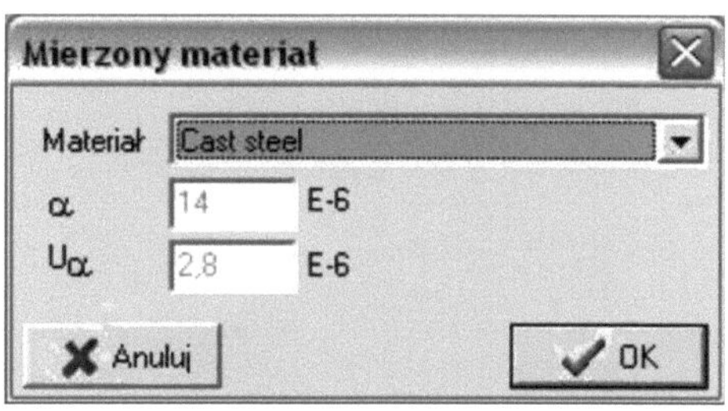

Fig. 3.55 Dialog box to choose the measured object material [16]

Stage 3. The next step is a choice of the group of characteristics from a toolbar (Fig. 3.56); then we indicate the particular characteristic for which the measurement uncertainty will be determined.

Stage 4. In the appropriate dialog box (Fig. 3.57) the user introduces measurement strategy parameters, which are coordinates of a few characteristic points of the measured object, used measuring stylus parameters, the number of sampling points, and the diameter or the thickness.

In a dialog box there is a picture of an exemplary object with the appropriate tolerance, required characteristic points, and coordinate system. The object is shown in three different orientations that can be changed with the appropriate button.

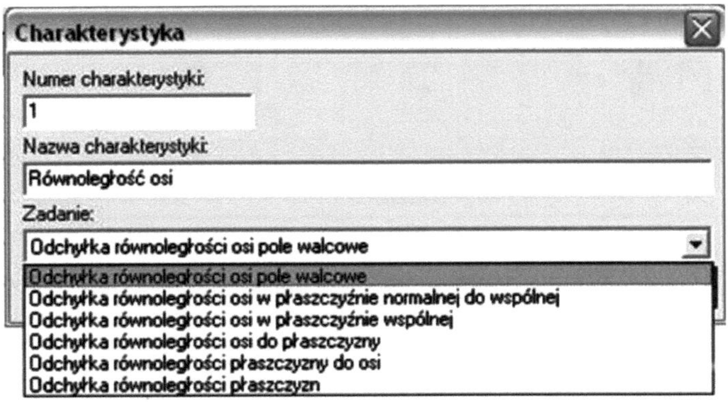

Fig. 3.56 Dialog box for characteristic choice [16]

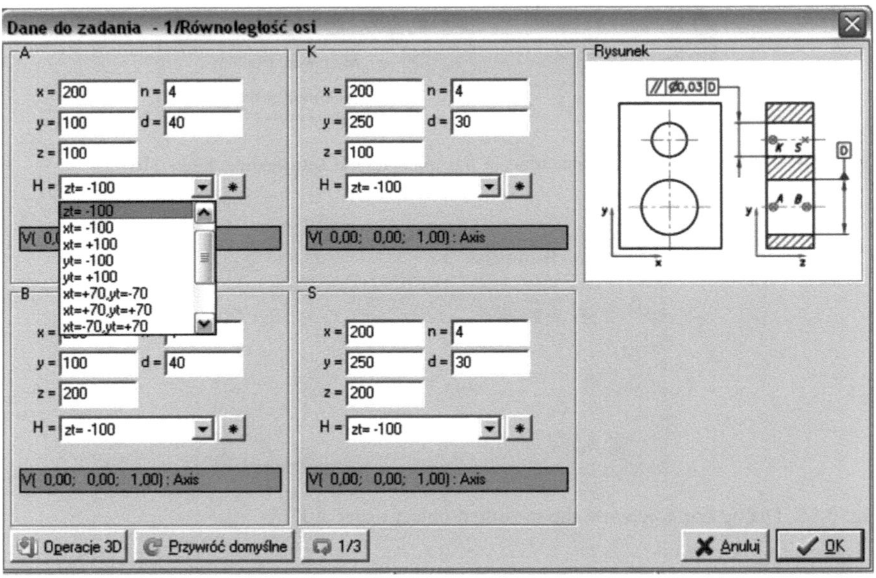

Fig. 3.57 Example of a dialog box for the data introduction needed for determination of the measurement uncertainty of axes parallelism deviation (cylindrical tolerance area) [16]

Stage 5. After the completion of all calculations it is possible to generate the report including the following information: headline with information about the enterprise, the measuring machine, and the measured object, and for every characteristic the characteristic ID and the result of uncertainty calculation.

It is possible to check whether all data are correctly introduced to the program. It is made by generating the expanded report, in which all data entered for every

No/Name: 1/Axes pararelium

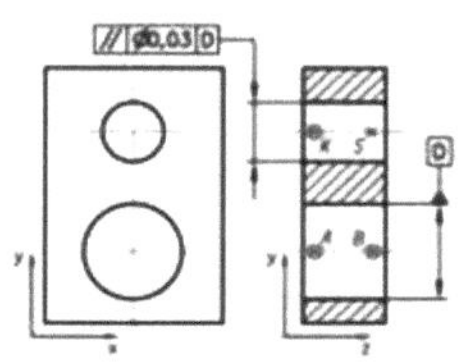

Pararelium axes deviation / cylindrical arene

-	X	Y	Z	Head	Direction vector	N	D
A	200	100	100	zt= -100	V(0,00; 0,00, 1,00) : Axis	4	40
B	200	100	200	zt= -100	V(0,00; 0,00, 1,00) : Axis	4	40
S	200	250	200	zt= -100	V(0,00; 0,00, 1,00) : Axis	4	30
K	200	250	100	zt= -100	V(0,00; 0,00, 1,00) : Axis	4	30

Uncertainty value 0.0079 µm

Fig. 3.58 Part of the expanded report presenting data for "axes parallelism deviation—cylindrical tolerance area" [16]

characteristic are presented together with the picture and the determined measurement uncertainty (Fig. 3.58).

This system is metrologically correct (type *B* method) and includes, similar to simulation programs, the most important error sources (geometric errors, probe errors, temperature errors). EMU software in the current stage enables the uncertainty evaluation for a huge amount of measuring tasks [16–18]. However, it is an offline system (an online evaluation of realized measurement uncertainty is not possible). According to [16–18], with its use it is currently possible to realize about 80 % of measuring tasks appearing in the technical documentation.

After presentation of methods for coordinate measurement uncertainty determination made in this chapter, it can be said that another, more universal, and easier method is necessary. It is also important to make possible the measurement uncertainty determination for every single point in the CMM measuring volume, and consequently, determination of measurement uncertainties for every realized measuring task. The solution is development of a computer (virtual) model of the coordinate system that permits us to calculate the uncertainty for all performed tasks using the simulation of the functioning real system. It should also take into consideration conditions in which this measuring task is performed. This issue is discussed in Chaps. 4 and 5 of this book.

References

1. Busch, K.: Überwachung von Koordinatenmessgeräten. VDI-B **751**, 51–62 (1989)
2. EASYTRAC Project—Easier and cheaper traceability in industry by up-to-date methods of calibration, EU Contract No. G6RDCT-2000-00188. Work package 6.3. Calibration of Free Form Parts on CMMs with Uncertainty Assessment according to the Method of Uncalibrated Objects and of Calibrated Objects—E.Savio, L.De Chiffre
3. Eumetron. www.eumetron.de

4. Feinmess GmbH & Co. KG. www.feinmess.com
5. Główny Urząd Miar—Central Office of Measures (Warsaw 1999) Guide to the Expression of Uncertainty in Measurement (GUM) ISO 1993/1995 with the addition to polish edition of Jaworski J.M
6. Hernla, M.: Messumsicherheit bei Koordinatenmessungen. Expert Verlag, Renningen (2007)
7. Hueser-Teuchert, D., Trapet, E., Wäldele, F., Wiegand, U.: Kalibrierung von Koordinatenmessgeräten mit Kugel – und Lochplatten Anleitung für DKD-laboratorien. Preprint PTB, Braunschweig (1993)
8. ISO/CD TS 15530-2—Geometrical product specifications (GPS)—Coordinate measuring machines (CMM): techniques for evaluation of uncertainty of measurement part 2: use of multiple measurement strategies in measurements of artefacts (under developments)
9. ISO/TS 15530-3 Geometrical product specifications (GPS)—Coordinate measuring machines (CMM): techniques for evaluation of uncertainty of measurement part 3: use of calibrated workpieces or standards (2004)
10. ISO/TS 15530-4 Geometrical product specifications (GPS)—Coordinate measuring machines (CMM): techniques for evaluation of uncertainty of measurement part 4: evaluating task-specific measurement uncertainty using simulation (2008)
11. ISO/TS 15530-5 Geometrical Product Specyfications (GPS)—Coordinate measuring machines (CMM): techniques for evaluation of uncertainty of measurement part 5: (Draft) GPS—Use of expert judgement
12. Jakubiec, W.: Analityczne wyznaczanie niepewności pomiaru we współrzędnościowej technice pomiarowej (Analilitical evaluation of the measurement uncertainty in coordinate measurement technique). Scientific Books, vol. 22. University of Bielsko-Biała, Bielsko-Biała (2008)
13. Jakubiec, W., Starczak, M.: Metodyka wyznaczania niepewności pomiarów współrzędnościowych (Methodology of evaluation of the coordinate measure¬ments uncertainty). VIth International Scientific Conference—Coordinate Measuring Technique: Scientific Books, vol. 10. University of Bielsko-Biała, Bielsko-Biała (2004)
14. Jakubiec, W., Płowucha, W., Starczak, M.: Modele błędów maszyny pomiarowej do oceny niepewności pomiaru (Models of measuring machine errors for assessment of measurement accuracy). Xth National, 1st International Scientific-Technical Conference Metrology in Production Engineering. Cracow University of Technology, Cracow (2003)
15. Jakubiec, W., Płowucha, W., Starczak, M.: The estimation of coordinate measurements uncertainty in industrial conditions. VIIth International Scientific Conference Coordinate Measuring Technique: Scientific Books, vol. 22. University of Bielsko-Biała, Bielsko-Biała (2006)
16. Jakubiec, W., Płowucha, W., Starczak, M.: EMU—oprogramowanie do analitycznego wyznaczania niepewności pomiarów współrzędnościowych (Software for analilitical evaluation of the coordinate measurements uncertainty). In: Sładek, J., Jakubiec, W. (eds.) Advances in Coordinate Metrology, pp. 83–89. University of Bielsko-Biała, Bielsko-Biała (2010)
17. Jakubiec, W., Płowucha, W., Starczak, M., Wizner, M.: Wdrażanie oprogramowania do analitycznego wyznaczania niepewności pomiarów współrzędnościowych (Implementation of the software for analilitical evaluation of the coordinate measurements uncertainty). In: Sładek, J., Jakubiec, W. (eds.) Advances in Coordinate Metrology, pp. 90–97. University of Bielsko-Biała, Bielsko-Biała (2010)
18. Jakubiec, W., Płowucha, W., Starczak, M.: Analytical evaluation of the coordinate measurements uncertainty. In: Sładek, J., Jakubiec, W. (eds.) Advances in Coordinate Metrology, pp. 169–176. University of Bielsko-Biała, Bielsko-Biała (2010)
19. JCGM.: Supplement 1 to the Guide to the expression of uncertainty measurement—Propagation of distribution using Monte Carlo method (2006)
20. Jednoróg, A., Dziuba, R.: Sprzęgnięcie współrzędnościowej techniki pomiarowej z systemami CAD/CAM (reverse engineering) (The connection of coordinate measuring technique with CAD/CAM systems (reverse engineering). IInd National Scientific Conference (with

international participation)—Coordinate Measuring Technique 22–24.04.96: Scientific Books, vol. 33, pp. 91–96. Lodz University of Technology, Bielsko-Biała
21. Koch, K.P., Peter, R., Weisig, S.: Koordinatenmessung mit einem Laser-tiangulationstaster. Feinwerktechnik & Messtechnik F&M 95, Heft 6 (1988)
22. Kowalski, M., Sładek, J.: Opracowanie wirtualnej wielowspółrzędnościowej maszyny pomiarowej z zastosowaniem do badań i korekcji błędów obiektów rzeczywistych i optymalizacji pomiarów (Development of virtual multi-coordinate measuring machine with the application for researches and correction of errors of real objects and for measurements optimization) PB 1367/T0795/08 KBN Research Grant (1995–1998)
23. Kunzmann, H., Schuster, H.J.: Temperature Measurements in Dimensional Metrology. PTB, Bericht F-17 Braunschweig (1994)
24. Metrosage. www.metrosage.com
25. Michawaki, H.: 3D measurement and CAD. Int. Jpn. Soc. Precis. **29**(2) (1995)
26. PN EN ISO 10360-2 Specyfikacje Geometrii Wyrobów (GPS) Badania odbiorcze i okresowe współrzędnościowych maszyn pomiarowych (CMM). Cześć 2 – CMM Stosowane do pomiaru długości (geometrical product specifications (GPS)—Acceptance and reverification tests for coordinate measuring machines (CMM)—Part 2: CMMs used for measuring linear dimensions)
27. PN-EN ISO/IEC 17025:2005—Ogólne wymagania dotyczące kompetencji laboratoriów badawczych i wzorcujących (General requirements for the competence of testing and calibration laboratories)
28. Poniatowska, M.: Spatial characteristic of geometric deviations of free-form surfaces determined in coordinate measurements. In: Sładek, J., Jakubiec, W. (eds.) Advances in Coordinate Metrology, pp. 267–274. University of Bielsko-Biała, Bielsko-Biała (2010)
29. Poniatowska, M., Werner, A.: Analysis of geometric deviations of free-form surfaces. In: Proceedigns of 10th International Symposium on Measurement and Quality Control, Osaka, Japan, 5–9 Sept 2010
30. Ratajczyk, E.: Optoelektroniczne sondy bezstykowe stosowane we współrzędnościowej technice pomiarowej (Opto-electronic, non-contact probes use in coordinate measurement technique) Mechanik nr 10, pp. 355–358 (1994)
31. Salsbury, J.G.: A simplified methodology for the uncertainty analysis of CMM measurements. Technical Paper, Conference on Precision Metrology/Applying Imaging and Sensoring, Society of Manufacturing Engineers, Indianapolis, IN, IQ 95–155:1–22 (1995)
32. Sato, O., Osawa, S., Kondo, Y., Komori, M., Takatsuji, T.: Calibration and uncertainty evaluation of single pitch deviation by multiple-measurement technique. Precis. Eng. **34**(1), 156–163 (2010)
33. Savio, E., De Chifre, L.: Uncertainty assessment using uncalibrated objects: calibration of freedom surfaces. Second International Workshop on CMM Calibration. Chech Metrology Institute, Prague (2001)
34. Savio, E., De Chiffre, L.: An artefact for traceable freeform measurements on coordinate measuring machines. Precis. Eng. J. Int. Soc. Precis. Eng. Nanotechn. **26**, 58–68 (2002)
35. Savio, E., Hansen H.N., De Chiffre, L.: Approaches to the calibration of freeform artefacts on coordinate measuring machines. CIRP Ann. **51**(1), 433–436 (2002)
36. Sładek, J.: Assessment of the accuracy of contact probe heads used in coordinate measuring machines. Doctoral Dissertation, Cracow University of Technology, Mechanical Department 1990
37. Sładek, J.: Modelowanie i ocena dokładności maszyn oraz pomiarów współrzędnościowych współrzędnościowych (Modeling and accuracy assessment of coordinate machines and measurements). Scientific Books, vol. 87. Cracow University of Technology, Cracow (2001)
38. Sładek, J.: The relationship between measuring machines performance specifications and their real capability to solve given measurement tasks. Metromeet 2005 International Conference on Industrial Dimensional Metrology Bilbao (Spain), April 7–8, 2005 Euskalduna Jauregia Invited Lecture

39. Sładek, J., Klecha, R.: Planowanie i interaktywne programowanie zadań pomiarowych we współrzędnościowej technice pomiarowej (Planning and interactive programing of measuring tasks in coordinate measuring technique). IIIth Scientific-Technical Conference Mechanotronika '97, Scientific Books, vol. 14, pp. 733–738. Warsaw University of Technology, Warsaw (1997)
40. Sładek, J., Kowalski, M.: Oprogramowanie QUINDOS jako baza do tworzenia zintegrowanych systemów zapewnienia jakości w budowie maszyn (QUINDOS software as a base for creating integrated systems for quality assurance in machine engineering) Proceedings of the seminar about quality maintenance systems in machine engineering, pp. 235–242. Cracow University of Technology, ITMiAP, Cracow (1994)
41. Sładek, J., Krawczyk, M.: Metody oceny dokładności pomiarów współrzędnościowych (Methods for coordinate measurements accuracy assessment) Pomiary Automatyka Kontrola (Measurement Automation and Monitoring), no. 9, vol. 53, pp. 478–481 (2007)
42. Sładek, J., Juras, B., Rewilak, J.: Wykorzystanie techniki współrzędnościowej do digitalizacji powierzchni i realizacja połączeń CAD-WMP-CAD (The coordinate technique application to surface digitalization and realization of CAD-WMP-CAD connections). In: Proceedings of VIth Scientific-Technical Conference Metrology in Production Engineering, pp. 225–240. Rzeszow University of Technology, Rzeszow (1995)
43. Sładek, J., Juras, B., Rewilak, J.: Programowanie zorientowane na cechy na przykładzie systemu QUINDOS (QUINDOS as an example of the software oriented to features). In: Proceedings of VIth Scientific-Technical Conference Metrology in Production Engineering, pp. 251–256. Rzeszow University of Technology, Rzeszow (1995)
44. Sładek, J., Rakoczy, R., Majdosz, R.: Realizacja i badanie powiązań informatycznych CAD WMP CAD z uwzględnieniem możliwości prognozowania dokładności pomiarów (Realisation and testing of the CAD CMM CAD IT couplings, considering the aspect of possibilities of measurement accuracy prognostication). In: IVth International Scientific Conference—Coordinate Measuring Technique, Scientific Books, vol. 53, pp. 209–221. Lodz University of Technology, Bielsko-Biała (2000)
45. Sładek. J., Kupiec, R., Gąska, A., Kmita, A.: Modelowanie czynników temperaturowych, wpływających na niepewność pomiaru, przy pomocy metody Monte Carlo (Temperature factors affecting the measurement uncertainty modeling with the use of Monte Carlo method) Pomiary Automatyka Kontrola (Measurement Automation and Monitoring) 01/2010
46. Sładek, J., Sitnik, R., Kupiec, M., Błaszczyk, P.: The hybrid coordinate measurement system as a response to industrial requirements. Metrol. Meas. Syst. **XVII** (1), 109–118 (2010)
47. Sładek, J., Rakoczy, R., Majdosz, R.: Research and realization of the CAD-CMM-CAD couplings, considering the aspect of possibilities of measure-ment accuracy prognostication. In: Proceedings of the 2nd International Conference on Surface Machining and Measurements of Sculptured Surfaces, pp. 71–84, Cracow 20–23 Sept 2000
48. Sładek, J., Błaszczyk, P.M., Kupiec, M., Sitnik, R.: The hybrid contact–optical coordinate measuring system—Elservier. Measurement **44**, 503–510 (2011)
49. Sommer, K.D., Kuhn, O., Weckenmann, A.: Use of MCM for uncertainty of non-linear and multivariate dimensional measurment task. In: 10th CIRP Conference on Computer Aided Tolerancing, Erlangen, Germany, Mar 2007
50. Trapet, E. Uncertainty analysis of measurements with CMMS using multiple measurement strategies. Report of EU Project EASYTRAC (2003)
51. Trapet, E., Wäldele, F.: Coordinate measuring machines in the production line influence of temperature on measuring uncertainties. In: Proceedings of the IV Congress International de Metrologia i Industrial, Zaragoza (Spain), pp. 229–242 (1989)
52. Trapet, E., Wäldele, F.: Rückführbarkeit der Messergebnisse von Koordinatenmessgeräten. VDI Berichte 1258/96
53. Trapet, E., Wäldele, F.: Substitution method to make traceable measurements with CMMs. Working paper ISO/TC3/WG10 (1996)
54. Trapet, E., Franke, M., Härtig, H., Schwenke, H., Wäldele, F., Cox, M., Forbes, A., Delbressine, F., Schellekens, P., Trenk, M., Meyer, H., Moritz, G., Guth, T.H., Wanner, E.:

Tracebility of Coordinate Measurements According to Method of Virtual Measuring Machine. PTB F-35, Braunschweig (1999)
55. Trapet, E., Savio, E., De Chiffre, L.: New advances in traceability of CMMs for almost the entire range of industrial dimensional metrology needs. CIRP Ann. Manuf. Technol. **53**(1), 433–438 (2004)
56. Treib, T.: Error budgeting—applied to calculation and optimisation of the volumetric error field of multiaxis systems. Ann. CIRP, **36/1/87**, 365–368
57. VDI/VDE 2617 Genauigkeit von Koordinatenmessgeräten Kenngrößen und deren Prüfung. Blatt7.Ermittlung der Unsicherheit von Messungen auf Koordinatenmessgeräten durch Simulation VDI/VDE- Richlinien ver. (2008)
58. Weber, H.: ISO/WD 15530 uncertainty assessment using calibrated workpieces. In: IVth International Scientific Conference—Coordinate Measuring Technique, Scientific Books, vol. 53, pp. 249–254. Lodz University of Technology, Bielsko-Biała (2000)

Chapter 4
Analysis of the Accuracy of Coordinate Measuring Systems

Abstract This chapter focuses on analysis of the coordinate system accuracy. It is assumed that the essence of a model of such a measuring system is that component errors overlap each other, determining the error vector for every point in the measuring volume. The description of sources of kinematic errors is presented as other methods for their identification with the use of plate standards, laser interferometers, or laser tracers together with the software description. The original matrix method (MM) for coordinate system error identification is described together with the connection of this MM method with the reproducibility error of the measuring point (REMP). Examples of research results are also presented in Annexes B, C, and D. The chapter mainly focuses on the issue of measuring tip contact with the measured object, and it is based on the author's original method for the identification of a probing system: contact probe errors with the use of a circular standard together with the system of software separation of components. The method of software correction of CMS accuracy is also presented. This method is based on determination of correction matrix components as a basis for software correction of errors of modern coordinate systems. Moreover, this chapter describes the basis for construction of virtual coordinate systems, including a system of articulated arm-coordinate measuring machine that uses Denavit–Hartenberg notation.

4.1 Sources and Causes of Coordinate Measuring Machine Errors

Every coordinate measuring system, as well as the classical machine, is a me-chanical construction, the imperfection of which is transferred to the effect of its operations, in this case to the measurement result. Solving a problem of measurement accuracy assessment, according to the description of the reproducibility error of measuring point reproducibility error of the measuring point (REMP) from Sect. 2.3 and its vector interpretation, the description of sources and causes of machine functioning errors as the measuring system, and in consequence to its

J.A. Sładek, *Coordinate Metrology*, Springer Tracts in Mechanical Engineering,
DOI 10.1007/978-3-662-48465-4_4

functional model is reached. Of course the mathematical model describing its action is also necessary. It has been assumed that the essence of the coordinate measuring system model is that component errors overlap each other, determining the vector of errors for every point in the measuring volume. The assumption has been made for both the classical coordinate machine as well as for the redundant system, the articulated arm–coordinate measuring machine. There are many publications describing kinematic errors characteristic for a classic coordinate machine [9, 10, 40–42, 25–27, 36, 56–59, 71, 86, 93, 99, 104, 106, 122–126, 134, 139, 140, 142, 143, 145, 146, 149, 156, 175–177, 187, 193, 196–198, 216–219]. The vector of errors contains three kinds/categories of errors:

- Known systematic influences
- Unknown systematic influences
- Random influences

Other division distinguishes two groups:

- Determined experimentally
- Estimated with other methods

The first group contains components that can be identified by measurement and their value is constant (does not change during the long term).

Some of them can be corrected because they bring known (additional) displacement to coordinates of measured points. Other measurable components may be described only statistically, for example, their influence on the measured point using probability distribution parameters. According to the division made by Trapet [186] for errors that are determined by the measurement, the following errors can be included.

M1: Systematic errors related to the process of tip contact with the object
M2: Random contact errors
M3: Errors related to the change of stylus (of probe head) and errors of the probe wrist (random errors)
M4: Systematic geometric deviations of CMM parts/groups (parametric errors of particular axes)

The last above-mentioned group of parametric errors of particular axes M4 is dominant among systematic influences that can be determined as a result of the measurement, and it can be possibly included in their later correction. This group consists of kinematic errors described by 21 components in particular axes of the machine (Fig. 4.1).

These components include:

6 translation errors (axis straightness) (*xty*, *xtz*, *ytx*, *ytz*, *ztx*, *zty*)
3 positioning errors (as special cases of translation errors) for every axis (*xtx*, *yty*, *ztz*), marked also as (*xpx*, *ypy*, *zpz*)
3 axis rotation errors (*xrx*, *yry*, *zrz*)
6 axis mutual rotation errors (*xry*, *xrz*, *yrx*, *yrz*, *zrx*, *zry*)
3 CMM axes perpendicularity errors (*xwy*, *xwz*, *ywz*)

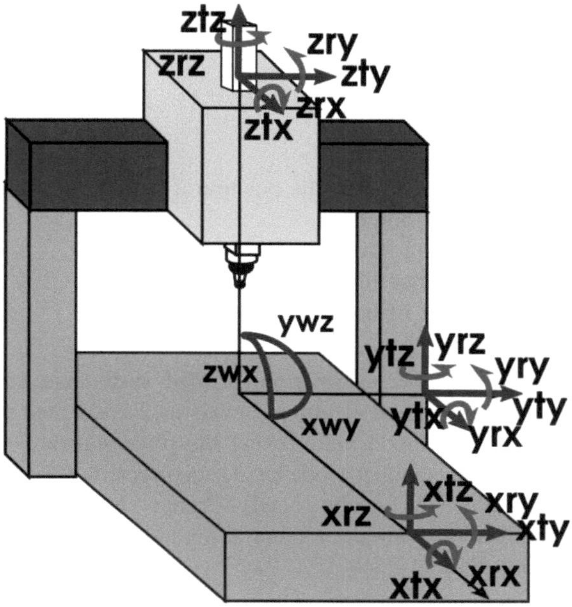

Fig. 4.1 CMM kinematic error components

Table 4.1 21 kinematic error components associated with particular axes of the machine

Axis direction	Rotary error around an axis			Translation error in direction		
	X	*Y*	*Z*	*X*	*Y*	*Z*
X	*xrx*	*xry*	*xrz*	*xtx*	*xty*	*xtz*
Y	*yrx*	*yry*	*yrz*	*ytx*	*yty*	*ytz*
Z	*zrx*	*zry*	*zrz*	*ztx*	*zty*	*ztz*

They are defined in Table 4.1.

Perpendicularity error of particular axes: *xwy*, *xwz*, *wyz*

$$A = \begin{bmatrix} 0 & -ywz - xrz & xwz + xry + yry \\ 0 & 0 & -ywz + xrx - yrx \\ 0 & xrx & 0 \end{bmatrix}$$

$$A_p = \begin{bmatrix} 0 & -xrz - yrz - zrz & xry + yry + zry \\ xrz - yrz - zrz & 0 & -xrx - yrx - zrx \\ -xry - yry - zry & xrx + yrx + zrx & 0 \end{bmatrix}$$

$$X = \begin{bmatrix} x \\ y \\ z \end{bmatrix} \quad X_p = \begin{bmatrix} x_p \\ y_p \\ z_p \end{bmatrix} \quad P = \begin{bmatrix} xtx + ytx + ztx \\ yty + xty + zty \\ ztz + xtz + ytz \end{bmatrix}$$

The dependence of the indication error from kinematic errors and measuring tip position errors [95] is presented below:

$$E = P + A \cdot X + A_P \cdot X_P \tag{4.1}$$

where

E	error vector in the position $X + X_P$
P	translation errors
A, A_P	rotation errors
X	reference tip position
X_P	vector of tip movement

It is possible to determine CMM systematic geometric errors applying laser interferometer and electronic levels or, as accepted in PTB according to the Novel method developed there, using the plate standard (ball or hole one). Systematic errors of the contact probe head (uncorrected) M1 and the errors associated with the stylus (or the probe head) change M3, identified in the process of its qualifications/correction, should be added in the form of the vector to every measuring point.

Random influences, which affect every single contact point separately (M2), need an additional discussion. For the described type of virtual CMM they are determined by statistical parameters, such as standard deviation. In the virtual model, errors are represented by the vector, value and direction of which are determined each time the measuring tip or probing direction changes. In the simulation model, after determining distribution parameters, this error can be generated from a given distribution (e.g., the Monte Carlo method).

The second group contains errors estimated with other methods. In the model of the virtual machine, as mentioned before, the other error components are included. However, they are not determined by direct measurement, but it has been assumed that their values would be appropriately estimated. According to [45] a group of these errors should contain:

- M5: Uncertainty of individual parametric errors, including the uncertainty of standard calibration or uncertainty of the system, for example, laser interferometer.
- M6: Long-term changes of individual systematical errors.
- M7: Temperature impacts on individual systematic errors.
- M8: The parametric errors model unconformity with more than one movement axis. These errors are called "elastic" [182] and they are expressed as the additional component of parametric errors specified separately for the arm construction, Fig. 4.2. They depend not only on the position on the axis on which they are identified, but also on the position on other axes. Sources of these errors are shown in Fig. 4.2.

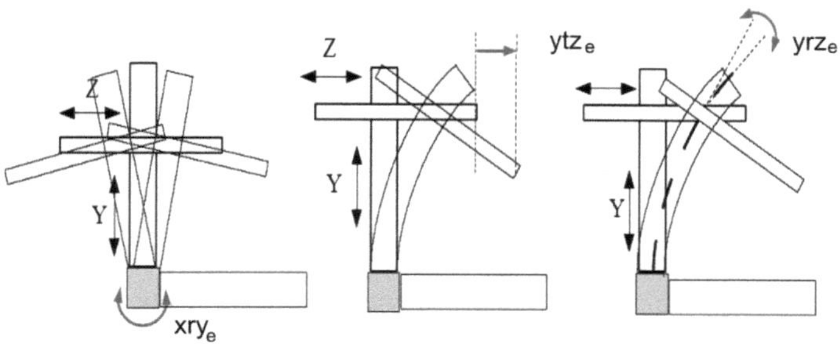

Fig. 4.2 Elastic errors and their sources expanding rotation errors of translation components *xrx*, yrz_e and *ytz* [67]

We can distinguish the following elastic errors characteristic for arm machines that supplement the rotation and translation errors.

Rotation error yrx_e: Dependent on the position of machine *Z* arm. For arm machines the error *xrx* is described by the equation:

$$xrx(x, z) = xrx(x) + wrx_e(z) \tag{4.2}$$

An error xrx_e is directly proportional to a displacement (of coordinate) along the *Z*-axis and it can be described with the function:

$$xrx_e(z) = A \cdot z \tag{4.3}$$

where *A* = a constant individually determined for particular coordinate machines.

Translation error yrx_e—is dependent on the machine position on the *Y*- and *Z*-axes. It may be described by equation:

$$ytz(y, z) = ytz(y) + ytz_e(y, z) \tag{4.4}$$

For ytz_e error it is not possible totally to separate components dependent on the movement in the *Y*- and *Z*-axes. The ytz_e error can be described by the equation:

$$ytz_e(y, z) + b(y) \cdot z \tag{4.5}$$

The *b*(*y*) function is implemented because producers can apply various methods for *Y* column strengthening, obtaining in the case of *b*(*y*) error a different error formula.

Rotation error yrz_e: Dependent on the position in *Y*- and *Z*-axes. It may be described by equation:

$$yrz(y, z) = yrz(y) + yrz(y, z) \tag{4.6}$$

$$yrz_e(y, z) = c(y) \cdot z \tag{4.7}$$

As in the previous case, the various construction solutions result in the different $c(y)$ function form.

In Eqs. (4.4) and (4.5) the distinction between the part dependent only on the position on the Y-axis and the elastic part is caused by different ways of their determination, as well as that $ytz(y)$ error is mainly caused by the track straightness error, and the $ytz_e(y, z)$ is dependent on the Y column shape change under the influence of machine Z-axis arm extension.

M9: Effect of the slow change of metrological characteristic of CMM (drift).

M5–M9: Not "pure" random components of the error, but unknown systematic components. They are approximated as constants in the case of perpendicularity errors or as linearly dependent location components in the case of rotation errors and translation errors.

The component errors mentioned above are usually estimated on the basis of experience acquired during the use of a certain machine or according to data delivered by producers. The issue of the parametric error component (coming from CMM geometry) estimation as well as estimation of other influences is the most difficult task that had been met in the case of analytical models use. It also constitutes a certain barrier in the wider spread of virtual machines built according to this concept.

4.2 Identification and Software Correction of Measuring Machine Errors

Specified errors are identified with the application of more and more perfect methods and measuring systems. CMM errors are identified for correction purposes (the accuracy improvement) and simulation purposes, in order to construct the measuring system model.

The development of the measuring device model should be preceded by analysis of the whole measurement process and should include all components that are the source of errors (they should be identified).

The issue that requires broader discussion is the error correction problem, realized at present in the case of CMM with the use of the computer-aided accuracy (CAA) matrix. Under this name we understand systems that are applicable for program corrections of geometric errors of coordinate measuring systems. Such a system was patented and implemented for the first time in the mid-1980s by the Zeiss Company [38] and up to now it has been well described in [7, 14, 15, 19, 25, 26, 30, 38, 82, 99, 102, 144, 207, 216]. However, practical application is the responsibility of particular producers. In the cases of coordinate machines and measuring arms (AACMM) information is given on the basis of information delivered by producers and the metrological practice and results of scientific works realized in LCM CUT.

The CAA correction system is based on identification of kinematic error components in chosen interpolation nodes, which are later corrected and entered into the measuring machine driver. During normal CMM operation (measurement performance) it is possible to correct the error in any position based on determined and remembered errors (deviations) in neighboring interpolate nodes. It is possible using linear interpolation and adding the calculated value to the measurement result [144].

There are many solutions of correction systems, differing in terms of volume and coverage factor of measuring area of the corrected machine. For machines of highest accuracies a large number of interpolation nodes are applicable. It means that, for example, for machines of highest accuracies the grid of interpolation nodes built in the CMM measuring volume has a 64-fold smaller raster scale than for machines of standard accuracy. The other method used by the Zeiss Company is a method based on the correction of machine deformation errors and the correction of the measuring stylus dynamic diffraction error, caused by probing forces and the acceleration. The machine drive system stores information about the stylus performance (working with different forces and speeds) which is obtained as a result of dynamic classification. In this way the drive system "predicts" the size of diffractions of dynamic machine elements and the stylus during measurement. Thus it is possible to adjust the machine motion path so that the diffraction of the measuring stylus would be as small as possible and at the same time to keep the measuring force on the same level during measurements. The supplement to these systems (closely cooperating with themselves) is a system of thermal error correction. It comprises a CMM construction with its elements allowing their free expansion without causing stresses independent of thermostable standards of movements Zerodur [212] or corrected with Renishaw expandability [136]. In the case of the Zeiss Company that system is called CARAT technology. There is also a different system, called adaptive compensation of temperature induced variations (ACTIV), developed by the DEA Group [21]. In this system, the machine is equipped with a number of temperature sensors located at appropriate points that systematically monitor thermal changes during performed measurements. Software algorithms calculate the structure deformations caused by temperature, therefore the measurement result can be corrected in realtime and given including current temperature changes. This system theoretically enables the coordinate measuring machine functioning in the range of temperature changes from 15 to 30 °C. The only restrictions refer to the temperature fluctuations, which cannot exceed 10 °C [21, 43] during a day.

Of course, in such varying conditions measurements are impossible because of uncorrectable deformations of the measurement object. In the case of stiff constructions (Leitz) made from natural materials (granite, steel) the temperature correction is also realized (both the scale correction as well as structural in the

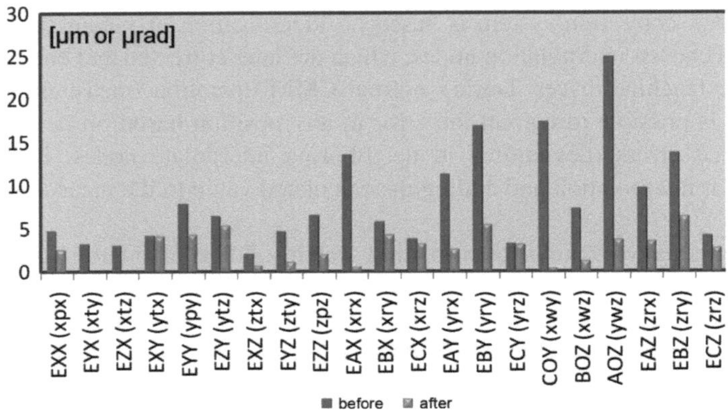

Fig. 4.3 Correction of geometric errors of measuring machine [27]

multidetector system) in relation to the CAA matrix, but due to the required high measurement accuracy, using good and stable air conditioning.

The effect of CAA correction in the area of kinematic components is presented in Fig. 4.3. Data used for the correction had been identified earlier with a laser tracer [27]. Darker bars indicate identified values of particular components of previous machine geometric errors, and bright-ashen bars after making the correction. As can be seen the components of these errors have been significantly reduced.

As a result of CAA correction systematic errors are reduced to residual errors which along with random errors create their own area of error of the measuring machine. Residual errors are errors of systematic character that are either impossible or too costly to be corrected. Such errors can be treated as errors of random character. This observation became a basis of a concept of the CMM model based on the usage of Monte Carlo simulation described in Sect. 5.2.4.

It can be noted that the process of error identification in order for their correction (CAA), as well as for development of CMM models, is technically the same and realized with the same methods. There are three methods of developing the particular geometric errors in chosen points of the machine measuring volume. Therefore it is assumed that it will be interesting for the reader to discuss them on the basis of the LCM CUT experience. The following methods are described below: the "traditional" method using the laser interferometer, Novel PTB method with the KalKom 4.1 software, and the method based on trilateration with use of both the laser tracer of the Etalon AG company, as well as the Leica LTD 840 laser tracker and TraCal software. The methods presented [29] are discussed in reference to determining the CAA correction matrix. In Annexes B, C, and D the description of kinematic error determination with a Renishaw ML 10 laser interferometer [136] and Tessa Microbevell electronic level [179], thermostable plate standard of Feinmess GmbH company made of Robax, the Novel method and Kalkom 4.1 software [108], and laser tracker LTD 840 of the Leica Company [72] and TraCal software [27] are presented.

4.2.1 Determination of Geometric Errors of the Measuring Machine Using the Laser Interferometer

Geometric errors of coordinate measuring machines are introduced in Sect. 4.1 and presented in Figs. 4.1 and 4.2. All these errors have an influence on the measuring tip location and affect the indication error. For measuring machines CAA correction matrix development requires (without including the contact system) the indication error E be expressed as a product of 21 components of geometric error vector k and **M** matrix (correction matrix) which presents the influence of each of these errors on x, y, z components of the indication error of the machine [17, 35, 51, 183]:

$$E = k \cdot M \tag{4.8}$$

where

$$E = [E_x, E_y, E_z]$$

$$k = [ywz, xwz, xwy, ytx, ypy, ytz, yrx, yry, yrz, xpx, \\ xty, xtz, xrx, xry, xrz, ztx, zty, zpz, zrx, zry, zrz]$$

$$M = \begin{bmatrix} -z & 0 & -y & 1 & 0 & 0 & 0 & z+z_t & -y_t & 1 & 0 & 0 & 0 & z+z_t & -y_t & 1 & 0 & 0 & 0 & z_t & -y_t \\ -z & 0 & 0 & 0 & 1 & 0 & -z-z_t & 0 & x+x_t & 0 & 1 & 0 & -z-z_t & 0 & x_t & 0 & 1 & 0 & -z_t & 0 & x_t \\ 0 & 0 & 0 & 0 & 0 & 1 & y_t & 0 & -x-x_t & 0 & 0 & 1 & y_t & -x_t & 0 & 0 & 0 & 1 & y_t & -x_t & 0 \end{bmatrix}^T$$

where

x, y, z indications of measuring systems

x_t, y_t, z_t length components of measuring stylus

The mathematical description presented in this form is a basis for the program correction of CMM accuracy, CAA.

CMM geometric error identification with the use of a laser interferometer belongs at present to standard operations in the case of the correction based on the determination of the CAA matrix as well as the CMM model development.

In the case of portal machine testing, the "rigid" model of kinematic errors (Sect. 4.1) is usually accepted which seems to be good enough for machines of "closed" construction, where the influence of diffractions associated with the pin-hole deflection is accepted as negligible. In machine indication error E particular component errors shown in Fig. 4.1 are cumulated. They are determined with the

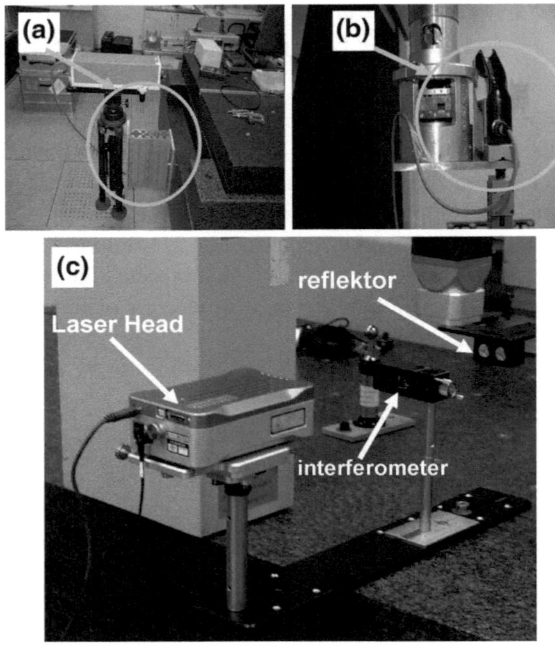

Fig. 4.4 Tests for determining CMM geometric errors **a** PMM 12106 machine, Hewlett-Packard laser interferometer, and Microlevel Tessa electronic level **b**, **c** with the use of Renishaw XL 80 interferometer connected with the machine table

use of an optic package of a laser interferometer and the electronic level presented in Annex B.

Of course the scheme illustrated in Fig. 4.1 does not represent the principle of real measurements, during which the laser probe head should be stiffly connected with the body of the tested machine. Figure 4.4 presents such measurements realized for the PMM12106 machine. Furthermore it is difficult to estimate axes' mutual rotation errors, that is, *xrx*, *yry*, *zrz*, with the laser interferometer. Therefore an electronic level is usually applicable for this task (Fig. 4.4b).

In order to use CAA accuracy correction, all possible identification components of the machine geometric errors should be determined in chosen points of the machine measuring volume. In the next stage measurement results should be saved in the memory of the measuring machine driver. Programs created by producers of measuring machines carry out this operation. In most cases, apart from the communication program, a dongle that gives access to the driver memory is also required. It is not available for the machine user by default due to security reasons. The example of the software used to create and send correction matrixes to the machine driver is the GeoComp program (Figs. 4.5 and 4.6) used in DEA Group machines. Values of geometric errors for particular axes (for assumed number of correction matrix nodes) obtained during research (e.g., using a laser interferometer) are loaded into this software. The number of nodes results from the planned (and the possible) machine accuracy. The software also includes the components of

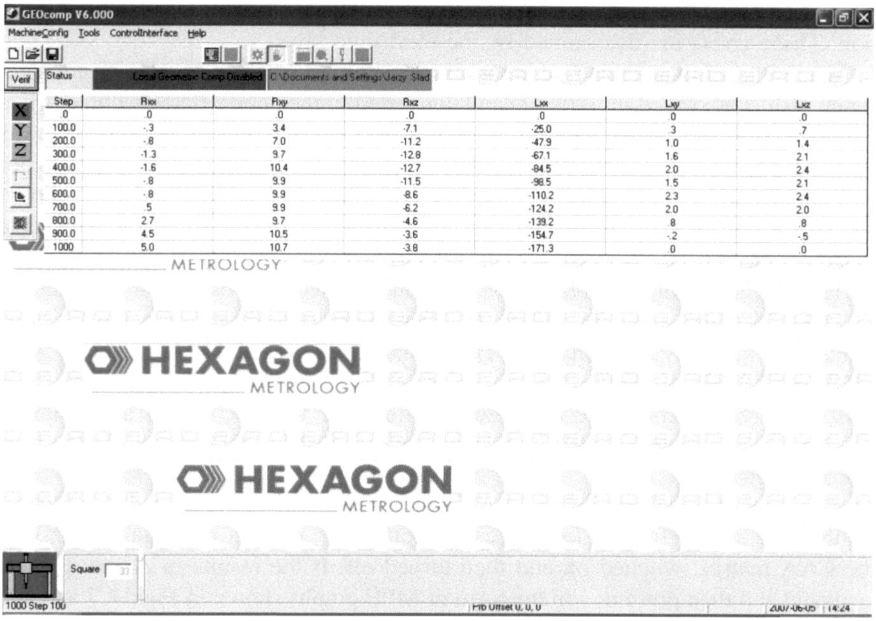

Step	Rxx	Rxy	Rxz	Lxx	Lxy	Lxz
.0	.0	.0	.0	.0	.0	.0
100.0	-.3	3.4	-7.1	-25.0	.3	.7
200.0	-.8	7.0	-11.2	-47.9	1.0	1.4
300.0	-1.3	9.7	-12.8	-67.1	1.6	2.1
400.0	-1.6	10.4	-12.7	-84.5	2.0	2.4
500.0	-.8	9.9	-11.5	-98.5	1.5	2.1
600.0	-.8	9.9	-8.6	-110.2	2.3	2.4
700.0	.5	9.9	-6.2	-124.2	2.0	2.0
800.0	2.7	9.7	-4.6	-139.2	.8	.8
900.0	4.5	10.5	-3.6	-154.7	-.2	-.5
1000	5.0	10.7	-3.8	-171.3	.0	.0

Fig. 4.5 Geocomp program: implementation of geometric errors associated with the *x*-axis [21, 43]

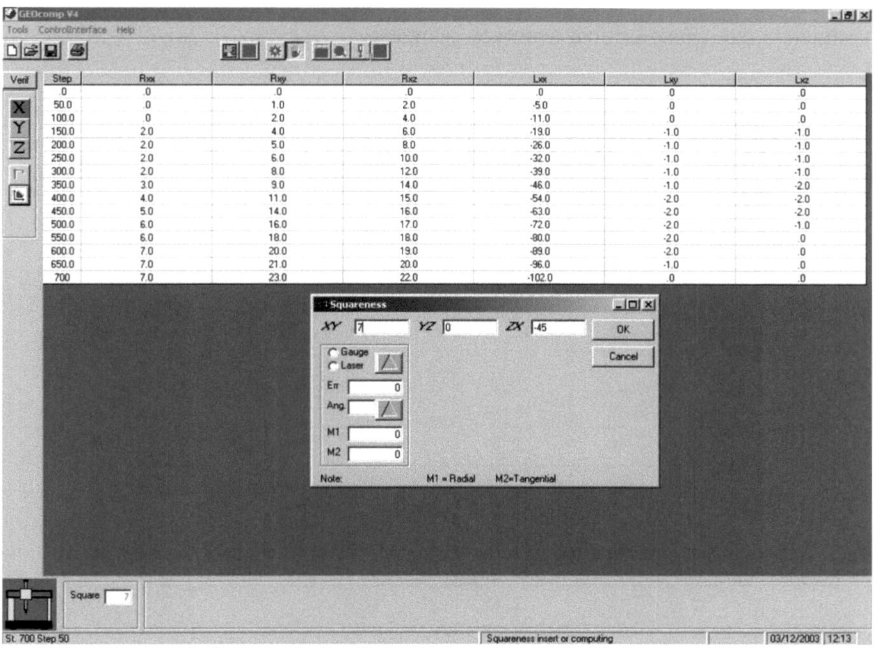

Step	Rxx	Rxy	Rxz	Lxx	Lxy	Lxz
.0	.0	.0	.0	.0	.0	.0
50.0	.0	1.0	2.0	-5.0	.0	.0
100.0	.0	2.0	4.0	-11.0	.0	.0
150.0	2.0	4.0	6.0	-19.0	-1.0	-1.0
200.0	2.0	5.0	8.0	-26.0	-1.0	-1.0
250.0	2.0	6.0	10.0	-32.0	-1.0	-1.0
300.0	2.0	8.0	12.0	-39.0	-1.0	-1.0
350.0	3.0	9.0	14.0	-46.0	-1.0	-2.0
400.0	4.0	11.0	15.0	-54.0	-2.0	-2.0
450.0	5.0	14.0	16.0	-63.0	-2.0	-2.0
500.0	6.0	16.0	17.0	-72.0	-2.0	-1.0
550.0	6.0	18.0	18.0	-80.0	-2.0	.0
600.0	7.0	20.0	19.0	-89.0	-2.0	.0
650.0	7.0	21.0	20.0	-96.0	-1.0	.0
700	7.0	23.0	22.0	-102.0	.0	.0

Fig. 4.6 Geocomp program: implementation of errors of the mutual perpendicularity of particular axes [21, 43]

perpendicularity errors of particular machine axes and errors of their mutual rotation. These errors are determined by the electronic level.

When values of machine geometric errors are uploaded to the GeoComp program, results are saved into the file and converted to the appropriate machine driver software form. When the process is complete, the CAA matrix created this way should be uploaded to the memory of the machine driver. Some software versions also enable online measurements meaning that values of measured geometric errors are transferred directly from the laser interferometer (its driving computer) and saved at once in the program that creates the matrix.

The procedure for geometric error component determination which was realized in LCM CUT is shown in Annex B. Tests were performed for a DEA Group standard accuracy machine. The maximum permissible error was: MPE = 0.0017 + 0.003/1000 × L mm, and the measuring volume was: 700 × 500 × 500 mm. The machine is compatible with PC-DMIS and Quindos software. Tests were carried out using the Renishaw Company ML10 laser interferometer.

In Annex B the results of standard accuracy machine calibration can be found. Calibration was performed for two cases, both times using a step gauge, first with the CAA matrix switched on and then turned off. If the results in Annex B were analyzed and then presented in the form of MPE graphs shown in Figs. 4.7 and 4.8, it would be noticed how important software correction is for modern CMM with free construction and how widely their base accuracy can be corrected.

It is also important to check the accuracy of software-corrected coordinate machines in the entire life cycle because, for example, changes of machine working parameters can cause critical accuracy decrease.

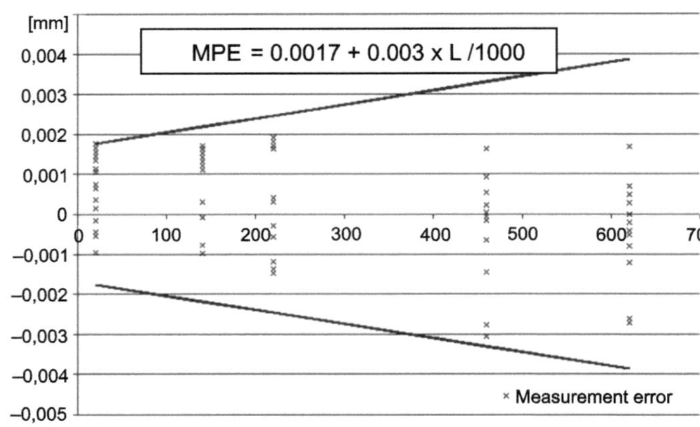

Fig. 4.7 Determined maximum permissible errors (MPE) for a standard accuracy machine; the machine had the CAA correction matrix turned on

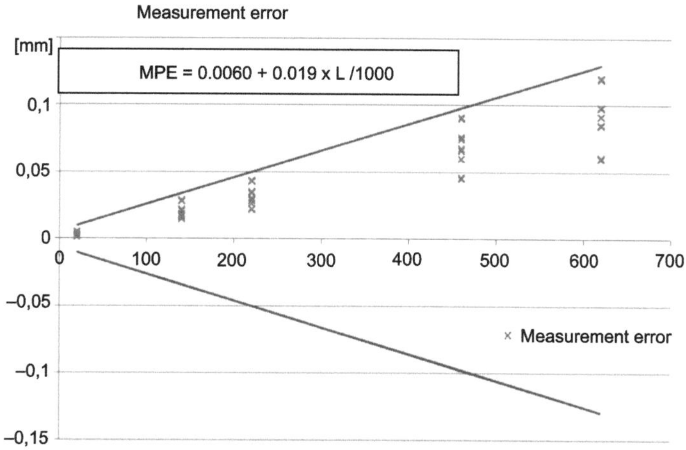

Fig. 4.8 Determined maximum permissible errors (MPE) for a standard accuracy machine; the machine had the CAA correction matrix turned off

4.2.2 PTB Method Using Plate Standard for Geometric Errors of Coordinate Measuring Machine Identification

In the mid-1990s in the PTB (*Physikalisch-Technische Bunde-sanstalt Braunschweig*) work was undertaken on the use of a two-dimensional standard in the form of a spherical plate to determine kinematic error components as the basis for the construction of a new CMM model. Authors of this method are: Dr. E. Trapet, Dr. K. Busch, and Prof. F. Waeldele from PTB, and H. Reisner from the Volkswagen laboratories in Saltzgiter.

It is worth noting that the first use of the plate standard for CMM error identification was realized at Technische Universitat Dresden in the late 1970s and 1980s [76, 178] and also at Cracow University of Technology in the 1980s [96, 97]. Works of W. Knapp from ETH Zurich on using circular standards for identification of the kinematic errors [52–56, 157] and author's works on this issue [93, 148–150] played a crucial role in adopting this concept. The complete method was formulated in PTB [190] and it is known in the literature as the Novel method [45, 68, 69, 183–186, 195]. In LCM CUT this method in the PTB version was implemented in the measurement practice and tested in the mid-nineties in the framework of the [65] project; the effects are presented in [60–66, 156, 162, 163, 201]. Moreover there are also many works describing the application of this method by other research centers [3, 127–129, 180, 213–215].

When performing measurements used in the PTB method the following principles should be respected according to [190].

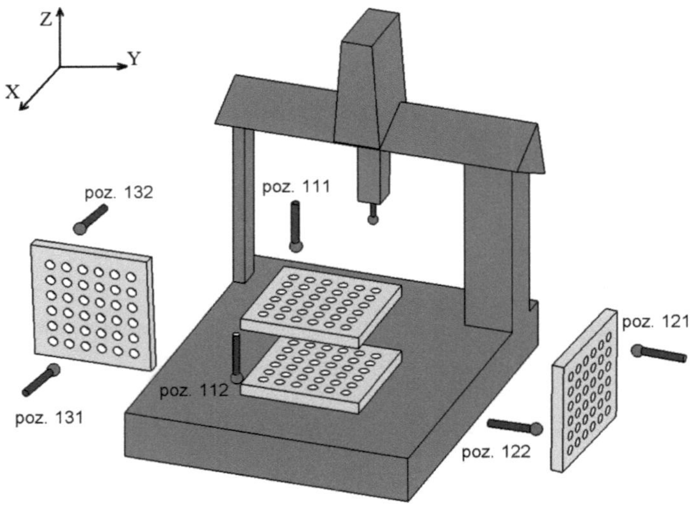

Fig. 4.9 Four required locations and six measurements used for error component determination with usage of the plate standard according to PTB method

- The standard element (hole/sphere) must be measured at the same points in which it was measured during the calibration.
- The sequence of the plate standard positions is fixed and precisely determined.
- The temperature of the plate standard should be stable and possibly close to 20 °C.
- The change of temperature during one complete measurement cycle for the plate standard in one position cannot be bigger than 0.5 K for steel. For invar, Zerodur, and Robax the temperature influence is not significant.
- If measurement conditions do not allow preserving the requirements specified above, it should be included in the error budget.

The plate standard used in this method is measured in three planes of the CMM coordinate system. Additionally in the vertical position two oppositely oriented measuring styli are used (positions 131 and 132, Fig. 4.9). In each of these locations the standard is set parallel to the appropriate planes of the machine coordinate system. To provide access to the measured standard dimensions, the probe head is equipped with the set of measuring styli put together in the star configuration, or an articulating probe head is used. A scheme showing the required locations of the plate standard in the measuring volume is shown in Fig. 4.9. A plate-holes standard made from Robax by the Feinmess Company tested on a PMM12106 machine in LCM CUT is presented in Fig. 4.10.

Based on differences between the measured and calibrated coordinates of the holes/spheres of the standard, errors of positioning, rotation, straightness, and squareness are defined.

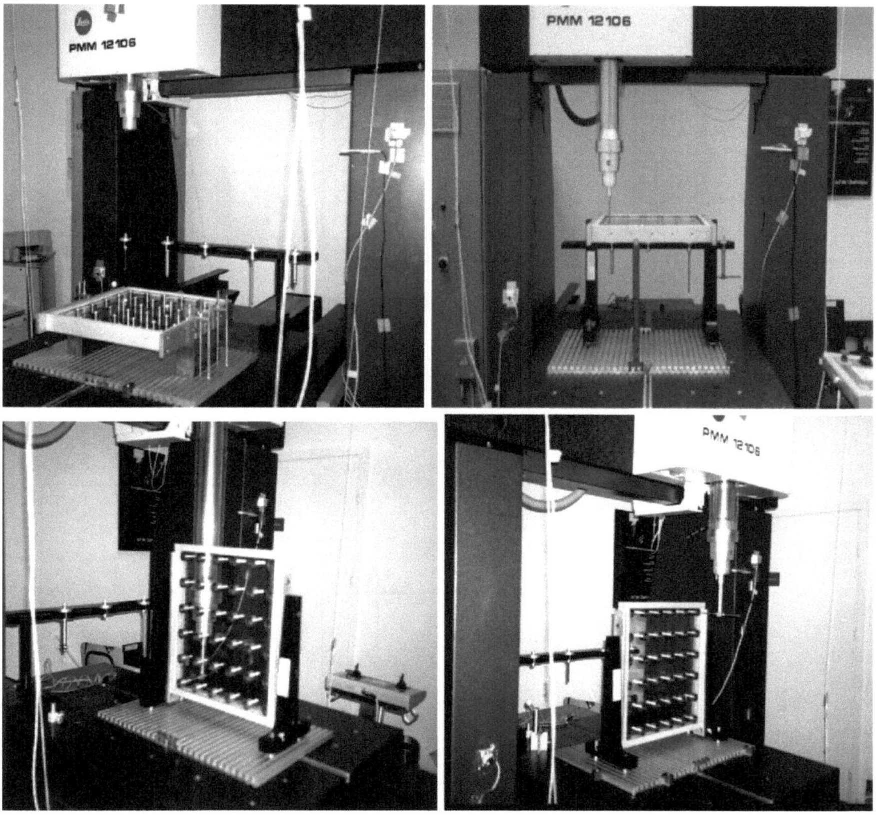

Fig. 4.10 Plate standard with holes, sandwich type, made from Robax, during LCM tests

A principle states that translation components of errors can be determined by a defined probe head in the certain baseline, and that slope and rotation errors may be defined by giving the distance to this line (compensatory line). It is obtained by moving the plate and by probe head orientation changes. To identify and eliminate systematic changes of standard dimensions in a given direction (drift), as well as to assess the results reproducibility, the measurement of the standard plate is performed twice in every position. Here a so-called snail method is applied, Fig. 4.11, in which the measurement starts from the center hole/sphere of the plate, and next goes on the spiral. During the second measurement the measurement of particular holes/spheres is carried out in the opposite direction.

When the CMM measuring volume is bigger than the plate measurement range, a chain of connections between the first plate location in a given plane with the next one is defined. It is made by performing overlapping measurements (Fig. 4.12). The possibility of such an enlargement of the error identification range is limited to two plate movements in one direction, counted from the base location.

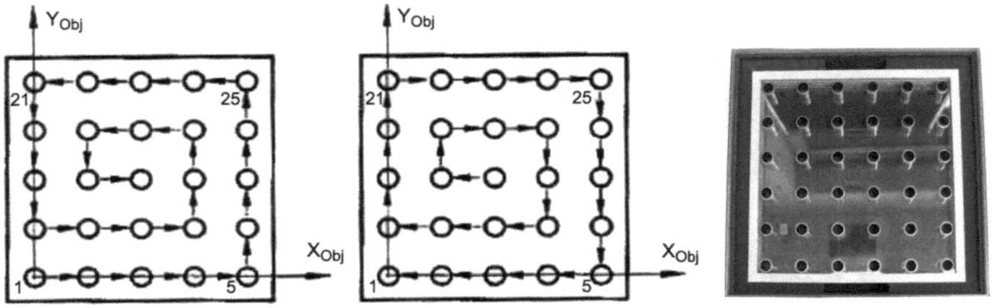

Fig. 4.11 Snail method of determining the coordinates of plate standard sphere centers

Fig. 4.12 Plate location for bigger measuring volume

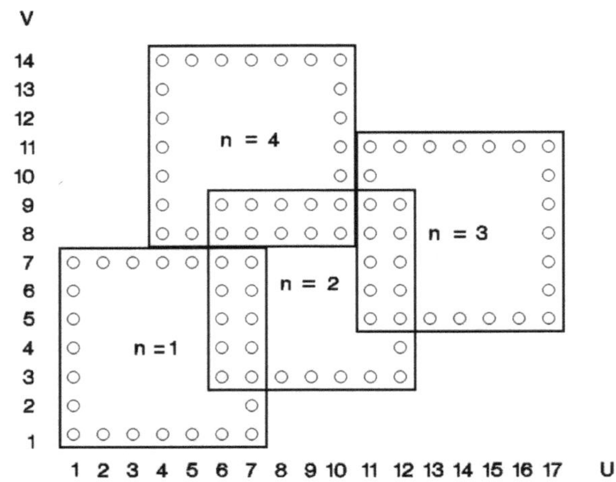

This restriction is related to the process of overlapping errors. Sufficient accuracy of plate positioning amounts to about 10 % of the distance between spheres. Coordinates of the center of each element on the standard plate are determined by measurement results.

After determining the coordinates of spheres or hole centers of the plate standard the analysis of the subsequent standard locations is performed. This analysis includes a comparison of the coordinates of centers obtained from measurements with nominal values designated during the calibration process. As a result direction and length of the error vectors for all standard positions are obtained (Fig. 4.13). These values are the basis for an analysis of 21 particular components of geometric errors of the checked machine. The measurement of the plate standard in a few locations in the CMM measuring volume enables the determination of positioning errors in every axis, straightness errors, axes mutual perpendicularity errors, and rotation errors.

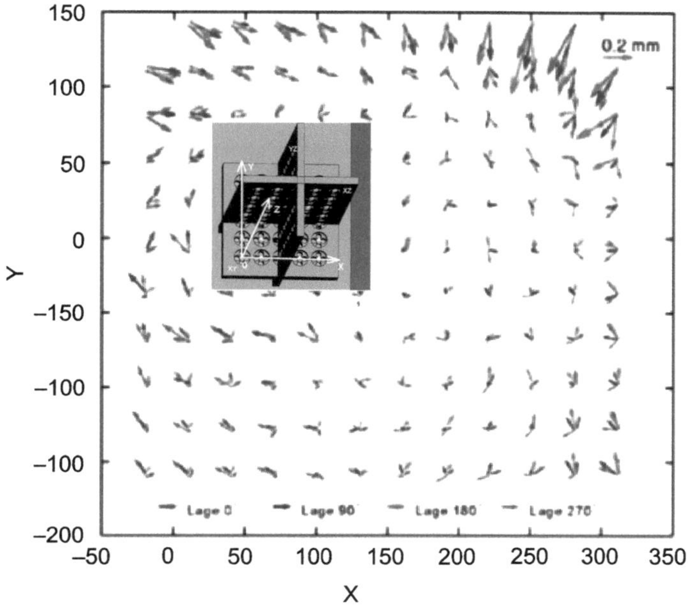

Fig. 4.13 Example of distribution and values of error vectors for position 11 in *XZ* plane according to measurement plan based on PTB method

A method of determining the individual error components for the method using the standard plate is presented in Fig. 4.14. A description of each error component is given below:

Positioning error (*xtx, yty, ztz*) (Fig. 4.14) is determined as a result of the intersection of the best-fit line, calculated from the measured points, with the reference line running through nominal points. Determined in this way, distances between points and standard lines represent positioning error.

Translation error (xty, *xtz, ytx, ytz, zty, ztx*) (Fig. 4.14b) is determined through the best-fit line calculation from measured points. Straightness error for a given point should be understood as a distance between this point and the determined line.

Rotation errors are determined individually for the case of axes rotation errors (*xry, xrz, yrx, yrz, zrx,* and *zry*) and in the situation of twist errors towards its own axis (*xrx, yry, zrz*).

Perpendicularity error (*xwy, xwz, ywz*) (Fig. 4.14c) is determined as the difference of the angle contained between two nominally orthogonal best-fit lines.

In the case of axes rotary error determination, differences of pairs of point coordinates are taken into account, as the result of the standard measurement in two locations in the *XY* plane, and for *XZ* and *YZ* planes as a result of determination of the same points from both sides of the plate by contrary oriented measuring tips. Rotation errors towards their own axis are determined through solution of an appropriate set of linear equations containing measurement results of the plate

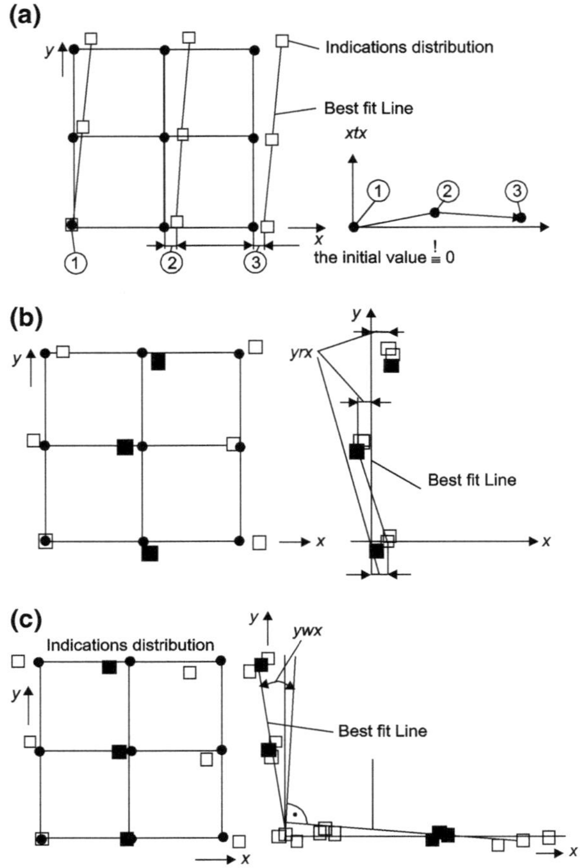

Fig. 4.14 CMM geometric errors determined during measurements with the plate standard usage: positioning errors (**a**), translation error (**b**), perpendicularity error (**c**) [141, 156]

standard in all six positions, because the measurement result of the standard in one location contains at least two rotation errors.

PTB–KalKom 3.1 software was used for the analysis and elaboration. Results obtained for the PMM12106 machine are shown in Fig. 4.15. In the abscissa axis, on which the measurement was carried out, a distance in mm is marked. The axis of ordinates describes deviation values in μm/m for positioning errors, and in μrad for rotation and perpendicularity errors. Apart from the graph, minimal and maximal values of the determined error are given. Throughout moving along the tested axis, values of errors for successive measured points are determined.

Particular locations are marked on the *X*-axis and values of errors determined for them on the *Y*-axis. The connection of these points creates the graph that afterwards is numerically smoothed with the use of spline functions. As a result an approximated diagram of changes of error values in the function of given axis coordinates is obtained.

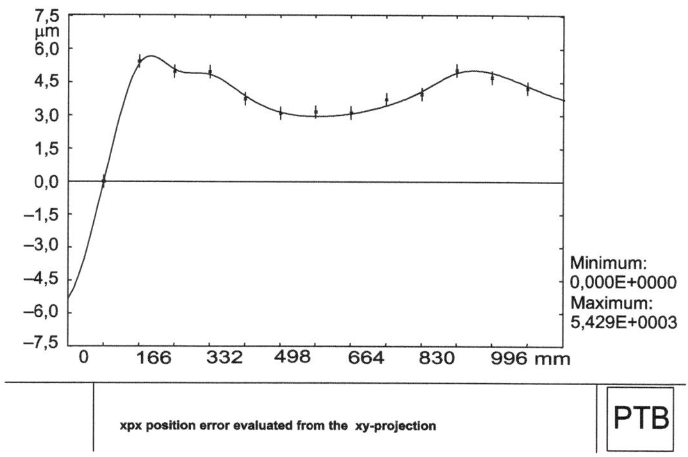

Fig. 4.15 Example of the graph developed as a result of analysis of geometric CMM errors with the KalKom program; translation error in the *X*-axis in *X*(*xtx*) direction; original printout—Kalkom 3.1 program

The procedure described above is also applied for other components of machine geometric errors.

Both in *XY* and *YX* planes the plate is measured in two locations: upper (position 111) and bottom (position 112) (Fig. 4.9). Therefore rotation errors of *X*-axis to *Y*-axis (*xry* error) and of *Y*-axis to *X*-axis (*yrx* error) are determined doubly twice, that is, *xryG*, *xryT*, and *yrxG*, *yrxT*. Because of their otherness and specificity, these errors require broader elaboration. *XryG* and *yrxG* errors form angles between the straight line drawn through points measured in the upper and bottom plate standard location, and the straight line determined from nominal points. *XryT* and *yrxT* errors associate with the torsion of the measurement system. It is caused by the difference in measuring tip angles during the measurements of the standard in upper and bottom positions. The errors discussed connected with the plate standard measurements in the *XY* and *YX* plane do not occur during the standard measurement in *XZ* and *ZX* planes (positions 122 and 121) (Fig. 4.9), because in both planes the measurement is performed only in one position. In the case of *xrzT* and *zrxT* errors the measurement is performed from two sides of the standard with two differently oriented measuring tips. Because of the small thickness of the standard and consequently small distances between measurements, the mentioned errors are negligibly small. Annex C contains the description of the KalKom 4.0 program together with exemplary results obtained in LCM CUT for the DEA machine. It shows the operation principles of the software that are so important for the method usage. Annex B contains a description and results of the correction matrix determination for the DEA Group standard accuracy machine, the same for which the CAA matrix had been determined with the use of the laser interferometer, enabling comparisons of the results obtained for various methods.

Values of CMM geometric errors determined with the described PTB method and developed on its base KalKom program [109, 141] are used both for the identification of kinematic errors to determine the CAA matrix, as well as for numerical error correction. It happens because the same error model is used by producers of measuring machines for the numerical correction of their accuracy. An example can be KalKmess (PTB) software, which directly generates data of the CAA/Zeiss matrix [109, 141]. The other important application of determined kinematic errors is their usage to build the accuracy functional model for a given CMM so-called virtual machine. In PTB such a virtual machine was developed in the form of MegaKal software, described in Sect. 5.2.1.

Contact Probe Head Error Identification

In the PTB concept, great significance was given to probe errors and their identification. Therefore a special system of this error assessment, focusing on their character was developed. The identification of probe errors is based on the measurement of the standard element—the reference sphere—with five measuring tips. Measurements are performed with these tips, which are later used in the uncertainty evaluation of the contact system or in the measurement process simulation. A program called TKal [109, 141] calculates probe head errors, comparing results of real measurements with calibration data of a standard element (sphere), calculating statistical parameters (standard deviation) for distances of all measuring points to the sphere determined center. The result is given as the sum of a systematic component (uncorrected) and a random one. A designated contact probe head error is given individually for every measuring tip. A graphical interpretation of results made by the TKal program, illustrating error distribution on the standard sphere is shown in Fig. 4.16 [109, 141]. Obtained data, similarly to the case of kinematic error identification, can be used for individual error assessment, as well as for virtual machine construction.

Assessing the PTB method, which uses the plate standard for kinematic error identification, one should pay attention to the accuracy of obtained results, which are based on the calibrated plate standard. The procedure of its calibration referred to the meter standard constitutes the primary system element. The procedure, described in Sect. 3.5, is based on application of the swing-round method and PKal program, in which the compensatory calculus is used, resulting in obtaining significantly smaller errors. Additionally the machine is used as a comparator, which also reduces errors. Currently this method together with the 4.0 version of KalKom software (Annex C) makes it possible to obtain expanded uncertainty ±0.5 μm (for the measuring volume of 0.7 m^3). For comparison, it should be said that for classic analytical methods (error budget) for the same conditions an uncertainty of ±0. 75 μm [65, 156] is determined in the best case. With the described method it is possible to determine not only 21 systematic components of errors of the stiff model, but also elastic components (described in Sect. 4.1). It is also possible to obtain further effects of the error assessment, such as the hysteresis and change of dimensions in a given direction as deviation from standard dimensions. It is very useful during permanent monitoring of the measuring machine, because it makes possible the determination of the value and the direction of error growth and,

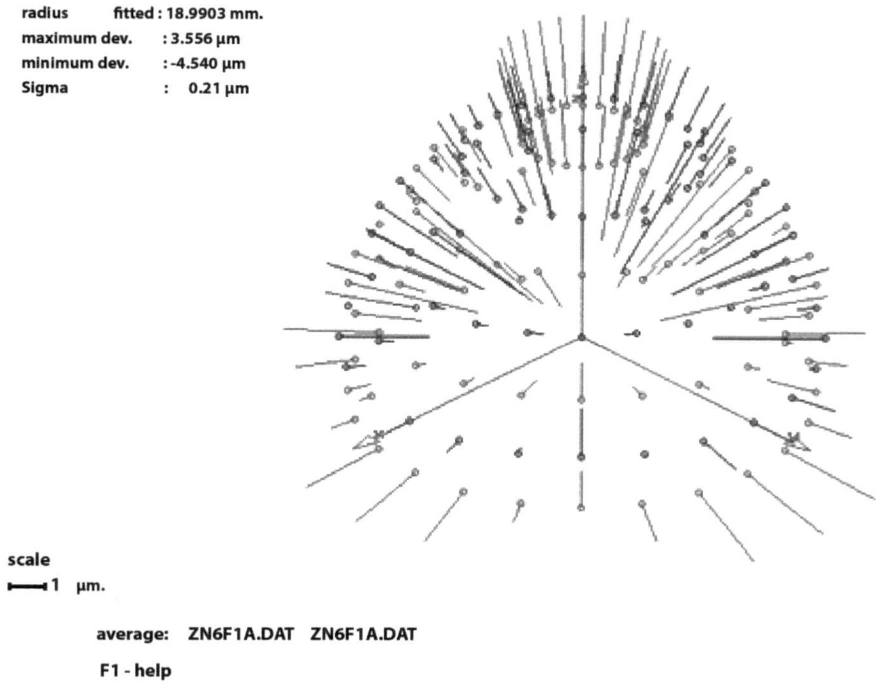

Fig. 4.16 Graphical interpretation of probe head errors (TKal program, developed in PTB)

consequently, their proper correction. It is also important to add that using the plate standard it is possible to determine the CMM maximum permissible error. Standard lengths are defined as combinations of the distances between sphere centers. This procedure is described in VDI/VDE 2617 Part 5 [192].

4.2.3 Identification of Geometric Errors Using Laser Tracker Systems and Multilateration Method

Laser tracker systems (LTS) are redundant systems enabling the determination of coordinates of a measuring point based on a distance (determined using an interferometer) from the mirror put in the retroreflector contact sphere to the interferometer (IFM), including angles of temporary location of the tracker system. LTS construction is presented in Fig. 4.17, which also shows its basic components. Describing LTS action it should be noted that laser light in the interferometer is divided into two parts. One passes through the moving mirror to the retroreflector, and afterwards comes back, where it is again divided in two parts. The first one goes to the photodetector and the second to the calculating unit. In the case of the retroreflector movement the difference of location is detected by the position sensor

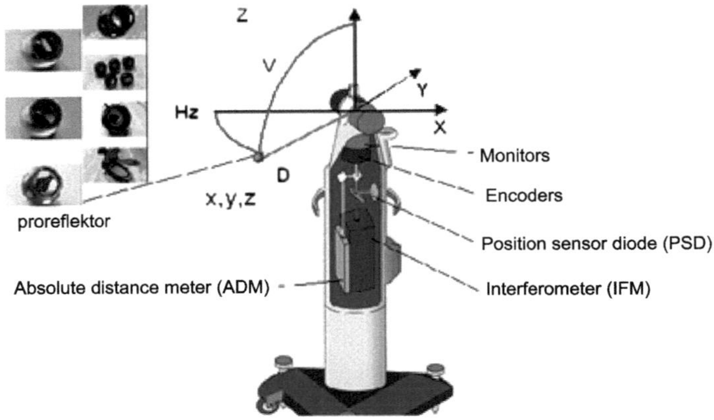

Fig. 4.17 Laser tracker system and its construction (Leica Company) [72]

(PSD). On the feedback principle the system aims to restore the previous state, which is a task for the motors and encoders system, and consequently moves the head tracking the retroreflector motion (Fig. 4.17). Therefore when the system determines the measuring point coordinates, it works in the same way as a manual measuring machine, designated as the contact of the retroreflector spherical area with the surface of measured object.

The determination of the retroreflector location coordinates can be performed keeping the appropriate connection (traceability) with identification of component errors of measuring machines in particular points of their measuring volume. Because of that the first ideas of LTS usage for large machine calibration were tested in the 1970s and described in [44]. However, the low accuracy of the first laser trackers, especially the resolution and the uncertainty of their angular encoders, held back wide use of this method for the mentioned purpose [85]. Only the improvement of technical solutions and the use of modern techniques of multilateration, as well as the PTB method described in the previous chapter, use of plate (sphere or hole) standards, and KalKom software as a basis of a technique for error determination, gave satisfactory results [83, 181, 189].

Multilateration techniques are successfully used, among others in geodesy, as well as in global positioning systems (GPS) techniques, used in satellite navigation. They consist of determining the position of a point from the known distance between the searched position and reference positions. In PTB multilateration has been used to determine CMM geometric errors using the virtual plate standard [207]. This method consists of the determination of the components of CMM kinematic errors in reference points. For this purpose, a retroreflector is mounted in the place of the machine probe head, the position of which is determined by LTS. The machine stops in reference points, which are regularly located on a rectangular path in a given plane of machine measuring volume. This path is repeated four times, and each time the position of the LTS changes. Then the reference points are

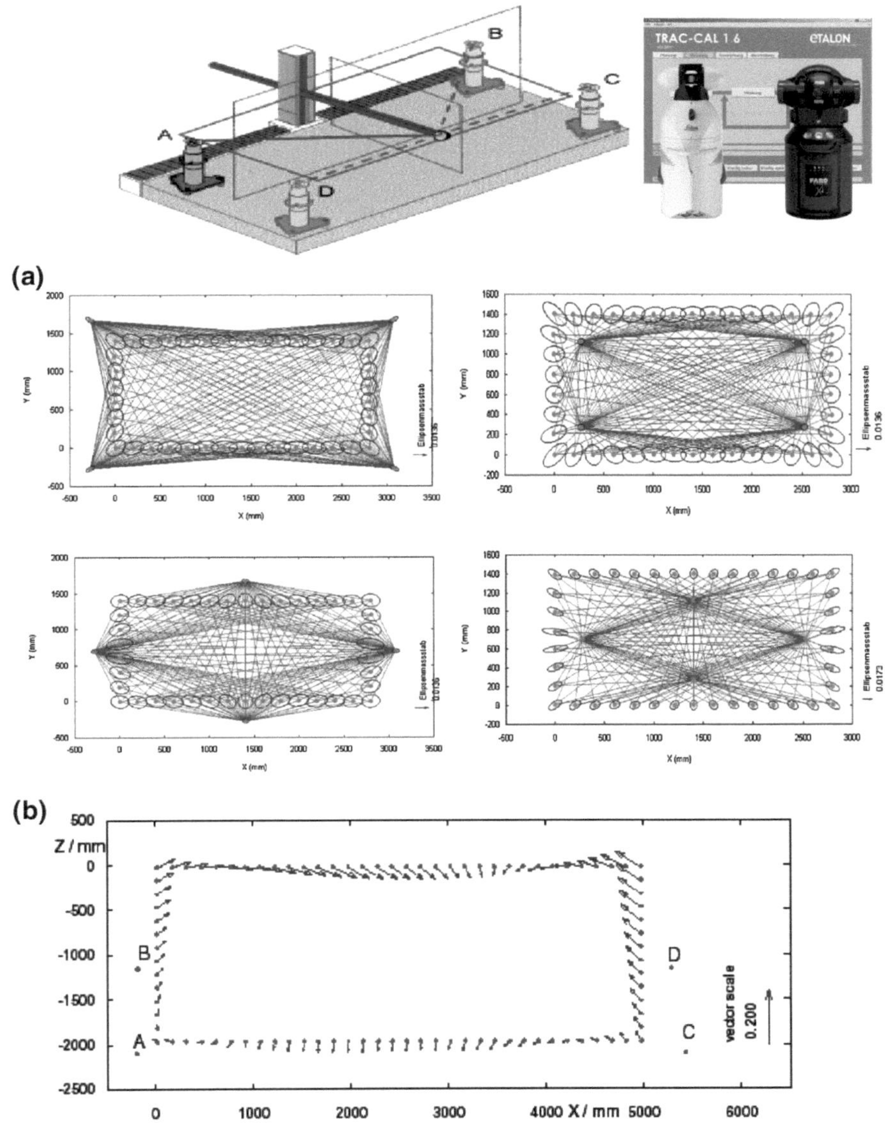

Fig. 4.18 Method of determining LCMM geometric errors with laser tracker and multilateration method using virtual plate standard. Charts of retroreflector movements and reference points of virtual plate standard for four locations of the laser tracker (**a**); error vectors in particular reference points for the selected location (**b**) [27, 207]

determined by the trilateration method based on the distances measured by the LTS including four different LTS positions. This process is repeated for three planes of the CMM measuring volume. The determination of kinematic errors is analogous to

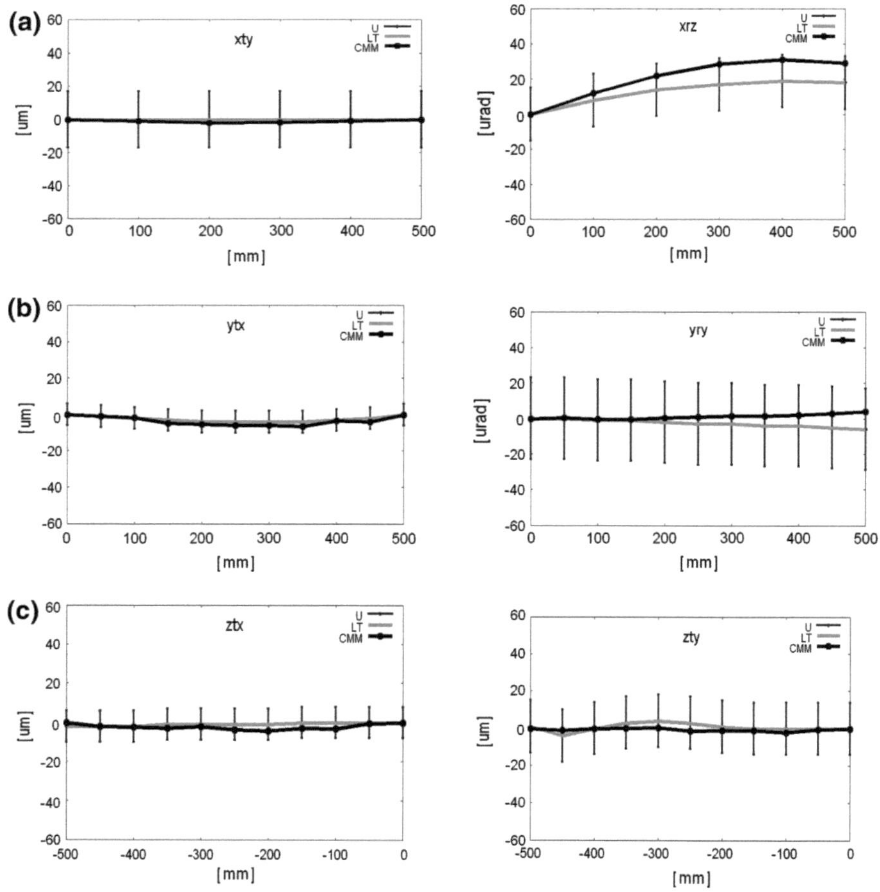

Fig. 4.19 Comparison of the determination of selected kinematic errors (**a**, **b**, **c**) using a laser interferometer (marked CMM) and laser tracer (marked LT)

that described in the previous chapter on the PTB method using the plate standard, but in this case the virtual plate standard as shown in Fig. 4.18 [144] is used.

For identification of CMM errors Trac-Cal software of the Etalon AG Company is used. This program is described in Annex D as well as the CMM error identification procedure with the use of the 840th LTD laser tracker of the Leica Company. Measurements were performed for the same Global Image machine of the DEA Group (average accuracy), for which kinematic errors were determined using a laser interferometer and plate standard as described in previous chapters. The CAA matrix correction was also determined using particular methods, allowing cross-comparison of all methods used.

Exemplary results of that comparison are illustrated in the graphs in Fig. 4.19, where the differences between the results obtained by the Renishaw ML10 laser interferometer (marked as CMM in the figures), and by errors determined by the laser tracker and TRAC-CAL software (marked LT) are demonstrated. The charts are shown for given components of kinematic errors together with calculated uncertainties of their determination. It can be generalized that obtained kinematic error components overlap and that differences are associated with the fact that the research was performed from "outside" the measuring machine system. The laser tracker was not rigidly connected with the tested machine, resulting in an obvious error propagation.

It was also noted that very important for the application of this method is the correction of these environmental conditions that have a direct impact on the uncertainty determination of error components. Additionally not only the number of base positions affects the uncertainty, but also the distance of the laser tracker in relation to the tested measuring volume. A significant reduction in uncertainty was obtained by setting the laser tracker at two different heights. The big advantage of this method of uncertainty estimation is the possibility of using simulation by the Monte Carlo method, depending on the configuration and retroreflector installation method. This allows us to plan the measurement in order to obtain uncertainty as low as possible. Another advantage is the shorter time of CAA matrix determination in relation to the method that uses the interferometer. For large machines it means several times shorter measurement time.

The barrier limiting the use of this method for more accurate measuring machines is the accuracy of the laser tracker. Therefore, under projects financed by the European Union, in PTB and NPL work was undertaken on the development of a significantly more accurate laser tracer system (the name refers to its use to ensure traceability) than a typical laser tracker (e.g., LDT 840) and intended only to measure the distance from the retroreflector to a special reference sphere with very small, of <50 nm, form of deviations.

As a result of this work a construction was developed, characterized by high accuracy. Due to, among others, constantly changing technological possibilities allowing the realization of the reference sphere with smaller and smaller shape

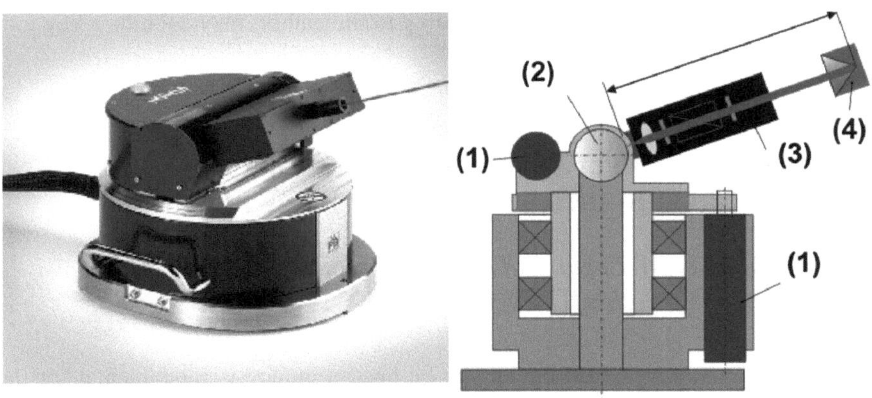

Fig. 4.20 Etalon AG Company laser tracer construction [27]

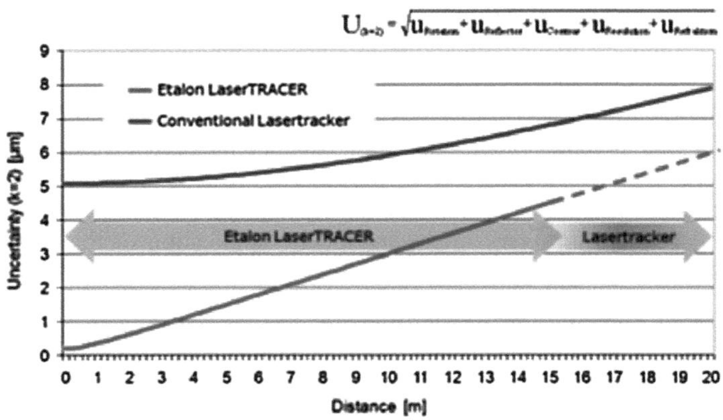

Fig. 4.21 Comparison of laser tracker and laser tracer in CMM accuracy research [27]

deviation, the accuracy of that solution keeps growing. The construction of the laser tracer is presented in Fig. 4.20. The laser beam that passes through the interferometer (3) is reflected from the retroreflector and then from the reference sphere (2), so that the distance between them is measured continuously. Because of encoders marked (1) it is possible to follow the movement of the retroreflector located in place of the machine probe head.

The implementation of this concept was delegated to Etalon AG, spin-off type, an innovative company established by the PTB. In recent years laser tracer construction has been widely developed, so nowadays it can be used to test even the most accurate systems. Currently, the length measurement uncertainty with this system is $U = 0.0002 + 0.0003\ L/1000$ mm. The comparison of accuracies of laser tracer and laser tracker is shown in Fig. 4.21.

The introduction of the laser tracer created a totally new quality not only in the CMM research area, but in most coordinate systems. Because of the accuracy and research possibilities, for example, based on the matrix method (MM) described in Sects. 4.4.3 and 4.4.4, this system, according to the author, may soon be a key tool in the studies of CMS accuracy.

4.3 Error Sources of Point Coordinates Contact Acquisition System—Probe Head Error Function

4.3.1 Analysis of Error Sources and Causes: Probe Head Error Function (PEF)

The issue of modeling CMM contact probe head accuracy is fundamental for the development of a complete CMM model, but to be able to build probe head models, an analysis of their functioning should first be performed. The author has realized

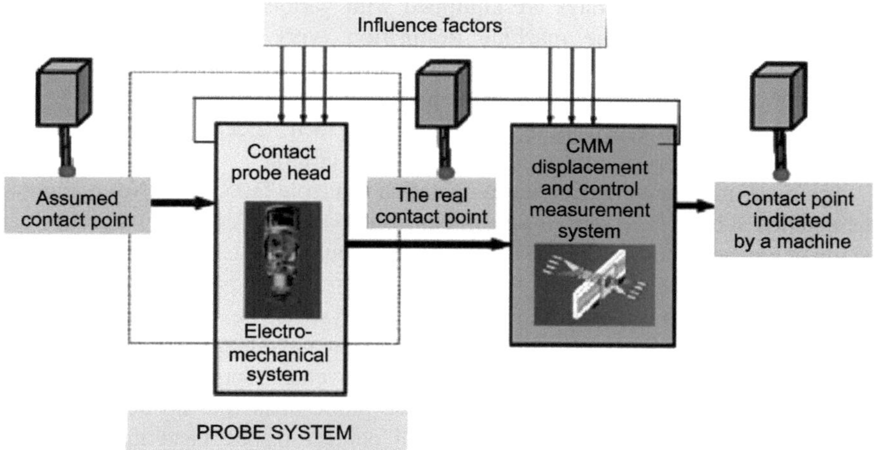

Fig. 4.22 Coordinates of CMM measuring point taking system

such works at Cracow University of Technology since the early 1980s in research projects [96, 97] the results of which are published in the papers [87–92, 94, 148].

Special research stations were built for this scientific work, and for the first time the so-called switching path for impulse probe heads was determined in order to base the probe head characteristic identification on the formulated probe error function (PEF) determined using circular standards [149]. The results of this work are published in [90, 91, 150–153, 161, 163, 166, 167]. These activities have also been continued in the research project [65], and another research project [154], described in [136]. The probe head research problematic is also dealt with in [98–100, 105, 106, 116, 118–121, 130, 194, 197, 199, 200, 202–206, 209–211].

The point coordinates acquisition system of CMM equipped with a contact probe head can be presented with the model shown in Fig. 4.22. Analyzing this scheme it can be said that from a theoretical point of view a faultlessly operating contact probe head should allow the determination of the real coordinates of the measuring tip contact point with the measured object. The nominal (assumed) coordinates of chosen contact point differs from its actual coordinates. This difference is caused by the irregularities in dimensions and form of the ending of measuring tip, interactions between measuring tip and measured surface, and above all—construction errors of the probe head. Finally, the probe head error manifests itself as the difference between the real coordinates of the probe contact with the measured surface and coordinates indicated by the machine displacement measurement system as a contact point of the measuring tip.

The difference of these coordinates corresponds to the contact probe head cumulative error. This error is the sum of a number of reasons, and it is one of the components of the REMP of CMM, as described in Chap. 2. Contact probe head errors are this reproducibility error component, the value of which does not depend on the location in the CMM measuring volume.

CMM produced nowadays are equipped with various types of contact probe heads. It depends on the machine destination (very accurate measurements, medium-accurate), measurement range, the level of automation, as well as on whether the producer has his own probe head construction solutions, or uses other companies' products (such as that of Renishaw [136]). Each probe head can be characterized and classified according to different criteria.

The most important criterion is the operation principle [105, 106]. Not going into the detailed subdivisions deriving from construction [96, 97, 149], two groups can be distinguished: contact probe heads with a built-in measurement system of measuring tip displacement and pulse probe heads (changeover).

For probes with an internal measurement system the measurement of tip deflection (probe heads of rotational measurement movements) is performed, or displacement in the direction of measurement movement (probe heads of translational measurement movements). In the pulse probe heads the tip deflection of a specified value causes the gap in electrical contact (probe head microcontact), which is taken as the impulse to the machine displacement measurement system informing about the contact of a tip with the measured surface. This signal can also be generated by a piezoelement action or strain gauge. In this case the microcontact action, described above, acts as a confirmation for the highly sensitive piezoelectric sensor (Zeiss Technology ATAC [212]) or strain gauge (Renishaw TP200 [136]).

The next criterion is the direction of measuring tip deflection (due to the inclusion of the angle between the direction of measurement movement and the normal to the surface at the measuring point). Due to the location of the measuring tip its deflection may occur in an undefined (pulse probe heads) and in normal to the surface direction (modern probe heads with built-in measurement system, measuring probe heads).

The criteria listed must be considered including the interaction of the contact probe head system and measuring machine. In this connection the method of determining coordinates of the measuring point must be taken into account. These coordinates may be determined either by summing the obtained measurement results of displacement of the measuring tip and the measurement of relative displacement of the whole contact probe head in the machine measuring volume (machine displacement measurement system); this is possible for probe heads with an internal measurement system, or only by a probe head relative displacement measurement system in the machine measuring volume (specific for pulse probe heads).

A reference point for the measuring point coordinate acquisition moment in systems using the tip and probe head displacement summation is the temporary absolute value of displacement. If the coordinate of the measuring point is taken as the machine displacement measurement system indication, the tip must be in time to take the point coordinate in a spatially defined reference point. This point can be referenced to the zero point of the tip displacement measurement system or to the point of electric contact closing and opening. It should also be distinguished whether it is taking coordinates of a single measuring point or if the tip during determination of the point coordinates moves along the outline of the measured surface (surface scanning).

To prepare a proper plan of research of errors that arise during contact with the measured surface process, a point coordinate acquisition model is necessary. Such a model should reveal most sources and causes that comprise a contact probe head error, and it also should allow determination of causes that depend on the solution used for head construction and are characteristic for a given head.

The probing process consists of four main phases (wherein the phase 3 is the most important from the point of view of the probe head operation aim):

1. Feed-in of measuring tip with a certain speed to the measured surface
2. Contact of a tip with the measured surface
3. Identification of the contact point location
4. Retraction from the measured surface

Although for the measuring probe phase 4 is more complex because the displacement measurement is carried out during the return movement of the tip after the contact with the measured surface, then specified phase 3, the identification of the contact point, occurs. The location of the tip ball is approximated to the state when the tip is not deflected, being on the edge of contact with the measured surface (Fig. 4.23).

Accordingly the determination of contact coordinates has a discrete character, even when surface scanning proceeds (the determination of coordinates is realized at intervals of the established scanning step). The contact process itself is performed by a combination of two elements that have significantly different surface curvatures. To obtain the contact of a measuring tip with the measured surface, tip feed-into this surface must be realized. In order to analyze the phenomena which occur in the process of measuring tip contact, a model, presented in Fig. 4.27 [149, 156], for taking the coordinate of a measuring point was constructed. This model has to illustrate the contact process, putting all phases in time. The contact process itself has been brought to a single axis, because the effects considered in the given measurement plane are affected mainly on the capture of the point coordinate in the measurement axis. The effects of influences in the contact area and of the probe head itself are shown in the form of values of appropriate displacements on the x-axis.

A detailed analysis of measuring probe head error causes is presented in Figs. 4.24, 4.25, 4.26 and 4.27, showing the characteristics of their operations in relation to the model from Fig. 4.23.

In Fig. 4.28 the characteristic of the electrocontact probe head operation is presented. The characteristic was based on research results presented in [90, 149]. In turn, characteristics shown in Figs. 4.25, 4.26 and 4.27 illustrate operation of the probe head with a built-in measurement system. Here are the results of research described in [65] and carried out by the author in PTB on the station based on a laser interferometer mounted on a PMM 12106 machine.

Therefore, the final value of the contact point coordinate is under the impact of two groups of factors: the first is related to the features of the measured object, and

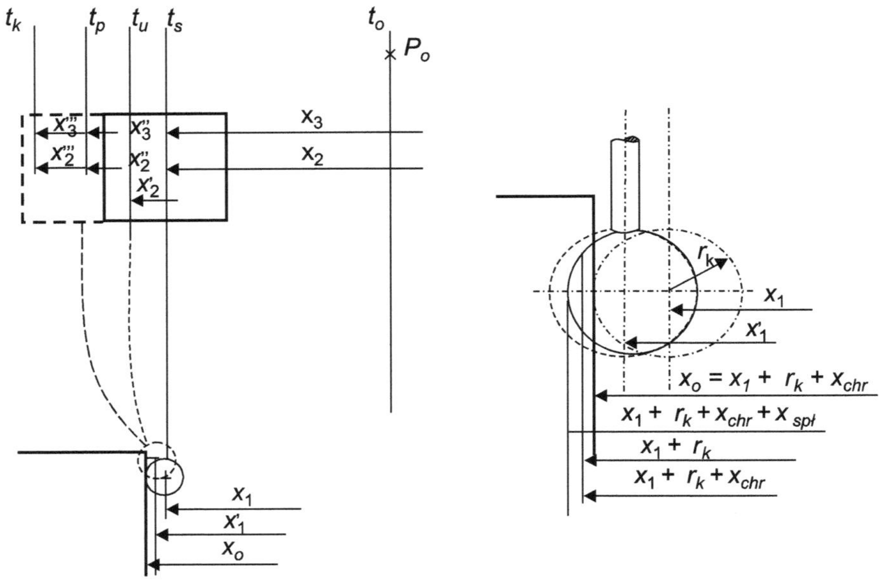

Fig. 4.23 Model of taking the measuring point coordinate realized by the contact probe head [156]: P_o reference point (e.g., the origin of coordinate system); x_o distance of measured object to the assumed reference point P determined along the measurement axis, the real value; x_1, the distance from the center of the measuring tip ball to Po at the time of the contact of measuring probe head, as an effect of calculation of deflection characteristic for the constant, established during configuration of the measurement force for measuring the probe head (close to zero) or for the pulse probe head at the time of microcontact opening (the contact impulse taken for the probe head with a piezoelectric transducer); x_2, distance of probe head neutral point to P where the neutral point, it is, for example, the center of rotation for the tilted probe head rotary or a point on its stationary element, and for the end of the stylus translational shift at the contact moment; x_3, the location indicated by CMM measurement systems along the axis of measurement, at the contact moment (theoretical indication); x_3'', the location indicated by CMM as a contact point; to, start of the probe head measurement movement; t, contact moment (the values related to it are x_1, x_2, x_3); t_u, the moment of ending of stylus deflection and stylus flattened values that correspond with it are x_1', x_2', x_3'; t_p, the moment of contact signal taking (for pulse probe head) and moment of the measuring movement end(for measuring probe head), corresponding values x_1'', x_2'', x_3''; t_c, the moment of termination of the tip back displacement, the starting moment for the measuring motion for the measuring probe head (this corresponds to a location $\approx x_1$); t_k, moment of termination of the tip displacement in the direction of the measured surface; tm, starting moment of the measuring motion (return) for measuring probe head, corresponding values: x_1''', x_2''', x_3'''; r_k, the radius of the spherical tip, including shape deviations in the contact point $r_k = r_{śr} + \delta_{\text{kulx}}$ where $r_{śr}$ is the average radius of the spherical tip; δ_{kulx}, shape deviation value (at the contact point); x_{chr}, the quantity depending on the state of the surface; x_{spl}, the quantity flattened due to Hertzian contact stress; x_p, the quantity depending on the displacement of the movable probe head assembly to the place of the contact signal taking; free path (switching for the pulse probe head) equals $x_{\text{prz}} = x_1' - x_1''$, $x_{\text{o}} - x_3''$, probe head error counted in the measurement axis

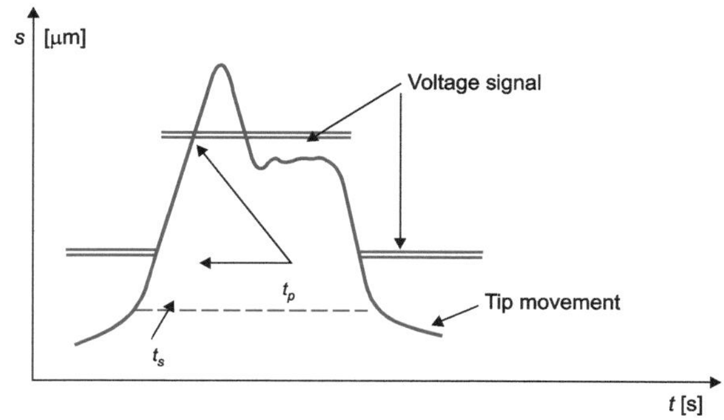

Fig. 4.24 Characteristic of the pulse-electrocontact probe head operation, measuring tip movement; markings consistent with the model of the contact probe head functioning in Fig. 4.23 [149, 156]

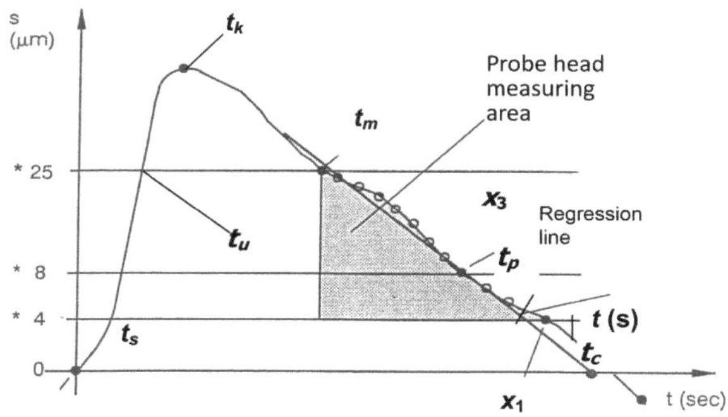

Fig. 4.25 Characteristic of the probe head with built-in measuring system operation, a single point coordinate determination (selected run of measuring tip movement); markings consistent with the model of the contact probe head functioning in Fig. 4.23 [156]

the second one to probe head construction and individual characteristics of the given specimen.

The first group includes:

x_r Condition of measured surface (roughness and shape deviations)
x_{cd} Tip ball contact deformations
x_d Stylus deflection under the measurement force
x_{sd} Influence of measuring tip shape deviation

The second group includes a construction solution and specific errors related to it:

x_s The way of switching for electrocontact pulse probe heads [88, 156]
x_j Nonlinearity and differentiation of the characteristic of inductive transducers in particular directions to measuring probe deflections appropriate for the

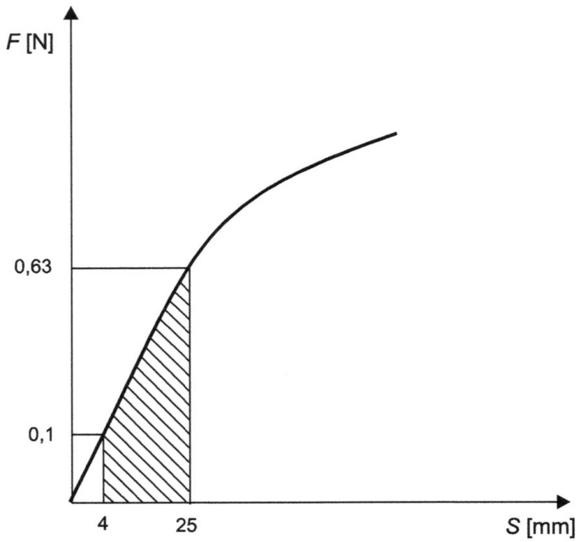

Fig. 4.26 Relation between measurement force F and displacement S of measuring tip for the probe head with built-in measurement system, with marked measurement area [156]

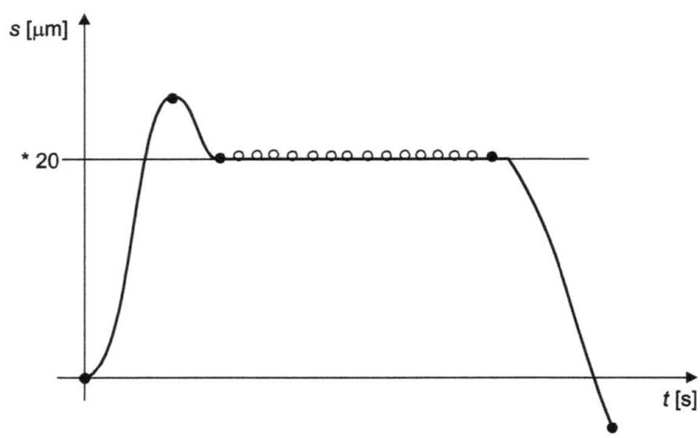

Fig. 4.27 Characteristic of the probe head with built-in measurement system operation, scanning measurement (selected course of measuring tip movement); marks consistent with model of the contact probe head functioning in Fig. 2.26 [156]

displacement directions of probe system (u, v, w) (described in detail in [65, 154]

x_u A reaction for uneven probe head load distribution and electronic balance system errors associated with it

x_{cs} Errors related to changes in sensitivity of the transducers for probe heads with variable characteristic (e.g., VAST type, Zeiss)

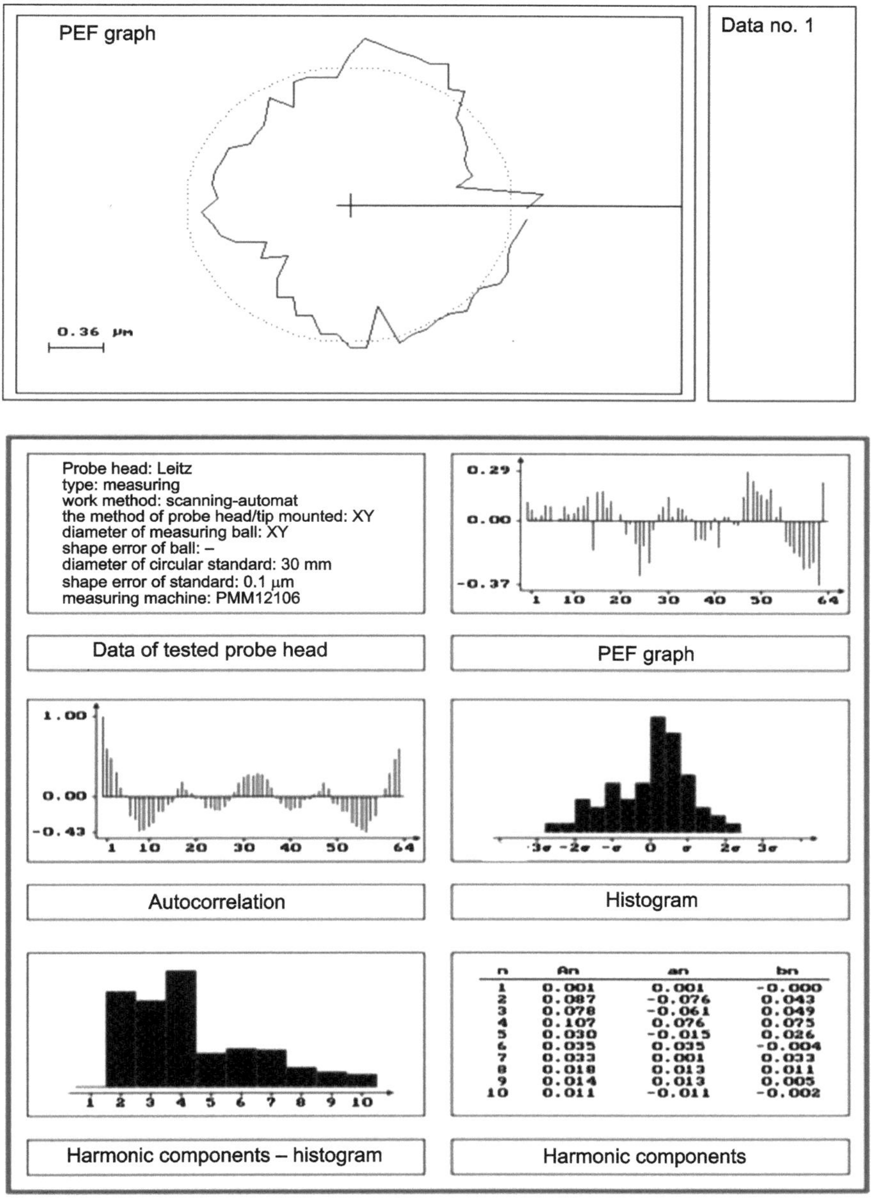

Fig. 4.28 Probing system error function (PEF) for head with built-in measurement system of PMM12106 Leitz machine and the result of harmonic analysis with distinguished values of particular components, PEF analysis program [149, 156]

x_{cod} Values of errors associated with the change of operation direction

After analysis of the adopted operation model and causes of partial error formation, the total probe head error PE (in the direction of a given measurement axis, e.g., x-axis) can be written as the sum of component errors:

$$\text{PE}_x = x_r + x_{\text{cd}} + x_d + x_{\text{sd}} + x_i + x_u + x_{\text{cs}} + x_{\text{cod}} \quad (4.9)$$

For all possible directions of probe head deflection error takes the form of a function dependent on the angle α between the direction of measurement and a normal to the measured surface, naming it according to [88] probe head error function, PEF. This function can be written in the form:

$$\text{PEF} = (\alpha, \text{PE}) \quad (4.10)$$

The markings used are consistent with the above description of the classification of error sources and their occurrence depends on the type of probe head construction solution.

4.3.2 *Contact Probe Head Error Tests*

Currently used standards and recommendations, such as ISO 10360 [48, 115] and VDI/VDE2617 [191], and publications based on them treat probing errors as random ones. This is a simplification, resulting from the lack of clear methods for accuracy assessment and identification of systematic errors, as well as unsuitable inference about their nature. However, the most common and universal method to probe head tests is assessment of one-, two- and three-dimensional reproducibility. Any type of measuring probe heads may be examined in this way and results obtained in such a manner may be compared.

Repeatability analysis consists of multiple contact of measuring tip with measured surface: flat, circular, or spherical standard (e.g., 25 measuring points, ISO 10360-2:2003 [48] or ISO 10360-5 s ed. 2010 [115] for the older version of VDI/VDE 2617) [191]. Maximum permissible errors determined in relation to the reference element are calculated for the contact point coordinates obtained. The area determined in that way does not allow a detailed determination of errors dependent on measuring probe head working specificity, and arising from the realized measuring task. This error is often inflated unnecessarily, and therefore it does not always correspond to reality.

In the years 1982–1985, under the R.I.3 program [96] and in the years 1986–1990, under CPBP [97], the author studied CMM contact probe heads. Then for the first time the direct method of determining the switching path (or more generally, free path) was developed. They were performed outside CMM on special measurement stations for a variety of construction solutions. The results obtained for electrocontact pulse heads led to the development of a measuring probe head operation model presented in Fig. 4.24, as well as the afore-mentioned free path. The results are published in [90, 91, 94, 96] and finally in [149] and [156]. It was found, however, that these methods can be used only under laboratory conditions and they are helpful especially for measuring probe heads and CMM designers, whereas for CMM users they are not very useful. They require not only special stations to be built, but also they have to be adapted each time to the tested type of probe head, which does not allow for direct comparison of results. However, work in this research direction was undertaken in the late 1980s at the Warsaw University of Technology, where research stations developed by Professors E. Ratajczyk, M. Dobosz, and A. Woźniak [22, 23, 118–121, 130, 208–211] were built. Another issue is direct measurements of the free path for probe heads with a built-in measurement system. They are possible and the author performed them under research on the measuring probe head of a PMM 12106 machine in the PTB laboratory in Braunschweig, using the HP laser interferometer system with the identification of the contact moment by a connection with the measuring probe head control system [65, 156]. Figures 4.25, 4.26 and 4.27 contain exemplary metrological characteristics of the probing system examined with this method. These studies have led to recognition of phenomena that occur at the contact moment, and thus complete the model presented in Fig. 4.28, so that it includes all types of contact probing systems.

In [54] attention is paid to the opportunities provided by studies based on standards of constant curvature. The idea of this research formulated in [96, 97] and realized in [149] is as follows. Measurements are taken on a standard of known dimensions and negligible shape deviation, respectively, to tested CMM resolution. As a result of the measurement, data are obtained, that related to the standard errors lead to the conclusion that obtained errors are associated with the measuring device. This principle can be used in probe head research. In this case, tests should be carried out in such a way that allows the elimination or significant reduction of other CMM errors.

Every measuring point located on the standard surface is theoretically burdened with the same error of random nature and the lack of systematic errors considered from a CMM resolution level, and coming from the standard used. The main role in this concept should be assigned to the information about the standard size referenced to the $x1$ value, known from the model of Fig. 4.23 and related to circular or spherical standards. The location of reference point P_0 should be associated with the center of the standard determined using compensation calculation. Therefore, the free path (or measuring path) identified indirectly is nothing but the apparent shape deviation d_{gi} in the measurement direction, calculated relative to a compensation circle or sphere. From the point of view of probing system error assessment the error components shown in Fig. 4.23 and listed in the formula (4.13) accumulate in a specific for PEF value given deflection direction α_i=.

The analysis of measuring tasks performed in [65, 97, 149] and [156] as well as recommended strategies, and above all of the calculation algorithms, leads to the conclusion that most modern contact probing systems operate by performing sampling at a particular plane, although the distribution of measuring points on the measured object surface is spatial. Most calculation algorithms are also built in this way, connecting spatial elements from measurements in certain planes. This is related both to the use of drive and control CMM systems, and to the construction of the measurement software. That is why the probing system error assessment may be performed in a certain plane. According to [97, 149] the best standard to use is a circular standard. It allows contact (in a given plane) in infinitely many directions of measuring stylus deflections.

Based on the results of work [90, 149, 156] deviation components d_{gi} from Eq. (4.13) were assumed as an error of the tested probe head occurring in a given direction of tip deflection α. Assumptions were done on the selection of standard diameter, its location in CMM measuring volume, as well as on the shape deviations of the measuring tip and used standard, to formulate probing error function PEF (a). Its domain is $<0, 2\pi>$in the given measurement plane. Use of the numerical values of PEF, summarized in tables, would be difficult, and that is why they are presented in the graphical form of a polar diagram (Fig. 4.28).

It should be assumed that for multiple measurements the average PEF value related to the compensation circle shows the error systematical part and the double standard deviation of the results obtained in the given direction shows the random part:

$$\mathrm{PEF}xy = d_{gi} + 2u' \tag{4.11}$$

On the basis of the above assumptions, the concept of PEF determination with use of circular standards was developed. However, in order to put it into practice and to allow determination of searching contact probing system error, many requirements and analyses related to the subject must be met. They are described in detail in [65] and [149]. These actions include:

- Choice of a reference element and probing system accuracy criteria
- Determination of the standard dimensions (shape and diameter deviations)
- Determination of the number of measuring points, their distribution, and the strategy of measurements realization

As a reference element (described above) a compensatory circle or a sphere, determined by the least squares method from all measuring points was adopted. Its diameter should be less than 30 mm, because then CMM kinematic errors [148–150, 152] can be omitted. Selection of the number of measuring points is also important, because it enables determination of the impact of changes in PEF value for each contact angle α, thereby allowing the probe head error analysis [149]. On the basis of years of research the author considers that the number 64 is correct and recommends that this number should not be less than 50.

It is also possible to give an area of maximum errors (PEF) as a value of range R:

$$R = \max \mathrm{PEF} - \min \mathrm{PEF} \tag{4.12}$$

It corresponds to the probe head errors designated in accordance with ISO 10360 [48] and VDI/VDE2617 [191], giving compliance of identified values with established maximum permissible values. In Figs. 4.28 and 4.29 exemplary measurement results of measuring the probing system, obtained by determining PEF and its analysis according to the concept set out in [149, 156] are presented.

Probing errors for a given angle of measuring stylus deflection can be described with the probing error function, which, as mentioned above, contains summed probing errors arising from a variety of reasons, such as the state of the measured surface (roughness and shape deviations), measuring tip contact deformation, measuring stylus deflection under the measurement force, measuring tip shape deviation, manner of switching (for pulse-electrocontact probing systems), non-linearity, and differentiation of characteristic of inductive transducers in particular directions of measuring probe head deflections, and so on.

Based on the performed studies and on PEF determination for different probing systems, it was found that this function at an assumed level of discretization of possible deflection directions in a given plane could be considered as signal carrying information about probing system errors. Therefore a harmonic analysis of this signal can be performed. This analysis can help to identify the nature of probe head errors, and to compare the value and the nature of errors for different types of contact probing systems. The research presented in [1, 148–150, 152] showed that this signal is a combination of a polyharmonic signal with random noise and therefore for particular deflection angles can be determined as systematic and random components of probing system error. The PEF signal is characterized by the following properties:

- It is a discrete signal describing the continuous phenomenon (there are possible infinitely many directions of tip deflection).
- It has a discretization step (2 $\Pi/64$) enforced in advance.
- Signal values are the average of several realizations.

Assuming that N values were determined at a distance equal to the h value in the interval of finite length T, the Fourier series for the $X(t)$ function has the form:

$$X_i = \frac{a_0}{2} + \sum_{n=1}^{N/2} a_n \cos(\frac{2\pi nih}{T}) + \sum_{n=1}^{N/2} b_n \sin(\frac{2\pi nih}{T}) \tag{4.13}$$

The estimator of the autocorrelation function for a signal can be determined using the formula:

$$\hat{R}_r = \hat{R}_x(rh) = \frac{1}{N-r}\sum_{n=1}^{N-r} X_n \times X_{n+r} r = 0, 1, 2, \ldots \tag{4.14}$$

The estimator of spectral density of power can be also determined in the form:

$$G_k = G_x(f_x) = \frac{1}{Nh}|X(f_k, T)|^2 = \frac{2h}{N}|X_k|^2 \tag{4.15}$$

The method of outputting the mentioned formulas can be found in [148–150, 152]. In Fig. 4.33 an exemplary graph of the probing system error function is presented together with the harmonic analysis described above, realized for the measuring probe system of a PMM12106 Leitz machine. Further research was performed in a similar way concerning the modeling of a contact probing system [186, 187]. Using PEF for the construction of a universal CMM model, experimentally obtained values in particular directions were added directly, and the maximum values of errors are the basis of operation of a random error generator.

In the described concept of contact error determination a random model was adopted, in which, according to Sect. 2.2, systematic and random parts are distinguished. The random part is represented by standard uncertainty $\mathbf{u}_{PEF}$.

When using different tips during a given measurement procedure the largest from obtained values is assumed as contact uncertainty.

The method of separation of random and systematic components in PEF is presented in papers [148–150]. It was based on observations of PEF harmonic analysis, when clear dominance of certain components was observed. Therefore, a concept of separation of these components in the form of a systematical part was developed—specified as a characteristic part of PEF (CPPEF)—and the random part, residuals function (RfPEF). The characteristic part is obtained as a combination of the minimum number of harmonics in such a way that after removing it from PEF the remaining result was a random part of probing system error. Randomness of this residuum is examined with a nonparametric series test. If to the values of the RfPEF function (di) (where $i = 1.64$ and da = average value) the sign (+) is assigned for $di \geq da$ and the sign (−) for $di < da$ then a certain sign string will be created. In this string the so-called series may be distinguished. A series is a sequence of identical symbols that follow or precede another symbol or do not precede any. Assuming that the number of symbols (+) is equal to the number of symbols (−) the randomness of the set of values RfPEF (di) may be hypothesized. It can be said that the number of series in the sequence will have a distribution that results from Table A.6 from [8]. The test of this hypothesis at the significance level $a = 0.05$ allows us to determine the maximum number of series for the range ($r_{n/2}$; $1 - \alpha/2$) and ($r_{n/2}$; $\alpha/2$), where $n = 64$ for the considered case. In accordance with Table A.6 [8] these numbers are: ($r_{n/2}$; $1 - \alpha/2$) = 54 and (rn/2; $\alpha/2$) = 77. If the series numbers obtained fall outside this range, then the hypothesis of RfPEF randomness will be rejected on an assumed significance level. In this case, the number of harmonic components forming RfPEF should be increased by one and for this amount it should be checked which compound composed of which harmonics sets RfPEF of random character according to the above test. In this example, PEF 4 harmonics of the highest value constitute the characteristic part. A nonparametric test gave a result of 56 series, therefore there is no evidence to reject the hypothesis of randomness of this discrete function [149, 156]. In Fig. 4.29 the results of the described analysis are presented, and a chart of PEF is shown together with separation of the characteristic part CPPEF, systematic component, and of residuals function RfPEF, random component.

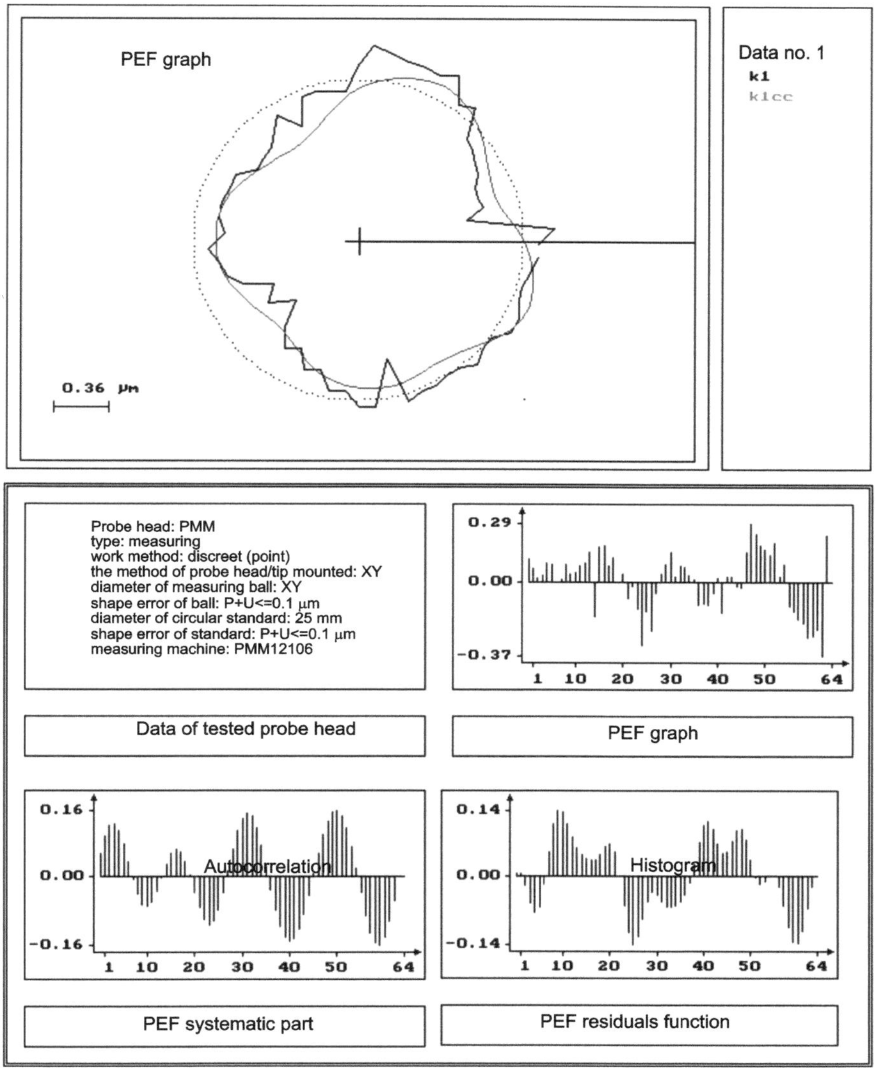

Fig. 4.29 PEF and separation of the characteristic part CPPEF, the systematic component also marked on the pie chart of PEF and residuals function RfPEF, random component, PEF analysis program [149, 156]

At present the development of control and calculate algorithms as well as the contact methods, for example, for a three-directional approach, allows adopting a spatial PEF description. The function given by Eq. (4.14) interpreted as a two-dimensional characteristic of a probing system [10, 11, 149, 156] was a kind of simplification, associated with the most common way of measuring tip deflection.

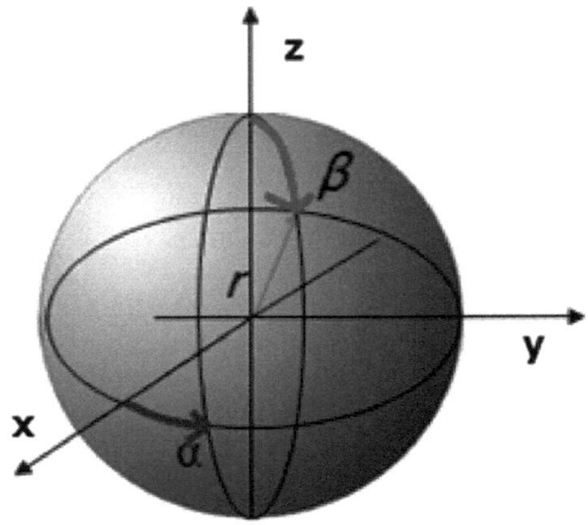

Fig. 4.30 Spherical coordinate system and the marks of PEF angles (α, β, PE)

However, for the general case its movement is spatial. In that situation, in order to completely describe measuring stylus deflection at least two deflection angles should be used [209, 210]. Therefore, the description of measuring stylus deflection may be given according to [34, 160] in a similar way to the description of spherical coordinates (Fig. 4.30).

Therefore in [34, 160] it was decided to adopt a PEF description in dependence on two angles and thus the description of the full three-dimensional probing system error volume. Eventually, the form of the probing error function is (4.20):

$$\mathrm{PEF} = (\alpha, \beta, \mathrm{PE}) \tag{4.16}$$

For probe heads of high accuracy harmonic analysis it has a special importance, because it allows us to realize not only the correction of its action, but to examine the interaction of standard and probing system, which may be crucial for the difficult and not yet resolved problematic of super accurate probing system [24, 46] calibration, because of the lack of a sphere standard with a shape deviation of 2.5 nm and the size deviation of 5 nm. A solution of this problem may be mapping of standards and usage of the fast Fourier transform (FFT). Such a study was carried out for the probe head of a PMM12106 machine, setting a discretization step equal to $2n/128$; the results are described in [160]. In this case, the PEF is obtained by measurements of a spherical standard. In these studies as a standard the reference sphere of diameter d = 24.9903 mm was used. Then a discrete PEF representation was determined using a fast Fourier transformation. As in the case of classical Fourier analysis, the PEF characteristic part is determined and can be regarded as a systematic part of probing system errors. Exemplary results are presented in Fig. 4.31. Work on the standards mapping as well as calibration technology tests are realized in LCM CUT and according to the author this is an important and forward-looking research direction in the field of contact probing system error examination and development of the methodology for the most accurate machine probing systems [160] calibration.

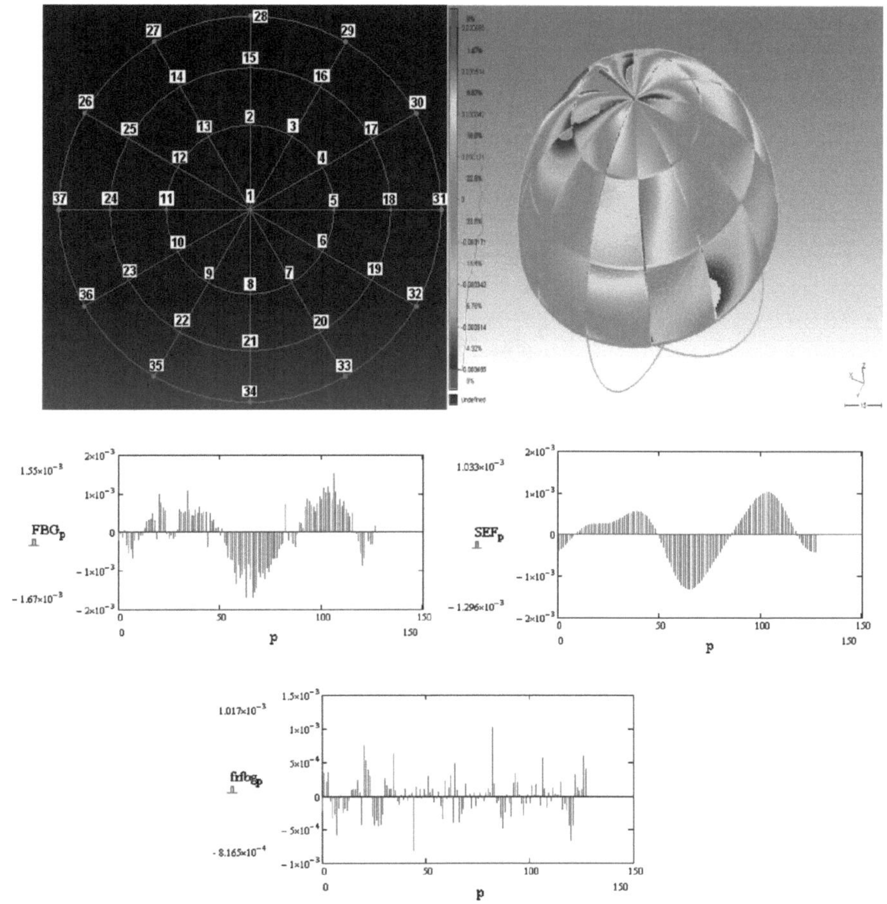

Fig. 4.31 PEF characteristic and FFT analysis results obtained for a reference sphere of $d = 24.9903$ mm diameter, and the shape deviation of 30 nm [160]

4.4 Matrix Method (MM) of CMM Accuracy Identification

4.4.1 *Idea of the MM Method*

According to the theoretical model described in Sect. 2.3.3, a method of CMM accuracy identification consistent with the model that was developed. The method is based on the assumption that point reproducibility error of REMP is a measure of coordinate system accuracy.

The first author's works dedicated to this issue covered the subject of standards usage to identify the location of the reference points in the CMM measuring volume

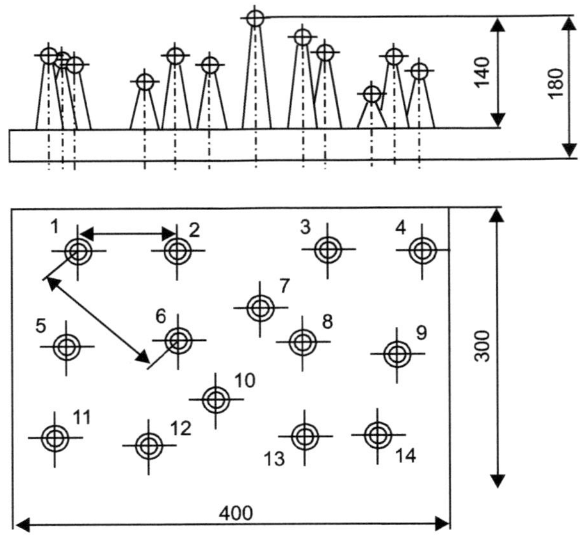

Fig. 4.32 Spatial plate standard with balls, used for CMM accuracy assessment [95], developed at the Cracow University of Technology

and focused on construction of standards representing the set of randomly distributed elements in the form of balls. Their location was determined by realizing the measurement on the precise CMM with a specially installed interferometry system for displacement measurement. The accuracy of standard ball coordinate determination was additionally increased by minimization of the mutual distance function of the reference elements, with the method of compensation calculation [95]. Figure 4.32 shows the developed spatial standard built according to this idea [95, 97].

Because of the difficulties in constructing the standards covering a larger area of CMM measuring volume, it was decided to build the standards generating a regular grid of reference points. The standard in the form of a cube, made of steel used for gauge blocks, with dimensions 49.9307 × 49.9706 × 49.9773 mm ± 0.5 μm was used here. The reference cube was placed at regularly distributed points in the CMM measuring volume, as described in [165]. However, determination of errors from the total CMM measurement area was difficult. The reason was a complicated procedure of standard displacement in CMM measuring volume and its compact dimensions. They were preferred because of the requirement that the dimensions of the basic element of standard shape were less than 50 mm (classic requirement that eliminates overlapping of long-term errors of the kinematic system and scale errors). Therefore, research was continued on finding the standard that would allow examination of a larger area of CMM measuring volume in one standard setting and would give the possibility of regular representation of a reference point set, so that the errors determined in these points designate the matrix characteristic for the tested machine, thus defining the original method. This method is called the matrix method [156].

4.4.2 Connection of MM Method with Reproducibility Error of Measuring Point (REMP)

The matrix method of coordinate system error identification is in its assumption different from the classical method based on control measurement of known length, taking as a reference the coordinate of a reference measuring point (consistent with the theory of coordinate measurements, where the only direct measurement is the determination of the coordinate of a measuring point described in Sect. 2.3). The first idea of this method has been formulated by the author under the research project CPP02.20 [97] in 1985–1989, but because of the lack of an appropriate standard simulating a grid of reference points and of low calculation possibilities of hardware it was not fully realized. Only the experiences gained during further research, and especially during use of plate standards with balls and holes on CMM, led to its refinement and realization. The essence of this method is to build a spatial grid of reference points. The use of a plate standard (holes or balls) placed in the measuring volume of the tested system determines the size and raster of the grid. The use of the laser tracer provides greater flexibility of its shaping. The difference between the coordinates of the centers of particular reference elements, obtained as a result of the standard calibration, and indicated by the tested CMM system, results in a CMM spatial error grid, that constitutes a base for the matrix model of error identification of a tested coordinate system. The matrix of errors corresponds to the distribution of reference points. Furthermore, it is the sum of two components: one associated with CMM errors dependent on the location in the measuring volume of the tested machine and the other independent of this location. The first one is mainly affected by errors (dependent on the kinematics and errors of systems of displacement measurements), and the second one by probing system errors (systematic and random). In the construction of the summary model it was assumed that the main responsibility for random error occurrence lies with the contact probing system. This assumption is possible in accordance with the results of research described in [65, 148–150, 152, 155, 156]. As a result of such a model construction, the set of vectors of reproducibility errors of measuring points was obtained (identified in the control area): the matrix model of coordinate system error identification, presented in Fig. 4.33.

Mathematically, the essence of the model, according to Eqs. (2.10) and (2.11), from Sect. 2.3, can be presented in the form of the following equation, assuming that for the standard the coordinate values correspond closely to the real ones:

$$\bar{\boldsymbol{P}}_s = \bar{\boldsymbol{P}}_r: \quad \bar{\boldsymbol{P}}_a = \left(\bar{\boldsymbol{P}}_p + \bar{\boldsymbol{P}}_{pr}\right) = \bar{\boldsymbol{P}}_s - \bar{\boldsymbol{P}}_m \tag{4.17}$$

The matrix was built based on the measurement results of 13 locations of the spatial plate standard with balls (or holes) in accordance with Fig. 4.37 and formula (4.17):

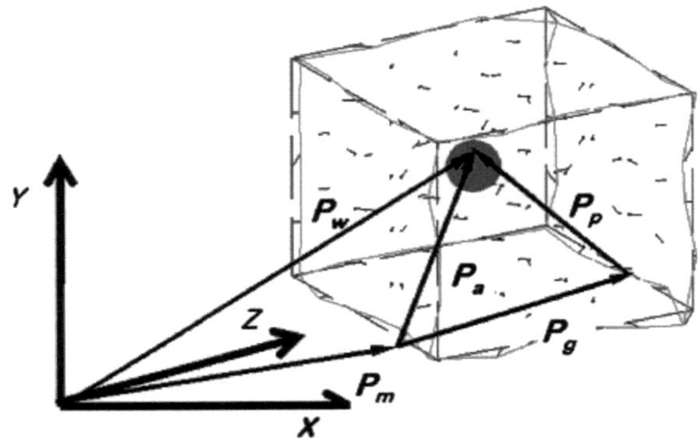

Fig. 4.33 Spatial grid of reproducibility errors of measuring points

$$\begin{bmatrix} P_{a(1,1,1,l)} & & \\ & \ddots & \\ & & P_{a(9,1,5,l)} \end{bmatrix} = \begin{bmatrix} P_{s(1,1,1,l)} & & \\ & \ddots & \\ & & P_{s(9,1,5,l)} \end{bmatrix} - \begin{bmatrix} P_{m(1,1,1,l)} & & \\ & \ddots & \\ & & P_{m(9,1,5,l)} \end{bmatrix} \quad (4.18)$$

where

$i,\ j = 1\ldots9,\ k = 1\ldots5,\ l = 1\ldots3$	indexes identifying the reference point location
$P_{m(i'j'k1)}, P_{m(i'j'k2)}, P_{m(i'j'k3)}$	$P_{i,j,k}$ coordinates of reference point determined by tested machine
$P_{s(i'j'k1)}, P_{s(i'j'k2)}, P_{s(i'j'k3)}$	coordinates $P_{i,j,k}$ of reference point, corresponding to real coordinates
$P_{p(i'j'k1)}, P_{p(i'j'k2)}, P_{p(i'j'k3)}$	coordinates of CMM position error vector in reference point
$P_{i,j,k}$	combined errors of kinematic system and errors of measurement systems
$\bar{\boldsymbol{P}}_{pr}$	vector of measuring probing system errors
$\bar{\boldsymbol{P}}_{a}$	vector of reproducibility errors of measuring point, the sum of position and probing system errors

Identifying errors in reference points (depending on the location, the position errors) the minimization of probe head error influence can be assumed, so that $\bar{\boldsymbol{P}}_{pr}$ can be omitted. An approximated relation is then obtained as

$$\bar{\boldsymbol{P}}_p \approx \bar{\boldsymbol{P}}_s - \bar{\boldsymbol{P}}_m \quad (4.19)$$

This assumption is possible due to the method of measurement realization, consisting of the regular distribution of measuring points on the surface of balls (or spheres) of the plate standard, and due to use of a four-point strategy, minimizing the influence of the contact angle contained between the measurement direction and a normal to measured surface on the probing system error. Thus it is possible to adopt the length of the vector (deviation) $\bar{\boldsymbol{P}}_p$ for a single reference point, $P_{i,jk}$.

The standard uncertainty of position error determination is calculated for error vectors in particular reference points. As the reference points are placed in different planes, the uncertainties are determined for different planes *XY, YZ, ZX* thus *UpXY, UpYZ, UpZX* values were obtained. From these values the characteristic maximum value for all reference points is selected.

$$Up_{\max} = \max Up_{pl} \tag{4.20}$$

The independent error from the location in the CMM measuring volume is the probing system error $\bar{\boldsymbol{P}}_{pr}$. It includes impacts from each tip, in the case of setting them into the form of a so-called star, or any position, in the case of use of an articulating probe head (e.g., type of Renishaw PH10).

The length of the $\bar{\boldsymbol{P}}_{pr}$ vector is determined separately, based on probing system error function PEF(a) defined in [149] and described in Sect. 4.3.1, which can be written as the formula:

$$P_{pr} = \mathrm{PEF}_{\mathrm{PRB}(i,l,r,k)}(\alpha_{\mathrm{i}}) \tag{4.21}$$

where $\mathrm{PEF}(\alpha_i) = d_{pri}$.

The probing system error function values defined in the spherical (or polar) system are given in the form of errors d_{pri}, and they are included in the area of expanded uncertainty of probing system error determination $U_{\mathrm{PEF\ PRB}\ i(l,r,k)}$.

where

P_{pr} probing system error vector length for the *i*th PRB *i*(*l*,*r*,*k*) measuring tip (Fig. 4.34) dependent on the measuring stylus length *l*, measuring tip radius *r*, measurement type (external or internal) *k*

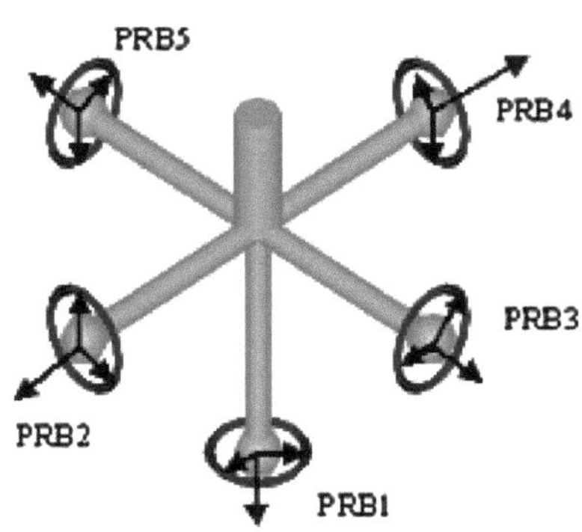

Fig. 4.34 The method of determining the probing system error individually for each tip

$U_{\text{PEF PRB }i(l,r,k)}$ expanded uncertainty of probing system error determination for measuring the tip marked PRB $i(l,r,k)$

The standard uncertainty of probing system error determination in the assumed model is experimentally determined separately for each tip. The approach direction for the ring and sphere standard in regularly distributed $N = 64$ points is taken into account. D_{GI} errors are calculated in relation to a compensatory circle or sphere according to [149]. Finally, the expanded uncertainty is defined as $U_{\text{PEF PRB }(i,l.r,k)}$. For the whole probe head (all tips) expanded uncertainty is given as relation:

$$U_{pr} = \max U_{\text{PEF PRB }i(l,r,k)} \tag{4.22}$$

However, it is recommended to give detailed test results for each measuring tip, due to the considerable variability of results. A detailed procedure has been developed in [148–150, 152, 155]. It is based on the probing system error model given in Sect. 4.3. According to this procedure it is possible to perform additional harmonic analysis in order to extract a systematic part of the identified characteristic, Fig. 4.29. Finally CMM measurement uncertainty established in the controlled area (cube) can be written using the formula:

$$U_{\text{MM}} = U_{pr} + U_p \tag{4.23}$$

whereby it should be understood as

$$M = A + B(l) \tag{4.24}$$

It is given as a concrete value related to the dimensions of the adopted test area −l, for example, $M = A + \text{B}\ (400)$, where $L = 400$ means the length of a cube side in mm. This cube is made of planes that run through spherical or hole elements of the used plate standard.

The adopted notation, despite the similarity to a known way of CMM accuracy (assessed by length measurement) expression, has a completely different nature; the L parameter is not a variable, but only a mark of the size of the test area. CMM accuracy error has been determined by REMP in the reference point and adopted for an established control area (cube) representative of the whole measuring volume of the tested machine. On the basis of the developed model, the defined uncertainty volume for a single point has the shape of a sphere (or more generally ellipsoid), consistent with theoretical considerations from Sect. 2.3, and publications. A CMM total accuracy error obtained by the method described above additionally allows construction of a detailed error map in the control area, allowing the realization of a comparative assessment of different machines, related to errors in particular reference points. Reference to the length standard through the calibration procedure for a plate standard with balls or holes is also provided, maintaining the relation between the calibration uncertainty and the expected uncertainty of the controlled machine. In the proposed method, the probing system is examined more broadly and more accurately than in other methods (e.g., ISO 10360-2:2003 [48]) and is similar to the examining method described in ISO 10380-5:2010 [115], because of the increased number of points and included diversity of error areas for particular tips. It is also possible to use a simple procedure for the assessment of the scanning probing system, which may be realized in a way similar to that recommended by the ISO 10360-4 standard [113], defining the tip movement paths during determination of the THP parameter.

On the basis of collected experience it is easier and more practical to determine the maximum error area for particular planes ME_{XY}, ME_{YZ}, ME_{ZX} and to use error values in reference points (error maps). This allows the precise determination of areas of defined accuracy characteristics, and because of that it is possible to plan the right measurement strategy. For comparative use and global assessment of given CMM–ME_{CMM} the largest identified error range (from all planes) should be adopted as the range for the whole examined volume (ME for CMM). The values obtained for U_{MM} and ME_{CMM} should be given on the control protocols, carried out on the basis of the proposed matrix method. Similarly, for the probing system it is also possible to determine maximum values ME_{PEF} for particular values of PEF defined for all specified tips and to choose from them the maximum value representative for the whole probing system ME_P. In the case of CMM calibration under real operating conditions (this is allowed because of the matrix method) it should be advised to determine the standard and expanded uncertainty in particular positions U_{MM} and afterwards to determine on their basis the range of maximum permissible errors for particular positions taking the probing system into account. The initial assessment, as well as the connecting and systematizing element, is based on the diagram shown in Fig. 4.35.

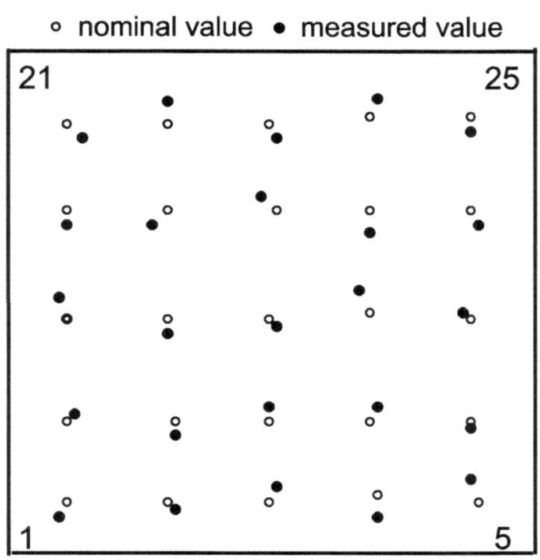

Fig. 4.35 Exemplary scheme of map error construction in reference points

4.4.3 Matrix Method Evaluation Based on Comparative Tests Carried Out on Accurate Measuring Machine

According to adopted assumptions the use of the matrix method is based on construction of a grid of reference points with use of a standard plate with balls or holes or with use of a laser tracker or laser tracer. The grid of points is made by determination of hole or ball centers of the plate standard and, in the case of laser trackers or laser tracers, coordinates of points in the CMM volume. However, when using plate standards, they should be placed in a specific way (Fig. 4.36).

In the case of a plate standard with 25 balls (5 × 5) there are five positions distinguished parallel to the *XY* plane, four planes parallel to the *YZ* plane, and four planes parallel to the *ZX* plane. The difference in the number of positions between the horizontal and vertical location is due to a greater number of measurements realized in normal metrological practice with a vertically oriented tip (PRB1). Vertical locations are designed in such a way that the plate standard was set exactly halfway between the holes placed on the horizontal locations. During measurements in these locations horizontally oriented styli are used (PRB2 and PRB4, Fig. 4.34). Due to measurement problems that can occur during measurements it is necessary to use a special strategy to compensate the effect of probe head imbalance by use of a balancing stylus and by performing the measurement of particular holes from both sides of the plate standard, as shown in Fig. 4.36.

In the MM measurement the strategy is based on the measurement of holes/balls (1–25) of the plate standard in each location shown above and maintaining their assumed order. The order of passing to the next hole/ball is significant. Due to the nature of determined errors and experiences gained from the use of this method the

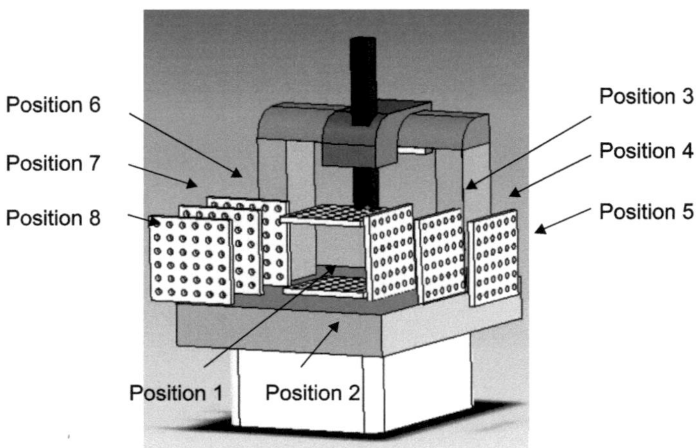

Fig. 4.36 Scheme of required positions of plate standard

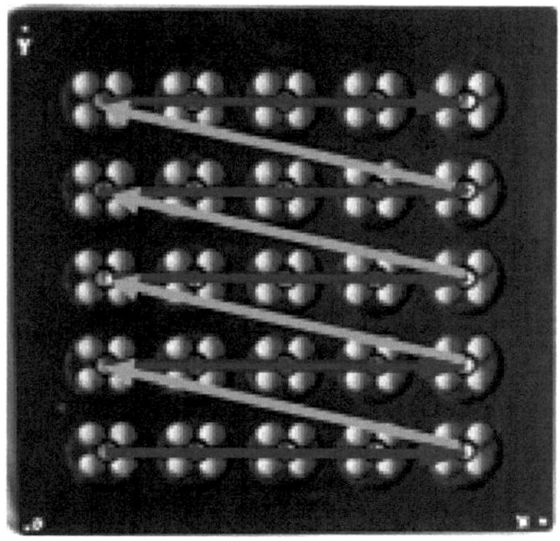

Fig. 4.37 Measurement scheme for snake method

"snake" method was finally adopted (Fig. 4.37), in which the direction of the probe head motion is set by hole numbers 1 through 5, 21 and to 25 (Fig. 4.36).

The research on MM use was carried out for an accurate PMM12106 Leitz CMM, maintaining conditions typical to realization of calibration or verification measurements, with temperature monitored during measurements. The temperature was measured close to each axis scale. The plate standard temperature and the air temperature in the CMM measurement area were also checked. A precise Almemo system from the Ahlborn Company, calibrated by the reference mercury thermometer of ±0.01 K uncertainty, was used during measurements. The method of the temperature monitoring and measuring was consistent with the one presented in Fig. 3.36 in Sect. 3.5, that presents the typical protocol of measurements performed by the matrix method.

During research shown in Fig. 4.36 a plate standard with Retter Company balls (25 balls (5 × 5) located at distances of 83 mm) was used. The standard was placed (as shown in Fig. 4.36) in 13 locations. To place a plate standard in these locations, special stable pillars had been made. The following plate standard locations during measurements are shown in Fig. 4.38. A grid of 325 reference points with known errors in each node, placed in the center of the measuring volume of the tested machine was obtained as a result of the measurements. Better results were obtained using a thermostable plate standard with holes, made of Robax (Fig. 4.38) by the Feinmess GmbH Company in accordance with the author's concept. To develop measurement results the matrix method Microsoft Excel application [80] was developed. The full protocol contains the results for all planes *XY*, *YZ*, and *ZX*. The developed method was verified. Therefore, accuracy tests of the same CMM, that is, PMM12106 were carried out in accordance with VDI/VDE 2617-5 recommendations [192] and the ISO 10360-2 standard [48]. For both methods the Microsoft Excel program "CMM test ISO/VDI" [47] was used. It is typically used to compile measurement results during CMM calibration by the Laboratory of Coordinate

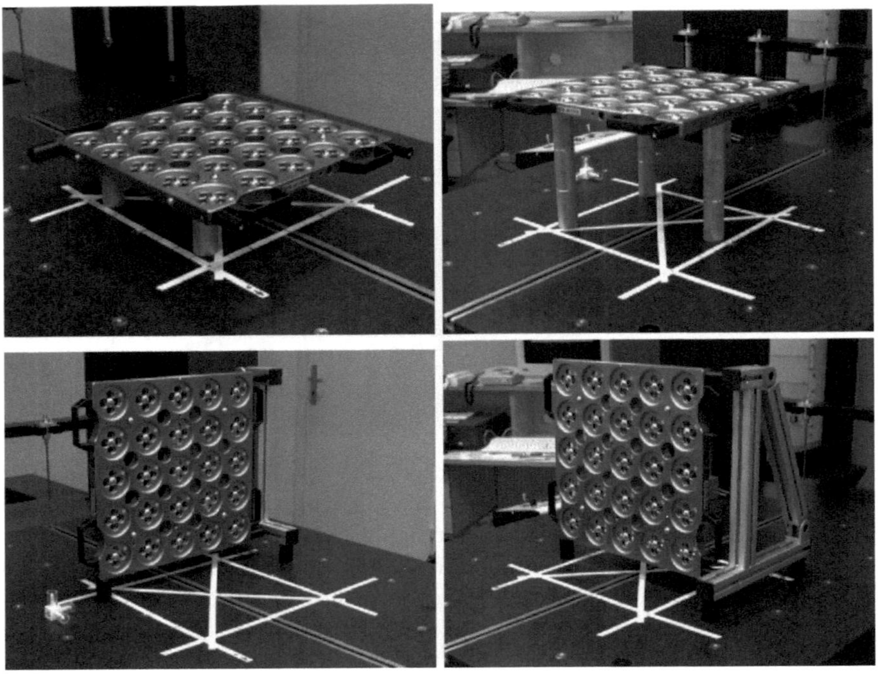

Fig. 4.38 Method of plate standard with balls placed during CMM error matrix identification

Metrology. In the ISO 10360-2 method [48] gauge blocks were used, and for probing system examination a reference sphere measured in 25 points was used, allowing the determination of the *R* parameter.

In addition, in the case of the VDI/VDE 2617-5 method [192] during measurement realization the same snake method was used as in the verified matrix method. In the case of the VDI/VDE (rev. 1986–1989) method [191] during the probing system test, in addition to the probe system accuracy parameter V_3 (taken into consideration when assessing CMM errors), the V_2 value was also determined based on measurement of 50 points, regularly distributed on the ring standard surface, separately for each probe tip (marked PRB1-5 as shown in Fig. 4.29), so that it would be possible to make a comparison with the matrix method. The characteristic feature of MM is error specification for each tip, with reference to each plane. That is why usage of vector diagrams—showing error directions in each plane—allows precise identification of errors in reference points including every used tip. This concept is confirmed by the new version of the ISO 10360-5:2010 standard [115] (e.g., ISO 10360-2 specifies errors only for a vertical tip). In the case of the VDI/VDE method, due to the similarity of measurement strategies and use of the same standard, obtained results can be easily compared. Exemplary measurement results are presented in Tables 4.2 and 4.3. Maximum error results obtained for the checked probing system are set separately.

Table 4.2 Summary of maximum errors method

	MM (matrix method) μm	Method VDI μm	ISO μm
ME_{CMM}	2.35	2.84	3.50
U_3—CMM uncertainty	1.12	1.57	1.88

Table 4.3 Results of probe head assessment

	MM (Matrix Method) PEF μm					Method VDI μm					ISO
	PRB1	PRB2	PRB4	PRB3	PRB5	PRB1	PRB2	PRB3	PRB4	PRB5	
PEF/ME_{probe}	0.52	0.57	0.60	0.59	0.54	0.56	0.55	0.58	0.56	0.55	0.40
U_{pr}—probe uncertainty	0.24	0.28	0.28	0.29	0.28	0.26	0.27	0.25	0.29	0.26	0.22

Values obtained for three different methods are similar. This means that measurements were performed correctly, and even though different methods were used, they fulfill the measurement reproducibility requirements [117], because the manner of obtaining the assessed value (length) for the ISO 10360-2 method [114] and [112] VDI/VDE2617 sheet 5 [192] is based on predetermined coordinates of the measuring points. Because the matrix method specifies errors separately for each tip, it is also helpful in planning optimal measurement strategies through the choice of the object location, number, and distribution of measuring points. The form of results accepted for MM presentation, in the form of a chart of uncertainty values and of a chart of error vector distribution separate for each examined plane, allows quick identification of subareas of higher accuracy, which is important in practice, because very often a CMM accuracy deficit is found while performing certain measuring tasks. Thanks to performed research it can be stated that the matrix method meets the requirements for CMM accuracy assessment and should be recommended both for the calibration process and for periodic testing. Due to the larger number of assessed values in the MM, extended uncertainty of CMM measurement better describes the accuracy of assessed CMM than the description based on maximum errors and it is consistent with the modern uncertainty theory adopted by ISO (GUM) [37]. Therefore, additionally, for comparison, the values of expanded uncertainty were calculated for both methods used in the MM verification.

In conclusion it may be said that this method fulfills set assumptions that were a premise to its development. The valuable advantage of the developed method is the possibility of its expansion in the direction of construction of simulation models used for measurement accuracy assessment. The further part of the work is dedicated to this subject. Additional research is required to establish the rules for extension of the control area, whereas standard dimensions are less than 1/3 of the CMM controlled area. This is especially important for large measuring machines.

The parameters describing the accuracy of the tested machine in accordance with the ISO 10360-2 standard [48] and VDI/VDE2617 recommendations [191] are listed below for comparison.

Accuracy according to VDI/VDE2617 sheet 5 [192] (measurement with use of plate standard):

$$\mathrm{MPE}_E = 0.8 + L/400 \tag{4.25}$$

The accuracy according to calibration carried out using Zerodur standards based on ISO 10360-2:

$$\mathrm{MPE}_E = 0.5 + L/400 \tag{4.26}$$

The accuracy according to the matrix method: with the use of a plate standard with holes, made of Robax:

$$\mathrm{MM} = A + \mathrm{S}(400) \tag{4.27}$$

M = 0.65 + 1.1 μm in spatial control area.

The obtained results are similar. Differences arise from the specificities of each method. In work [84] MM was developed for large machines using a laser interferometer and plate standard, as described in Sect. 4.4.4.

4.4.4 Use of Matrix Method for Error Identification of Large Measuring Machines (LCMM)

In the case of large machines the identification of real parameters of machine accuracy is a difficult task. The difficulty is caused by several factors including measurement ranges and technical problems in the construction and use of standards [5, 6, 12, 13, 28, 31, 32, 40, 50, 79, 103, 107, 111]. Therefore, under works [168] and [157], the concept of the use of the matrix method was planned and realized. Its use in the case of LCMM is a challenging task. This is because the measuring volume of tested LCMM can be even several times larger than the measurement range of the largest plate standards, which in the classic version of the method determines the grid of reference points. Therefore, in the concept of using this method for LCMM the division of the whole measuring volume into smaller control subspaces was adopted. In these subspaces identification and assessment of errors of reproduction of reference point position using a plate standard are realized [84, 164, 169–171].

However, it should be noted that the division of the volume into smaller ones in the final stage requires the obtained errors to be identified $\vec{p}_p$ in relation to the whole volume of the checked machine (according to formula (4.20)). This creates a need for precise definition of relations between the subspaces. Two methods can be used here:

1. The first involves the use of determined earlier geometric errors of tested LCMM. Because of that it is possible to define position reproducibility error between tested subspaces.
2. The second method uses a system of a leading standard, so it allows for defining the position of the plate standard with balls independently from the machine.

In the case of determining both machine kinematic errors as well as errors of a standard leading system, it is necessary to use the laser interferometer or laser tracker system for that purpose. Because of wide availability and economic considerations a laser interferometer [157, 168] was used. The concept of division of the tested LCMM measuring volume into control subspaces [84] allowed implementation of the matrix method in the form of model tests carried out for portal machines with small dimensions (DEA Global Image type Clima 07.07.05, Fig. 4.39a, and Global Performance 07/05/05) with the same accuracy equation MPEE = 1.7 + L/333 μm (Fig. 4.39b) with the use of a specially constructed plate standard, which allows obtaining the error grid with dimensions of 165 × 165 × 165 mm and is twice as small as the typical plate standards. The standard meets the requirements necessary to develop a simulation model due to the characteristic error variability of LCMM with portal construction. The reference elements are 25 balls with a diameter of 12.5 mm placed in 5 rows for which the sampling was performed on the sphere of every ball.

Both machines used in model tests were equipped with a standard measuring probing system type TP200 from the Renishaw Company and with universal measuring tips from the same company. All measurements of the plate standard

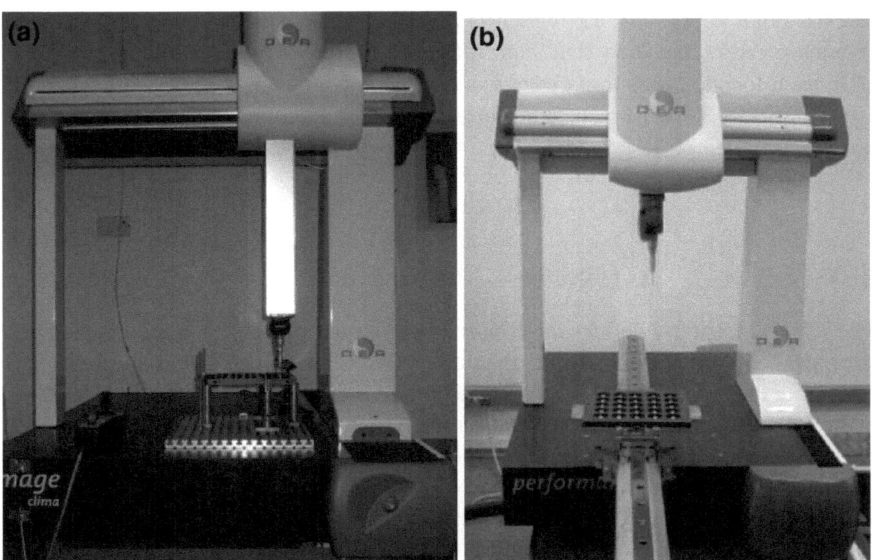

Fig. 4.39 Global image machine (**a**); global performance machine (**b**) [84, 157]

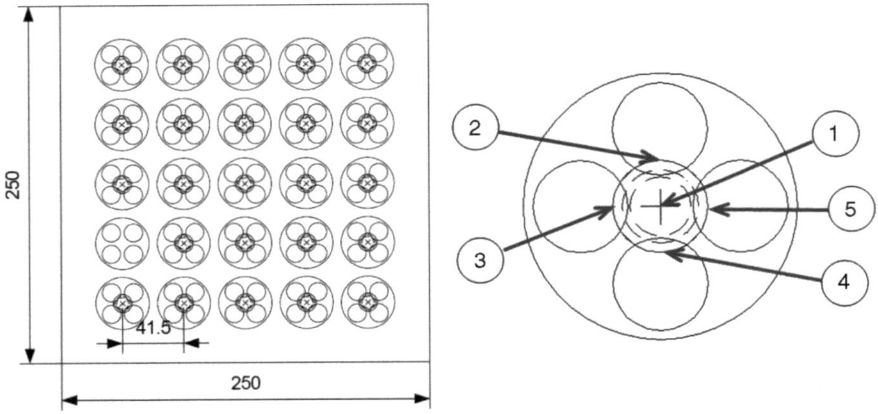

Fig. 4.40 Standard for model tests with the strategy of the reference point determination which is the center of the ball [84]

with balls were realized with the use of the metrological program PCDMIS CAD++. Construction of the reference point grid involves the ball measurement, and the machine measuring tip contact is performed at five points regularly distributed on each, as shown in Fig. 4.40. Reference elements were measured in an automatic cycle, and in each case the standard measurement strategy was consistent with the one presented in Fig. 4.37.

A method was also developed for reducing the amount of plate standard positions from 13 to 3 mutually perpendicular and parallel to main planes of the machine, that is, *XY*, *XZ*, *YZ*, which is possible because of the long-term nature of geometric errors and their dominant influence on the error value $\vec{p}_p$.

In each location, position errors were determined, and x and y deviations can be used to obtain three-dimensional information of each error $\vec{p}_p$ by appropriate summing, which is described by Eq. (4.30):

$$\vec{p}_{pkji} = \frac{1}{2}\begin{bmatrix} x_{xy(5(j-1)+i)} + x_{xz(5(k-1)+i)} \\ y_{xy(5(j-1)+i)} + x_{yz(5(k-1)+j)} \\ y_{xy(5(k-1)+j)} + y_{xz(5(k-1)+i)} \end{bmatrix}, \quad i,j,k = 1\ldots5 \tag{4.28}$$

where k_{ji} increasingly defines the consecutive number of measuring points along the error grid, up to the point with the index 555. Thanks to that it is possible to determine the $5 \times 5 \times 5$ vector error grid, and to prepare its visualization [157, 168].

In the first phase of tests the measuring volume was divided and measured in control subspaces, which in no point overlap with each other, and vector grids of point position indication errors in local systems, which created in this way did not have a common area. In this case their efficient combination assumes usage of external devices that allow obtaining information on the behavior of position errors between tested grids, where it is not known how their values change due to lack of common points.

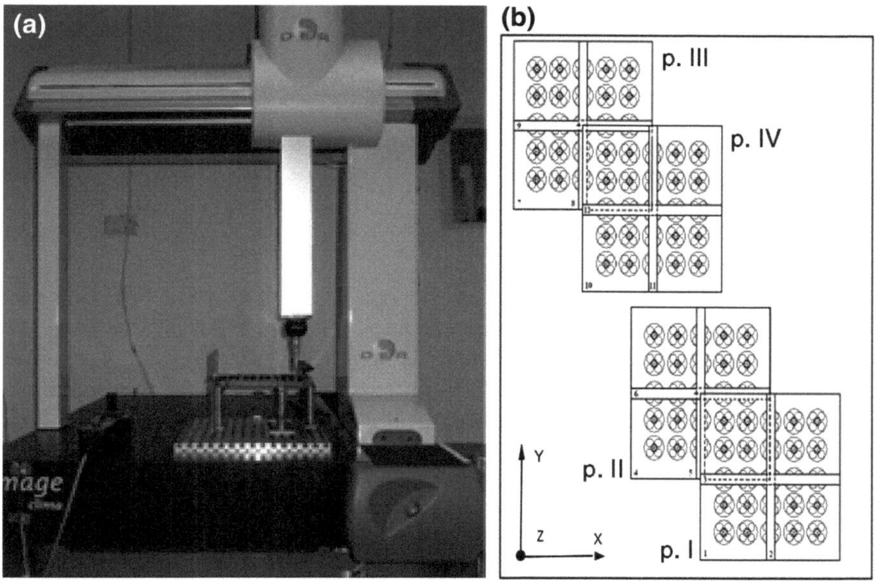

Fig. 4.41 Measurement station (**a**); location of control subspaces on global image machine (**b**) [84, 171]

In the first method, realization of the subspace connection assumes the use of machine geometric errors. For their determination the laser interferometer ML10 of the Renishaw Company was used, equipped in optic set to position, straightness, and angle measurements.

A plate standard with balls was moved along the diagonal and based always on the same balls. For each of the four subspaces determined, local coordinate systems were constructed, two overlapping each other and two separated from each other, where error grids were built (Fig. 4.41).

Subsequently, for error grids $\vec{p}_p$ constructed in this way and identified as local coordinate systems, it was necessary to define the error between their beginnings for each of them in the following locations; analytically determined four position errors of machine $\vec{E}_{pm}$ were used for this purpose, calculated for each of the initial balls of grid of index$_{111}$ for which the position error values $\vec{p}_p$ obtained from the measurement performed by the machine in accordance with Eq. (4.30) were equal to zero.

Because measurements were performed (according to the assumption) only using the reference tip, the position error equation took the form (4.29):

$$\vec{E}_{pm} = \vec{P} + A\vec{X} \tag{4.29}$$

where
P translation errors matrix
A rotation errors matrix

The definition of the connection between subspaces (e.g., I and III) can be written with modified Eq. (4.29):

$$\vec{E}_{V(I-III)} = \vec{E}_{pm}\left(P_0^{III}\right) - \vec{E}_{pm}\left(P_0^{I}\right) \tag{4.30}$$

For control subspace connection in the second method a Rexroth Company linear guide was used, which was mounted on the pipe profile from the Schucko-Design Company. An assembly with a mounted plate standard was moved through the linear guide on tracks. The system that ensured the movement and orienting of the standard in three measurement planes is presented in Figs. 4.42 and 4.45.

As a result of preliminary tests several methods of control subspace connection were distinguished. During test realization it had been noticed that there are several other optimal methods, which largely depend on the size of the examined measuring volume, therefore it was decided to connect the control subspaces by a method using one universal plate standard with holes or balls. The measuring volume created may contain control subspaces that do not overlap (without a common area) and overlap each other (with a common area).

German PTB was the first research center that used the method of the common area ("on the tab"). The method was used to determine CMM geometric errors. The enlargement of volume was done by specific displacement of the plate standard, the orientation point of which in the next location of the standard overlaps another point (just one point is enough, although the number of three or more allows a better determination of the relation). Speaking of orientation points, they refer to given centers of balls, reference points of the plate standard, which should coincide with the centers of balls in the next standard location. Standard displacing in such a manner each time allows construction of a local coordinate system (under such a system it is easy to identify and calculate the position indication errors). The beginning of such a system is set in a place where the orientation point in the previous location of the plate standard was determined (Fig. 4.42). It must be remembered that by moving the standard in this way, it is created each time for a new group of errors and axes of local systems in the next locations are parallel to each other. In view of PTB works, the discussed assumption can be accepted, when by moving the standard translational errors are generated the mostly long-term nature of machine kinematical errors is assumed, and thus it is taken into account that a new beginning of a local coordinate system must be within ±10 mm distance in relation to the position of particular reference elements. The above assumptions determine the number of next locations of the standard, the maximum number of method application up to five times. It should be noted here that with the increasing number of standard displacements the uncertainty of the whole measurement also

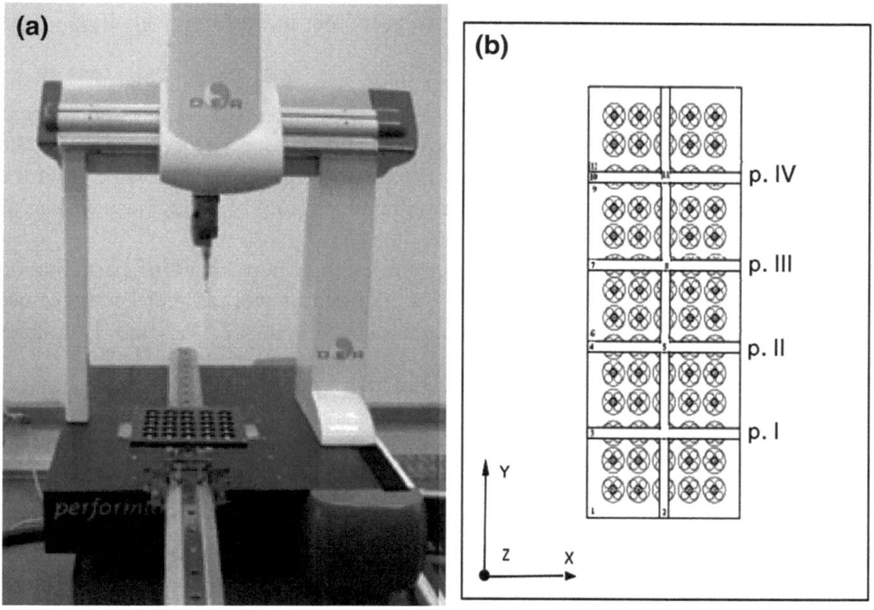

Fig. 4.42 Measurement station (**a**); location of control subspaces on global performance machine (**b**) [84, 171]

increases, in accordance with the principle of error propagation. The advantage of such connections of control subspaces lies in the possibility of obtaining the largest from grids that is easy to implement.

Finally, the connection of error $\vec{p}_p$ determined for the first subspace for the start of the second will be:

$$\vec{E}_{V(I-II)} = \vec{E}_{\mathrm{pp}}\left(P_0^I\right) \tag{4.31}$$

If the plate standard with balls is positioned by the leading system, a linear guide, it is first necessary to know its geometric tolerances, which significantly influence the identification of the plate standard location, and hence values of obtained position errors. It is worth mentioning that during plate standard displacement from the first position to the next one, external factors such as deviations of guide construction start to exert an influence, and what is most important, the positioning truck rotation errors and system vibrations also are added, therefore their values must be determined and taken into account. The errors of position indication $\vec{E}_{pi}$ [157] were finally calculated taking into consideration linear guide deviations and laser interferometer errors.

The developed method of LCMM accuracy assessment with the use of a laser interferometer and linear guide for determining the KWP position required an analysis of component factors, which have a decisive influence on obtained result uncertainty. The realization of assessment of particular uncertainties finally allowed defining the conditions of its application.

Because the positioning of the ball standard is performed with the use of the system mounted on trucks, which move along the guide (Fig. 4.46) and using a laser interferometer, the analysis must take into account the influences of both the standard rotation errors (due to backlash between the truck and the slide) and the impact of translation errors (associated with the uncertainty of slide construction) as well as errors of interferometer indication.

Based on determined component uncertainties their percentage of participation on determined position indication errors was defined as well as real vector values for checked control subspaces (for orientations KWP in *XY*, *XZ*, and *YZ* planes during the proper tests) [84, 157].

In accordance with the established order of work, at first an uncertainty and KWP correction error analysis in the *XY* plane was made. Indication results were influenced by the following factors determined by Eq. (4.32):

$$u_{XN} = \sqrt{\left(\frac{\delta x_N}{\delta \alpha}\right)^2 \cdot u_\alpha^2 + \left(\frac{\delta x_N}{\delta x_T}\right)^2 \cdot u_{x_T}^2 + \left(\frac{\delta x_N}{\delta x_{\mathrm{nom}}}\right)^2 \cdot u_{x_{\mathrm{nom}}}^2 + \left(\frac{\delta x_N}{\delta y_{\mathrm{nom}}}\right)^2 \cdot u_{y_{\mathrm{nom}}}^2} \quad (4.32)$$

where

u_{XN} indication of real coordinate *X* uncertainty
u_α indication resulted from KWP rotation uncertainty
u_{x_T} indication resulted from KWP translation uncertainty
$u_{x_{\mathrm{nom}}}$ indication of *X* value of checked machine uncertainty
$u_{Y_{\mathrm{nom}}}$ indication of *Y* value of checked machine uncertainty

On this basis, it was possible to determine the uncertainty for every coordinate of the plate standard ball. Because standard rotation has the greatest impact on the resulting values, uncertainties of ball coordinates with numbers 1, 5, 21, and 25 for both the *X* and *Y* coordinates in local coordinate systems were analyzed:

$$u_{XN} = \sqrt{(y_{\mathrm{nom}} - \alpha \cdot x_{\mathrm{nom}})^2 \cdot u_\alpha^2 + (1)^2 \cdot u_{x_T}^2 + \left(1 - \frac{\alpha^2}{2}\right)^2 \cdot u_{x_{\mathrm{nom}}}^2 + (\alpha)^2 \cdot u_{y_{\mathrm{nom}}}^2} \quad (4.33)$$

where

x_{nom} nominal *X* coordinate from plate standard measurement
y_{nom} nominal *Y* coordinate from plate standard measurement
α rotation angle of standard plate

Similarly, in the case of uncertainty for real coordinates after including translation and rotation effects the formula for the coordinate *Y* indication uncertainty is obtained:

$$u_{YN} = \sqrt{\left(\frac{\delta y_N}{\delta \alpha}\right)^2 \cdot u_\alpha^2 + \left(\frac{\delta y_N}{\delta L}\right)^2 \cdot u_L^2 + \left(\frac{\delta y_N}{\delta x_{\text{nom}}}\right)^2 \cdot u_{x_{\text{nom}}}^2 + \left(\frac{\delta y_N}{\delta y_{\text{nom}}}\right)^2 \cdot u_{y_{\text{nom}}}^2} \quad (4.34)$$

where

u_{YN} indication of Y real coordinate uncertainty

u_L indication of the position by laser interferometer uncertainty

$$u_{YN} = \sqrt{(x_{\text{nom}} - \alpha \cdot y_{\text{nom}})^2 \cdot u_\alpha^2 + (1)^2 \cdot u_L^2 + \left(1 - \frac{\alpha^2}{2}\right)^2 \cdot u_{x_{\text{nom}}}^2 + (\alpha)^2 \cdot u_{y_{\text{nom}}}^2} \quad (4.35)$$

The above equations allow us to determine real (after correction) position indication errors $\vec{p}_p$. They were at first determined for the XY plane for the average mean of indications from three series for all four positions, enlarged 2000 times, shown in Fig. 4.43 [84, 157].

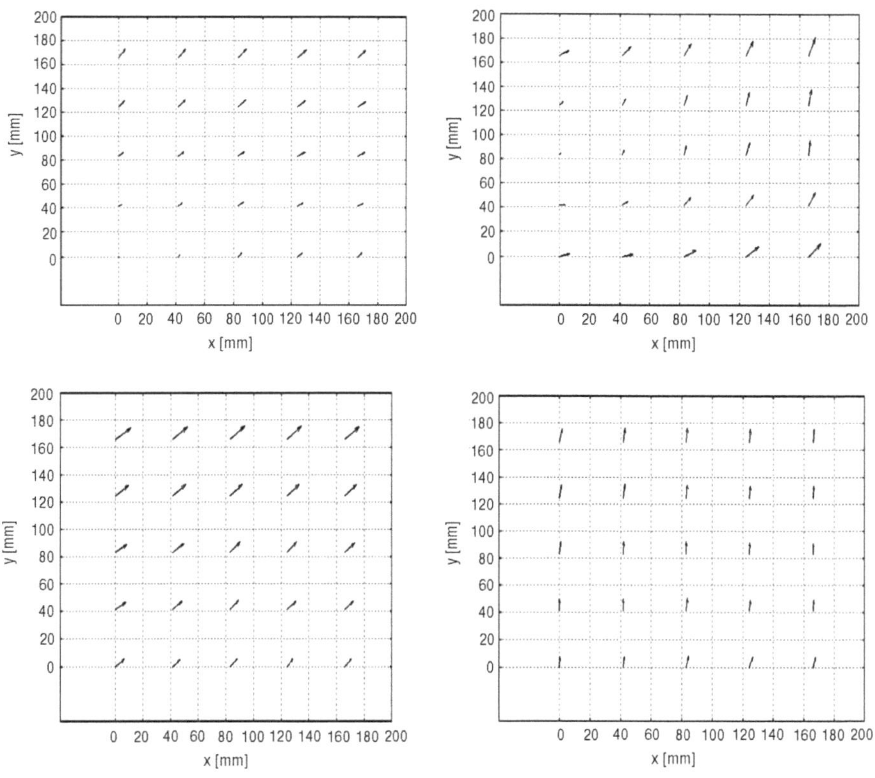

Fig. 4.43 Position indication errors $\vec{p}_p$ in the XY plane for four positions (enl. = 2000×) [84, 157]

In the same way the indications correction in the *YZ* plane was performed, where there is also the impact of errors of standard rotation and its displacement. In the end the correction of indication errors in the *XZ* plane was determined due to the influence of rotation errors only.

Model research performed in LCM CUT on DEA portal machines of similar dimensions allowed adaptation of the matrix method for large machines, assuming the division of checked machine work volume into smaller control subspaces. The machine accuracy assessment was performed by finding the connection between measuring points locally determined with KWP, which were then evaluated globally. These activities were possible due to the use of assistive devices such as a laser interferometer and a linear guide. Thanks to them the location of the standard in the machine volume (DEA performance) was accurately determined, and for DEA Global Image they allowed learning the tested machine errors.

One of the final activities was to create vector grids of position reproducibility errors, which define the spatial nature of errors: identify maximum errors of tested control subspaces. Vector error grids were determined based on results from KWP measurements (the impact of rotation and standard twisting during its movement along the linear guide was included, as well as its construction errors). Each vector $\vec{p}_p$ was calculated according to Eq. (4.28). Thanks to Gnuplot V.4.2 software determined errors were graphically reproduced (Fig. 4.44).

Accuracy equations were finally determined according to the two developed methods. Accuracy of Global Image Clima CMM given by the producer is described by equation:

$$\mathrm{MPE}_{E\mathrm{Image}} = 1.7 + 1/333\,\mu\mathrm{m} \tag{4.36}$$

In both cases, measurements were performed with the use of a reference sphere of 15.786 mm diameter. It allowed determination of the accuracy component of contact system U_{pr} of 1.62 μm. Substituting for the U_p parameter the maximum error made in the subspace of 165 × 165 mm size a value equal to 3.2 μm was obtained, which allowed us to write the accuracy equation according to the principle of the matrix method:

$$U_{mm} = U_g + U_p = 1.62 + 3.2(165)\,\mu\mathrm{m} \tag{4.37}$$

Then vectors of control subspace connection were determined, finally obtaining the following form of the accuracy equation for the grid 600 × 600 mm in accordance with the matrix method,

$$U_{(\mathrm{MM})\mathrm{Image}} = 1.62 + 5.5(600)\,\mu\mathrm{m}. \tag{4.38}$$

Then the second developed method was verified, using for this purpose the developed station to move KWP along the guide on the Global Performance machine with measuring volume 500 × 700 × 500 mm and specified by producer Eq. (4.39):

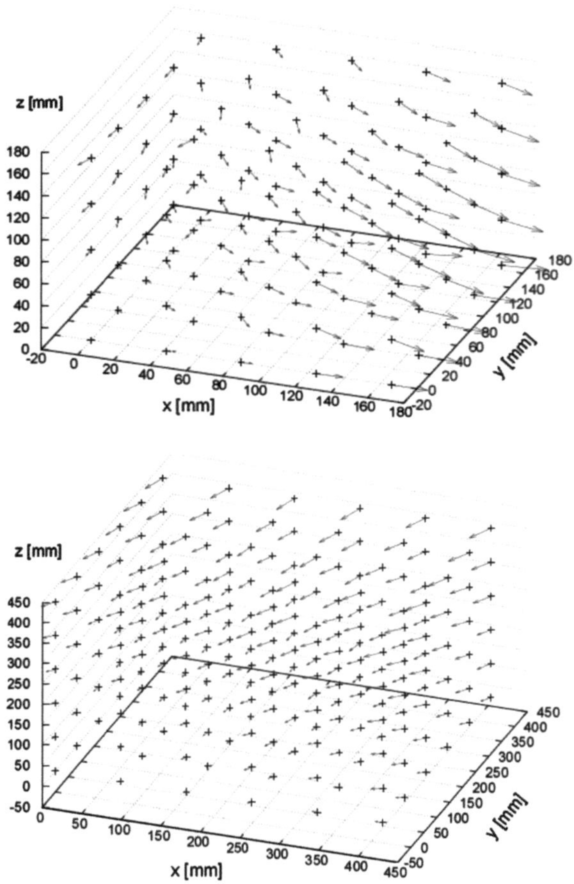

Fig. 4.44 Exemplary vector grids of reproducibility of position errors in subspace 1 in the local system, left and global one, right (enl = 2000× = [84, 157])

$$\mathrm{MPE}_{E\,\mathrm{Performence}} = 1.7 + L/333\,\mu\mathrm{m} \tag{4.39}$$

The next step in this case was the determination of the maximum vector value $\vec{p}_p$, which was equal to 3.4 μm, and the error of the contact system, equal to 1.67 um. The maximum error including the control subspaces of 600 × 600 mm connection into one area determined the component *U*p at 7.2 um. Such a high value was associated with backlash between the driving elements of the station and a guide used. However, it should be distinguished that this method is considerably less time consuming than the method previously verified.

Verification tests for the real large volume machine (LCMM) were performed on a DEA Global Image (Fig. 4.45) with measuring volume of 2000 × 3000 × 1500 mm, so it can be included in the group of large machines. Similarly, as previously discussed, this machine is equipped with a articulating probing system and pulse head type TP20.

Fig. 4.45 Verification of the developed method on LCMM global image type [84, 157]

The maximum permissible error is given by the producer with Eq. (4.40):

$$\mathrm{MPE}_{\mathrm{EImage}} = 4.5 + L/200\,\mu\mathrm{m} \tag{4.40}$$

As a result of verification measurements performed on the measurement station shown in Fig. 4.45 a contact system error equal to 4.3 μm was determined.

The maximum value of machine indication for tested subspaces was 6.7 μm and because of that the machine accuracy equation presented by formula (4.41) has the form:

$$\mathrm{MPE}_{E\mathrm{Image}} = 4.5 + L/200\,\mu\mathrm{m} \tag{4.41}$$

The accuracy determined by the connection of subspaces with the use of a linear guide is:

$$U_{\mathrm{MM}} = 4.3 + 21.4L/500\,\mu\mathrm{m} \tag{4.42}$$

In conclusion, the obtained results confirmed the possibility of the matrix method usage for LCMM and proved that it is useful for a task that is especially difficult from the metrological point of view, that is, the identification of LCMM accuracy. It also confirmed its compliance with the theory of coordinate measurements.

Use of a laser tracker or laser tracer to build a reference point grid is an obvious and actual solution. The number of points in that case is greatly increased, improving the accuracy of LCMM error identification. The multilateration method must also be used here (Sect. 4.2.3).

In Fig. 4.46 the idea of laser tracker use combined with the matrix method for LCMM is presented.

In Fig. 4.47 the concept of the matrix method use in the task of identification of portal machine errors is also presented, but for error grid determination the more accurate laser tracer by Etalon AG (Sect. 4.2.3) was used. Here, similarly as for LCMM, the error grid is determined based on the multilateration method in

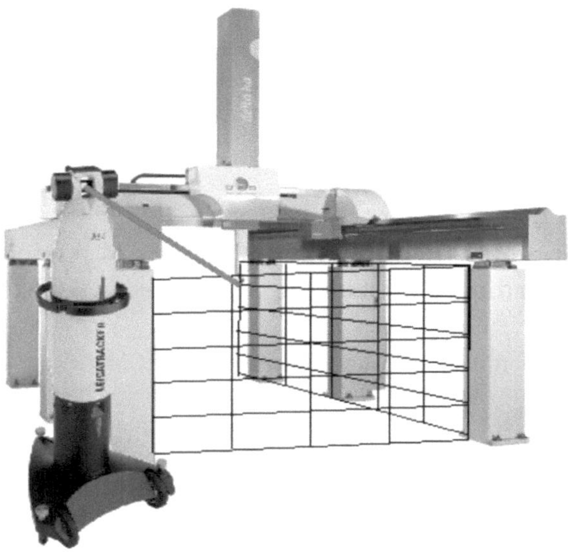

Fig. 4.46 Concept of laser tracker systems usage for the realization of the method for LCMM

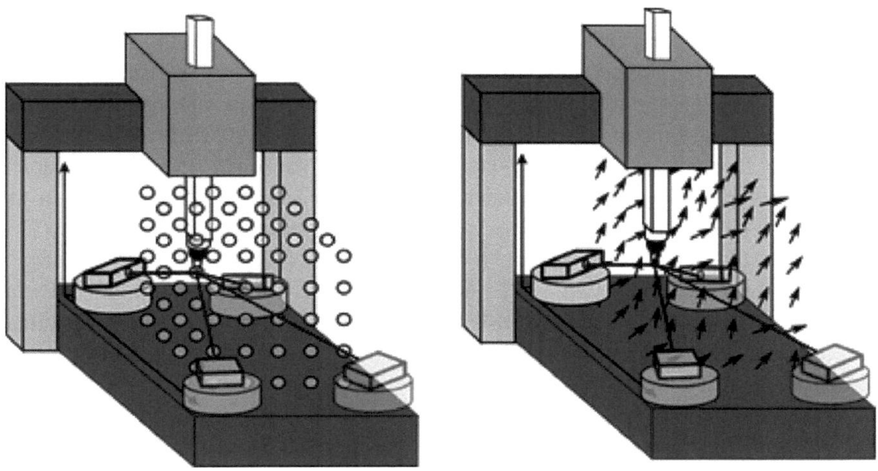

Fig. 4.47 Concept of building a reference point grid for CMM with the use of a laser tracer

accordance with the presented scheme. Different from the method using standards there is no limit on the number of reference points, and therefore the assessment of CMM volume with the MM method can be realized in accordance with the concept of the method and identify the whole machine error field. The accuracy of the laser tracer allows testing all currently produced CMMs. The method of displacement of determined reference points corresponds to locations of virtual plate standards in accordance with the method of CMM error identification described in Sect. 4.2.3. It should be distinguished that different from classical methods of error identification in the MM components of geometric errors are not determined, but a grid of reference points and vectors of indication errors, which in the author's opinion is the most promising method and may be developed for every coordinate system. Research on this method with the use of an Etalon laser tracer is realized under development project [158] in LCM CUT.

4.5 Modeling and Identification of Errors of Articulated Arm Coordinate Measuring Machines (AACMM)

Articulated arm coordinate measuring machines (AACMMs) are modern mobile measuring systems with rotary kinematic pairs. Their characteristic feature is redundancy, that is, the possibility of achieving the same contact tip position from an infinite number of positions and configurations of a particular arm's elements. There are manual systems and, as already mentioned, mobile ones used for industrial measurements for the purpose of verifying the correctness of assembly. They are also used in less accurate measurements of elements in mechanical, furniture, and household goods production. Equipped with a contactless scanner AACMM became an efficient tool in reverse engineering, for example, in industrial design helping in digital modeling of products. They were originally developed for these tasks in the early 1980s. Only at the beginning of the twenty-first century, when they were equipped with accurate encoders and modern metrological software, did they gain wide recognition and become one of the most important coordinate measurement systems. In fact thousands of AACMMs already found application in small and medium-sized production plants (MSP), where a quick dimension verification is needed and the accuracy of element realization is in the range of several micrometers. These systems have an impressive bibliography [33, 39, 49, 58, 70, 75, 77, 101, 110, 131–133, 135, 137, 138, 147, 172–174, 188].

The system used in LCM CUT presented in Fig. 4.48 is an articulated arm coordinate measuring machine, ROMER Omega 2025, with the accuracy specified according to the producer by equation MPE = 0.068 mm in accordance with ASME [2]. However, the monitoring of AACMM accuracy is more and more often realized in accordance with ISO 10360-2, which also is specifically recognized in VDI/VDE 2617 Blatt 9 recommendations [193]. Such an approach is justified because these systems are often used to complement CMM measurements justifying the use of the

Fig. 4.48 Articulated arm coordinate measuring machine ROMER Omega 2025 in LCM CUT

same calibration procedure. As in the case of CMM, also for AACMM the development of procedures for measurement accuracy assessment is of utmost importance. Of course there are classic methods, such as the comparative and multiposition method (described in Sects. 3.1–3.3), but nowadays it is more important to build a simulation system of online accuracy assessment of realized measurements.

4.5.1 Idea of AACMM Errors Model

On the basis of previous experience in coordinate systems model building (Sects. 4.1–4.4) work was started on developing a kinematic model of articulated arm coordinate measuring machine KmAACMM as a basis for virtual arm development which could be applied to online measurement accuracy assessment. Another application of this model is the development of accuracy correction system AA–CAA (articulated arm CAA). The model was realized under [101] and [159].

The first task, as described above, was to develop a kinematic model of AACMM. In addition to the work of LCM CUT these issues were also the subject of several works [32, 39, 137, 138]. In the case of the model realized in LCM CUT a basis of the adopted metrological concept is a vector model of measurement accuracy based on REMP in accordance with Sects. 2.3 [149, 156]. AACMM is a redundant device, so in this case the Denavit–Hartenberg D–H notation [20] is used to connect each rotary joint with the local coordinate system of the device and for determination of the transformation sequence of neighboring coordinate systems leading to the construction of the device kinematics model as a sum of these transformations. The calculation of the position and orientation of the measuring stylus tip in relation to the reference system for a given set of configuration coordinates is possible by realization of a forward kinematic task. This task can be treated as a representation of the kinematic chain position description from configuration coordinate volume to Cartesian coordinate volume.

Determination of all possible sets of angles and linear displacement values (configuration coordinates) in the movable connections, and the achievement of given positions and orientations of the measuring stylus are possible by performing the inverse kinematic task.

Therefore if it is assumed that the coordinates obtained from the calculation of a forward kinematic task are nominal coordinates and coordinates read from the AACMM interface (encoders) are given points, a kinematic analysis of AACMM can be carried out, and thus construction of a model of any articulated arm coordinate measuring machine. Thanks to that a wide field of research on AACMM opens, which includes:

- AACMM error field identification
- AACMM accuracy correction task realization (AA CAA)
- Building a virtual AACMM with the task of measurement simulation for calculating the online uncertainty of a realized measuring task

In the last case, not only the use of the kinematic model must be assumed, but the model of AACMM random errors should be prepared, taking as the description basis the REMP formulated in Sect. 2.3, including the error of orientation of measuring stylus tip included in formula (2.11) and shown in Fig. 2.6. To made the model complete, it is also important to determine the vector random field $U(\mathrm{Pa})$ around the measuring stylus tip at the moment of contact and taking of the coordinates of the determined measuring point, written with Eq. (2.13), and shown in Figs. 2.7 and 2.8.

The first kinematic model of AACMM developed in LCM CUT was created for articulated arm coordinate measuring machine Sigma, model: 2025 manufactured by ROMER. It has six elements connected with themselves with six rotary pairs. Axes of rotary connections are marked with letters from A to F shown in Fig. 4.49. The articulated arm presented (Fig. 4.49) can be mounted to a special base or to a mounting plate. Sigma 2025 AACMM data:

Accuracy Test in Accordance with ASME [2]
Test A (ball test) ± 0.017 mm
Test B (cone test) ± 0.024 mm
Test C (spatial test) ± 0.034 mm

Angle Measurement Systems (Encoders):
Incremental encoders are zeroed always in the same positions, by “transition” through “zero” positions during the start-up and preparation of the arm to the measurement. The resolutions of encoders were defined on the basis of the author’s experience and consultations with ROMER:

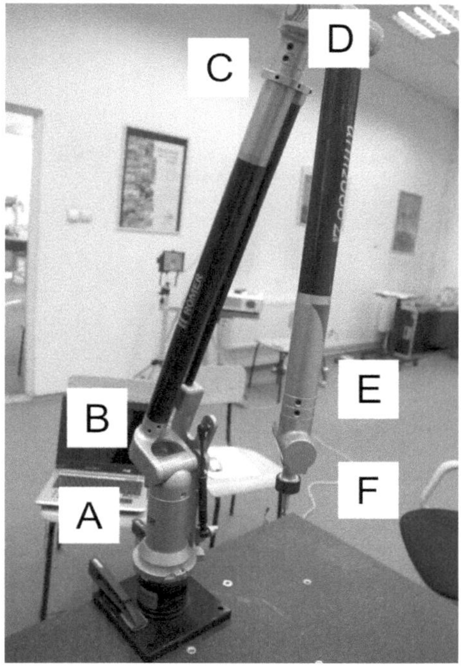

Fig. 4.49 AACMM ROMER Sigma2000 in LCM CUT with markings of particular *A*, *B*, *C*, *D* segments

A axis encoders:	1,280,000 points per rotation
B, C, D, E, F axes encoders:	640,000 points per rotation

The counterweight that maintains the C segment can be also mounted to the arm, making it possible to perform measurements with one hand. This solution introduces an additional stress to the C rotation axis, increasing the measurement errors. Therefore, during measurements necessary to gather the data needed for the kinematic model of the arm, all tests were carried out without the counterweight. The Sigma articulated arm coordinate measuring machine is equipped with a rigid contact measuring probing system with the measuring ball (steel and ruby) and the tip to point measurements. AACMM may also cooperate with scanning probing systems and with the fork head for tube measurements. All probing systems are automatically recognized by the arm software, so they do not require qualification each time. Tests were performed with a contact rigid probing system with a ruby ball diameter of 6 mm. The arm connects to the computer through a wireless network, using the producer's Romosoft software and GDS module (since 2010, ROMER [43] has introduced a new RDS module replacing the previous one; Figs. 4.50 and 4.51).

The arm startup and configuration procedure is performed in the GDS module, where it is possible to read the current values from encoders and coordinates of the measuring tip center, and also the measuring stylus direction–stylus orientation (unfortunately at present it is not possible to save data from GDS into the file). It is presented as three unit vector projections of the stylus direction on the axes of the main coordinate system.

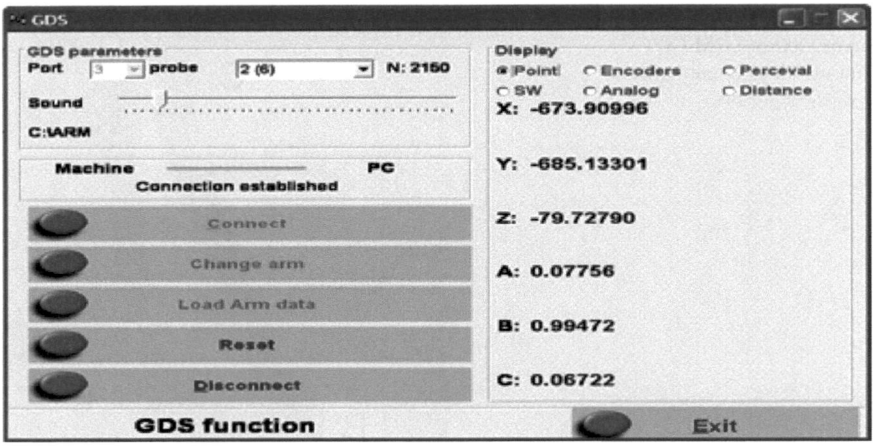

Fig. 4.50 GDS module window showing the position of measuring tip (*XYZ*) and the orientation of measuring stylus (*ABC*)

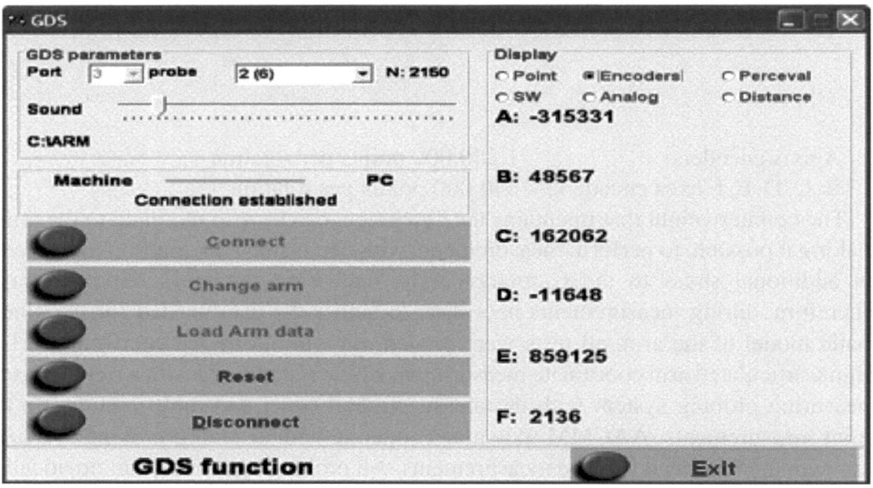

Fig. 4.51 GDS module window showing indications of current values of encoders

4.5.2 AACMM Kinematic Description: Denavit–Hartenberg Notation

For a description of the articulated arm coordinate measuring machine kinematic as a redundant system Denavit–Hartenberg notation (D–H) [20] was adopted. It links the local coordinate system defined for each joint in the transformation chain with neighboring coordinate systems of the device. According to the position and

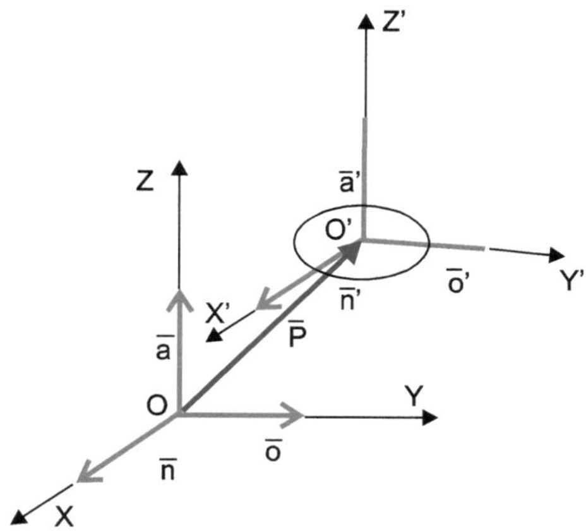

Fig. 4.52 AACMM kinematic description (as a rigid body) in the Cartesian volume

orientation vector notation, it is assumed that particular AACMM segments are rigid bodies. Then their location in the reference system is described by the position and orientation vectors [101, 159, 173]. Marking the Cartesian reference coordinate system by *xyz* letters with the beginning at the point *O* and unit vectors *i*, *j*, *k* of the *x*, *y*, *z* axes to AACMM as a rigid body the *x'y'z'* system is assigned with the beginning at the point *O'* and unit vectors *i'*, *j'*, *k'* (Fig. 4.52).

The O′ point location in relation to the *xyz* system is described by the location vector:

$$\bar{p} = \bar{n}p_x + \bar{o}p_y + \bar{a}p_z \tag{4.43}$$

where p_x, p_y, p_z are its components along particular axes of the reference system, that is:

$$\bar{p} = [p_x + p_y + p_z] \tag{4.44}$$

To minimize the number of elements describing the rotation, the representation is introduced [101, 159]. The rotation is represented by unit vector ***k*** of the rotation axis, described in the reference system *xyz*, together with the rotation angle θ. In the literature, this method is known as an axis/angle representation [18]. To determine the resultant rotation matrix $R_k(\theta)$, which is presented as an assumption of elementary rotations around the axis of the reference system, the following sequence of transformations in relation to the reference system *xyz* is performed (as shown in Fig. 4.53).

- Rotation by the $-\gamma$ angle around the *z*-axis, and then by the $-\beta$ angle around the *y*-axis (because of this transformation the *k* vector becomes compatible with the *z*-axis).
- Rotation by the θ angle around the *z*-axis.

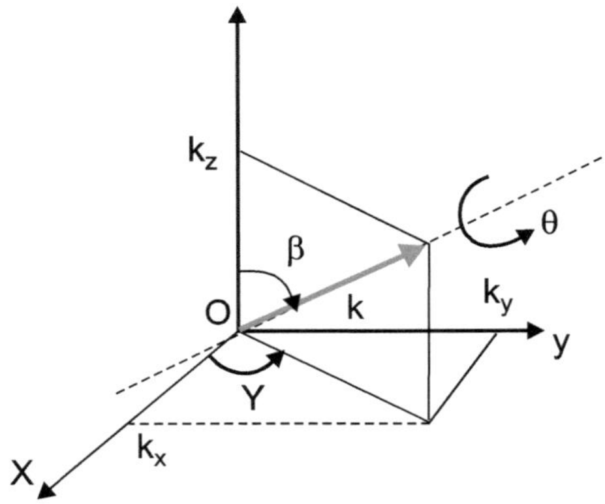

Fig. 4.53 Rotation around any axis

– Rotation by the β angle around the y-axis, then by the γ angle around the z-axis (because of this transformation the k vector obtains primary sense).

This sequence is described by the equation:

$$\mathrm{Rot}(k,\theta) = R_k(\theta) = R_z(\gamma)R_y(\beta)R_z(\theta)R_y(-\beta)R_z(-\gamma) \tag{4.45}$$

Angles β and γ are determined on the basis of components of unit vector $k = [k_x, k_y, k_z]^T$:

$$\sin\gamma = \frac{k_y}{\sqrt{k_x^2 + k_y^2}} \quad \cos\gamma = \frac{k_x}{\sqrt{k_x^2 + k_y^2}} \quad \sin\beta = \sqrt{k_x^2 + k_y^2} \quad \cos\beta = k_z \tag{4.46}$$

The final formula is:

$$\mathrm{Rot}(k,\theta) = \begin{bmatrix} k_x^2(1-\cos\theta)+\cos\theta & k_xk_y(1-\cos\theta)-k_z\sin\theta & k_xk_z(1-\cos\theta)+k_y\sin\theta \\ k_xk_y(1-\cos\theta)+k_z\sin\theta & k_y^2(1-\cos\theta)+\cos\theta & k_yk_z(1-\cos\theta)-k_x\sin\theta \\ k_xk_z(1-\cos\theta)-k_y\sin\theta & k_yk_z(1-\cos\theta)+k_x\sin\theta & k_z^2(1-\cos\theta)+\cos\theta \end{bmatrix} \tag{4.47}$$

Because, in practice, it is interesting to determine the vector k and the angle θ, knowing the R rotation matrix:

$$R = \begin{bmatrix} n_x & n_y & n_z \\ o_x & o_y & o_z \\ a_x & a_y & a_z \end{bmatrix} \tag{4.48}$$

and according to [1]:

$$\theta = \arccos\left(\frac{n_x + o_y + a_z - 1}{2}\right) \quad k = \frac{1}{2\sin\theta}\begin{bmatrix} a_y - o_z \\ n_z - a_x \\ o_x - n_y \end{bmatrix} \tag{4.49}$$

In the case of axis/angle representation the rotation is described by four parameters, that is, the angle and three components of the rotation axis unit vector [4]. D–H notation identifies four parameters: θ_i—rotation around the z_{i-1} axis, λ_i—displacement, offset along the z_{i-1} axis, l_i—distance between z_{i-1} and z_i axes measured along the common perpendicular axis, that is, the x-axis, ai rotation angle of the z_i-axis in relation to $z_{i\text{-}1}$, as the relative rotation to the x-axis (Fig. 4.54). In the case of a rotational pair the angle-θ_i is variable and offset λ_i is constant [16, 18, 20, 73, 74, 81].

Homogeneous transformation matrices, corresponding to listed parameters, can be presented in the form of Eqs. (4.50)–(4.53).

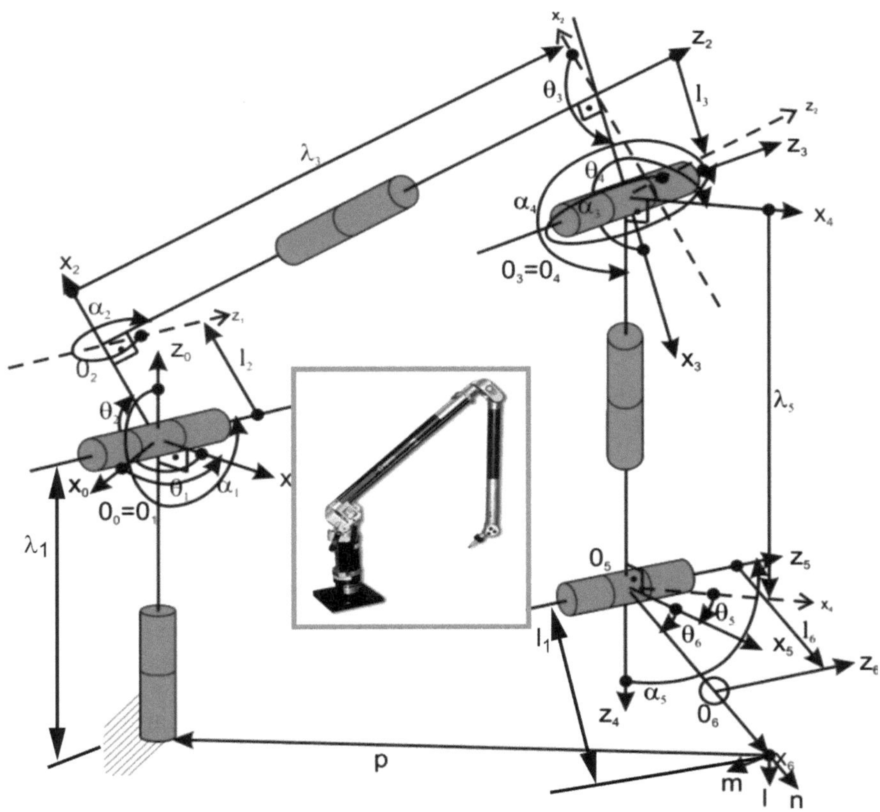

Fig. 4.54 AACMM kinematic scheme, in D–H notation [101, 159]

$$A(z_{i-1},\theta_i)=\begin{bmatrix}\cos\theta_i & \sin\theta_i & 0 & 0\\ -\sin\theta_i & \cos\theta_i & 0 & 0\\ 0 & 0 & 1 & 0\\ 0 & 0 & 0 & 1\end{bmatrix} A(z_{i-1},\lambda_i)=\begin{bmatrix}1 & 0 & 0 & 0\\ 0 & 1 & 0 & 0\\ 0 & 0 & 1 & -\lambda_i\\ 0 & 0 & 0 & 1\end{bmatrix} \tag{4.50}$$

$$A(x_i,l_i)=\begin{bmatrix}1 & 0 & 0 & -l_i\\ 0 & 1 & 0 & 0\\ 0 & 0 & 1 & 0\\ 0 & 0 & 0 & 1\end{bmatrix} A(x_i,\alpha_i)=\begin{bmatrix}1 & 0 & 0 & 0\\ 0 & \cos\alpha_i & \sin\alpha_i & 0\\ 0 & -\sin\alpha_i & \cos\alpha_i & 0\\ 0 & 0 & 0 & 1\end{bmatrix} \tag{4.51}$$

By multiplying these matrices the position and orientation matrix can be obtained which transforms the i-th coordinate system in the $(i-1)$ system in the form:

$$A_i^{i-1}=\begin{bmatrix}\cos\theta_i & -\cos\alpha_i\sin\theta_i & \sin\alpha_i\sin\theta_i & l_i\cos\theta_i\\ \sin\theta_i & \cos\alpha_i\cos\theta_i & -\sin\alpha_i\cos\theta_i & l_i\sin\theta_i\\ 0 & \sin\alpha_i & \cos\alpha_i & \lambda_i\\ 0 & 0 & 0 & 1\end{bmatrix} \tag{4.52}$$

where

$$A_i^{i-1}=\begin{bmatrix}B & & & p\\ 0 & 0 & 0 & 1\end{bmatrix} B_i=\begin{bmatrix}n_i & o_i & a_i\end{bmatrix}=\begin{bmatrix}n_{ix} & o_{ix} & a_{ix}\\ n_{iy} & o_{iy} & a_{iy}\\ n_{iz} & o_{iz} & a_{iz}\end{bmatrix} p_i = \begin{bmatrix}p_{ix} & p_{iy} & p_{iz}\end{bmatrix}^T \tag{4.53}$$

where
B_i orientation matrix,
P_i position matrix [16, 18, 50].

The forward kinematic consists in calculation of the position and orientation of the coordinate system $(z+1)$ that is next in relation to the reference system.

Using the definition of a homogeneous transformation matrix they can be presented in the form of:

$$A_i^{i-1}=A_1^0A_2^1A_3^2\ldots A_i^{i-1} \tag{4.54}$$

The inverse kinematic task is based on Eq. (4.54). The left side of this equation defines the location and orientation of the measuring stylus relative to the basic system and it is given; that is, the location and orientation of the measuring tip in a certain temporary position is known. The right side is the product of homogeneous matrices determined for each node from the base of AACMM to their measuring stylus [101, 159]. This task involves the determination of generalized locations (i.e., connects the particular AACMM degrees of freedom) and is hard to realize due to

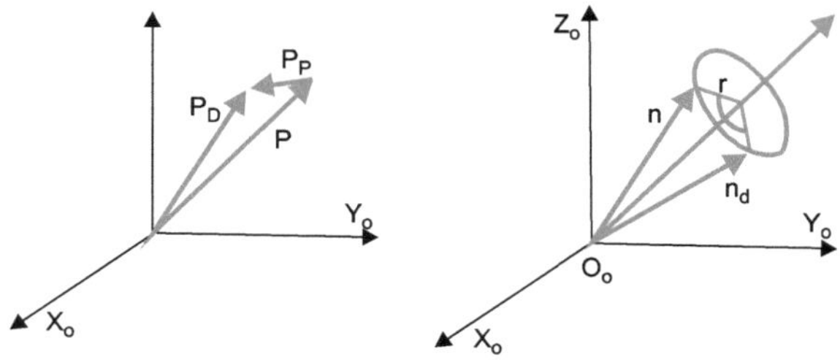

Fig. 4.55 Location error (**a**); orientation error definition (**b**)

the nonlinearity of the equations, for which finding a clear solution is just difficult. In the case of redundant systems, such as AACMM, the answer can have an infinite number of solutions.

The location error is expressed by the equation:

$$\overline{p_p} = \overline{p_d} - \bar{p} \tag{4.55}$$

This is illustrated in Fig. 4.55a, where the vector $\overline{p_l}$ is the difference between the vector $\overline{p_d}$ (determined as a result of the simulation), and $\bar{p}$ indicated by the machine.

In the case of determining the orientation error (Fig. 4.55b) the formula (4.54) is used, and as a result, a measuring stylus orientation matrix determined relative to the basic system is obtained. Finally, the formula for the calculation of the rotation errors must be found, that is, the difference between the set rotation and the actual one. In general, to determine the rotation three vectors ($\bar{\boldsymbol{o}}\,\bar{\boldsymbol{a}}\,\bar{\boldsymbol{n}}$) can be used, defined by formula (4.54). The set orientation is marked by R_d, and the actual one by R. In the case of the software (AACMM) used, only the $\bar{\boldsymbol{n}}$ vector can be obtained directly which is consistent with the location of the measuring tip; therefore in Fig. 4.55b unit vectors o, a, are omitted to make the drawing more clear [101, 159]. The orientation error can be calculated based on formula (4.56):

$$\bar{e} = \bar{k} \sin\theta \tag{4.56}$$

A method for the orientation determination is presented in Fig. 4.55b. The basic coordinate system is also marked (as in Fig. 4.53), vectors $\bar{n}$ and $\overline{n_d}$, unit vector $\bar{\boldsymbol{k}}$, and the rotation angle θ around it. It was also assumed that vectors $\overline{n_d}$ and $\bar{n}$ have their beginnings in the point o (Fig. 4.51a). Then this error must be expressed by orientation matrices R_d and R. Because $\bar{\boldsymbol{k}}$ is a unit vector designating the rotation axis and the angle θ is the equivalent rotation angle between $\boldsymbol{n}\ \boldsymbol{o}\ \boldsymbol{a}$ and $\boldsymbol{n_d}\ \boldsymbol{o_d}\ \boldsymbol{a_d}$ systems, that is, between $\bar{n}$ and $\overline{n_d}$ vectors, then the orientation error is calculated by the formula:

$$\mathrm{Rot}(k, \theta) = R_d R^T \tag{4.57}$$

This formula describes the transformation that must be performed to obtain the coverage of R i Rd systems, and in the analyzed case—$\bar{n}$ i $\overline{n_d}$ vectors. Performing this transformation, the right side of the equation must be determined (4.54) by writing vectors in the form of matrix (3 × 1) according to the vector and matrix calculus. In order for the two systems or vectors to overlap, they must meet the following assumption according to vector calculus:

$$R_d R^T = I \tag{4.58}$$

Marking the set vector by $\bar{n} = \begin{bmatrix} n_x & n_y & n_z \end{bmatrix}$, and by $\overline{n_m} = \begin{bmatrix} n_{xm} & n_{ym} & n_{zm} \end{bmatrix}$, the vector of the actual AACMM tip location obtained directly from the AACMM software system, and after their multiplication, a matrix (3 × 3) is obtained:

$$nn_m^T = \begin{bmatrix} n_x n_{xm} & n_x n_{ym} & n_x n_{zm} \\ n_y n_{xm} & n_y n_{ym} & n_y n_{zm} \\ n_z n_{xm} & n_z n_{ym} & n_z n_{zm} \end{bmatrix} \tag{4.59}$$

The next step is to present the formula (4.58) using formulas (4.59), left side, and (4.59), right side. By equating the sum of diagonal terms of both matrices and using the assumption:

$$k_x^2 + k_y^2 + k_z^2 = 1 \tag{4.60}$$

it was given:

$$(1 - \cos\theta)(k_x^2 + k_y^2 + k_z^2) + 3\cos\theta = n_x n_{xm} + n_y n_{ym} + n_z n_{zm} \tag{4.61}$$

and after simplification:

$$\cos\theta = \frac{n_x n_{xm} + n_y n_{ym} + n_z n_{zm} - 1}{2} \tag{4.62}$$

Subtracting out the diagonal terms (1.2) and (2.1) of both matrices we obtain:

$$k_x k_y (1 - \cos\theta) + k_z \sin\theta - k_x k_y (1 - \cos\theta) + k_z \sin\theta = n_y n_{xm} - n_x n_{ym} \tag{4.63a}$$

then after simplification:

$$k_z \sin\theta = \frac{n_y n_{xm} - n_x n_{ym}}{2} \tag{4.63b}$$

Next by subtracting out the diagonal terms (3.1) and (1.3) we obtain:

$$k_x k_z(1-\cos\theta) + k_y \sin\theta - k_x k_z(1-\cos\theta) + k_y \sin\theta = n_x n_{zm} - n_z n_{xm} \tag{4.63c}$$

where after simplification we obtain:

$$k_y \sin\theta = \frac{n_x n_{zm} - n_z n_{xm}}{2}. \tag{4.63d}$$

Next by subtracting out the diagonal terms (2.3) and (3.2) of both matrices we obtain:

$$k_y k_z(1-\cos\theta) + k_x \sin\theta - k_y k_z(1-\cos\theta) + k_x \sin\theta = n_z n_{ym} - n_y n_{zm} \tag{4.63e}$$

and after simplification it is:

$$k_x \sin\theta = \frac{n_z n_{ym} - n_y n_{zm}}{2} \tag{4.63f}$$

Using the formula:

$$\sin^2\theta + \cos^2\theta = 1 \tag{4.64}$$

we get

$$\sin\theta = \sqrt{1 - \frac{1}{4}(n_x n_{xm} + n_y n_{ym} + n_z n_{zm})^2} \tag{4.65}$$

According to [150], an orientation error formula can be given in the following form, which is expressed by rotation pseudo-vector:

$$\bar{e} = \bar{k}\sin\theta = \begin{bmatrix} k_x \sin\theta \\ k_y \sin\theta \\ k_z \sin\theta \end{bmatrix} = \begin{bmatrix} \frac{n_z n_{ym} - n_y n_{zm}}{2} \\ \frac{n_x n_{zm} - n_z n_{xm}}{2} \\ \frac{n_y n_{xm} - n_x n_{ym}}{2} \end{bmatrix} = \frac{1}{2}\begin{bmatrix} n_z n_{ym} - n_y n_{zm} \\ n_z n_{xm} - n_x n_{zm} \\ n_y n_{xm} - n_x n_{ym} \end{bmatrix} \tag{4.66}$$

It is a searched relation for calculating the orientation error. It also should be noted that the restriction for the rotation angle θ implies the following conditions for the scalar product $n^T n_m \geq 0$. Of course if vector $\bar{n}$ corresponds to a vector $\overline{n_m}$, then the orientation error is equal to 0.

4.5.3 Kinematic Model (KmAACMM) Parameter Identification

In order to identify parameters of the real AACMM for construction purposes of the arm model, it is necessary to carry out measurements of real geometric dimensions of particular arm segments. Data given by the producer in accordance with Fig. 4.57

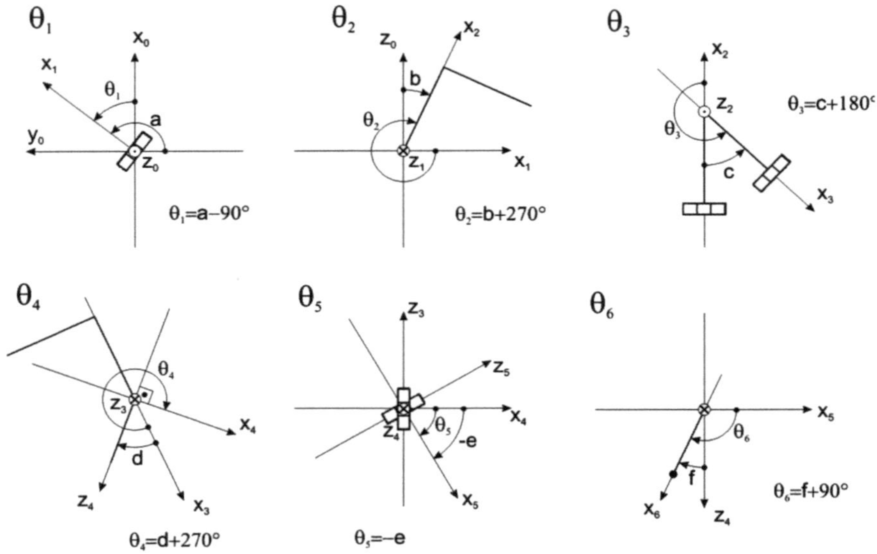

Fig. 4.56 Relations between model configuration angles (θ_1, θ_2, θ_3, θ_4, θ_5, θ_6), and angles calculated from encoder readings (*a*, *b*, *c*, *d*, *e*, *f*)

turned out to be not very accurate, making it impossible to rely on them during calculations of forward kinematics. After analysis it turned out that the arm software takes zero point associated with its base—mounting—in a place other than specified in the documentation, that is, significantly higher. Therefore it was necessary to perform the measurement of arm segments on a PMM12106 machine and to correct the given dimensions of particular segments, such as λ_1 dimension, Fig. 4.54. Angles θ_1 – θ6 (Fig. 4.54) gain variable values corresponding to location angles of one segment in relation to the previous one, and constitute configuration variables of building the model. These are directed angles, and clearly defined, a model description in D–H coordinates. The values of these angles are determined by dividing the values obtained from the encoder by its resolution. The starting point is determined by the point of the encoder zeroing, and the turn by the producer.

Observations of each encoder reading result in the relations between configuration coordinates and angles read from encoders shown in Fig. 4.56.

However, it should be noted, that "zero" angles for encoders do not need to comply with each rotating pair configuration presented in Fig. 4.56. These are zero positions in the project assumptions. Their real (zero) positions depend directly on the encoder mounting in arm joints and therefore may be subjected to additional error in relation to the project assumptions.

This error can be so big that it should be taken into account. Thus, the angles $\theta_1 - \theta_6$ should be considered as the sum of the angle calculated from encoder readings and the angle of "zero movement," the value of which must be additionally determined:

$$\theta i = \Theta i + \xi i \tag{4.67}$$

For the analysis of the model correctness and to verify the accuracy of AACMM indications a set of reference points in the arm measuring volume was used. Each point indication of each encoder was read, and also the point coordinates and the orientation of a stylus. It is a set of several points of the arm work volume obtained for various configurations of arm kinematics. These data were collected shortly after arm periodic correction and read for each point of the articulated arm location, so that configuration variables (encoders) are the only factor that affects the location and orientation of reference points. That is why the AACMM model is based on data from encoders, the position, and the orientation of its measuring tip.

The kinematic model, and the next model based on it (e.g., metrological model), as described above, must include the accurately determined geometric lengths of the particular segments to avoid errors resulting from summation of errors of particular segments.

Data for KmAACMM construction must be obtained at locations with a fixed next arm segment. It is also important to determine encoder zero points, because according to what is stated above, they can insignificantly change their position (it refers also to the whole encoder). It has a huge impact on the geometric accuracy of the model built on the basis of vectors of a given beginning and ending, and forming a closed polygon. Beginnings and endings of these vectors are not determined by the length of particular elements, but rather by centers of encoders. To determine lengths of each element, encoder indications are read, as well as coordinates and the measuring tip direction which is provided directly from the producer software. Readings must be taken so that they cover the whole measuring volume, in a series of 17 readings. Such series should be at least eight (four on the same level in each quadrant of the circle, i.e., the measuring volume, and then four on the level above, about 30–60 cm, depending on the AACMM model), as shown in Fig. 4.57. In order to determine zero points of encoders and to verify their indications the counterweight should be removed and elements B and C, and afterwards D and E, should be blocked (Fig. 4.49). This operation should be repeated at least 25 times. The dispersion of encoder indications must be within the measurement uncertainty of each encoder (e.g., for ROMER Sigma 2025 arm indications cannot

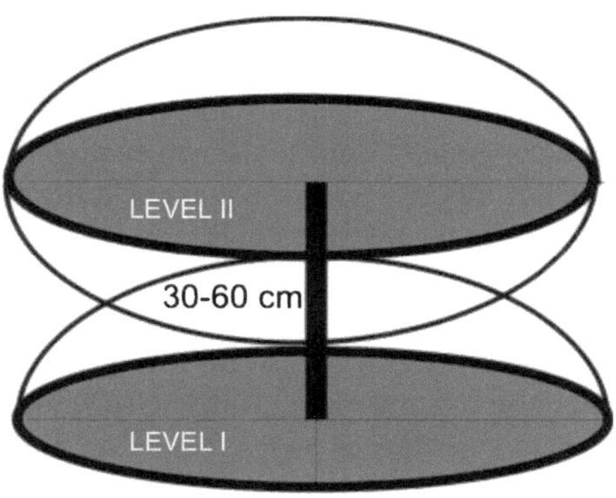

Fig. 4.57 AACMM position during parameter identification

exceed 5 points, i.e., about 7′). For KmAACMM construction 17 variables are necessary (that is why one series consists of 17 measurements, as much as is needed to build a system of equations), and these are displacements of encoders zero, six lengths for the x variable and six lengths for the z variable. Then systems of equations with 17 variables are set for each unknown length as the average from the obtained results, obtaining specified geometric lengths of particular segments, together with the uncertainty of their determination. Calculation of AACMM geometric parameters is made, as shown above, based on readings from encoders for particular locations and orientations of the measuring tip. Of course, for a single configuration of arm segments, there are an infinite number of possible solutions; that is why a system of equations, which will indicate enough positions, should be created, so that only one solution will correspond to each configuration. Because the initial measurements showed that most of the articulated arm model parameters are different from the data given by the producer, the previously accepted model should be verified, especially values of angles and distances taken as zero. An important advantage of the determination of model parameters in this way is its dependence on the current arm calibration corrections. This is very important, because according to AACMM producer recommendations (ROMER in this case) they must be corrected every six months due to the time propagation of indication errors or when exceeding the maximum permissible error during periodic calibration performed by an accredited laboratory. Of course, the need for a correction performance (by the producer) entails the necessity to "correct" the model. The advantage of the determination of model parameters based on data from Romosoft software is that corrected parameters can be calculated relatively quickly by substituting data saved in an Excel file. In order to use the model, for example, as a correction system of arm accuracy (articulated arm computer-aided accuracy AA CAA), or as the basis for the planned metrological model, virtual AACMM, the data must be online saved into a file in a format accepted by the model. Thanks to this solution for the AACMM model the accuracy can be adopted such as declared, for example, by the producer, on the basis of tests of A, B, and C types [2].

Calculation of such an extensive system of equations originally chosen for this task, the Mathcad system, turned out to be difficult. A major restriction is the accuracy of calculations. Therefore, in system building many more equations describing the Z coordinate values were used. What is more, these equations were much simpler and more of them could be used without overloading the program. The key to getting the optimal result turned out to be a continuous verification of error location for all available AACMM configurations and addition of locations with the greatest error to the solved system of equations. The results obtained on the basis of different sets of equations differ from each other by a few hundredths of a millimeter for distances and by 0.0001 radian for angles.

Due to the described difficulties connected with the use of the Mathcad program it was necessary to develop an integral system realizing the kinematic model of KmAACMM. Such software was developed [159] in programming environment C ++, as an integral model KmAACMM. In Fig. 4.58 the program is presented as well as hardware dependencies that led to KmAACMM model development.

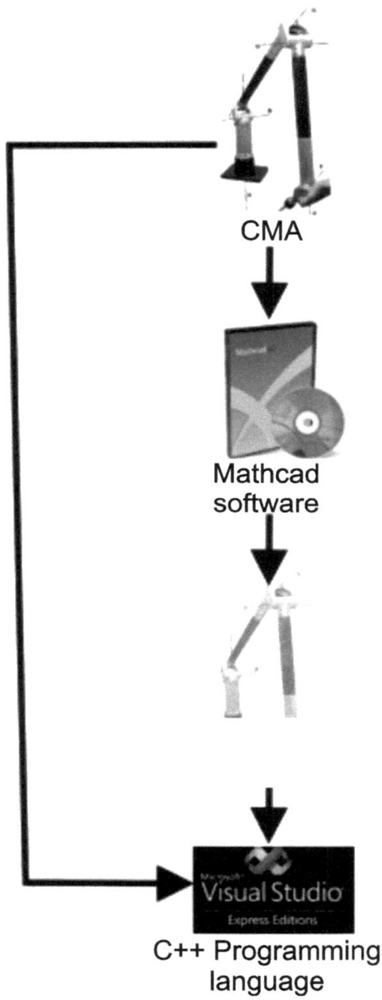

Fig. 4.58 Order of AACMM kinematic model KmAACMM program realization

This software, based on the data from encoders, determines the *x*, *y*, *z* coordinates and may simulate AACMM real work or correct its accuracy.

4.5.4 Visualization and Correctness Assessment of Kinematic Model (KmAACMM)

The best way to assess the correctness of the kinematic model, and also a way to visualize the model behavior together with its comparison with the real object, turned out to be the 3-D model of the articulated arm coordinate measuring machine

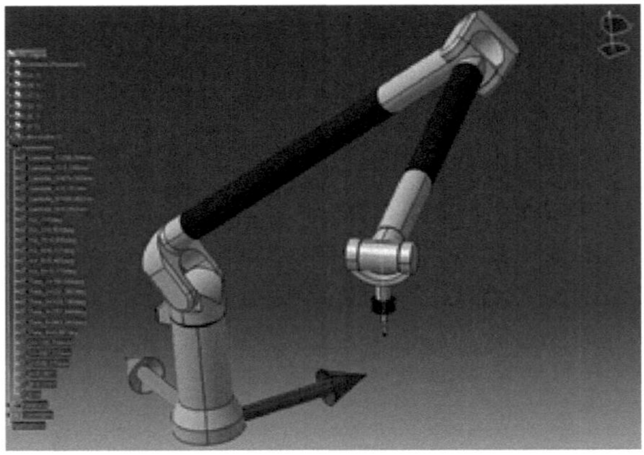

Fig. 4.59 AACMM parametric model in the Catia v5 AACMM R17 [101, 159]

built in the Catia V5 R17 program [47, 78, 101, 159]. The Catia program also allowed attaching the results obtained during measurement, and generated using a kinematic model KmAACMM, in the form of a table, and parameterized all dimensions included in the D–H model. All this makes KmAACMM verification easier (Fig. 4.59).

Thanks to the DMU Kinematics module the behavior of the model and the real arm can be compared; error analysis is also available due to comparison of received and nominal orientation. This allowed the isolation of error source components and their corrections. For articulated arm coordinate measuring machines whose elements—tubes–are made of aluminum, a methodology to identify the thermal influences on parameters of the built kinematic model was developed. For the arm, for which thermal influences were tested, a "temperature distribution identification frame" was created; that is, temperature sensors were distributed in its measuring volume, as well as on its component elements (Fig. 4.60). The temperature was measured and read by a set sequence for whole measuring volume, at the time of arm indication testing. The procedure was repeated for the other temperature which is changed by the air-conditioning system. The obtained results for the assumed temperature range are the basis of changes in the developed kinematic model by their introduction to an appropriate software module of KmAACMM. In the case of the ROMER Sigma arm, made of carbon fiber (tubes connecting the particular rotary elements), for which the thermal expansion is low (in the temperature range of 16–30 °C), a clear effect on the accuracy was not obtained, so the developed system was not used for identification of thermal influences.

In Figs. 4.61 and 4.62 developed models of KmAACMM for two articulated arms are presented: ROMER's Sigma and Multi Gauge. These models respond identically as real articulated arm coordinate measuring machines and repeat given locations of kinematic sets allowing online model assessment.

Verification results developed for the orientation error determination task for the ROMER Sigma modeled arm are posted in Fig. 4.63. All results are in the range of 0.6–0.8 μm.

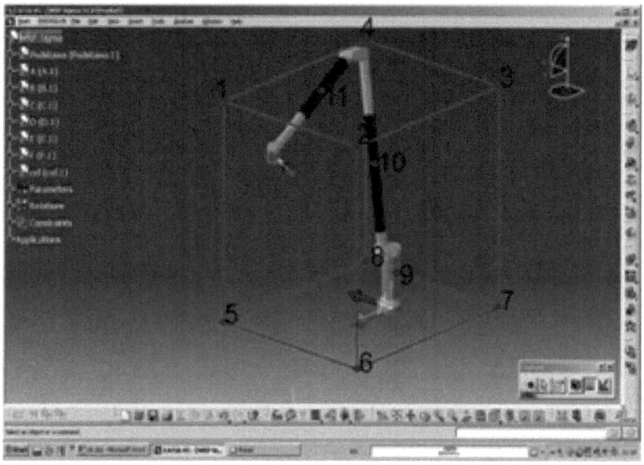

Fig. 4.60 Distribution of temperature sensors [101, 159]

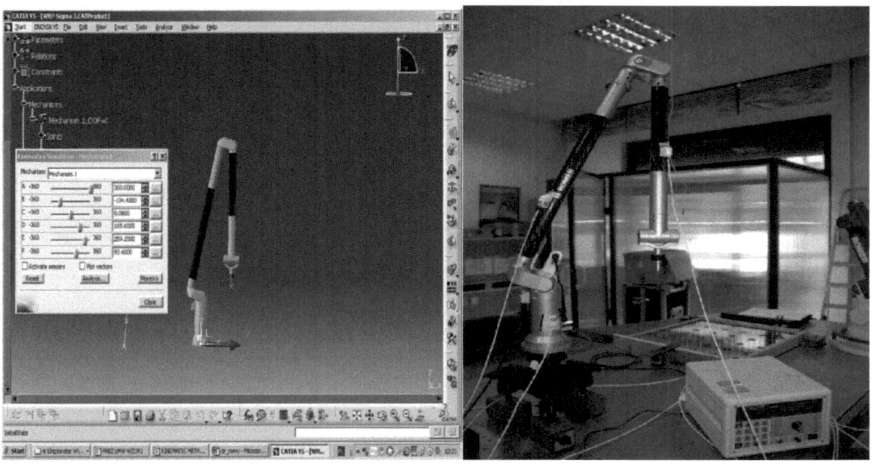

Fig. 4.61 Kinematic model in Catia v.5 program in DMU Kinematics module for ROMER Sigma arm [101, 159]

Another procedure showing a part of the verification tests realized under [101, 159], was the development of correction procedure AA CAA for the Multi Gauge arm. Tests verifying the developed correction system AA CAA consisted of determination of reference point coordinates in three-ball sockets located on the standard made of carbon fiber (ball-bar type), for which the calibration procedure was carried out on PMM12106. In Fig. 4.64 the indication error determination results are shown both for an uncorrected system and after performing the correction involving the model KmAACMM activation. Significant reduction of

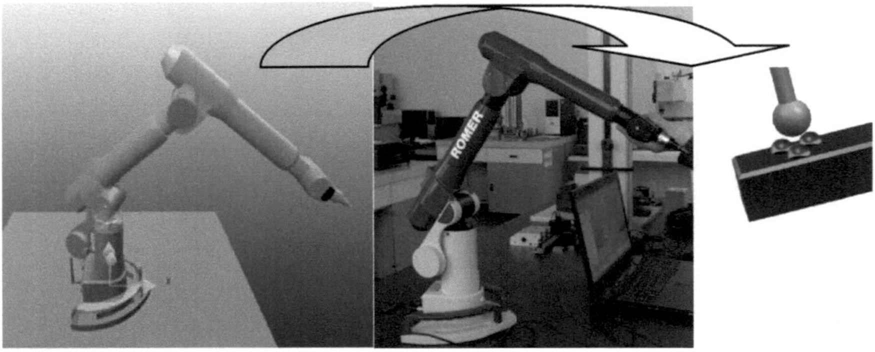

Fig. 4.62 Kinematic model of the Multi-Gage arm. View of reproduction particular arm segment arrangement during real and simulated measurement; view of a standard with three-point sockets mounted on a standard of ball-bar type to test the indication error of modeled arm [101, 159]

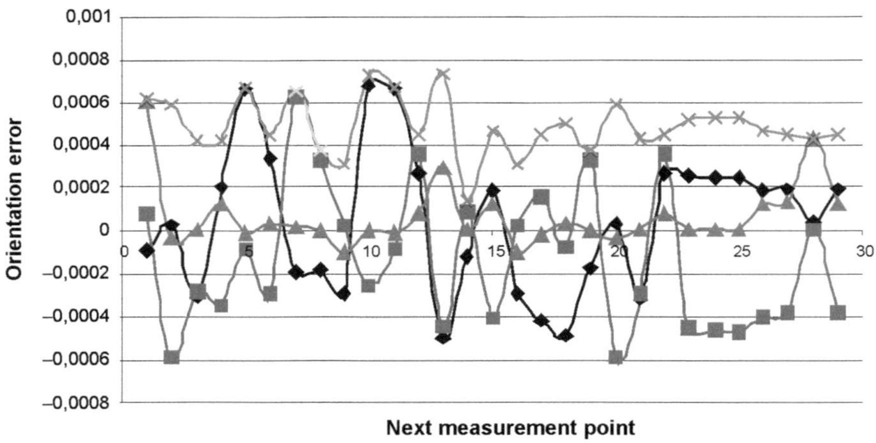

Fig. 4.63 Orientation error graph (*filled diamond* variable *x*, *filled square* variable *y*, *filled triangle* variable *z*, *multiplication sign* orientation error): values given in mm [101, 159]

indication errors was obtained. This confirms that the model was properly developed.

The developed AACMM kinematic model (KmAACMM) in the form of integral software KMAACMM C++ [208] can be easily implemented in all types of AACMM. However, to make the system operate in realtime, it is necessary that producers provide an indication of the encoder position under the given configuration, the location of a particular arm's segments, which should be sent online to developed software. The built model has significant scientific importance; it shows and disseminates the concept of CAA AA system construction and enables its

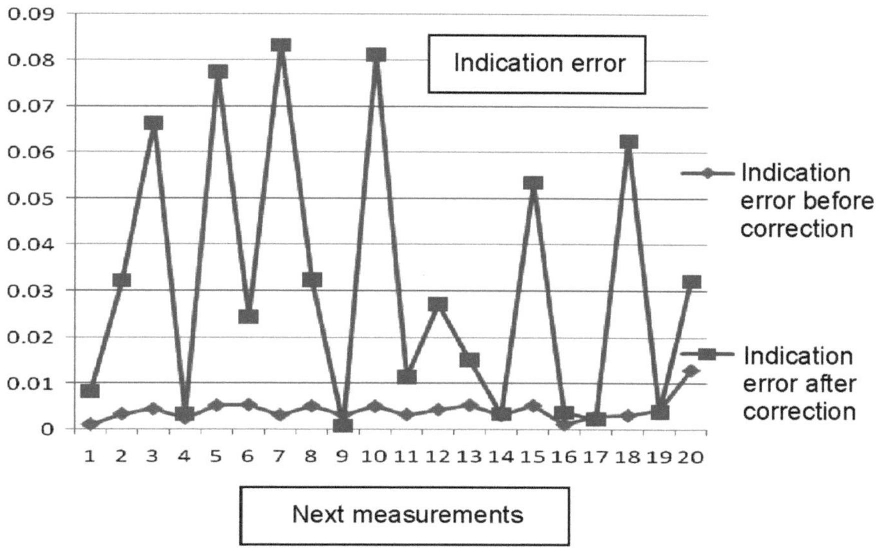

Fig. 4.64 Chart of indication error before and after correction of AACMM model [101, 159]

implementation by any producer. It also has significant importance for the development of arm calibration procedures. However, the most important use of the developed model is first of all a construction of a metrological model, a virtual AACMM with application to the online assessment of a realized measurement accuracy task. The simulation system developed in LCM CUT is described in Sect. 5.2.6.

References

1. Adamczak, S., Janecki, D., Stępień, K.: Investigating the accuracy of cylindricity deviation measurment on CMM's using the cross-correlation function. VIIth International Scientific Conference Coordinate Measuring Technique. Bielsko-Biała 2006. University of Bielsko-Biała. Scientific Books 22
2. ASME B89.4.22 2004: Methods for Performance Evaluation of Articulated Arm Coordinate Measuring Machines
3. Balsamo, A., Pedone, P., Ricci, E., Verdi, M.: Low-cost interferometric compensation of geometrical errors. CIRP Ann. Manufact. Technol. **58**(1), 459–462 (2009)
4. Barker, L.K.: Modified denavit-hartenberg parameters for better location of joint axis systems in robot arms. NASA Technical Paper 2585, INASA-TP–2585 19860018481 (1986)
5. Bartscher, M., Busch, K., Schwenke, H., Trapet, E., Wäldele, F.: Artefact based traceability concept for large coordinate measuring machines (CMMS). In: Proceedings of the 9th METROLOGIE Conference, pp. 158–161, Bordeaux (Frankreich), ISBN 2–909430–88X, (1999)

6. Bartscher, M., Busch, K., Franke, M., Schwenke, H., Wäldele, F.: New artifacts for calibration of large CMMs. In: ASPE, Proceedings of the 5th Annual Meeting, pp. 542–546, Scottsdale (Arizona) (2000)
7. Belforte, G., Bona, B., et al.: Coordinate measuring and machine tools self-calibration and error correction. Ann CIRP **36**(1), 359–364 (1987)
8. Bendat, J.S., Piersol, A.G.: Random Data: Analysis and Measurement Procedures. Wiley, New York (1971)
9. Bosch, J.A.: Coordinate Measuring Machines and Systems. Marcel Dekker Inc., New York, Basel, Hong Kong (1995)
10. Breyer, K.H.: Längenmessunsicherheit. VDI-B **378**, 21–28 (1980)
11. Breyer, K.H.: Prüfen von Mass, Form und Lage mit Koordinatenmessgeräten. VDI-B 529/84
12. Brudekin, M., Lee, E.S.: A hole-plate artifact design for the volumetric error calibration of CMM. Int. J. Adv. Manuf. Technol. **17**, 508 (2001)
13. Brudekin M., Lim C.K.: Rapid volumetric calibration of coordinate measuring machines using a hole bar artefact. Proc. Instn. Mech. Engrs. Part B: J. Eng. Manuf. 216, 1083–1093 (2002)
14. Busch, K.: Überwachung von Koordinatenmessgeräten. VDI-B **751**, 51–62 (1989)
15. Busch, K., Kunzmann, H., Wäldele F.: Numerical Error Correction of CMM. In: Proceedings of International Symposium on Metrology for Quality Control in Production, Tokyo, pp. 270–282 (1984)
16. Chablat, Ph, D., Wenger, P., Majou, F., Merle J.P.: An interval analysis based study for the design and the comparison of 3-DOF parallel kinematic machines. Int. J. Rob. Res. **23**, 6.15–624 (2004)
17. Castro, H.F.F., Burdekin, M.: Evaluation of the measurement uncertainty of a positional error calibrator based on a laser interferometer. Int. J. Mach. Tools Manuf **45**, 285–291 (2005)
18. Craig, J.J.: Wprowadzenie do robotyki, Mechanika i sterowanie" (Intoduction to robots: Mechanics and Control) Wydawnictwo Naukowo –Techniczne (Scientific and Technical Publishing) Warsaw (1995)
19. Curran, E., Phelan, P.: Quick check error verification of coordinate measuring machines. J. Mater. Process. Technol. **155–156**, 1207–1213 (2004)
20. Denavit, J., Hartenberg, R.S.: A kinematic notation for lower-pair mechanisms based on matrices. Trans. ASME J. Appl. Mech. **23**, 215–221 (1955)
21. DEA, www.dea.it
22. Dobosz, M., Woźniak, A.: Metrological feasibilities of CMM touch trigger probes. Part II. Experimental verification of the 3D theoretical model probe pretravel. Measurement **34**(4), 287–299 (2003)
23. Dobosz, M., Woźniak, A.: CMM touch trigger probes testing using a reference axis. Precision Eng. **29**(3) 281–289 (2005)
24. Donker, R., Widdershoven, I., Spaan, H.: ISARA 400: Alarge measurement volume ultra-precision CMM—advances in coordinate metrology, monograph. In: Sładek, J., Jakubiec, W. (eds.), University of Bielsko-Biała pp. 50–56 (2010)
25. Duffie, N.A., Yang, M.; Generation of parametric kinematics error-correction functions from volumetric error measurements. Ann. CIRP **34**(1), 435–438 (1985)
26. Duffie, N.A., Malmberg, E.: Error diagnosis and compensation using kinematics models and position error data. Ann. CIRP **36**(1), 355–358 (1987)
27. ETALON AG, www.etalon.de
28. Estler, W.T., Edmundson, K.L., Peggs, G.N., Parker, D.H.: Large-scale metrology—an update. CIRP Ann. Manuf. Technol. **51**(2), pp. 587–609 (2002)
29. Feinmess GmbH & Co. KG, www.feinmess.com
30. Flack, D.: CMM verification. Measurement Practice Guide no.42. National Phisical Laboratory (2001)
31. Forbes, A.B., Peggs, G.N.: A large reference artefact for cmm verification. In: Ford, D.G., Postlethwaite, S.R. (eds.) Laser Metrology and Machine Performance III, pp. 394–400. Computational Mechanic Publications, Southampton-Boston (1997)

32. Furutani, R., Kamahora, K.: Development of a new artifact for the calibration of large scale instruments. Measurement **30**, 139–143 (2001)
33. Furutani, R., Ozaki, M.: Uncertainty of calibration of 2D planar coordinate measuring machine. IXth International Scientific Conference, Coordinate Measuring Technique, Bielsko-Biala—Ustroń,14th–16th of April 2010
34. Gąska, A.: Modeling of accuracy of coordinate measurement with use of monte carlo method. Ph.D. Thesis Dissertation, Cracow University of Technology, Faculty of Mechanical Engineering. Thesis Advisor—Prof. J.A. Sladek (2011)
35. Geus, D.A.: Comparison of measurement data from 3D-measurement. In: Ossana, P.H., Postrednik, D., Durakbasa, N.M. (eds.) Metrology for quality control in production. In: Proceedings of 6th ISMQC IMEKO Symposium, pp. 181–187, Wien 8–10 Sept 1998
36. Gawlik, J., Kowalski, M., Ryniewicz, A.: Analysis of measurement of precision machines guides. In: Kunzmann, H., Waeldele, F., Wilkening, G., Corbet, J., MacKeow, P., Weck, M., Huemmler, J. (eds.) 4th International Conference on Ultraprecision in Manu-facturing Engineering. In Progress in Precision Engineering and Nanotechnology, vol. 1, pp. 319–322. PTB-Braunschweig u. Berlin P.u.Oe–Verlag (1997)
37. Główny Urząd Miar—Central Office of Measures (Warsaw 1999) Guide to the Expression of Uncertainty in Measurement (GUM) ISO 1993/1995 with the addition to polish edition of J. M. Jaworski
38. Halupczok, E.: Definicja dokładności współrzędnościowej techniki pomiarowej na przykładzie maszyny UPMC CARAT (Definition of coordinate metrology accuracy on the example of UPMC CARAT machine),Carl Zeiss Industrielle Meßtechnik GmbH, Esslingen
39. Hamana, H., Tominaga, M., Ozaki, M., Furutani R.: Calibration of articulated arm coordinate measuring machine considering measuring posture. 10th International Symposium on Measurement and Quality Control, Osaka Japan, September 5–9 (2010)
40. Hansen, H.N., De Chiffre, L.: An industrial comparison measuring machines in Scandinavia with focus on uncertainty stastements. Precision Eng. **23**, 185–195 (1999)
41. Herbst, C, Tutsch, R.: Simulation of micro coordinate measuring machine for parallel measurement of microstructures. In: Sładek, J., Jakubiec, W. (eds.) Advances in coordinate metrology, pp. 15–24. University of Bielsko-Biała (2010)
42. Hernla, M.: Unsicherheiten einfach abschaetzen. QZ Qualitaet und Zuveraessigkeit, **45** (2000)
43. Hexagon Metrology, www.hexagonmetrology.pl
44. Hocken, R.: Tree-dimensional metrology. Ann. CIRP **26**(2), 378–382 (1977)
45. Hueser-Teuchert, D., Trapet, E., Wäldele, F., Wiegand, U.: Kalibrierung von Koordinatenmessgeräten mit Kugel – und Lochplatten Anleitung für DKD-laboratorien. Preprint PTB, Braunschweig (1993)
46. IBS Precision Engineering, www.ibspe.com
47. Investigation CMM ISO/VDI. Microsoft Excel Application Cracow University of Technology Lab of Coordinate Metrology (2000)
48. ISO 10360.: Part 2—CMMs used for measuring linear dimensions, 2n edn, pp. 2–01 (2003)
49. Iwasińska, O., Ratajczyk, E.: Badanie dokładności maszyn hybrydowych na przykładzie współrzędnościowej maszyny pomiarowej ScanMax (Hybrid machines accuracy tests on the example of ScanMax coordinate measuring machine) IVth International Scientific Conference—Coordinate Measuring Technique, Lodz University of Technology Scientific Books, branch in Bielsko-Biała, no. 53, Bielsko-Biała, pp. 106–114 (2000)
50. Jakubiec, W., Starczak, M.: Analysis of inspection accuracy of large CMMs by means of bar-ball gauges. Second International Workshop on CMM Calibration. Czech Metrology Institute, Prague (2001)
51. Jakubiec, W., Malinowski, J.: Metrologia wielkości geometrycznych (Metrology of geometrical quantities) WNT (Scientific and Technical Publishing), Warsaw (2007)
52. Knapp, W.: Test of the tree-dimensional uncertainty of machine tools and measuring machines and its relation to the machine errors. Ann. CIRP **32**(1), 459–464 (1983)

53. Knapp, W.: Circular test for tree-coordinate measuring machines and machine tools. Prec. Eng. **5**(3), 115–124 (1983)
54. Knapp, W., Control of the accuracy of multi-coordinate measuring machine. UNO/ENE Seminar on Present Use and Prospects for Precision Measuring Instruments and Engineering Industries. Dresden 20–24 Sept 1982
55. Knapp, W.: Anwendung eines zylindrischen Normales zur Kontrolle eines Mehrkoordinaten-Messgeräten. PTB, Bericht Me **38** (1983)
56. Knapp, W.: Vorschlag zur Abnahme von Drei-Koordinaten. Meßgeräten. Dissertation ETH-Zürich (1984)
57. Knapp, W., Bucher, T.: Comparison of different artefacts for interim coordinate measuring machine checking: a report from the swiss standards committee. Precision Eng. **13**, 277–282 (1991)
58. Koperska, A., Ratajczyk, E.: Porównanie testów dokładności współrzędnościowych ramion pomiarowych (Aticular Arm Coordinate Mesuring Machines accuracy tests comparison). In: Sładek, J., Jakubiec, W. (eds.) Advances in Coordinate Metrology. University of Bielsko-Biała, Bielsko-Biała (2010)
59. Kowalski, M.: Analiza pomiarów na współrzędnościowych maszynach pomiarowych obarczonych błędami kształtu (Analisys of measurements done with the use of coordinate measuring machines affected by form errors). Doctoral Dissertation, Cracow University of Technology (1993)
60. Kowalski, M.: Model wirtualnej wielowspółrzędnościowej maszyny pomiarowej i jej zastosowanie (Model of virtual multi- coordinate measuring machine and its application). In: Proceedings of VIth Scientific-Technical Conference Metrology in Production Engineering, pp. 218–223, Rzeszow University of Technology (1995)
61. Kowalski, M.: Interpretacja wyników badań dokładności WMP realizowanych płytami kontrolnymi (Interpretation of CMM accuracy tests results carried out by control panels). In: Proceedings of VIIth Scientific-Technical Conference. "Metrologia wtechnikach wytwarzania maszyn", Mechanika 63, vol. I, pp. 275–284. Kielce University of Technology, Kielce (1997)
62. Kowalski, M.: Wybrane aspekty realizacji modelu wirtualnej współrzędnościowej maszyny pomiarowej (Chosen aspects of realization of the virtual coordinate measuring machine model). Polish Academy of Sciences—KBM, Scientific Book no. 63, Works of Institute of Machine Technology and Production Automation, pp. 117–126. Cracow University of Technology (1999)
63. Kowalski, M.: Zastosowanie wirtualnej WMP do wyznaczania i minimalizacji błędów pomiarowych. (The use of virtual CMM for determintaion and minimization of measurement errors). In: Proceedings of VIIIth Scientific-Technical Conference Metrologia w technikach wytwarzania maszyn (Metrology in Machine Production), vol. II, pp. 401–408. Szczecin–Międzyzdroje 13–15 IX (1999)
64. Kowalski, M., Ryniewicz, A.: Metodyka badań temperatur maszyn współrzędnościowych (Coordinate machines temperature research methodology). In: Proceedings of VIIIth Scientific-Technical Conference Metrologia w technikach wytwarzania maszyn, vol. II, pp. 393–400. Szczecin–Międzyzdroje 13–15 IX (1999)
65. Kowalski, M., Sładek, J., et al.: Opracowanie wirtualnej wielowspółrzędnościowej maszyny pomiarowej z zastosowaniem do badań i korekcji błędów obiektów rzeczywistych i optymalizacji pomiarów (Development of virtual multi- coordinate measuring machine with the application for researches and correction of errors of real objects and for measurements optimization). Final Report of KBN Research Project PB 1367/T07/95/08, Cracow University of Technology, not published (1998)
66. Kowalski, M., Sładek, J., Rakoczy, R.: Opis wirtualnej współrzędnościowej maszyny pomiarowej (Decription of virtual coordinate measuring machine). In: IIIth International Scientific Conference—Coordinate Measuring Technique, pp. 71–80. Lodz University of Technology Scientific Books, branch in Bielsko-Biała, no. 44, Bielsko-Biała (1998)

67. Krawczyk, M.: Metoda oceny dokładności Współrzędnościowych Maszyn Pomiarowych z zastosowaniem Laserowego Wzorca Stopniowego (Coordinate measuring machines accuracy assessment method with the use of laser step standard). Ph.D. Thesis Dissertation. Cracow University of Technology, Faculty of Mechanical Engineering (Thesis Advisor - Prof. J.A. Sladek) (2005)
68. Kunzmann, H., Trapet, E., Wäldele, F.: A uniform concept for the calibration and periodic inspection of coordinate measuring machines using reference object. Ann. CIRP **36**, 561–564 (1990)
69. Kunzmann, H., Trapet, E., Wäldele, F.: Results of the international comparison of ball plate measurements in CIRP and WECC. PTB, Preprint Braunschweig (1994)
70. Kurtz, R., Hayward, V.: Multiple—Goal Kinamatic optimization of a parallel spherical mechanism with actuator redundancy. IEEE Trans. Robot. Autom. **8**(5) 1992
71. Lau, K.C., Hocken, R.J.: Tree and five axis laser tracking systems. US Patent No. 4714339 (1987)
72. Leica, www.metrology.leica-geosystems.com
73. Lenarcic, J., Thomas, F.: Robot Kinematics, Theory and Applications. Kluwer Academic Publisher, Dordrecht, The Netherlands. ISBN 1–4020-0696–9 (2002)
74. Lenarcic, J., Stanisić, M.M.: Advances In Robot Kinematics. Kluwer Academic Publisher, Dordrecht, The Netherlands. ISBN 0–7923–6426-0 (2002)
75. Lin, S.-W., Wang, P.-P., Fei, Y.-T., Chen, C.-K.: Simulation of the errors transfer in an articulation-type coordinate measuring machine. 11 November 2004, Accepted: 4 April 2005, Published online: 8 December 2005, Springer-Verlag London Limited (2005)
76. Lotze, W.: Prüfkörper für Koordinatenmessgeräten. Feingerätetechnik 30, und VDI-Z 123, pp. 154–155 (1981)
77. Lotze, W.: Form testing by means of the universal CMM ScanMax. In: IIIth International Scientific Conference—Coordinate Measuring Technique, pp. 82–90. Lodz University of Technology Scientific Books, branch in Bielsko-Biała, no. 44, Bielsko-Biała (1998)
78. Martinetz, T., Ritter, H., Schulten, K.: Learning of visuomotor-coordination of a robot arm with redundant degrees of freedom. Proceedings of the INT. Conference on Parallel, Processing in Neural Systems and Computers (ICNC). Elsevier, Dusseldorf (1990)
79. MESTRAL Project Artefacts and methods to establish traceability of large CMMs EU PROJECT (SMT4-PL97–2330) Final Report
80. Metoda Macierzowa (Matrix Method MM). Microsoft Excel Application Cracow University of Technology Lab of Coordinate Metrology (2000)
81. Morecki, A, Knapczyk, J., Kędzior, K.: Teoria mechanizmów i manipulatorów" (The theory of mechanisms and manipulators). Wydawnictwo Naukowo- Techniczne, Scientific and Technical Publishing (2001)
82. Moustafa, S.; Gerwien, N.; Haertig, F.; Wendt K.: Comparison of error mapping techniques for coordinate measuring machines using the plate method and Laser Tracer. In: XIX IMEKO World Congress Fundamental and Applied Metrology, Lisbon, Portugal, 6–11 Sep 2009
83. Muralikrishnan, B., Sawyer, D., Blackburn, C., Phillips, S., Borchardt, B., Estler, W.T.: ASME B89.4.19 Performance Evaluation Tests and Geometric Misalignments in Laser Trackery. J. Res. Natl. Inst. Stand. Technol. **114**, 21–35 (2009)
84. Muzyka-Żmudzki Michał: Metoda oceny dokładności Wielkogabarytowych Współrzędnościowych Maszyn Pomiarowych (WgWMP) przy zastosowaniu małych wzorców (Large Coordinate measuring machines (LCMM) accuracy assessment method with the use of small standards). Ph.D Thesis Dissertation, Cracow University of Technology, Faculty of Mechanical Engineering 2009 (Thesis Advisor - Prof. J.A. Sladek realized as a PhD grant no. 1261/T02/2007/32)
85. Nakamura, O., Goto, M., Toyoda, K., Tanimura, I., Kurosawa, A.: Development of coordinate measuring systems with tracking laser interferometers. Ann. CIRP **40**(1), 523–530 (1991)

86. Nawara, L., Kowalski, M.: Analiza błędów w pomiarach przestrzennych. Mechanik 12/80, pp. 674–679
87. Nawara, L., Kowalski, M.: The investigations on selected dynamical phenomena in the heads of multi-coordinate measuring devices. Ann. CIRP **33**(1), 373–376 (1984)
88. Nawara, L., Kowalski, M.: Analysis of the random component of multicoordinate measuring machines and metrological robots position error. Ann. CIRP **36**(1), 373–377 (1987)
89. Nawara, L., Sładek, J.: Influence of surface state on the measuring accuracy measuring machine. In; 7th International Conference on Production Engineering, pp. 696–671, Tokyo Japan (1984)
90. Nawara, L., Sładek, J.: Investigations of measuring heads errors influence on the measuring accuracy of multi-coordinate machine. In: 7th International Conference on Automated Inspection and Product-Control, pp. 305–312. Birmingham, England (1985)
91. Nawara, L., Sładek, J.: Analise der von den Tastkäpfen den in elastischen Fer-tigungssystemen eingesetzten Mehrkoordinaten Messmaschinen Herrühren-den Fehler. In: International Conference TITAN 85, pp. 301–309, Bucuresti Romania IX (1985)
92. Nawara, L., Kowalski, M., Sładek, J.: Badanie maszyn pomiarowych 3-koordynatowych ze względu na dokładność. PTMiU (Examination of 3-coordinate measuring machines due to accuracy. PTMiU), Quarterly PAN (Polish Academy of Sciences) pp. 141–150 (1983/2)
93. Nawara, L., Kowalski, M., Sładek, J.: The influence of kinematics errors on the profile shapes by means of CMM. Ann. CIRP **37**(1), 398–401 (1989)
94. Nawara, L., Kowalski, M., Sładek, J.: Badanie głowic pomiarowych stosowanych w maszynach wieloosiowych (Examination of probe heads used in multiaxial machines). Proceedings of Interuniversity Conference of Metrologists, Poznań (1984)
95. Nawara, L., Kowalski, M., Sładek, J.: Pewna metoda oceny błędów maszyn pomiarowych (Sure method for assessment of measuring machines errors). Mechanik no. 1/87, pp. 14–19
96. Nawara, L., Kowalski, M., Sładek, J.: Badanie maszyn pomiarowych 3-koordynatowych ze względu na dokładność i wskaźniki techniczno-ekonomiczne (Examination of 3-coordinate measuring machines due to accuracy and technical-economic indicators). Works of ZMiKT PK under the R.I.3 problem, stages 1–5, 1980–1984
97. Nawara, L., Kowalski, M., Sładek, J.: Badania metrologiczne maszyn pomiarowych i urządzeń ze szczególnym uwzględnieniem dokładności geometrycznej, pozycjonowania i strategii pomiarów (Metrology tests of measuring machines and devices with special include of geometrical accuracy, positioning and measurement strategy). Subtopic 1, Works of ZMiKT PK under CPBP no. 02.20, 1985–1989
98. Neumann, H.J.: Koordinatenmesstechnik. im Industriallen Einsatz. Verlag Moderne Industrie, 86895 Landsberg /Lech 2000
99. Neumann, H.J., et al.: Praezisionmesstechnik in der Fertigungs mit Koordinatenmessgeraeten. Expert Verlag (2005)
100. Nocuń, M.: VAST- Nawigator skaning trzeciej generacji) VAST- Nawigator third generation scanning) Mechanik 2, pp. 140–142 (2004)
101. Ostrowska, K.: Accuracy assessment method for measurements done with the use of articulated arm coordinate measuring machines. Ph.D. Thesis Dissertation, Cracow University of Technology, Faculty of Mechanical Engineering (Thesis Advisor - Prof. J.A. Sladek) (2010)
102. Ozono, S., Takamasu, K.: Data processing and calibration of coordinate measuring machine. Int. J. Japan Soc. Prec. Eng. **29**(2), 105–108 (1995)
103. Pedone, P., Romano, D.: Designing small samples for form error estimation with coordinate measuring machines. Precision Eng. **35**(2), 262–270 (2011)
104. Pfeifer, T.: Koordinatenmesstechnik für Qualitätssicherung. VDI Verlag GmbH, Düsseldorf (1992)
105. Pfeifer, T., Bambach, M., Fürst, A.: Ermittlung der Messunsicherheit von 3D-Tastsystemen. Technische Messen tm, 1979/2 und 4, pp. 47–52 und 161–169

106. Pfeifer, T., Bambach, M.: Definition und Prüfung von Kriterien zur Bestim-mung systematischer und zufälliger Fehler von Dreikoordinatenmessgerä-ten. Forschungsberichte des Landes Nordhein-Westfalen No. 2856
107. Philips, S.D., Sawyer, D. et al.: A novel artifact for testing large coordinate measuring machines. Precision Eng. J. Int. Soc. Precision Eng. Nanotechnol. **25**, 29–34 (2001)
108. Physikalisch-Technische Bundesanstalt, www.ptb.de
109. Physikalisch-Technische Bundesanstalt. Kalkom v.3.1 i v 4.0, v 4.1. Software to determining kinematic errors accordingly to Novel method
110. Piratelli-Filho, A., Lesnau, G.R.: Virtual spheres gauge for coordinate measuring arms performance test. Measurement **43**(2), 236–244 (2010)
111. Plath, H.H.: Checking accuracy of 3D-CMM by use of different types of calibrated artefacts —a comparative report of industrial experiences. In: First International Workshop on Coordinate Measuring Machines Calibra-tion Prague (Czech Republic), pp. 36–42, June 1–2 1999
112. PN- EN ISO 10360 Specyfikacje Geometrii Wyrobów (GPS) Badania odbiorcze i okresowe współrzędnościowych maszyn pomiarowych (CMM) (Geometrical Product Specifications (GPS) –Acceptance and reverification tests for coordinate measuring machines (CMM))
113. PN- EN ISO 10360 Specyfikacje Geometrii Wyrobów (GPS) Badania odbiorcze i okresowe współrzędnościowych maszyn pomiarowych (CMM), Cześć 4. CMM stosowane w trybie pomiaru skanowanego (Geometrical Product Specifications (GPS). Acceptance and reverification tests for coordinate measuring machines (CMM), Part 4: CMMs used in scanning measuring mode)
114. PN- EN ISO 10360 Specyfikacje Geometrii Wyrobów (GPS) Badania odbiorcze i okresowe współrzędnościowych maszyn pomiarowych (CMM), Cześć 2. CMM Stosowane do pomiaru długości - wydanie trzecie, 2009 (Geometrical Product Specifications (GPS). Acceptance and reverification tests for coordinate measuring machines (CMM), Part 2: CMMs used for measuring linear dimensions, 3rd edn (2009)
115. PN- EN ISO 10360 Specyfikacje Geometrii Wyrobów (GPS) Badania odbiorcze i okresowe współrzędnościowych maszyn pomiarowych (CMM), Część 5. CMM z zespołem głowic pomiarowych wielotrzpieniowych (Geometrical Product Specifications (GPS). Acceptance and reverification tests for coordinate measuring machines (CMM), Part 5: CMMs using single and multiple stylus contacting probing system) 2n edn (2010)
116. Poniatowska, M., Werner, A.: Probe radius compensation and fitting errors in cad based measurements of a free-form surfaces: a case study., XIX IMEKO World Congress Fundamental and Applied Metrologypp.1821–1826. Lisbon, Portugal, 6–11 Sept 2009
117. Polski Komitet Normalizacji. Międzynarodowy słownik metrologii—Pojęcia podstawowe i ogólne oraz terminy z nimi związane (VIM)" - PKN-ISO/IEC Guide 99:2010 (International vocabulary of metrology. Basic and general concepts and associated terms (VIM)) (2010)
118. Ratajczyk, E.: Sondy stykowe stosowane w maszynach i robotach pomiarowych (Contact probes use in measuring machines and robots) Mechanik no. 8–9, pp. 297–301 (1994)
119. Ratajczyk, E.: Optoelektroniczne sondy bezstykowe stosowane we współrzędnościowej technice pomiarowej (Opto-electronic, non-contact probes use in coordinate measurement technique) Mechanik no. 10, pp. 355–358 (1994)
120. Ratajczyk, E.: Konfiguracje sond pomiarowych w układzie współrzędnościowej maszyny pomiarowej (Measuring probes configurations in coordinate measuring machine) Mechanik no. 1, pp. 19–22 (1995)
121. Ratajczyk, E.: Sondy stykowe—własności metrologiczne i metody ich badania (Contact probes—metrology features and methods of their examination) Mechanik no. 2, pp. 57–61 (1995)
122. Ratajczyk, E.: Metody kompleksowe umożliwiające wykrywanie źródeł błędów współrzędnościowych maszyn pomiarowych (Complex methods for coordinate measuring machine errors sources detection). In: Proceedings of VIIIth National Metrology Confrence, Scientific book 4, pp. 259–266, Warsaw University of Technology, Warsaw (1995)

123. Ratajczyk, E.: Wyznaczanie błędów geometrycznych współrzędnościowych maszyn pomiarowych dla wirtualnego modelu ich kompensacji (Determination of coordinate measuring machines geometrical errors for virtual model of them compensation) III Szkoła -Konferencja nt. Metrologia wspomagana komputerowo MWK'97. In: Proceedings of the IIIth Conference Computer aided metrology, vol. 3, pp. 175–182. Zegrze neer Warsaw 19–22 May 1997
124. Ratajczyk, E.: Metody sprawdzania dokładności współrzędnościowych maszyn pomiarowych. Pomiary długości, prostoliniowości i prostopadłości. (Coordinate measuring machines accuracy verification methods. Length, straightness and perpendicularity measurements. Part 1, Mechanik no. 3, pp. 141–146 (1999)
125. Ratajczyk, E.: Metody sprawdzania dokładności współrzędnościowych maszyn pomiarowych. Pomiary pozycjonowania i błędów rotacyjnych. (Coordinate measuring machines accuracy verification methods. Positioning and rotary errors measurements) Part 2, Mechanik no. 4, pp. 277–275 (1999)
126. Ratajczyk, E.: Metody sprawdzania dokładności współrzędnościowych maszyn pomiarowych. Zastosowanie wzorców płytowych (Coordinate measuring machines accuracy verification methods. Plate standarts use) Part 3, Mechanik no. 5–6, pp. 411–416 (1999)
127. Ratajczyk, E.: Procedury sprawdzania dokładności współrzędnościowych maszyn pomiarowych wykorzystujące wzorce płytowe (Coordinate measuring machines accuracy verification procedures. using plate standarts). In: Proceedings of VIIth Scientific-Technical Conference Metrologia w technikach wytwarzania maszyn, vol. II, pp. 419–428. Szczecin University of Technology, Szczecin (1999)
128. Ratajczyk, E.: Charakterystyka oprogramowań komputerowych do wyznaczania błędów pomiaru długości za pomocą wzorców płytowych (Characteristic of computer softwares for determination of length measurement errors with the use of plate standards). In: IVth International Scientific Conference—Coordinate Measuring Technique, pp. 171–185. Lodz University of Technology Scientific Books, branch in Bielsko-Biała, no. 53, Bielsko-Biała (2000)
129. Ratajczyk, E.: Wzorce płytowe w sprawdzaniu doraźnym współrzędnościowych maszyn pomiarowych (CMM) (Plate standards in coordinate measuring machine (CMM) verification) Przegląd Mechanika, pp. 154–159. LXIV, Scientific book 9S (2005)
130. Ratajczyk, E.: Współrzędnościowa Technika Pomiarowa (Coordinate Measuring Technique). Warsaw University of Technology, Warsaw (2005)
131. Ratajczyk, E.: Ramiona pomiarowe – budowa, parametry techniczne, zastosowania (Articuled Arm Coordinate Measuring Machines—construction, technical parameters, applications) – Mechanik no. 12, pp. 1051–1054 (2008)
132. Ratajczyk, E.: Współrzędnościowe Ramiona pomiarowe i ich testy dokładności (Articuled Arm Coordinate Measuring Machines and their accuracy tests) Przegląd Elektrotechniczny no. 5 (2008)
133. Ratajczyk, E.: Ramiona pomiarowe – testy dokładności (Articuled Arm Coordinate Measuring Machines – accuracy tests) Mechanik no. 2 (2009)
134. Ryniewicz, A., Kowalski, M., Gawlik, J.: Research into rotational errors of coordinate measuring machine. In: 6th ISMQC IMECO Symposium. Metrology for Quality Control in Production. IX, Technical University of Vienna, Wien (1998)
135. Ratajczyk, E., Zawadzki, M.: Accuracy test of measuring arms—is it possible to compare ASME and ISO standard requirements. Współrzędnościowa Technika pomiarowa – problemy i zastosowania (Coordinate Measuring Technique—problems and applications), pp. 137–146. University of Bielsko-Biała, Bielsko Biała (2008)
136. Renishaw Sp. z o.o., www.renishaw.com.pl
137. Santolaria, J., Aguilar, J.J., Yagüe, J.A., Pastor, J.: Kinematic parameter estimation technique for calibration and repeatability improvement of articulated arm coordinate measuring machines. Precision Eng. **32**, 251–268 (2008)

138. Santolaria, J., Yagüe, J.-A., Jiménez, R., Aguilar, J.-J.: Calibration-based thermal error model for articulated arm coordinate measuring machines. Precision Eng. PRE–5655
139. Schultschik, R.: Componente of the volumetric accuracy. Ann. CIRP **24**(1), 223–229 (1977)
140. Schüssler, H.H., Schäfer, A., Schmidt, R.: Praktische Erfahrungen bei der Ab-nahme von Koordinatenmessgeräten hinsichtlich der Längenmessunsicher-heit nach der Richtlinie VDI/VDE2617. Blatt 2.1 VDI-B 529 (1984)
141. Schwenke, H.: Bedienanleitung MegaKal, KalKom, TKal, PKal, KalKmess, PTB, Braunschweig 1996–1999
142. Schwenke, H., Trapet, E., Wäldele, F.: Pushing the limits of accuracy of coordinate measurements. In: Kunzmann, H., Wäldele, F., Wilkening, G., Corbet, J., MacKeow, P., Weck, M., Huemmler, J. (eds.) Progress in Precision Engineering and Nanotechnology, vol. 1, pp. 323–326. PTB-Braunschweig u. Berlin P.u.Oe–Verlag (1997)
143. Schwenke, H., Neuschaefer-Rube, U., Pfeifer, T., Kunzmann, H.: Optical methods for dimensional metrology in production engineering. CIRP Ann. Manuf. Technol. **51**(2), 685–699 (2002)
144. Schwenke, H., Franke, M., Hannaford, J., Kunzmann, H.: Error mapping of CMMs and machine tools by a single tracking interferometer. CIRP Ann. Manuf. Technol. **54**(1), 475–478 (2005)
145. Schwenke, H., Knapp, W., Haitjema, H., Weckenmann, A., Schmitt, R., Delbressine, F.: Geometric error measurement and compensation of machines—an update. CIRP Ann. Band **57**(2), 660–675 (2008)
146. Schwenke, H., Schmitt, R., Jatzkowski, P., Warmann, C.: On-the-fly calibration of linear and rotary axes of machine tools and CMMs using a tracking interferometer. CIRP Ann. Manuf. Technol. **58**(1), 477–480 (2009)
147. Shorya, A., Slocum, A.H.: Constraint-based design of parallel kinematic xy flexure mechanisms. J. Mech. Des. **129**(8), 816–830 (2007)
148. Sładek, J.: Badanie głowic stykowych stosowanych w maszynach pomiarowych wieloosiowych (Examination of contact probes use in multiaxial measuring machines). In: Proceedings of Conference: Postępy w technologii maszyn, Częstochowa 1–2 X, pp. 48–51 (1989)
149. Sładek, J.: Assessment of the accuracy of contact probe heads used in coordinate measuring machines. Ph.D. Thesis Dissertation, Cracow University of Technology, Faculty of Mechanical Engineering (1990)
150. Sładek, J.: The examination of multicoordinate measuring devices (MMD) by means of circular standards. In: Proceedings of International Conference MATE. Automation 92" 18–19th February 1992, Budapest, Hungary, pp. 333–339
151. Sładek, J., Kowalski, M.: Problems of robotization of measuring—control operation in automated manufacturing systems. In: IFAC Workshop Automatic Control For Quality and Productivity, ACQP 92", pp. 288–289, Istanbul, Turkey 3–5 June 1992
152. Sładek, J.: Metoda badań głowic stykowych stosowanych w wieloosiowych urządzeniach pomiarowych i OSN (Method of examination of contact probes use in multiaxial measuring instruments and CNC machines). Works of Institute of Machine Technology at Wroclaw University of Technology, no.49, Proceedings of Conference: Podstawy technologii maszyn 91, pp. 371–372
153. Sładek, J.: Einsatz des Prüfkörpers mit einem einfachen Formfehler für Prüfen der Tastsysteme von Dreikoordinaten Messmaschinen und Messrobotern. In: 3 Internationales Symposium von DAAAM, Budapest, Ungarn 4–6 November 1992
154. Sładek, J.: Opracowanie systemu oceny dokładności pomiarów oraz metody identyfikacji własnego pola błędów współrzędnościowych maszyn pomiarowych (Developing of a system for assessing the accuracy of measurements and of a method for identification coordinate measuring machine's errors field). Final Report of the Research Project funded by the KBN under contract No. PB-0921/T07/99/1

155. Sładek, J.: errors identification and measurement accuracy assessment of coordinate measuring machines (CMM). Adv. Technol. Mach. Equip. Postępy Technologii Maszyn i Urządzeń, Quarterly PAN (Polish Academy of Sciences) **23**(4), 113–136 (1999)
156. Sładek, J.: Modelowanie i ocena dokładności maszyn oraz pomiarów współrzędnościowych (Modeling and accuracy assessment of coordinate machines and measurements). Cracow University of Technology Scientific Books, Mechanika no. 87, Cracow (2001)
157. Sładek, J.: Metoda oceny dokładności Wielkogabarytowych Współrzędnościowych Maszyn Pomiarowych (WgWMP) przy zastosowaniu małych wzorców (Method for accuracy assessment of Large Coordinate Measuring Machines (LCMM) with the use if small standards) Ministry of Science and Higher Education grant no. 1261/T02/2007/32 (2007)
158. Sładek, J.: Badania rozwojowe nad wzorcowaniem (z wykorzystaniem Laser Tracera) i wyznaczanie niepewności pomiarów dla systemów współrzędnościowych (Development researches on calibration (with the use of Laser Tracer) and determination of measurements uncertainty for coordinate systems) Grant no. 0869/R/T022010/10 National Centre of Research and Development (2010–2013)
159. Sładek, J.: Metoda oceny dokładności pomiarów realizowanych redundantnymi systemami współrzędnościowymi (RSW) (Method for accuracy assessment of measurements done with the use of redundant coordinate systems (RCS)) Research Project No. N505 255935 (2008–2010)
160. Sładek, J, Gaska, A,: Modelling of CMM probe head errors. In: Proceedings of 10th ISMQC–TC14 IMEKO, Osaka, Japan, 5–9 Sept 2010
161. Sładek, J., Kowalski, M.: Planung des Messablaufs der Geometrischen Gro-essen für Automatisierte Fertigungssysteme. In: 4 Internationales Symposium von Donau Adria Assoziation für Automation und Messtechnik, DAAAM, Automation und Metrologie: Mensch, Technik, Umwelt: TU Brno Fakultät Maschinenbau, Brno Tschechien, pp. 267–277, 16–18 Sept 1993
162. Sładek, J., Kowalski, M.: Stoch, identyfikacja dokładności sumarycznej WMP przy użyciu wzorców (Total CMM accuracy identification with the use of standards). In: Scientific Conference—Coordinate Measuring Technique, pp. 89–97. Lodz University of Technology Scientific Books, branch in Bielsko-Biała, no. 19, Bielsko-Biała (1994)
163. Sładek, J., Kowalski, M.: Anwendung des Virtuellen normales zur Schätzung der Genauigkeit bei Mehrkoordinaten-Messungen. In: 6 Internationales Sympo-sium von Donau Adria Assoziation für Automation und Messtechnik, DAAAM, Automation und Metrologie: Mensch, Technik, Umwelt: TU Kraków, Fakultät Maschinenbau, Kraków 24–26 Okt. 1995, pp. 187–188
164. Sładek, J., Krawczyk, M.; Modeling and assesment of large CMM's accuracy. XIII IMEKO World Congres—Metrology in the 3rd Millenium, Cavat, Dubrovnik, Croatia (2003)
165. Sładek, J., Rakoczy, R.: Wykorzystanie koncepcji sztucznych sieci neuronowych do modelowania współrzędnościowej maszyny pomiarowej (WMP) przy zastosowaniu wzorca przestrzennego (The use of the concept of artificial neural networks for modeling the coordinate measuring machine (CMM) with the use of spatial standard). Works of Institute of Machine Technology and Production Automation, Polish Academy of Sciences, Committee on Machine Building, Technology Bases Section, Scientific Book no. 63, Cracow University of Technology, pp. 193–206 (1999)
166. Sładek, J., Kowalski, M., Rewilak, J.: Komputerowe wspomaganie operacji planowania pomiarów wielkości geometrycznych (Computer-aided planning operation of geometrical quantities measurements) Mechanotronika 92, scientific book 2, pp. 141–143. Warsaw University of Technology, Warsaw (1992)
167. Sładek, J., Kowalski, M., Juras, B.: Robotyzacja wybranych operacji pomiarów wielkości geometrycznych (Robotics of given geometrical quantities measurements oparations). In: Proceedings of Conference: Metrologia w budowie maszyn, Poznań (1992)
168. Sładek, J. et al.: Opracowanie systemu oceny dokładności pomiarów realizowanych na wielkogabarytowych współrzędnościowych maszynach pomiarowych (The system of accuracy assessment of measurements done with the use of large coordinate measuring

machines development). Grant 5 T07D 038 24. Ministry of Science and Higher Education (2003–2005)
169. Sładek, J., Kowalski, M., Ryniewicz, A., Juras, B., Muzyka-Żmudzki, M.: The analysis of the possibility of the assignment of the reference points for the large CMM's. In: VI International Scientific Conference Coordinate Measuring Technique, Bielsko-Biała, pp, 270–278. University of Bielsko-Biała, Scientific books (2004)
170. Sładek, J., Kowalski, M., Ryniewicz, A., Juras, B., Muzyka-Żmudzki, M.: The method of the identification of the orientation of control standard platter for the assessment of large coordinate measuring machine's (LCMM's) accuracy. In: Measurement 2005, 5th International Conference, Smolice Slovakia (2005)
171. Sładek, J., Kowalski, M., et al.: Identification method of the CMM measurement subspace localization based on location errors. VII Int Sci Conf Coordinate Measuring Tech. **22**(7), 291–301 (2006) (Bielsko-Biała)
172. Sładek, J., Ostrowska, K., Sokal, G., Kmita, A.: Wzorcowanie współrzędnościowych Ramion Pomiarowych (WRP)"(Calibration of Artifical Arm Corrdinate Measuring Machines (AA CMM)), Acta mechanica et automatica, vol.1 no.2 (2007)
173. Sładek, J., Ostrowska, K., Gąska, A.: Wirtualne współrzędnościowe ramie pomiarowe (WWRP) (Virtual articulated arm coordinate measuring machine (VAACMM), Pomiary Automatyka Kontrola (Measurement Automation and Monitoring) 01/2010
174. Sładek, J., Ostrowska, K., Gąska, A.: Modeling and identification of errors of coordinate measuring arms with use of metrological model. Measurement **46**, 667–679 (2013)
175. Sładek, J. Ostrowska, K., Gacek, K., Bryndza, M.: Designation of operator impact on errors of measurements realized by coordinate measuring arm. In: Sładek, J., Jakubiec, W. (eds.) Advances in Coordinate Metrology, Monograph. University of Bielsko-Biała, Bielsko-Biała (2010)
176. Stahl, R.: Praktische Erfahrungen bei der Abnahme von Koordinaten-messgeräten hinsichtlich der Komponenten der Messabweichungen nach der Richtlinie. VDI/VDE2617, Blatt 2.1 VDI-B 529 84
177. Slocum, A.H.: Precision machine design. Prentice Hall (1992)
178. Teichmann, U.: Metrologische probleme bei der pruefung von koordinatenmessgeraeten. Feingeraetentechnik (1978)
179. Tesa Technology, www.tesabs.ch
180. Tikka, H.: Calibrating a boring plate and checking a 3-Dimensional Measuring Machine by that One. IMECO-Vortrag S.8.1 1982 Berlin Preprint V/II
181. Takatsuji, T., Goto, M., Kirita, A., Kurosawa, T., Tanimura, Y.: The relationship between the measurement error and the arrangement of laser trackers in laser trilateration. Meas Sci Technol (2000)
182. Trapet, E.: Large coordinate measuring machines—state of art and future for geometry error correction, performance, verification and calibration. In: Proceedings of 1st International Workshop on CMM Calibration, Published by The Czech Metrological Institute, Prague Czech Republic, June 1–2 1999
183. Trapet, E.: Uncertainty analysis of measurements with cmms using multiple measurement strategies. Report of EU Project EASYTRAC (2003)
184. Trapet, E., Wäldele, F.: A uniform concept for calibration, acceptance test and periodic inspection of co-ordinate measuring machines using reference objects. Ann. CIRP **39**, 561–564 (1990)
185. Trapet, E., Wäldele, F.: A reference artefact based method to determine the parametric error components of coordinate measuring machines and machine tools. Measurement **9**, 17–22 (1991)
186. Trapet, E., et al.: Tracebility of Coordinate Measurements According to the Method of the Virtual Measuring Machine. European Commission SMT- EDG XII/C/5,Project No. Mat.1 CT94-0076, final report
187. Trapet, E., Franke, M., Härtig, H., Schwenke, H., Wäldele, F., Cox, M., Forbes, A., Delbressine, F., Schellekens, P., Trenk, M., Meyer, H., Moritz, G., Guth, Th., Wanner, E.:

Tracebility of coordinate measurements according to method of virtual measuring machine. PTB F–35, Braunschweig (1999)
188. Tyka, A, Góralczyk, A.: Problemy realizacji nadzorowania niektórych hybrydowych współrzędnościowych maszyn pomiarowych Problems of monitoring of some hybrid coordinate measuring machines) Monograph, Współrzędnościowa Technika pomiarowa – problemy i zastosowania (Cooridnate Measuring Technique – problems and applications) University of Bielsko Biała. Bielsko Biała. pp. 21–28 (2008)
189. Umetsu, K., Furutnani, R., Osawa, S., Takatsuji, T., Kurosawa, T.: Geometric calibration of a coordinate measuring machine using a laser tracking system. Meas. Sci. Technol. **16**, 2466–2472 (2005)
190. VCMM.: User Manual, PTB Braunschweig (2002)
191. VDI/VDE 2617 Genauigkeit von Koordinatenmessgeräten Kenngrößen und deren Prüfung—Blatt 2.1 –Leitfaden zur Anwendung von DIN iSO 10230-2 zur Messung von Laengenmassen Beuth Verlag Berlin u. Köln (2005)
192. VDI/VDE 2617 Genauigkeit von Koordinatenmessgeräten Kenngrößen und deren Prüfung. Blatt 5.1 Überwachung mit Kugelplatten (2009)
193. VDI/VDE 2617 Genauigkeit von Koordinatenmessgeräten Kenngrößen und deren Prüfung. Blatt 9. Annahme und Bestätigungsprüfung von Gelenkarm Koordinatenmessgeräten VDI/VDE- Richlinien ver (2009)
194. Warnecke, H.J., Dutschke, W.: Fertigungsmesstechnik. Springer, Berlin (1984)
195. Wäldele, F., Hagelmann, R., Trapet, E.: Coordinate metrology. PTB-Bericht F–11, Braunschweig (1992)
196. Weckenmann, A.: Übersicht über Abnahme und Überwachungsverfahren für Mehrkoordinaten-Messgeräte. VDI-B **378**, 1–10 (1980)
197. Weckenmann, A.: Koordinaten-Messtechnik, Überblick, Entwicklung, Anwendung. VDI-B **408**, 15–24 (1981)
198. Weckenmann, A.: Kengrossen für die Angabe der Genauigkeit von Koordinaten-Messgeräten. Technische Messen 50 1983/5, pp. 179–184
199. Weckenmann, A.: Notwendigkeit und Möglichkeit der Normung von Antaststrategie und Auswertegrundsätzen. VDI-B **529**, 209–227 (1984)
200. Weckenmann, A.: Koordinatenmesstechnik—Flexible Strategien für Funktion und fertigungsgerechtes Prüfen Carl Hanser Verlag München, Wien (2012)
201. Weckenmann, A., Petrovic, N.: Comparation of cmm length measurement tests conducted by a step guage and a ball plate. VII Int. Sci. Conf. Coordinate Measuring Tech. **22**(7), 373–380 (2006) (Bielsko-Biała)
202. Weckenmann, A., Goch, G., Springborn, H.D.: Korrektur der Taststiftbiegung bei Messung mit Mehrkoordinaten-Messgeräten. Feinwerktechnik u. Messtechnik 87 1979/1, pp. 5–9
203. Weckenmann, A., Henrichowski, M., Mordhorst, H.J.: Design of gauges and multipoint measuring systems using coordinate measuring machines data and computer simulation. Prec. Eng. **13**, 203–207 (1991)
204. Weckenman, A., Geus, D., Gubesch, A.: High precision form and dimension measurements with tactile measuring machines. In: IVth International Scientific Conference—Coordinate Measuring Technique, pp. 255–264. Lodz University of Technology Scientific Books, branch in Bielsko-Biała, no. 53, Bielsko-Biała (2000)
205. Weckenmann, A., Estler, T., Peggs, G., McMurtry, D.: Probing systems in dimensional metrology. CIRP Ann. Manuf. Technol. **53**(2), 657–684 (2004)
206. Weckenmann, A., Jiang, X., Sommer, K.-D., Neuschaefer-Rube, U., Seewig, J., Shaw, L., Estler, T.: Multisensor data fusion in dimensional metrology. CIRP Ann. Manuf. Technol. **58** (2), 701–721 (2009)
207. Wendt, K., Schwenke, H., Waldele, F, Krawczyk, M, Kniel, K.: Error mapping of large CMM's by sequential multi-lateration using a lasertracker. Materials of Conference. EUSPEN (2001)
208. Wirtual WRP—simulation software for determining uncertainty of measurements done with the use of articulated arm coordinate measuring machine AACMM

209. Wozniak, A.: Dokładność głowic stykowych maszyn współrzędnościowych (Accuracy of contact probes of coordinate machines) Scientific books Mechanika no. 235—Warsaw University of Technology (2010)
210. Woźniak, A., Dobosz, M.: Metrological feasibilities of CMM touch trigger probes. Part I. 3D theoretical model of probe pretravel. Measurement **34**(4), 273–286 (2003)
211. Woźniak, A., Dobosz, M.: Influence of measured objects parameters on CMM touch trigger probe accuracy of probing. Precision Eng. **29**(3), 290–297 (2005)
212. Zeiss Europa, www.zeiss.pl/imt
213. Zeleny, V.: Coordinate measuring machines calibration in the czech republic. In: Ossana, P.H., Postrednik, D., Durakbasa, N.M. (eds.) Metrology for Quality Control in Production, Proceedings of 6th ISMQC IMEKO Symposium., Wien, Austria, 8–10 Sept 1998, pp. 743–749
214. Zeleny, V.: Calibration of coordinate measuring machines with respect to standards. In: V International Scientific Conference Coordinate Measuring Technique. Bielsko-Biała, Lodz University of Technology Scientific Books, branch in Bielsko-Biała **18**(3), 32 (2002)
215. Zeleny, V., Steskal V.: The newest ways of CMM calibration. In: IVth International Scientific Conference, Coordinate Measuring Technique, Lodz University of Technology Scientific Books, branch in Bielsko-Biała, no. 53, Bielsko-Biała, pp. 284–292 (2000)
216. Zhang, G., Veale, R., Charlton, T., Borchardt, Hocken R.: Error compensa-tion of coordinate measuring machines. Ann. CIRP, **34**(1), pp. 445–447 (1985)
217. Zhang, G., Ouyang, B., Lu, B., Hocken, R., Veale, R., Donmez, A.: A Displece-ment Method for Machine Geometry Calibration. Ann. CIRP **37**(1), 515–518 (1988)
218. Zhang, G.X., LI, X.H., Lin, Y.B., Liu, S.G., Liu, X.L., LI, X.F., Guo, J.B., Qiu, Z.R., Zhao, S.Z., Jiang, C.Z., Fan, Y.M.: A study on the optimal design of laser-based multi-lateration systems. CIRP Ann. Manuf. Technol. **52**(1), 427–430 (2003)
219. Zhuang, H., Motaghedi, S.H., Roth, Z.S., Bai, Y.: Calibration of multi-beam laser tracking systems. Robot. Comput. Integr. Manuf. **19**(4), 301–314 (2003)

Chapter 5
Simulation Methods for Assessing Accuracy of Measurements

Abstract This chapter presents simulation methods for assessing the accuracy of coordinate measurements, which are considered as a continuously tested alternative to the classic methods described in Chap. 3. It describes the bases for modeling measuring systems. It also presents the concept of CMM error identification, which is based on the measurement of reference objects, as a basis for the construction of a virtual measuring machine. Moreover, it reports the concept of virtual CMM as developed by the National Metrology Institute of Germany (in German *Physikalisch-Technische Bundesanstalt*, *PTB*), the basis for the ISO/TS 15530-4:2008 standard, using plate standards and models developed in the Laboratory of Coordinate Metrology at Cracow University of Technology (LCM CUT). These include a universal model based on the identification of component errors, other models based on further developments of the matrix method (MM), and on the use of artificial neural networks and the Monte Carlo method-based model. In the last case, both the concept of testing of so-called residual errors and the new principle for the development of virtual CMM for systems using CAA software correction is given. In addition, the description of the virtual model of a coordinate measuring machine for an articulated arm, which is described in Chap. 4, is presented. Here, the rules for the application of measuring system simulators for the task of forecasting coordinate measurement accuracy are also investigated. This chapter discusses the results of research aimed at verifying the correctness of the developed models and the software used to implement virtual CMM and the virtual articulated arm CMM in practice.

5.1 Introduction to Modeling of Measurement Devices and Systems

Chapter 3 offers a description of the classical methods of uncertainty estimation. From it, it turned out that these methods do not meet the requirements of universality thus they cannot be used outside of specialized accredited laboratories of industrial concerns, and of research units or national measurement institutes. Due to

J.A. Sładek, *Coordinate Metrology*, Springer Tracts in Mechanical Engineering,
DOI 10.1007/978-3-662-48465-4_5

constantly increasing requirements, the industry demands effective solutions allowing the online assessment of the accuracy of realized measurements. It is therefore necessary to develop a different, more universal, and easier to implement method that allows an estimation of measurement uncertainty for each performed measuring task. This kind of solution is the development of simulation models of measuring systems. Such a model according to [76, 78] could be developed for each system used to measure geometric quantities, especially for the most complex systems, such as coordinate systems. The essence of simulation model operation is a virtual performing of given tasks and estimation of measurement uncertainty using the *A* method. Because of that, the calculation of uncertainty of complex measurement procedures becomes easier, and in some cases there is only one method that allows uncertainty estimation. Simulation techniques are used in various fields of science and technology. Examples of this include the Monte Carlo method and the use of artificial neural networks. These have also been implemented in coordinate metrology [8, 17, 20, 53, 54, 78, 86, 88, 89, 95, 96, 106, 108, 110]. The Monte Carlo method is also included as a supplement to GUM [116]. Overall, the growing importance of simulation methods in metrology is related to the rapid development of numerical methods. Simulation methods are significantly useful when it is impossible to find an analytical solution or when the measurement model is complicated and the uncertainty estimation for many error sources is difficult or impossible. In these cases a mathematical model can be developed at first, and afterwards a functional model of the analyzed system can be built by using the appropriate simplifications. Uncertainty estimation using simulation is based on the classical model of the uncertainty budget of the measuring process, and its evaluation is based on multiple numerical repetitions of signal way and statistical estimation of given uncertainty, in accordance with Fig. 5.1.

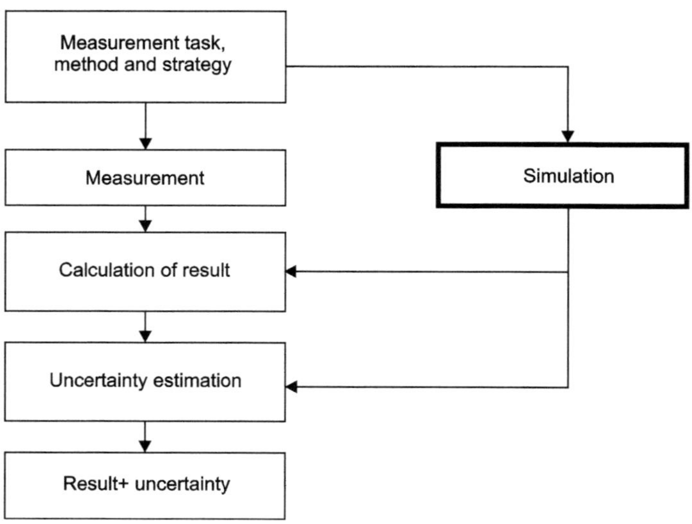

Fig. 5.1 Simulation method scheme for estimating uncertainty

This approach has the following advantages.

- It allows the connection of any algorithm and the ability of data filtering.
- The mathematical model does not need to be linear.
- The problem of measuring simulation of many measuring points is possible to be realized by using the data structure.
- Correlations between error components can be shown graphically.
- It is possible to include all, also asymmetric probability distributions.
- Once built, the complete simulation model of a measuring instrument can be used to calculate the uncertainty for all measuring tasks performed with this instrument.

The development of the measuring instrument model must be preceded by an analysis of the whole measuring process considered as a signal analysis. All the important variables affecting the course of measurement should be identified, to make it possible to develop a model of the measuring process in the form of simulation modules. In the final stage of work it is necessary to determine the size and variability of the input data as a basis for subsequent simulation. The variability of the process and the response of the model to interference allow the assessment of the quality of a given model in the form of difference assessment.

5.1.1 Construction of the Model

1. The measurement procedure description, signal analysis
2. Localization of significant quantities influencing the measurement course
3. Measurement process model in the form of simulation modules
4. Designation of sizes and variability of input data

5.1.2 Simulation

1. Measurement process simulation
2. Assessment of difference of model and real system

To better present the idea of systems of virtual measurements built in the field of coordinate metrology, and to encourage their development, an example of a simulation model of a simple measuring instrument, the caliper, is presented by [76, 78].

The measurement process for this instrument can be described as a signal course between functional blocks (modules; Fig. 5.2). The signal that goes through the particular stages comes under transformations, until the final measurement result of the measured quantity (e.g., diameter) is obtained. Each course in the form of a signal can be also individually developed for a given measuring task.

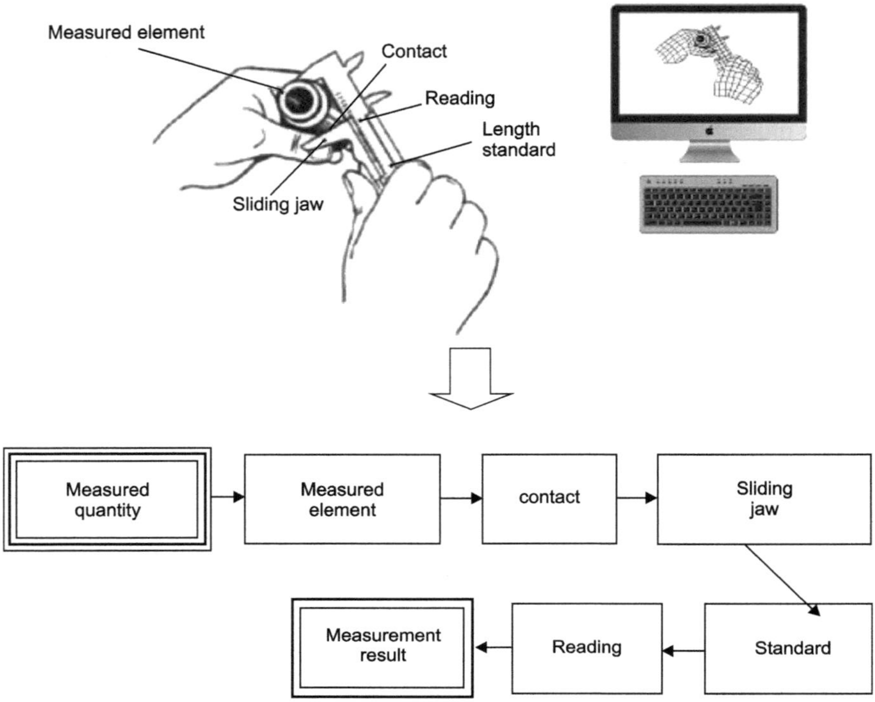

Fig. 5.2 Signal course for the measurement process on the caliper example

Functional blocks are often presented as: sensor, transducer, comparator, reading device. The transmitter can also contain other functional components, as well as transmission elements or data processing algorithms. The measuring signal course used for measurement uncertainty estimation must include all measurement process components, which are error sources. In the signal course components such as the impact of the measured object in the form of, for example, surface roughness, the shape of the contact with the object, or the clearances between the sliding jaw and slide, the formula, and result reading with the use of the vernier scale should be included. The quantities affecting the possibility of process full description also need to be identified.

5.1.2.1 Localization of Significant Quantities Affecting the Measurement Course

In the second step the affecting quantities are identified. The procedure is analogous to that given in the 14253-2 standard, where the input dataset in the form of a list is used by the metrologist performing the measurement as a checklist. This stage of the model construction is shown in Fig. 5.3.

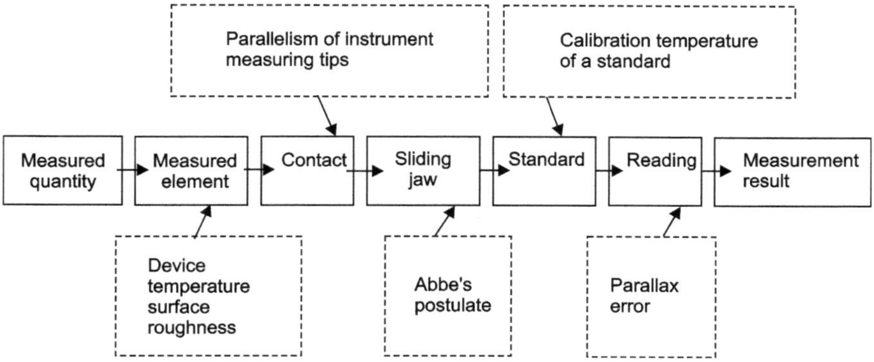

Fig. 5.3 Localization of quantities affecting the signal course on the caliper example [76]

But before the measurement is presented in the form of simulation modules the assessment of particular uncertainty components should be performed in order not to overlook essential components, and therefore not to simplify the model significantly.

5.1.2.2 Measurement Process Model in the Form of Simulation Modules

At this stage of construction of the simulation model, the measurement course is reconstituted by putting together modules analogous to the previously developed measurement process using simulation software and library of simulation modules. Thanks to them, it is possible to include quantities influencing the measurement result easily.

In Fig. 5.4 the simulation model for the caliper and the measurement process realized with this measuring tool as a basis of simulation software is shown. The

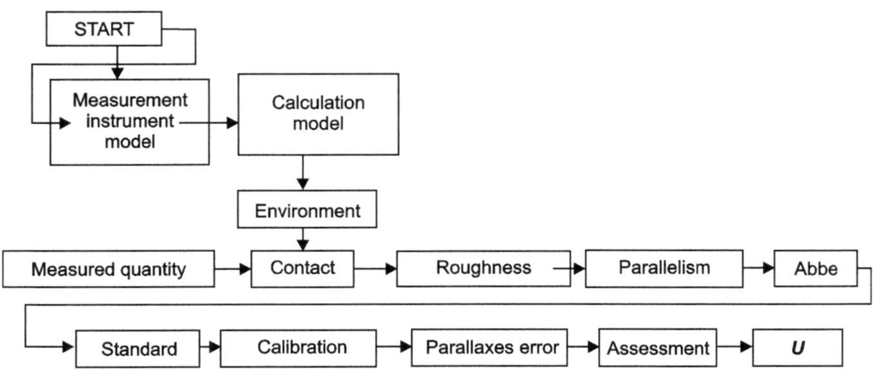

Fig. 5.4 Simulation model of the caliper measurement process described in the form of a signal flow [76]

measurement process is presented in the form of modules, “black boxes”. The proper measurement process, from input data to the final measurement result, is shown in the form of a loop that is realized by three blocks: the instrument module, the calculation module, and the module simulating environmental conditions. The purpose of this loop is to allow multiple measurement repetitions in order to obtain a significant number of random samples required for the calculation of uncertainty.

In each of the modules, it is possible to give and change the necessary parameters characteristic for input data. Thus, for example, to simulate the effect of thermal expansion of a measured object, parameters such as coefficient of material thermal expansion value, and its uncertainty, the temperature difference between the measured object and the environment, and value of the uncertainty attributed to this difference are given.

5.1.2.3 Determination of Input Data

Quantification of input data is a very important step in uncertainty determination. Using knowledge, practice, or statistical assessment, the user may specify how particular input data affect the uncertainty components. Through the use of modular structure it is possible to disable particular modules and to analyze the impact of a single quantity on the uncertainty of the whole process, proceeding in accordance with the ISO 14253-2 [70] standard. As recommended, in the first step the general estimation should be performed, which afterwards helps in accurate assessment. By turning on particular modules, only the ones that significantly affect the uncertainty can be selected.

5.1.2.4 Simulation of the Measurement Process

After the construction of the measurement process model and quantitative evaluation of the impact of particular components, the right simulation stage takes place. It consists of a large number of simulated random samples carried out by random generators for identified probability distributions. In the first course a nominal value (the measurement result without measurement error influence) is determined. Simulation modules are activated in the next course. The simulation course has the structure of a loop shown in Fig. 5.5, where N is the number of independent random samples of simulation, and P is the number of points in the measurement process. The P value during, for example, roughness or shape deviation measurement, is large and is contained mostly in the range from 1000 to 8000, and during the length measurement done on the caliper $p = 2$ [76]. The random sampling number N should be chosen in a way to obtain reasonable confidence of a simulation result.

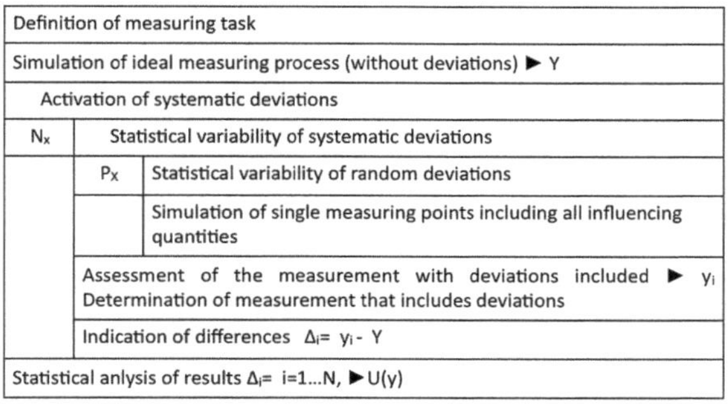

Fig. 5.5 Structure of simulation loop [76]

5.1.2.5 Assessment of Differences Between Simulated Measurement Results and a Value Obtained in the Real Process of Measured Quantity

The difference $\Delta_i = y_i - Y$ between the simulated measurement results y and the true value of measured quantity Y, represents the deviation of the measurement result and after N simulation repetitions may be used for statistical calculations. Simulation is comparable to the real measurement if the uncertainty of the standard calibration is $U = 0$ and under a large number of repeated measurements including "probable" conditions for their implementation. The simulation technique meets the same GUM requirements of experimental uncertainty estimation [135]; see Sect. 3.4.1.

5.1.3 Comparison of Uncertainty Budget with Simulation

Assessing the simulation technique in uncertainty estimation is related to the classical error budget. In Table 5.1 the implementation stages of the classical uncertainty budget and simulation are given.

Table 5.1 Comparison of uncertainty budget with simulation [76]

Uncertainty budget according to GUM	Simulation
Mathematical model of measurement process $y = f(x_1, x_2, \ldots, x_N)$	Formulation of the measurement process in the form of the signal course
The calculation of sensitivity coefficient $C_i = \frac{\partial f}{\partial x}$,	Identification of all significant distortions
Standard uncertainty establishment $u(x_i)$	Reproduction of the process in the simulation software, the use of a modular structure
Determination of correction coefficients $r(y_i, y_j)$	Quantification of the impact size for each of the modules
Calculation of combined standard uncertainty $u_c(y)$	Establishment of the confidence level
Calculation of combined expanded uncertainty $U = k \cdot u_c(y)$	Numerical simulation for the disturbed and undisturbed measurement and statistical evaluation of differences

If models and input quantities are consistent, then the uncertainty values estimated by both methods are similar.

As for the caliper, the measurement on CMM can be presented as a signal course (Fig. 5.6). The model of component uncertainty identification is described in accordance with the signal course in the measurement process. The signal that begins formulating the measuring task for a given object is distinguished here, including its thermal expansion and characteristic shape deviations, internal stresses, and so on. Then the point identification by the contact of the measuring tip with the measured object should be taken into consideration. This contact is associated with participation of errors of: the contact probe head (systematic and random components), machine geometry [errors of kinematic, hysteresis, dynamic, temperature gradient effect, standard scales (temperature, linearity error)], and

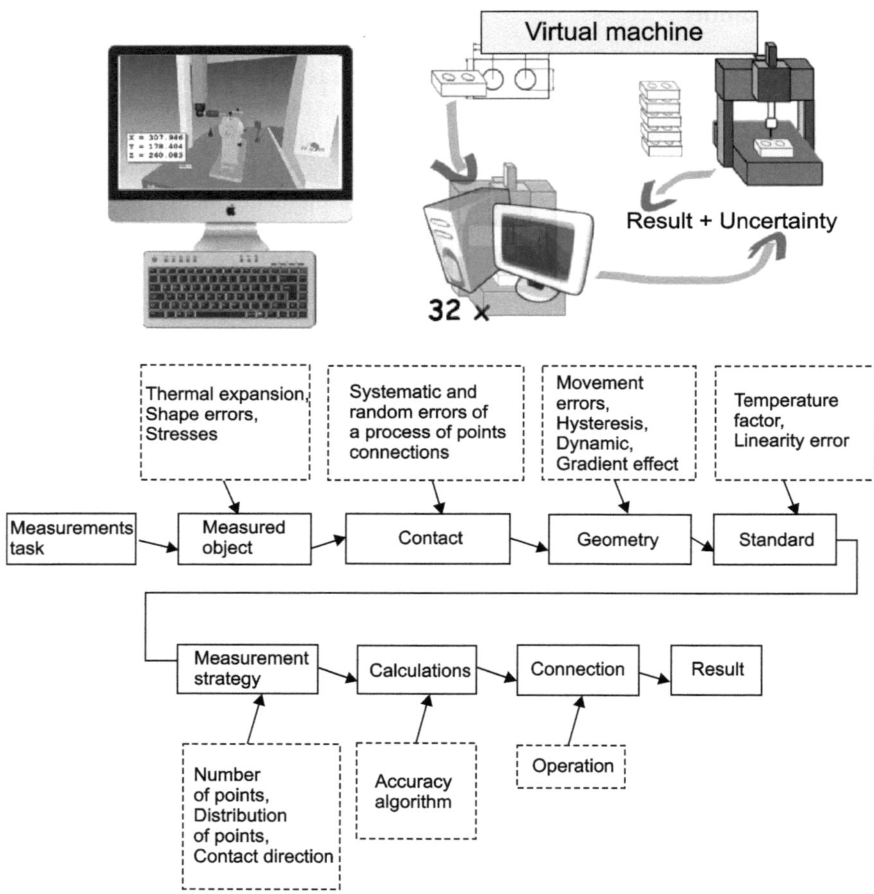

Fig. 5.6 Measurement on CMM as a signal course and influential quantities in the form of component uncertainties

participation of the measurement strategy (the number and manner of distribution of measuring points, the direction of the contact). After identification of measuring point calculations with the participation of software (for which algorithm errors occur) are performed, connections and relations of shape elements are defined, and finally a measurement result is given.

5.1.4 Model of Measurement Process

The signal course for CMM is the basis for development of the measurement process model. There are various simulation modules that realize the particular components of the process. So in the "task" section we have these simulation modules: G3, shape element generator; G9, module that simulates translation; and G10, rotation. In the "machine" section we can distinguish: TM6, module related to thermal expansion of an object; TM9, module that simulates the workings of contact probe head; and TM10, module that simulates the machine geometry, its kinematic errors; and a module that reflects the influences of thermal expansion for all three machine scales. In the "calculate" section we can distinguish simulation modules: A1, which consists of six compensatory (reference) elements and A7, the module for connection and relation of shape elements. At the end, there is the U section, "uncertainty," in which, on the basis of programmed simulations, the standard uncertainty is calculated. In Fig. 5.7 the measurement process model for CMM is shown, and in Table 5.2 particular sections and modules of simulation software are presented.

Modeling of complex measuring systems in a modular form is a concept that can be used specifically for coordinate systems. The implementation of this task with the use of different concepts, for both machines and articulated arms, is described in the next sections of this chapter.

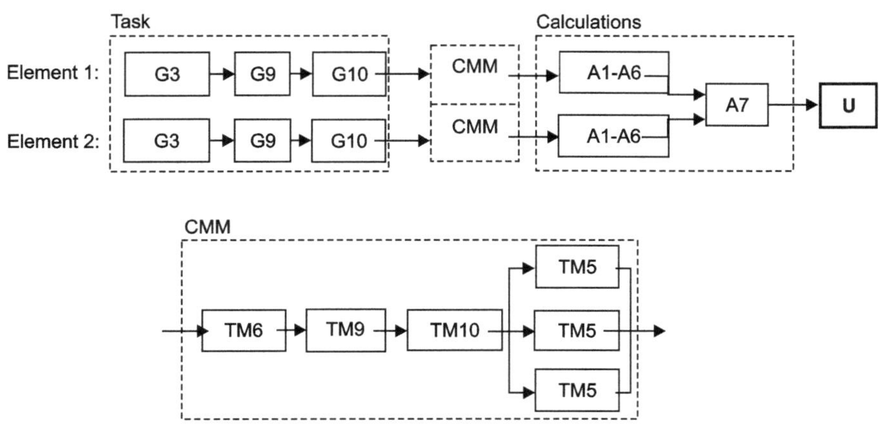

Fig. 5.7 Model of the measurement process of exemplary geometric characteristic (e.g., parallelism deviation) for CMM, including particular modules in the simulation software [76]

Table 5.2 Sections and simulation software modules for CMM

G3	Shape element generator	TM10	Machine geometry
G9	Displacement	TM5	Thermal expansion of scales
G10	Rotation	A1-6	Calculation of compensation elements
TM6	Thermal expansion of an object	A7	Connection and relation of shape elements
TM9	Contact system	U	Uncertainty

5.2 Simulation Models of Coordinate Measuring Systems

On the basis of the current literature [1, 2, 8, 9, 24, 25, 33, 36, 37, 41–46, 52–54, 63, 68, 76, 77, 82–84, 87–89, 91, 92, 96, 97, 99, 101–103, 107, 109, 113, 114, 123–128] it can be stated that the issue of the accuracy assessment of any measurement carried out with the use of a coordinate measuring system (CMS) can find an effective solution on the basis of simulation of measurements using its model, a virtual measuring system that has to reproduce the metrological behavior of the real machine or the other coordinate measuring system with the highest fidelity.

The use of virtual CMS led to re-creating the impact of not only the method of measurement realization, but also of the conditions of its carrying out. The effect of random errors is included by multiple simulations and definition of distributions of particular errors. The first work in this scope was undertaken at Cracow University of Technology in the mid-1980s, focused on CMM modeling, and was carried out under the research program CPBP 02.20 [61], but the issue of determination of particular errors in an experimental way still remained unresolved. In particular, it concerns kinematic errors. The situation changed with the development in Physikalisch Technische Bundesanstalt Braunschweig of a method for identification of these errors, with the use of plate (balls or holes) standards. That allowed the effective use of a virtual machine [126]. A simplified representation of a virtual machine for CMM is presented in Fig. 5.8, where the essence is the use of a material standard for which the identified errors are not the standard's but the errors of tested CMM.

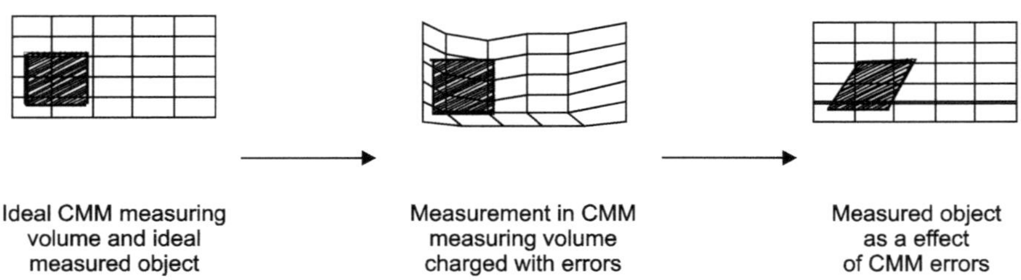

Fig. 5.8 Concept of CMM error identification based on measurement of standard objects as a basis for construction of virtual measuring machine [89]

A virtual machine system developed in the PTB was installed in the mid-1990s in the Laboratory of Coordinate Metrology at Cracow University (LCM CUT) and on its basis, virtual models of several measuring machines were built. At the same time in the LCM CUT a new computer model realizing virtual CMM under the project [44] described in Sect. 5.2.2 was developed. The versatility of this model is that it allows the identification of CMM component errors with any available method. Measurements can be performed here with a laser interferometer and level, and it is also possible to use plate standards. Another model, also built in the LCM CUT, namely the model based on the matrix method of CMM error identification in reference points with the use of artificial neural networks, described in [15, 88–91, 93, 94] is presented in Sect. 5.2.3. Further work on the modeling of coordinate systems was focused on modeling residual errors for measuring machines equipped with software error correction systems (CAA). The Monte Carlo method described in Sect. 5.2.5 [18] was used here. An important result of work on virtual systems was the development in the LCM CUT of a model for redundant systems, such as articulated arm coordinate measuring machines, where for the modeling of their kinematics Denavit–Hartenberg notation was used. Such a system is described in Sect. 5.2.4. [64, 92, 104]. The use of coordinate measuring system models in metrological practice is still a difficult task, despite development of the ISO/CD 15530-4 standard project [33] and recommendations of VDI/VDE 2617-7 [128] concerning the use and validation of these models. The implementation of commercial simulation software and its accreditation, for example, for calibration laboratories, is carried out mainly on the basis of the PTB model. Furthermore, for several years the information has been given by the Metrosage Company [58] offering the PUNDIT/CMM system, but there are no closer data on implementations of this software.

5.2.1 Virtual Measuring Machine PTB

Using the method of identification of CMM kinematic errors presented in Sect. 4.2. 2, a system that allows building a virtual measuring machine was created in the PTB. This system is based on the MegaKal [56, 75] software; the concept of the system is shown in Fig. 5.9, and the scheme and operation principle are shown in Fig. 5.10. The main objective of the operation of the virtual measuring machine PTB is to determine the measurement uncertainty in online mode for measuring tasks performed on CMM. The PTB model is based on the experimental identification or on the estimation of the following parameters: CMM kinematic errors, systematic and random CMM probe head errors, and changes in environmental conditions, which influence the uncertainty of simulated measurements.

The simulation of a measuring task with the estimation of its uncertainty requires data analogous to those requested for a real measurement such as profile parameters (size and location). The strategic elements of the simulated task (distribution of measuring points, configuration of measuring tips, position in machine volume, etc.) are also needed.

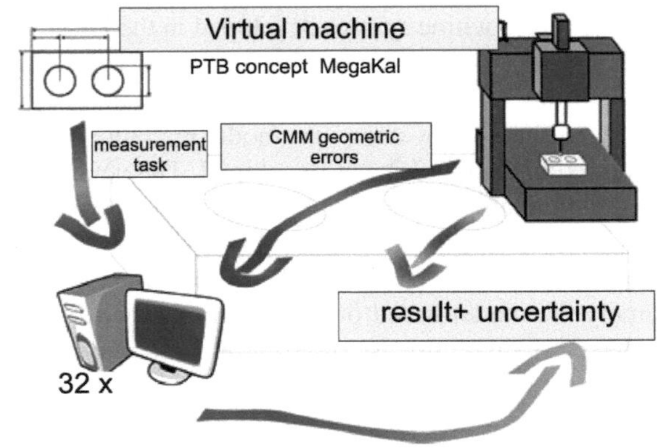

Fig. 5.9 Concept of virtual machine developed in PTB

The model presented in Fig. 5.10 uses the methodology described in Sect. 4.2.2 to determine the various error components involved in CMM model building. The calculations providing information to the MegaKal software are carried out by separate programs (modules) forming the system. These modules in the form of individual programs are shown in Fig. 5.11. The virtual machine (MegaKal) operates in accordance with the scheme described in Chap. 4. First, the measuring

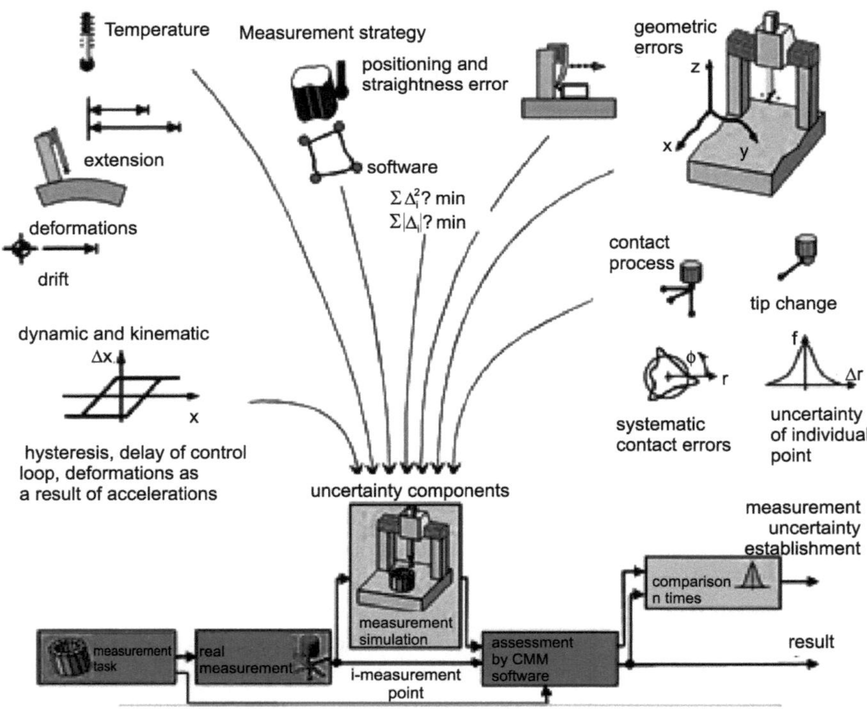

Fig. 5.10 Scheme and operation principle of virtual CMM developed in PTB [126]

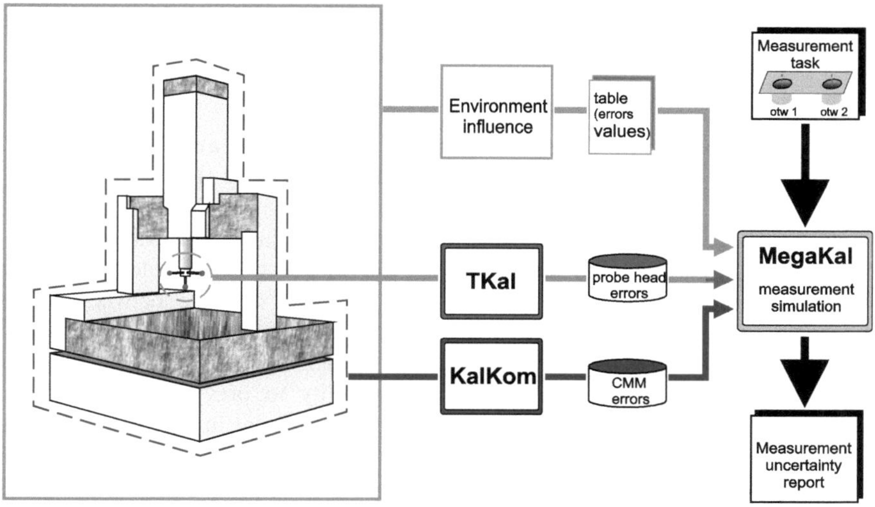

Fig. 5.11 Scheme of relations between programs in PTB MegaKal system, with marked: KalKom, basic program for the identification of geometric CMM errors; TKal, program for the identification of measuring probe head errors; MegaKal, measuring task simulation and uncertainty calculation system [89]

task (geometry, position, orientation in CMM measuring volume, distribution of measuring points, and configuration of measuring tips) should be described. Next, the measuring points for a given profile are generated automatically and the measurement is simulated. During this simulation, a group of errors characteristic for a simulated CMM is appointed to all measuring points, taking into consideration estimated uncertainties and adopted parameters of their distributions [126]. The measurement is repeated many times in a virtual way and thanks to the built-in metrological software, by randomly changing coordinates of particular points the uncertainty of a given characteristic is calculated. The system allows us to carry out the measurement at any point in the machine volume, thus the uncertainty of measurement of a given task can be determined for any location. It is also possible to determine the area in which this uncertainty takes the best (smallest) value.

In Fig. 5.12 an example of a simulation of a simple measuring task is presented. The task is to determine the distance between cylindrical hole axes (precisely: the

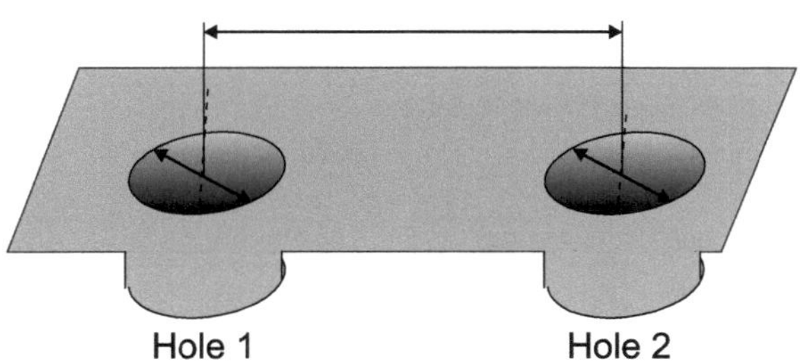

Fig. 5.12 Virtual measuring task: the calculation of the distance between the centers of two cylindrical holes [89]

distance between the points of intersection of hole axes with the front plane) and to give the accuracy of its determination as a measurement uncertainty.

5.2.1.1 Task of Distance Measurement Between Holes 1 and 2, with Determination of Measurement Uncertainty (Fig. 5.12)

Component uncertainties of hole 1 diameter and hole 2 diameter are evaluated first. Then a relation between generated elements (the intersections of hole axes with a plane) is created in order to calculate afterwards the distance between points obtained in this way. For specified points and tasks a simulation for multiple measurements is carried out in order to calculate their uncertainty.

When generating elements, there is a possibility to preview them. Exactly speaking, these generated points that create a given element in measuring volume together with the simulation of the contact points of the selected measuring tip can be previewed. The description and creation of the relationship between the elements is realized through metrological software. When the simulation process is complete, the program generates a report that is shown in Table 5.3.

Table 5.3 MegaKal

Report of a Simulated Measurement				
DEA MISTRAL 10705 basic $MPE_E = 3.5 + L/250$				
Evaluation: **7.9.1996, 12:46**				
User: **Jerzy Sladek**				
Head error: M1.TCH/−				
Parametric errors: M1.KDT				
Uncertainty coefficients: M1.UNS				
Macro: **TWOHOLMKR**				
Comment: **SIMULATION OF MEASUREMENT OF HOLE DIAMETERS AND THE DISTANCES BETWEEN THEIR CENTERS**				
Uncertainty coefficients				
XRX: 1.00 µrad/m XRZ: 0.20 µrad/m XRX: 0.30 µrad/m XTX: 4.40 µm/m				
YRX: 5.00 µrad/m YRZ: 2.30 µrad/m YRY: 0.20 µrad/m YTY: 1.20 µm/m				
ZRX: 0.30 µrad/m ZRY: 2.00 µrad/m ZRZ: 0.30 µrad/m ZTZ: 3.50 µm/m				
XWY: 0.00 µrad/m XWZ: 0.00 µrad/m YWZ: 0.00 µrad/m				
Measurements Result: Distance Point-1–Point-2				
Parameter	Nominal	Bias	Standard deviation	Uncertainty U_{95}
Point 1				
X	400.00000	−0.00875	0.00182	0.01083
Y	400.00000	−0.00022	0.00049	0.00080
Z	100.00000	−0.00127	0.00075	0.00231

(continued)

Table 5.3 (continued)

Measurements Result: Distance Point-1–Point-2				
Parameter	Nominal	Bias	Standard deviation	Uncertainty U_{95}
Point 2				
X	600.00000	−0.00890	0.00944	0.01186
Y	400.00000	0.00045	0.00068	0.00104
Z	100.00000	0.00016	0.00076	0.00122
Distance	200.00000	−0.00028	0.00093	0.00187
Head error: M1.TCH/+				
Comment: results with probe head use				
Distance	200.00000	−0.0009	0.00158	0.00319
Measurements result: cylinder 1				
Parameter: center of cylinder	Nominal	Bias	Standard deviation	Uncertainty U_{95}
Point				
X	400.00000	−0.00837	0.00181	0.01044
Y	400.00000	−0.00022	0.00049	0.00078
Z	75.00000	−0.00121	0.00059	0.00200
Direction: XY				
Axis: XZ	90.00000°	3.13″	0.11″	3.26″
YZ	90.00000°	0.03″	0.12″	0.18″
Space rotation			0.11″	3.27″
Diameter	100.00000	0.00023	0.00089	0.00179
Head error: M1.TCH/+				
Comment: results with probe head use				
Diameter	100.00000	0.0017	0.00124	0.00249
Measurements result: cylinder 2				
Parameter center of cylinder	Nominal	Bias	Standard deviation	Uncertainty U95
Point				
X	600.00000	−0.00856	0.00270	0.01151
Y	400.00000	−0.00036	0.00049	0.00092
Z	75.00000	−0.00001	0.00058	0.00079
Direction: XY				
Axis: XZ	90.00000°	2.77″	0.11″	2.89″
YZ	90.00000°	−0.80″	0.12″	0.97″
Space rotation			0.11″	3.05″
Diameter	100.00000	0.00015	0.0011	0.00223
Head error: M1.TCH/+				
Comment: results with probe head use				
Diameter	100.00000	0.0008	0.00138	0.00276

The example was carried out for the model of the DEA Mistral 10705 machine developed in the Laboratory of Coordinate Metrology at Cracow University of Technology. This is a mid-accuracy machine made of light metals. According to the producer's data $MPE_E = 3.5 + L/250$. The machine worked in the air-conditioned laboratory with the Split system, having the thermal stability of ± 1 K. During the measurement, the temperature was registered in constant intervals, and then included in the form of mean values in the virtual machine system. In the example, results were varied by including or ignoring the impact of probe head errors. The measuring probe head TP2 6 W made by Renishaw was used. When analyzing simulation results a dominant influence of probe head errors can be observed, especially in the field of random errors. In addition to the ability to simulate a variety of measuring tasks, MegaKal has also a function that allows us to simulate the measurement of many known lengths in the whole measuring volume. A three-dimensional graph showing the distribution of length measurement errors together with the equation describing the uncertainty of simulated measurement is obtained as a result. This corresponds to the typical procedure for CMM accuracy assessment through a control length measurement (ISO 10360-2 standards). MegaKal can be used to formulate a given CMM maximum error equation. Figure 5.13 shows the result of a simulation procedure of length measurement, performed for the DEA Mistral 10705 machine model, together with its simulated equation of maximum errors. One of the main advantages of the PTB method is the

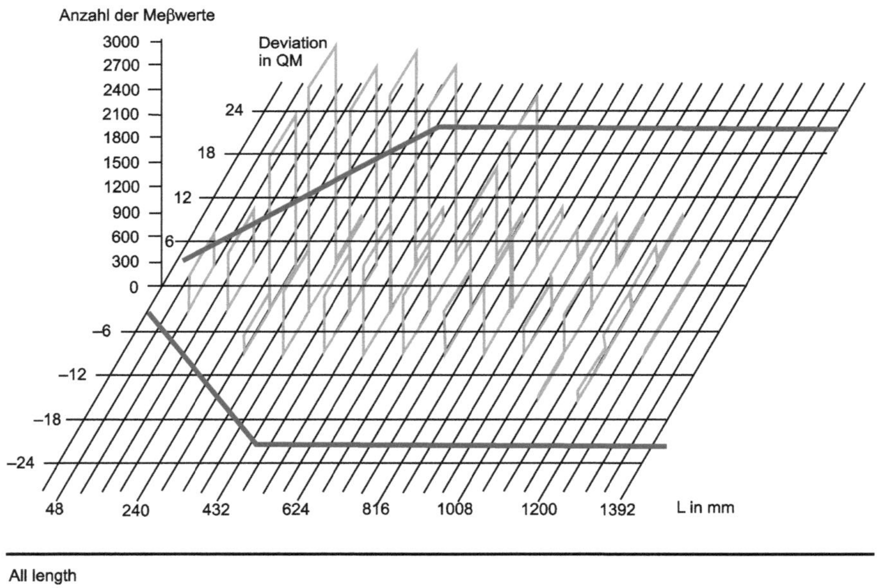

Fig. 5.13 Length measurement errors obtained by simulation for the DEA Mistral machine (accuracy provided by the producer $MPE_E = 3.5 + L/250$) model (MegaKal program) with marked area of determined maximum errors [89]

use of a plate standard to identify the kinematic errors (KalKom). The uncertainty estimation process is done by acceptance of component uncertainties according to the scheme shown in Fig. 5.10. This method assumes a dominant contribution of probe head errors in the shaping of the random error range, but associates systematic errors mainly with kinematics. This is confirmed by a measuring task simulation example with the MegaKal system (Table 5.3), and extensive test results contained in [42]. In effect, the system works very well under laboratory conditions, if they are known, and under ensured stable external conditions, and if the variability of particular parameters is controlled. Using the MegaKal PTB system in the LCM at Cracow University of Technology very good results for the accurate machine operating under strict air-conditioning conditions were achieved. However, the use of the system in conditions similar to an industrial one (laboratory air conditioned within ± 2 K limits) became problematic, especially when the average accuracy machine, made of light alloys, was modeled. Then both the identification of kinematic errors and also the uncertainty estimation became difficult. Therefore, under the research project [44], an effort aiming at the development of a universal metrological CMM model correcting the faults of the PTB model was undertaken.

In conclusion it should be noted that in order to use this system under industrial conditions, it is necessary to use a different approach to assess the impact of component uncertainties (marked in Sect. 4.1 as M5–M9). The external conditions of the CMM room should be monitored in a long-term manner, including not only the temperature measured at several points (as, e.g., in Sect. 3.5), but also vertical and horizontal thermal change gradients. This makes it possible to determine not only changes in the range of individual components, but also their uncertainties. Thereby it becomes possible to determine their effect on model parameter changes. But the use of the PTB system in the LCM CUT finally was modified so that for calculations the metrological software of the modeled machine was used and the module of the virtual machine installed as operating in a real measurement background. A similar principle was the basis for the development of Quindos (Leitz Messtechnik) and CALIPSO (Zeiss) [127] software options which was carried out under a program financed by the European Union [126] (Fig. 5.14).

Currently, software with the virtual machine option can be installed by two specialist laboratories: FEINMESS [12] (for Quindos system, Leitz Company) and Eμmetron [10] (for the CALIPSO system, Zeiss Company) [71, 128]. These options have been installed in several laboratories, including the Messzentrum QFM at the University in Erlangen-Nuernberg [11] which obtained the DKD-1 accreditation for calibration measurements with the use of a virtual machine in the version with the CALIPSO software. However, this model belongs to solutions used mainly in laboratories with very stable thermal conditions in which temperature changes are not higher than ± 0.2 K. Attempts to use it for machines operating in industrial conditions did not bring satisfactory results.

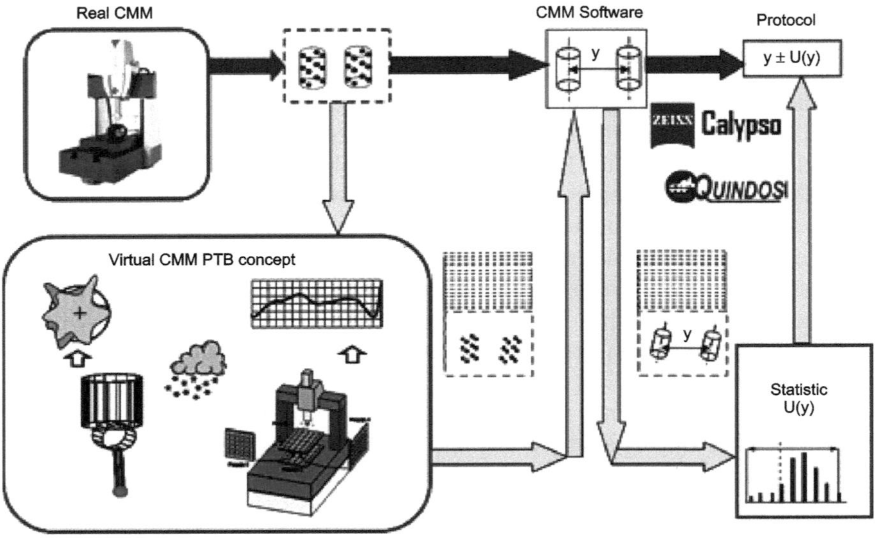

Fig. 5.14 PTB virtual machine system cooperating with the modeled CMM metrological software

5.2.2 Universal Model of Coordinate Measuring Machine: Virtual CMM CUT

Initial work on the construction of VCMM in order to use it to determine the accuracy of measurements has been carried out from the mid-1980s at Cracow University of Technology and used to evaluate the circular contours [41, 61]. During continuation of the project [44] the purpose was to develop a universal CMM model for which component error data could be determined using various possible methods (laser interferometer, electronic level, or plate standard). Differently from the PTB model, the new CMM model should consider various possibilities of component error identification. Especially for large machines, where a laser interferometer is deployed, the method using the material standard shown in Fig. 4.17 cannot be used. Moreover, the new model should be suitable for laboratories with a lower thermal stability (±2 K), which makes it necessary to determine experimentally the values of the component errors at given points. The evaluation of their dispersion should also be performed. Similarly as in case of the PTB model, MegaKal, the built model is equipped with its own metrological software that performs the basic functions of shape measurements. The created measuring machine model is based on software overlapping of CMM component errors, according to the accepted model using the theory of the vector error field (described in Sect. 2.3). It is also based on determining for each point of the measuring volume of a real measuring machine an appropriate error vector that is the sum of its all components:

- Kinematic errors
- Measuring probe head errors
- Errors of the measuring system displacement
- Errors resulting from the impact of measured object
- Errors related to temperature changes

In addition to the included CMM error vector, as in the PTB model, the virtual CMM should allow the assessment and optimization of measurement accuracy in the dependence of set requirements through the analysis of many possible options and strategies of measurement realization or the impact of an object location in the machine volume. It should also allow the determination of the minimum and maximum errors or their ranges for different measuring tasks. In developing the CMM model [44], the authors analyzed the kinematic errors. The form deviations of slides are the main reason. Thereby the measuring tip will move through the spatial curve. The path of the measuring stylus tip movement is a consequence of the overlapping of linear and rotation movements in particular axes. In general, it is assumed that the movement of the stylus tip is described by 18 components of summary error and 3 components of measuring machine coordinate axes squareness error. In the case of an error model that assumes a rigid mechanical structure of the CMM, the model shown in Fig. 4.1 can be assumed.

According to the assumption, which was adopted during development of this model, the identification of component errors can be realized by using any possible system, including the recently introduced laser tracer.

The simulation process of a coordinate measuring machine measurement runs according to the scheme presented in Fig. 5.15. It is based on the parameter values of all CMM component errors that have already been determined experimentally and introduced to the program. Additionally, data concerning types of probe heads and measuring styluses used and information on the measurement process course are given. In order to determine the measurement uncertainty, the coordinates of measuring points of a given measured object generated or measured before on real CMM are introduced into the system. Next the distribution of points on the element surface (determination of the sampling strategy) and setting of an object in the measuring volume of a simulated CMM (definition of location and orientation) are defined. Initially, the wanted element's parameters are calculated by the program with the use of expansion calculus (the Gauss least squares method) on the basis of coordinate values of all the points of the given geometric object. The first set of parameters is obtained. Then, to the same surface coordinates of points (measured by CMM), second coordinate values of machine error vectors are added. From the thus-received coordinate values of points, corrected by listed components, the metrological characteristics of the measured element are determined. In order to include random influences the virtual measurement is performed 32 times. During simulation random errors are generated. They change with each measuring point with a certain probability distribution, usually set in the range of ±2 d. In this area, contact probe head errors with their determined uncertainty, together with permissible changes in environment temperature and changes in the machine

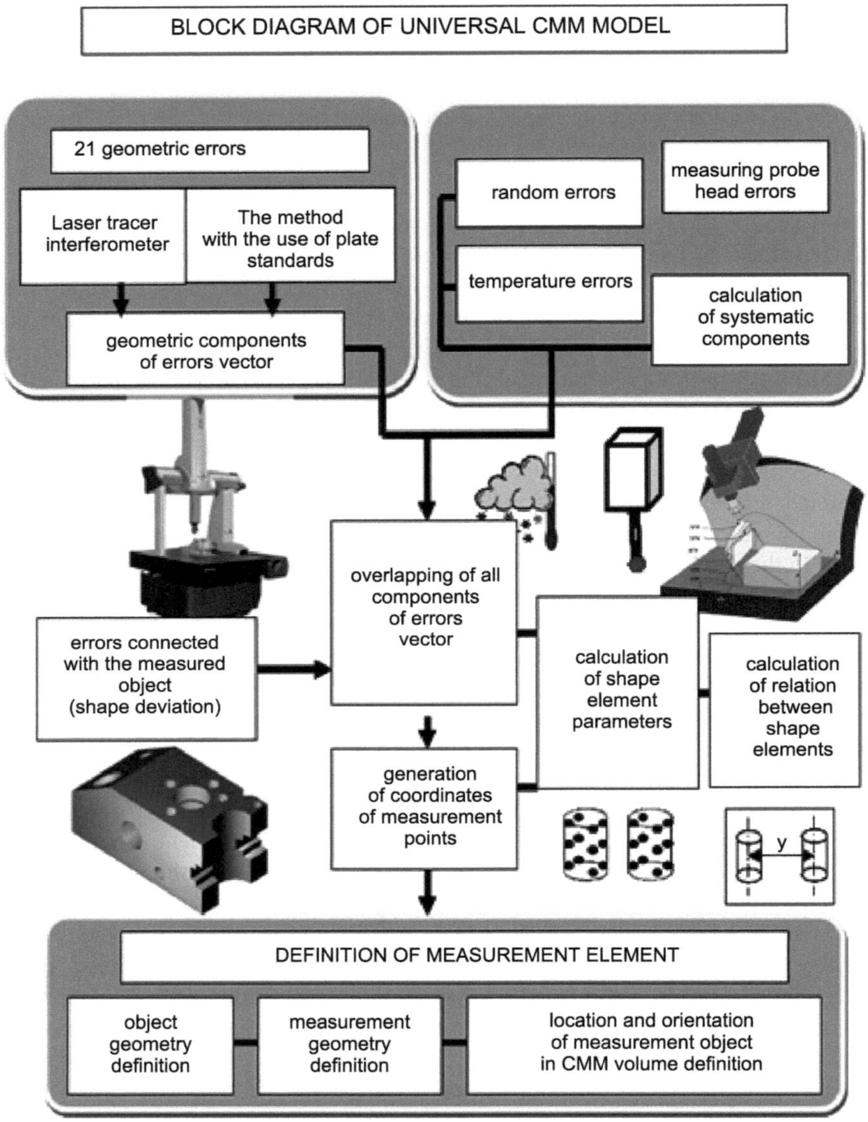

Fig. 5.15 Block diagram of CMM model developed in the LCM at Cracow University of Technology [44]

temperature distribution are included. Sequentially obtained sets of parameters are saved after each simulation and finally their values are used to perform some statistical operations. Final results include calculated average values of geometric parameters of the object together with their expanded uncertainty as a measure of the measurement accuracy.

When the system is used as a research system, for example, for the analysis and optimization of measurement strategies, we use ideal point coordinates generated by the program for measured surfaces. Then a set of nominal points is obtained, for which the errors calculated for particular points using the developed error model are added (summing of following deviation vectors in the following point). This corresponds to obtaining them as during the real measurement on CMM.

The CMM model—in the form of virtual CMM CUT software [134] written in the C/C++ programming language—is composed of many interdependent subprograms and procedures that together carry out a simulation of the real CMM measuring system operations. Algorithms are written in particular modules in the form of functional blocks, corresponding to the theoretical concept presented in Fig. 5.7 and Table 5.2. Communication with the user is performed via a communication interface that allows the interactive determination of the program course. Main procedures and algorithms, focusing on the metrological software module, are described in detail in [44].

An important advantage of the model was its flexibility because of the possibility to use direct measurement data. Under work on [91] the virtual CMM CUT model had been significantly modified, because for the realization of this project an open CMM model that was able to operate as a reference was necessary. This system was the only one allowed to receive information about the kinematic errors and CMM probe head errors identified by various methods. Therefore the specially written, proprietary metrological software, which had been used thus far, was changed. Currently, this system operates with Quindos and PCDMIS software and also due to the interface compatible with I++ DME it may be connected to any software that uses this interface. The scheme of universal system Universal Virtual CMM CUT [44] is shown in Fig. 5.16. This software is still a reference for other virtual systems [85, 86, 89] developed in the LCM CUT.

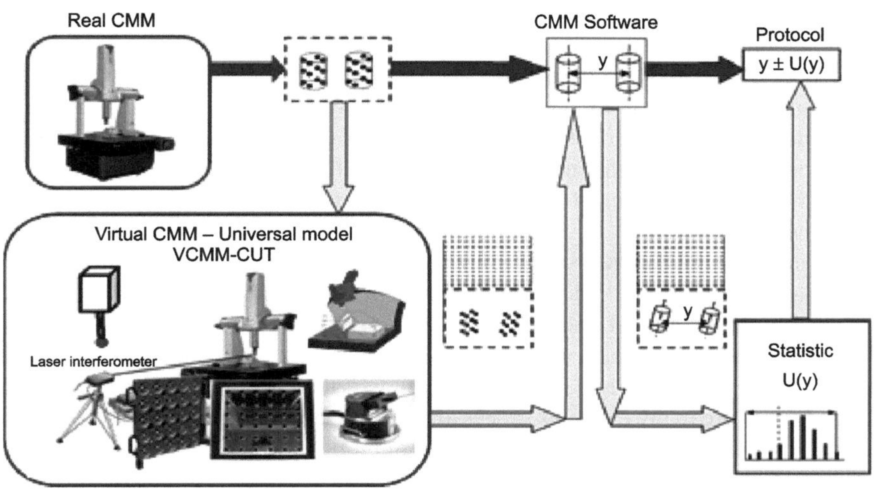

Fig. 5.16 Scheme of a universal virtual CMM CUT system

This system, however, due to the need of identifying component errors, was difficult to use and required both a lot of experience and expensive research equipment. In addition, tests especially for industrial conditions were very long term and difficult. Therefore another simpler system allowing for modeling errors and their identification was necessary. Such a solution was a model using artificial neural networks.

5.2.3 Virtual CMM Based on Artificial Neutral Networks

The critical analysis of CMM models described in Sects. 5.2.1 and 5.2.2 was a sufficient premise for searching other methods for their construction. The nature of CMM errors, a large number of components, and their not fully understood mutual interaction in various measurement conditions contributed to investment in research on methods being more universal and simpler in use. In addition, a deep knowledge of mathematical description of CMM errors and its preparation should not be a purpose for the virtual machine constructor. The purpose of the virtual machine construction is first of all a simulation of the real measurement under repeatability conditions, to enable the later assessment on the basis of the uncertainty of realized measurement. The construction of the mathematical model based on kinematic errors is very difficult [41, 44, 89, 123–126] and requires far-reaching simplifications. Another aspect of this issue, which poses many difficulties, is (as shown in Sect. 3.5) the estimation of component uncertainties coming from various error sources. Additionally, in assessment of measurement accuracy, CMS accuracy standards of known calibration uncertainties are usually used. Therefore the concept of a new method that better meets the described above requirements based on knowledge of a set of reference point coordinates was developed. When analyzing work [27, 39, 59, 119, 120] it was noted that features of many models based on the use of neural networks converged with the problems that occur during CMM model construction. The most important are the possibility of predicting the values and direction of error vectors for any points in the CMM measuring volume, the ease of model development due to taking new components of the summary error (the probe head and temperature influence), and the possibility of modeling accuracy in the global form of measurement uncertainty without the necessity of particular error components estimation.

Artificial neural networks are informatic-mathematical tools and their origin dates back to the 1940s. They were used then for the first time as a calculation model. The first artificial neuron model was created in 1943 by McCulloch and Pitts. For many years, they were not in use due to the low processing power of computers. It was a problem because according to their origins they were a tool in which there were many elements processing provided information without requiring parallel programming, but using only the learning process [55]. Thus they could be used to try to resolve the discussed issue, which was realized in [89]. Possible applications of neural networks in signal processing are still not fully explored, and

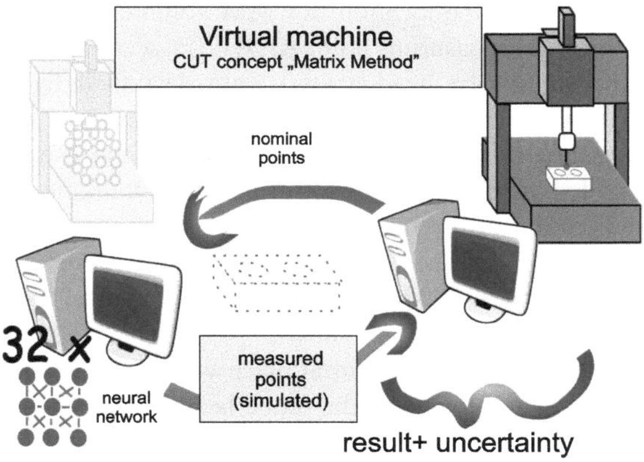

Fig. 5.17 Use of neural networks for modeling CMM errors

it seems that they will be deciding the progress in information technology for many years. Therefore the approach based on the analysis of signals described in Sect. 5.1 opens up new possibilities for measuring system modeling with use of this tool.

Due to analysis in [83] assumptions that motivate the application of artificial neural networks for modeling CMM errors were formulated:

- Rejection of analysis of various CMM errors sources
- Opportunity to build a model based on errors in chosen reference points
- Possible small number of measuring points required to build a CMM model
- Use of standards to build a set of reference points
- Possibility to predict errors occurring beyond the reference points
- Possibility to use the model to correct CAA matrix errors of modeled CMM

Formulated assumptions allowed the application of neural networks to construct a virtual machine. The scheme of this concept based on signal analysis is discussed in Sect. 5.1.

The use of artificial neural networks to build a CMM simulation model including data, which are achieved through the application of the matrix method described in Sect. 4.4 is presented in Fig. 5.17.

5.2.3.1 CMM Neural Model Associated with Matrix Method (MM) of Error Identification

The development of a new CMM model based on a neural network was carried out under the work [89]. The plate standard was used to create a reference point grid in which the error discrete field characteristic for a given CMM was determined. For creation of a reference point grid the matrix method was used, described in Sect. 4.4 , as shown in Fig. 5.35. Errors were identified in reference points of a regular grid created by the hole centers of a plate standard. They were used to model errors dependent on the location in CMM volume, and were described by reproducibility error of the measuring point REMP (in accordance with Sect. 2.3). Errors of point gathering (probe head errors) were described by the probe head error function (a) (defined in Sect. 4.3). Random errors appeared mainly in the contact process. As an assumption, the system was based on the use of metrological software of the modeled machine. The measuring task simulation process was performed in the background of the real measurement.

In Fig. 5.18 an operation scheme of a model using artificial neural networks, **Virtual Neuro CMM** and a plate standard used to determine the REMP error field of a modeled CMM are shown.

In this concept the central place was taken by the neural model of the error grid and the neural model of the probe head. As the modeled object under the work [86] the PMM12106 machine working with the metrological software Quindos was adopted, as described in further works included in Sect. 5.2.3.6 and published in [15, 93, 94, 105]. Models for other machines of the DEA Group using PCDMIS metrological software were built as well.

The **Virtual Neuro CMM** system has been designed so as to allow operation in a given range of temperatures. Therefore there is the need to repeat tests for a selected range of temperatures with a fixed interval between the temperature limits.

In the developed model in accordance with the algorithm of the developed system shown in Fig. 5.19, the obtained data in the form of nominal coordinates of the measuring point are associated with the error calculated by the neural network depending on the location. In that way the error vector in a given point is

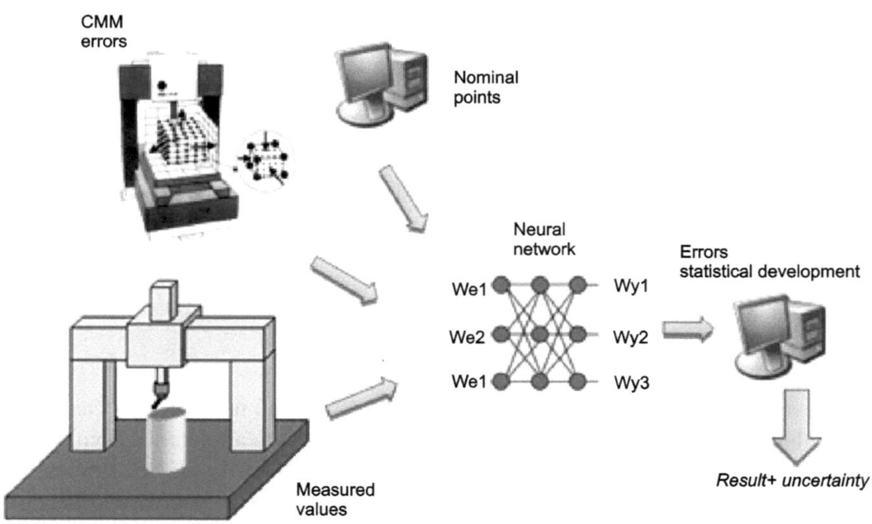

Fig. 5.18 Operation scheme of the model based on determination of the field of reproducibility errors of measuring point REMP

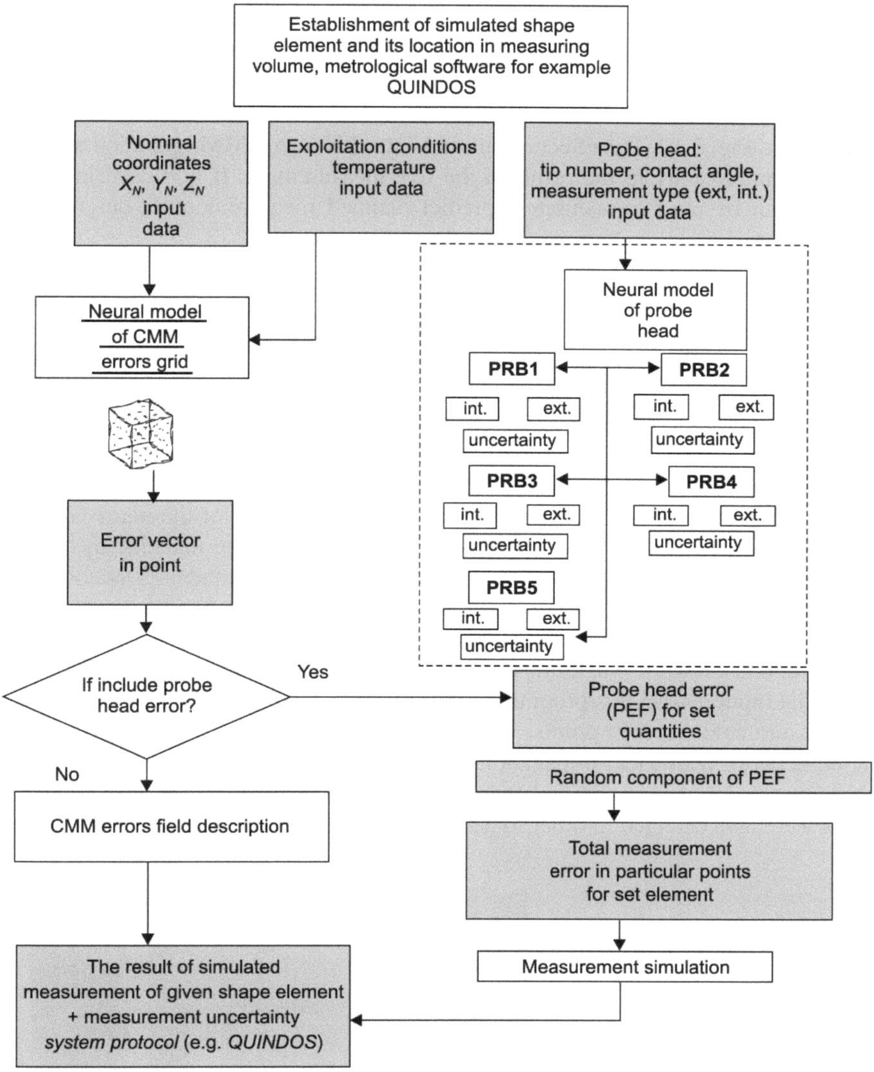

Fig. 5.19 Virtual Neuro CMM system algorithm [89]

formulated, and then the decision about including the probe head errors is made. This alternative is needed due to the possibility of the use of the developed model to correct the matrix (CAA) of the modeled CMM. If we decide to include probe head errors in the model (Fig. 5.19), then data for a proper tip should be selected depending on the specific measuring task. The HA CMM machine works with a variety of styli and for each one it is necessary to collect the necessary data. The model is based on the use of the contact system in the so-called star system, shown earlier in Fig. 4.29. Therefore, the probe head neural model was constructed in this way. All information about summed errors in a given measuring point go back to Quindos software where appropriate measurement simulation and calculation of geometric characteristic value and the uncertainty of its measurement is performed, as shown in Fig. 5.19. The effect of online **Virtual Neuro CMM** operation adds the simulated uncertainty to the result of the real measurement. If it is used in offline mode it can be used to accurately predict planned measurement as described in Sect. 5.2.3.6.

5.2.3.2 Construction of a Grid that Models Errors Dependent on the Location in CMM Volume

The characteristic of CMM accuracy based on the grid of reference points, and on errors determined in them was adopted as the basis for the construction of a neuron network that metrologically models the operation of the machine. A grid of 325 reference points in accordance with the conditions of the use of the matrix method (described in Sect. 4.4.2) was built exactly in the center of the machine measuring volume, which is the area chosen most frequently for measurements. Data obtained in the form of errors in node points, after conversion to the correct format, are the basis for the learning set for the neural network (Fig. 5.20).

The neural network was built using the NeuroShell2 v 4.0 program [30, 116–118]. The input data in this program are coordinates of the standard and output data are measurement errors in points. The best results were obtained for the three-layer neural network with back-propagation of error and with various forms of activation. The network has been modified many times, and the learning set and activation forms were also changed in order to get its optimal construction. The scheme of this

Fig. 5.20 Reference point grid and thermostable hole plate from Robax with a coefficient of thermal expansion +0.5^m/m/K of the "sandwich" construction type made according to the author's project by the Feinmess GmbH company and calibrated in the Laboratory of Coordinate Metrology at Cracow University of Technology with the swing-round method used for CMM model tests

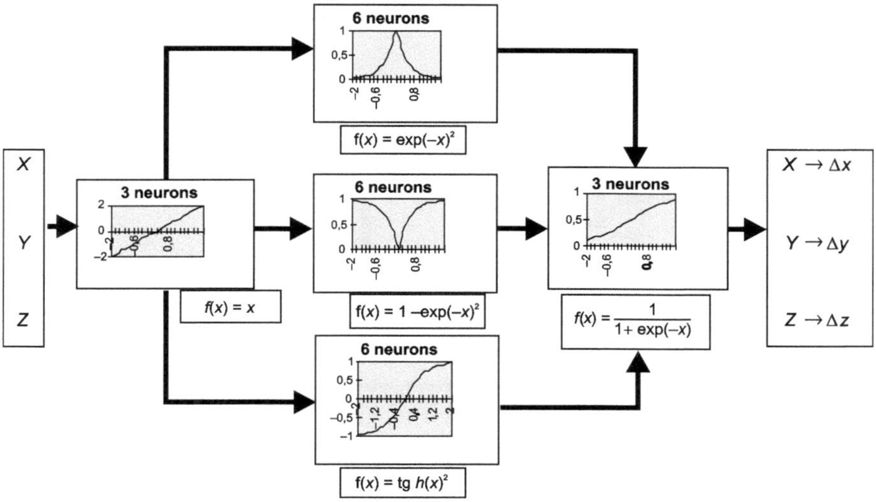

Fig. 5.21 Neural network construction used in CMM virtual model including adopted activation functions

network is shown in Fig. 5.21, also showing the finally accepted activation forms and numbers of neurons for each layer.

The division of the dataset on learning, testing, and checking is realized by the NeuroShell2 program. It is only needed to define the percentage division of these data. This division has also been done experimentally and was different for various datasets (dependent on their frequencies).

The NeuroShell program allows us to determine the structure of the network and of particular coefficients used for network learning, the modules of which are marked as Slab $1 - n$, as shown in the program window posted in Fig. 5.22.

5.2.3.3 Neural Model of Contact Probe Head Errors

An important element of the model is the contact probe head module. During modeling 10 artificial neural networks (five for external measurements and five for internal measurements) were developed. For the probe head, as well as for the model of errors independent of locations of kinematic sets, networks were built experimentally. Finally their construction was adopted analogous for errors independent of the location. The differences relate only to the number of neurons in each layer, presented in Fig. 5.23. Similar actions were taken with the percentage division of learning, testing, and checking data. In order to identify probe head errors, according to the MM method and on the basis of [80, 81] a probe head error function was defined (Sect. 4.3.1). The PEF systematic part was determined by assuming mean values for each of 64 locations from 32 repetitions as most probable

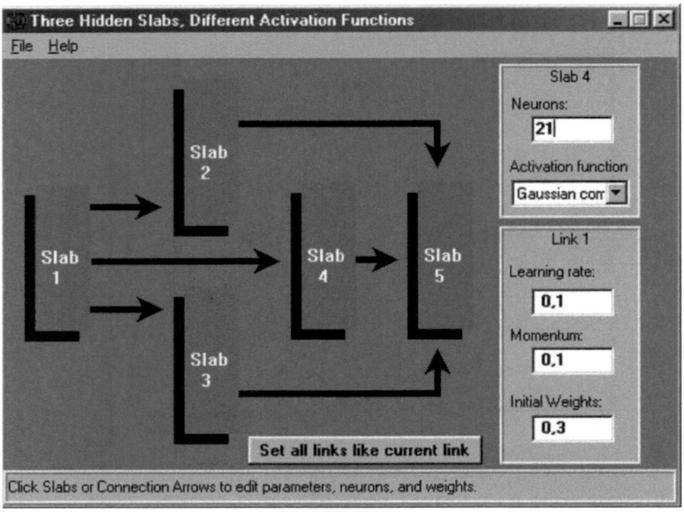

Fig. 5.22 NeuroShell software window showing the neural network construction

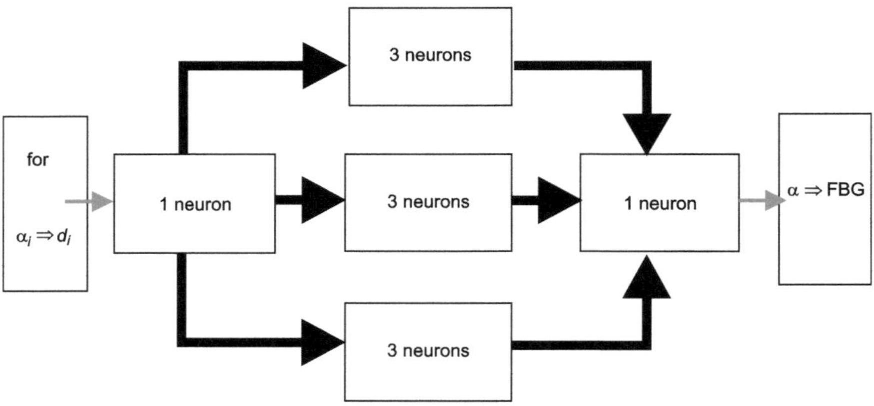

Fig. 5.23 Construction of neural network modeling contact probe head errors

values modeling real values of PEF. Measurements were performed in a polar system for standard ring Ø 25 (Fig. 5.24) and as a result five datasets were obtained, from which each contained 64 (accepted number of measuring points) pairs: angle-deviation (PEF value). In Fig. 5.25 measurement results of the ring standard for the PRB1 tip are shown in the polar system giving the minimum, maximum, and average values from each point set for the given angle of approach (contact) of the tip. In the developed model an external measurement was also distinguished, for which the construction of the calibration dataset was similar to that for the internal measurements (of a standard ring). A different standard was used: a sphere with a

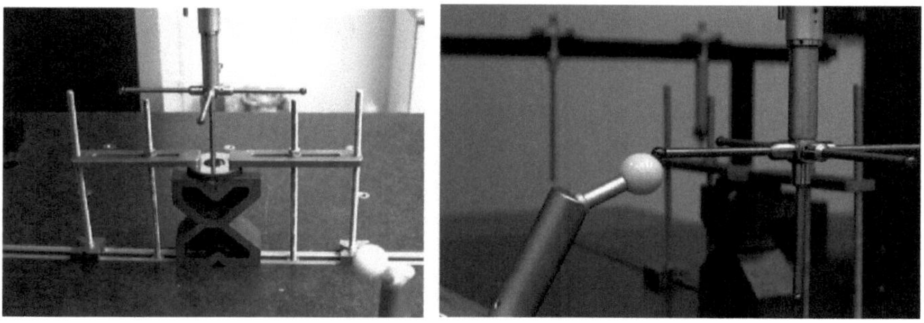

Fig. 5.24 Measurement of the ring and reference sphere, the external and internal measurement in order to identify probe head error function (PEF) [89]

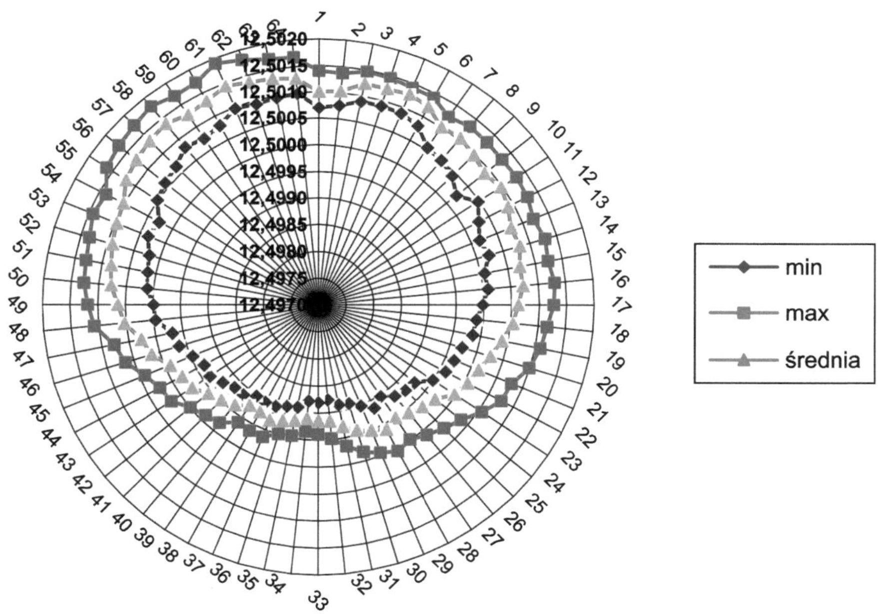

Fig. 5.25 Example values of PEF for the PRB1 tip of PMM12106 machine probe head as data to build a neural network for this probe head

diameter of Ø 25 (Fig. 5.24). In that way a total number of 10 neural networks (five for the internal and five for the external measurement) were obtained for the modeled probe head, as the next element of the CMM neural model. In Fig. 5.26 the process of fitting the network to the probe head learning data is shown graphically.

The last part of the CMM virtual model is a simulator of the random part of the probe head error function. To build the simulator, according to [80, 89], 32

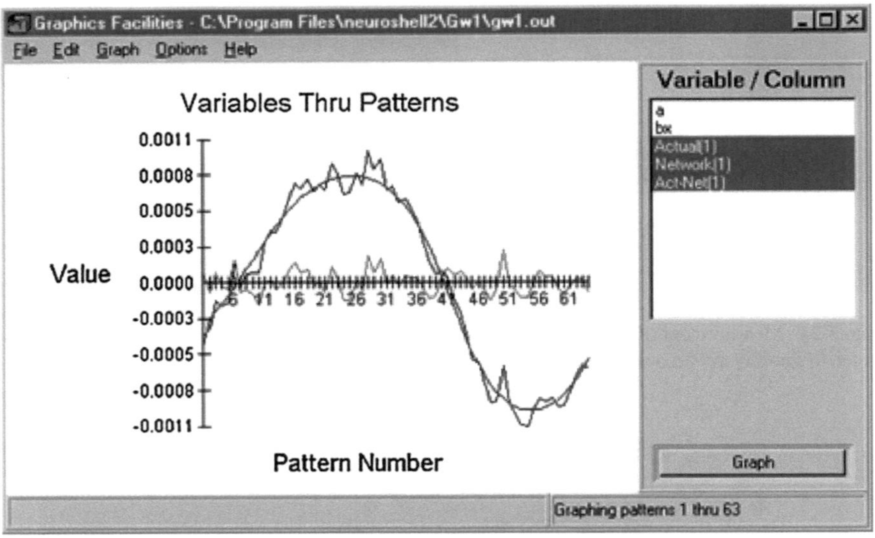

Fig. 5.26 Graphical presentation of quality of network matching to learning data for the probe head

measurements of the reference ring in 64 points were made. It was assumed that the division on 64 points models the variability for all points (for any contact angles) well enough. In Fig. 5.27 the obtained values of standard deviation of the repeated 32 measurements for a given approach angle (64 angles) of PRB1 tip (as shown in

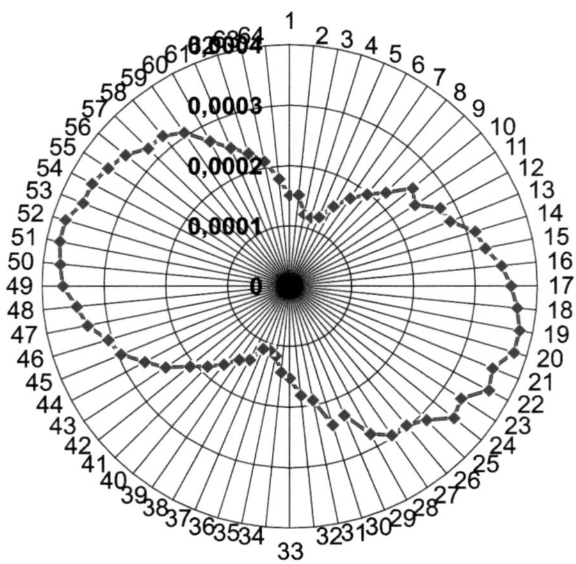

Fig. 5.27 Example values of standard deviation for the PRB1 tip of PMM12106 machine probe head as data to build a simulator of PEF random part

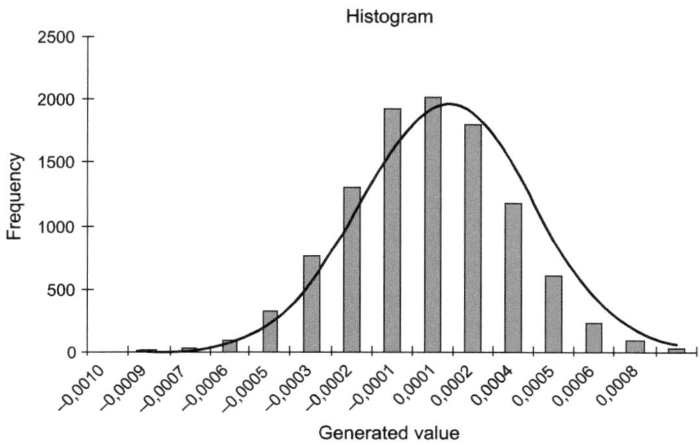

Fig. 5.28 Histogram of simulated distribution adopted for the modeling of probe head random errors [89]

Fig. 4.34) are shown in a polar system. However, to cover the range of any tip deflections (not only for the 64 deflection directions) it is necessary to model possible values of standard deviations for any intermediate points. This task was realized by an additional neural network. The construction of this network is analogous to a network prepared for modeling probe head systematic errors.

The obtained results allowed building an experimental distribution, which was tested for compatibility with a normal distribution. The histogram built on the base of 10,000 simulations is shown in Fig. 5.28. Finally for random errors of the contact probe head a normal distribution generator was developed.

In the network learning process the data pairs were adopted: a point on the standard surface and associated with it the angle between the measurement direction and a normal to the measurement surface and the standard deviation determined for a given location. As a result the network modeling changes of the standard deviation dependent on the contact direction of the measuring tip were obtained. The operation of the complete simulator consists of two parts: the first sets the value of the standard deviation, and the second starts the generator of the normal distribution for simulation of random part of the probe head error.

5.2.3.4 Simulation of Measuring Tasks—Virtual Neuro CMM CUT Application

The primary purpose of the Virtual Neuro CMM, which is the work in the metrological software background and calculation of uncertainties of realized measurements using the simulation method, requires writing a computer program that realizes the algorithm given in Fig. 5.37. For this purpose the Microsoft programming tool Visual C++ 4.0 was selected. The central assumption was that the program has to cooperate and use the calculation procedures of metrological software of modeled CMM. Because of that the part responsible for error simulating was developed as a program working in a test course starting with the command line with appropriate parameters. Due to this, the application can work in a hidden course, and also can be started by other programs. The file with program output data was constructed in a manner to be read by metrological software, in this case by the Quindos NT system. The scheme of cooperation of written application with this system is compatible with the algorithm presented in Fig. 5.19. The input data are downloaded by the program, and therefore the coordinates of nominal points are processed in the program by the function that supports the network modeling machine errors. The result consists of errors in machine volume points obtained as a result of the operating network. Thanks to the tip name, access paths to files containing modeling networks are found. The argument for the function that supports network modeling a probe head for a given tip is the angle of approach (contact) with the measured surface, which is downloaded from the metrological software. The parameters that can be configured before, during program starting, are: the name of the file with information about points, which may be simulated and the name of the output file in which points calculated by the virtual machine will be saved. Other data required for calculations, such as the number of repetitions and access paths to files in which information about appropriate neural networks is collected, are in the initialization file, vcmm.ini.

The algorithm of proceeding during performing the cycle of CMM error simulations in a given point of its measuring volume is shown in Fig. 5.29, and in Fig. 5.30 application screens from Virtual Neuro CMM CUT C++ are shown [89, 133]. According to it, the input file contains in a single line the following information:

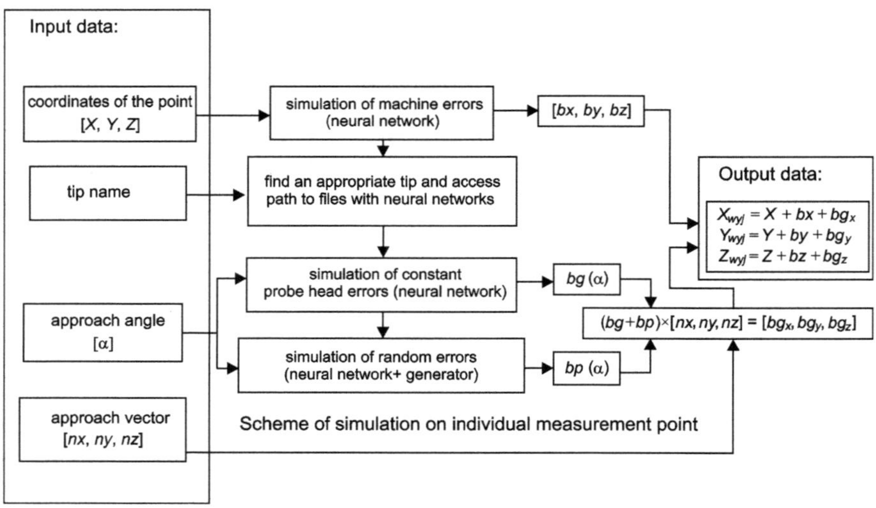

Fig. 5.29 Simulation scheme of individual measuring point [133]

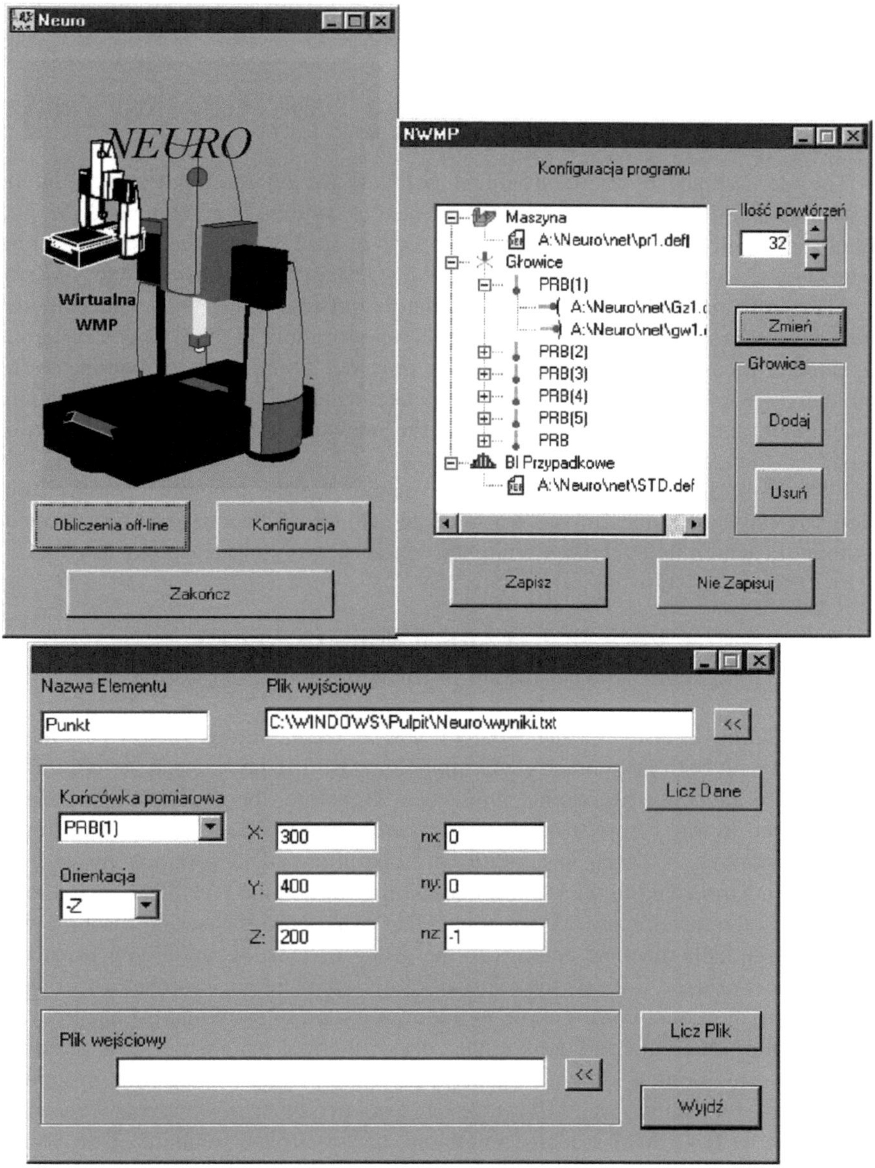

Fig. 5.30 Application of Virtual Neuro CMM CUT C++ [133]

- Coordinates of the point (x, y, z)
- Contact angle (approach) a
- Approach vectors (nx, ny, nz)
- Name and marking of the tip

An exemplary line of that file has a form:

123.49232 108.55050 500.00000 340.00000 −0.94 −0.34 0.00 PRB(1).

The file accepted by the metrological software (in the described case, Quindos system) can also be freely modified and adapted to conditions of a given CMM, and more specifically, to the software that supports it. In this case, the use of different metrological software, that is, of the PCDMIS system was confirmed and verified.

The simulation and uncertainty calculation process itself takes place in the metrological program. Therefore, the simulation procedure in the Quindos system is repeated n times (assumed $n = 32$), each time obtaining: a set of parameters of simulated geometric elements or relations occurring between them, the standard uncertainty, and next **the uncertainty of the measurement result** in the form of expanded uncertainty.

In the fully automatic course, the user cannot see the program operation, but observes only the effect in the form of an additional line in the measurement protocol in which its uncertainty is given together with the measurement result.

5.2.3.5 Research Verification of Virtual Neuro CMM CUT Application: Determination of Measurement Uncertainty

When planning tests of Virtual Neuro CMM it was assumed that error sources associated with both its geometry and applied measuring probe head should not be disturbed by unidentified external influences. Therefore, the construction of learning sets, and testing of prepared system required tests realized for a high-accuracy machine working in strict air-conditioning conditions. The adoption of such an assumption is reflected in the high reproducibility of obtained results, which allows us to verify the developed CMM model. The PMM12106 measuring machine was used in tests. This machine belongs to the group of most accurate devices of this kind in the world, so it meets the assumptions adopted above. Another assumption adopted during model construction, in addition to that of stable CMM accuracy, is the control of external conditions, especially of thermal parameters. It is provided by the air-conditioning system together with a temperature monitoring system, a description of which is given in Sect. 3.5. In tests two types of standards were used. In the case of HA CMM machine modeling, the steel plate standard of the Retter Company with a coefficient of thermal expansion of 11.5 μm/m/K was used. It was thermally stabilized by 48 h near the measurement place, obtaining the maximum change in standard temperature $\Delta T_W = 0.1$ K. Another standard that was used was the thermostable standard made of Robax with a coefficient of thermal expansion equal to 0.5 μm/m/K shown in Fig. 5.31. This standard is used especially to build Virtual Neuro CMM models for machines constructed of light metals, as shown in Fig. 5.47.

Fig. 5.31 Thermostable standard of Feinmess Company made of Robax, during tests on HA CMM machine in the Laboratory of Coordinate Metrology at Cracow University of Technology. The distribution of thermal sensors to monitor thermal changes can be also seen. The system used is Almemo 5591 of the Ahlborn Company [89]

The developed Virtual Neuro CMM model was tested in several stages. The first tests were carried out for the individual measuring points distributed along the diagonal of the modeled measuring volume (these tests are described in [89, 98]). As a reference system MegaKal software was adopted with the modifications described in Sect. 5.2.1. It was used together with the Quindos system. Their results are given in Table 5.4. In Fig. 5.32 comparative test results obtained for both Virtual Neuro CMM and for the MegaKal–PTB system are presented.

Table 5.4 Comparison tests results for classical and neural model of virtual CMM

	Nominal coord.			Virtual Neuro WMP			MegaKal		
	X	Y	Z	BX	BY	BZ	BX	BY	BZ
1	100	100	550	−0.00236	0.00029	−0.00243	−0.00346	0.00188	−0.00178
2	200	180	500	−0.00236	0.00029	−0.00243	−0.00248	−0.00033	−0.00293
3	300	260	450	−0.00205	−0.00031	−0.00242	−0.00254	−0.00090	−0.00294
4	400	340	400	−0.00108	0.00035	−0.00226	−0.00148	0.00003	−0.00212
5	500	420	350	−0.00034	0.00055	−0.00081	0.00010	0.00066	−0.00054
6	600	500	300	−0.00017	0.00091	0.00098	−0.00003	0.00085	0.00101
7	700	580	250	−0.00034	0.00063	0.00136	−0.00051	0.00058	0.00170
8	800	660	200	0.00016	0.00019	0.00157	0.00123	0.00036	0.00222
9	900	740	150	0.00073	0.00004	0.00188	0.00076	0.00084	0.00140
10	1000	820	100	0.00097	−0.00001	0.00195	−0.00041	0.00029	0.00024
11	1100	900	500	−0.00243	0.00046	−0.00068	−0.00216	−0.00056	−0.00103

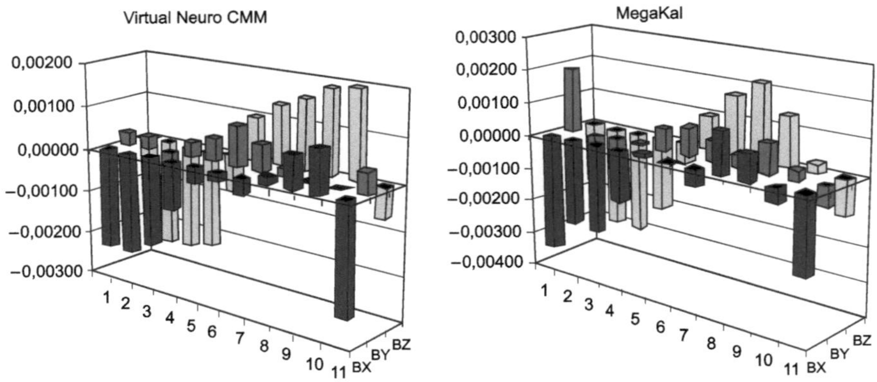

Fig. 5.32 Comparison of point coordinate values for two different CMM models [89]

Comparative test results have confirmed the adopted methodology on which the construction of learning set is based, especially in the part of the area on which the reference point grid was built. Results obtained for particular points allowed carrying out model tests based on real measurements for typical, appropriately selected, shape elements, presented in Fig. 5.33.

In Fig. 5.33 the reference ring of diameter 104.9998 mm, reference sphere of diameter 99.9996, and a reference cylinder of diameter 100.2498 and length 315 mm (dimensions were obtained as a result of calibration measurements) are presented.

An adopted method of verification of a virtual model operation based on real measurements is presented in Fig. 5.34. The nominal base locations of NPT (nominal point) points (error-free) are obtained by their generating. It is carried out by Quindos software for the nominal diameter and location of the object obtained for initial measurements, as this is done in practice, by realizing the measurement for the object using the automatic generation of the measuring point function. Then a cycle in accordance with the adopted strategy is performed, and using CNC control of measuring machine and automatic measurements for generated measuring points (after selecting the appropriate measuring tip) APT (actual point) coordinates are obtained.

Measurements are repeated 32 times in order to calculate the uncertainty (type A). Obtained results are used later as a base comparative material. In order to generate NPT points the virtual machine Virtual Neuro CMM is started. In this process, for each NPT point the location of APT Virtual is calculated with the generation of random parts in order to obtain for each point the reproducibility error of the measuring point (REMP). Thus the calculated APT Virtual point is transmitted to the Quindos system, and here the right simulation (of selected shape element for a given set of points) takes place. The simulation is repeated for new (32) sets in the same way as it is for measurements on the real CMM. In the final phase, results of the expanded uncertainty and obtained values of particular shape element parameters are compared.

Fig. 5.33 Standards used in neural virtual model tests

To perform research verification of the developed model two cycles of tests were planned. The first cycle was designed to assess the metrological compliance of the model and the real object at given points in the CMM measuring volume. It was realized for selected shape elements such as reference cylinder, sphere, and ring.

The proposed shape elements have been selected for model tests because they are the most frequent elements in measurements, and provide the best illustration of measured characteristics such as length, diameter, coordinates of a center, or the form error. However, they are mostly standards with stable dimensions and with negligible form errors; others would distort the assessment of obtained results. During model testing and performing real measurements on modeled CMM, measured and simulated elements were placed in three locations along the diagonal of the CMM measuring volume. Measurements were performed using 8-mm tips with the length equal to 100 mm, and of the stylus thickness of 5 mm, marked with the symbol PRB1 (Figs. 4.29 and 5.35). The use of such a tip came from years of

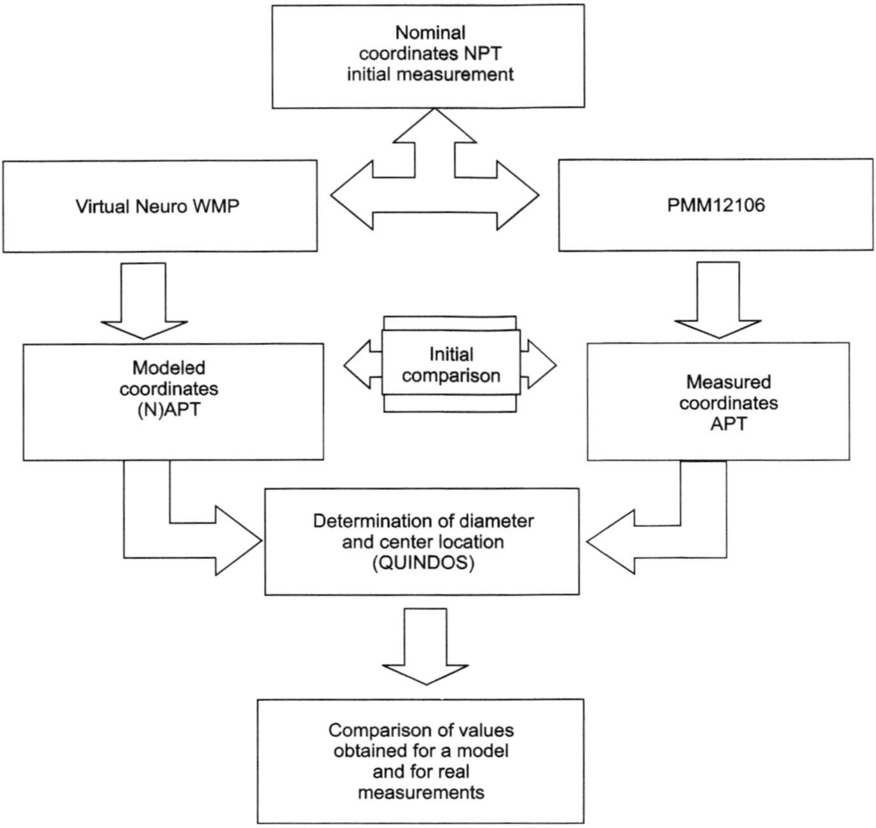

Fig. 5.34 Method of research verification of neural CMM model in comparison with the real object [86]

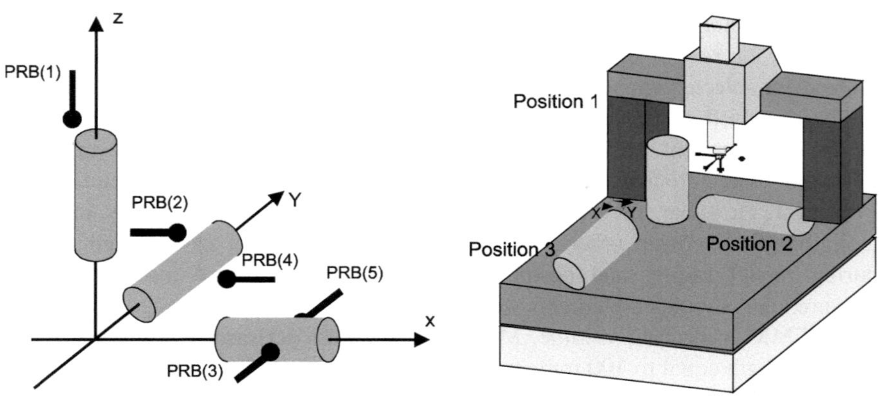

Fig. 5.35 Location of cylindrical standard in the direction of tested axes, with marking of used measuring tips [89]

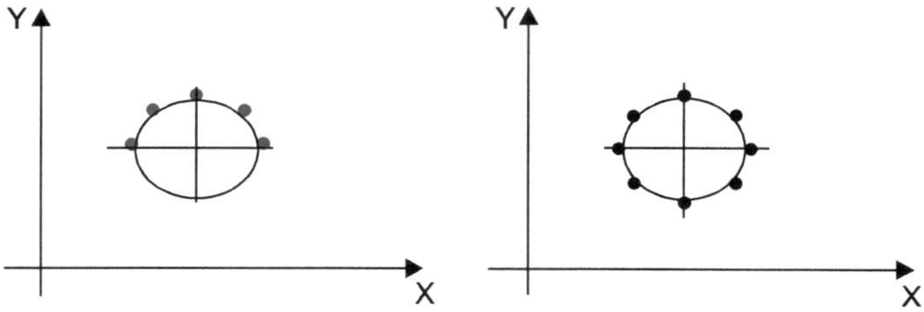

Fig. 5.36 Points distribution on the surface of the tested cylinder. Marks: 5 pt—5-points distributed over the half of the circle, 8 pt—8-point strategy distributed over the whole circumference, 32 pt—32-point strategy distributed over the whole circumference

experience (tips with these dimensions give the most stable results in the case of HA CMM machines). In this cycle the determination of the center and diameter of the ring and sphere and of their form errors, and diameter and straightness of the cylinder were planned. Points were distributed regularly for each measured contour. The sphere was measured in 9 points: 4 points on the equator, 4 on the parallel and 1 on the pole. The circle in 8 points, and the cylinder in 32, distributed on 4 circles, were located as far as possible from each other. Dimensions of tested shape elements were so chosen that the range of measurement movements allowed including REMP components depending on the location in the CMM volume. Assumed standards dimensions are greater than 83 mm, of the basic dimension of the reference point grid adopted during the development of CMM model, so that during measurement realization points with errors "additionally built in" by the grid were included (Fig. 5.36).

The developed Virtual Neuro CMM model was compared to other existing models of the same CMM, that is, the MegaKal system: PTB and Virtual CUT universal model. The results of tests carried out for this verification cycle are presented in Figs. 5.37, 5.38 and 5.39.

The second cycle of verification of the developed model was carried out under conditions for which there is a significant impact of the probe head and the measurement strategy. It is consistent in its idea with actual guidelines for simulation of model testing described in VDI/VDE 2617—part 7 [128] and in Sect. 4.2.5.4 of this book. When these tests were carried out, the above requirements had not been formulated yet, but their validity from the methodological point of view was confirmed later by these guidelines. Their purpose was to see whether they interact in a similar way on the measurement result for the developed model, as in the case of real measurements. Therefore, tests with the use of a reference cylinder (cylindrical perpendicular standard with a nominal diameter of 100.250 mm, and a length of 315 mm) were performed. An object of measurement was a circle located in the middle part of the cylinder placed along the particular CMM axes: *X*, *Y*, *Z* as shown in Fig. 5.35. Because of that, PRB3 and PRB5 tips were used for measurements of the cylinder placed in the *Y*-axis direction, PRB2 and PRB4 tips were used for the measurements of the cylinder placed in the *Y*-axis direction, and the PRB1 tip was

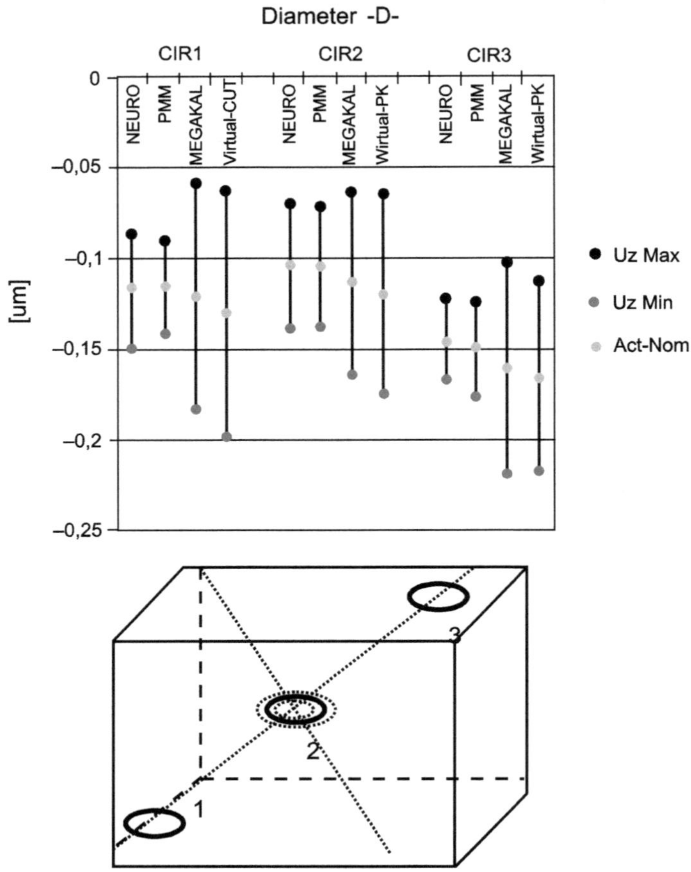

Fig. 5.37 Results of the model verification: Virtual Neuro CMM based on comparative measurements [89]. *Neuro* tested Virtual Neuro CMM model, *PMM* results of measurements on real HA CMM, *MegaKal* CMM model developed in PTB, *Virtual CUT* Universal CMM model—development of the LCM CUT

used for the cylinder placed vertically on a CMM table (in the Z-axis direction). The adoption of such cylinder locations also allowed checking if the model correctly responded to the inevitable changes in the accuracy in each axis. Additionally, such locations forced the use of tip changes during measurements (cylinder arranged in the X- and Y-axis direction) and the verification of the model compatibility with real measurements for this important case. Measurements were performed with the use of two tips: appropriately PRB3, PRB5, and PRB2, PRB4 in one measurement cycle. Both Quindos software and the developed Virtual Neuro CMM model allow the realization of such a measurement cycle (of course after the necessary calibration of measurement tips). In these locations measurements with the use of a variety of strategies based on the changeable number and distribution of measuring

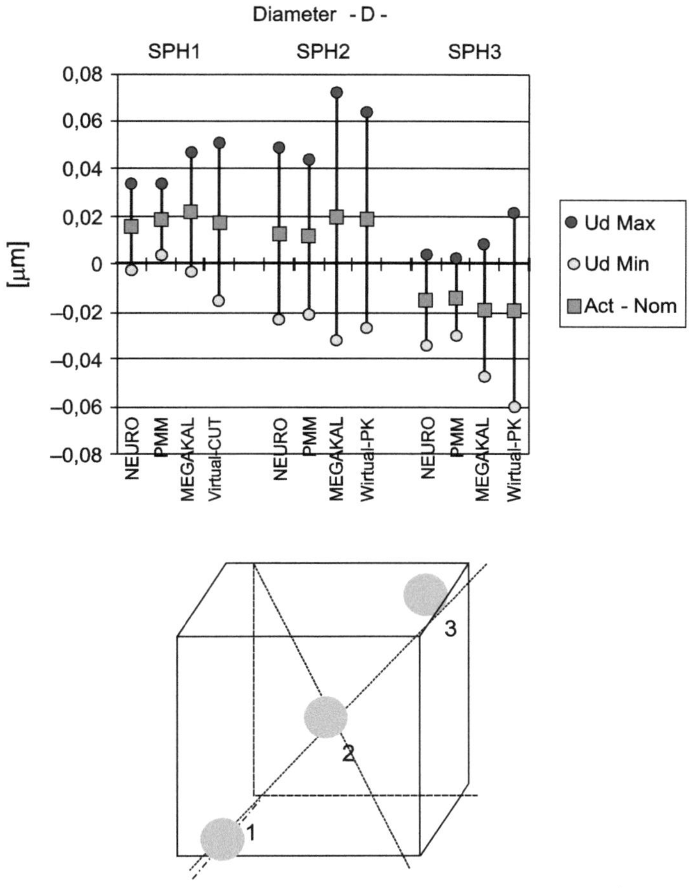

Fig. 5.38 Results of the model verification: Virtual Neuro CMM based on comparative measurements [89]: *Neuro* tested Virtual Neuro CMM model, *PMM* the results of measurements on real HACMM, *MegaKal* CMM model developed in PTB, *Virtual CUT* Universal CMM model —development of the LCM CUT

points were carried out. Measurements were taken for the least favorable, from the point of view of obtained accuracies, a 5-point strategy with the point distribution on the semicircle, and for the regular distribution first of 8, and then of 32 points. The least measurement strategy is the most favorable from the point of view of reducing the PEF impact and also compensation algorithms, which allows obtaining the best accuracy [38, 44, 60, 80, 89, 91, 129, 130].

Tested element—the circle. The distribution of test elements in the CMM measuring volume, marks **CIR1**, **CIR2**, **CIR3**.

Tested element—the sphere. The distribution of test elements in CMM measuring volume, marked as **SPH1**, **SPH2**, **SPH3**.

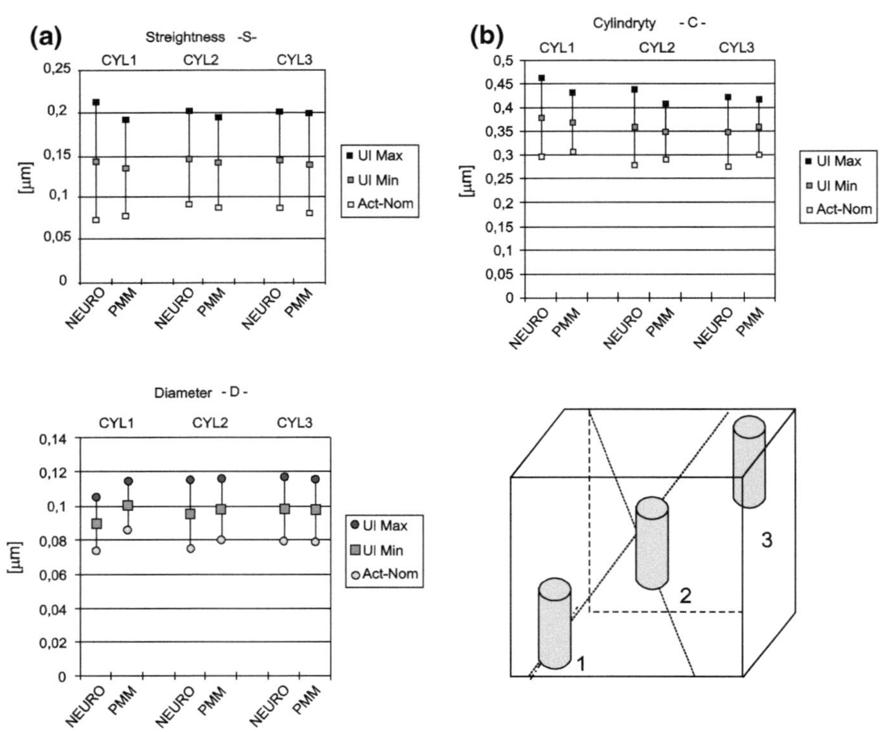

Fig. 5.39 Results of the model verification: Virtual Neuro CMM based on comparative measurements [89]: *Neuro* tested Virtual Neuro CMM model, *PMM* the results of measurements on real HACMM

Tested element—the cylinder. The distribution of test elements in CMM measuring volume, marked as CYL1, CYL 2, CYL 3.

The contact effect of the PRB1 tip with a curved surface was tested for cylinder positions in *X*- and *Y*-axes with the use of 3- and 5-point strategy (the direction of measurement movement, *Z*). The location of the center, the diameter, the shape deviation of the measured circle (of course, for the higher than 3-point measurement strategy) were under assessment. The point distribution in the case of described strategies is shown in Fig. 5.36 (with no results for the 32-point strategy, because they are analogous for the 8-point one presented). In Fig. 5.40 an example of a protocol of real measurements and simulations based on a neural CMM model is shown. In addition to results such as the center coordinates and diameter the protocol gives an expanded uncertainty obtained for realized measurements. The results of the circle and sphere measurements are presented. The results of the comparative tests described above are shown in Figs. 5.41, 5.42 and 5.43.

During measurements the most difficult task was to provide the strict measurement repeatability conditions. Hence the temperature monitoring system

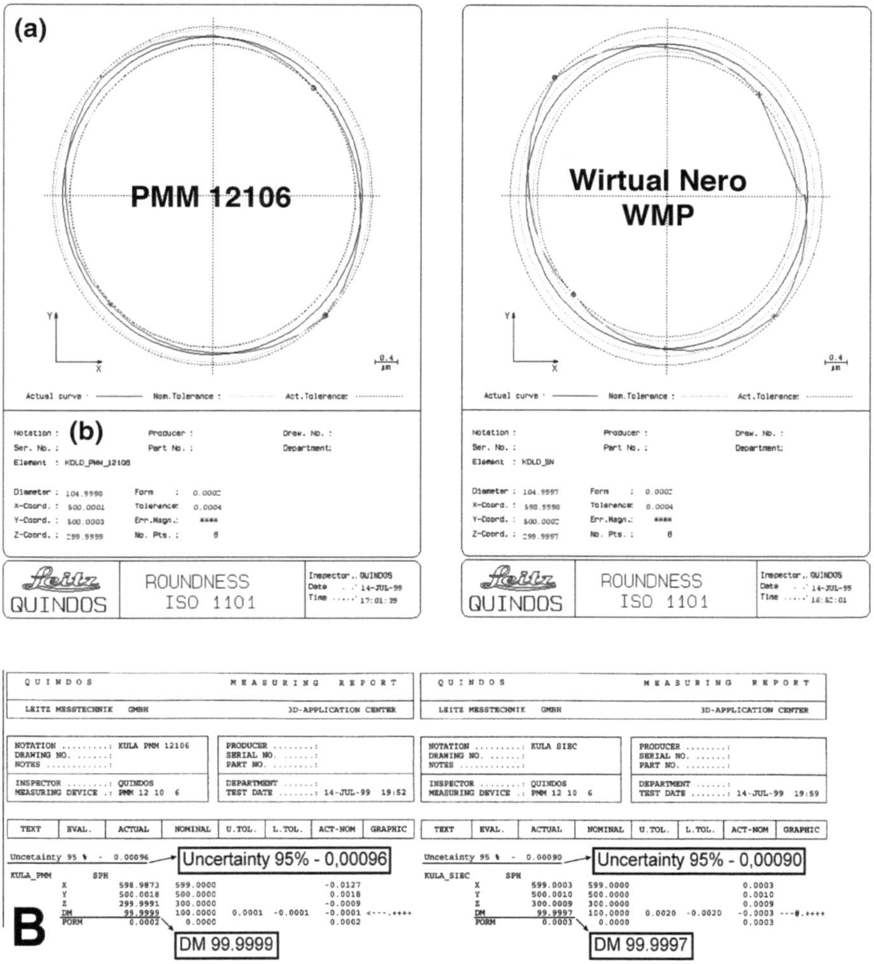

Fig. 5.40 Example protocol from realized real and virtual measurements of circular standard. Quindos system with Virtual Neuro CMM model: [89]: **a** reference ring measurement—the chart of measurement errors, **b** protocol of reference sphere measurement with the determined measurement uncertainty

(described in Sect. 3.5) was used, and all measurements were carried out without the presence of the operator, to not interfere with the thermal stability conditions. Each time after switching the location of the standard there was a pause of about two hours and only then were measurements started, running them remotely from the outside of the machine room to get good indications repeatability.

During analysis of the obtained results it was assumed that the most important was to compare the results obtained for the developed model and for real

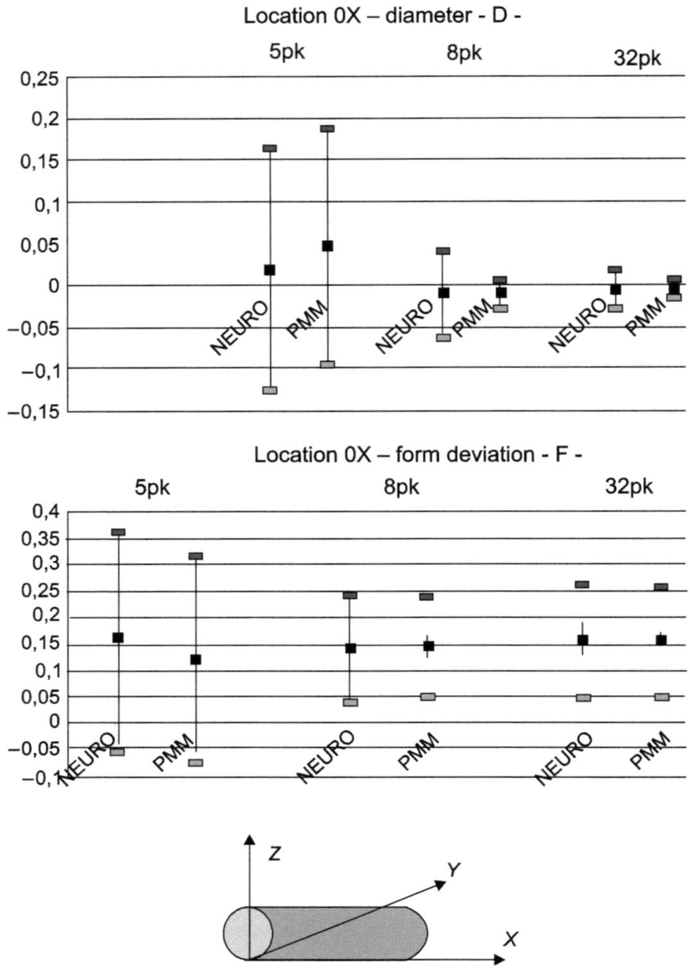

Fig. 5.41 Results of the model verification: Virtual Neuro CMM based on comparative measurements. The tested element type, the circle placed in the *X*-axis direction, the circle measurement in the *ZY* plane: [89] *Neuro* tested Virtual Neuro CMM model, *PMM* the results of measurements on real HACMM; Points distributions: 5 pt—5-point strategy, 8 pt—8-point strategy, 32 pt—32-point strategy

measurements. However, it was also necessary to make comparisons with other models used thus far. The range of obtained uncertainties is also important; it lies within the limits close to CMM resolution, indicating a very high repeatability and very good quality of protection against unforeseen external influences (mainly temperature). When analyzing results in two cycles (shown in Figs. 5.37, 5.38 and 5.39, and 5.41, 5.42 and 5.61), a high compatibility in the controlled parameters field for the Virtual Neuro CMM model with real measurement results performed

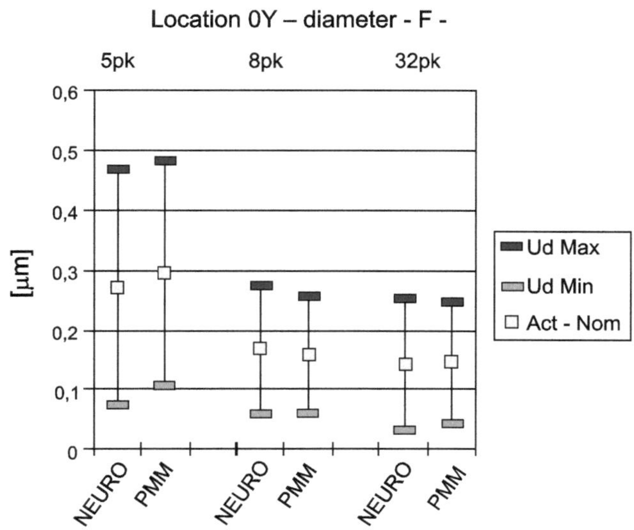
Location 0Y – diameter - F -
5pk
8pk
32pk
0,6
0,5
0,4
0,3
0,2
0,1
0
[µm]
Ud Max
Ud Min
Act - Nom
NEURO
PMM
NEURO
PMM
NEURO
PMM

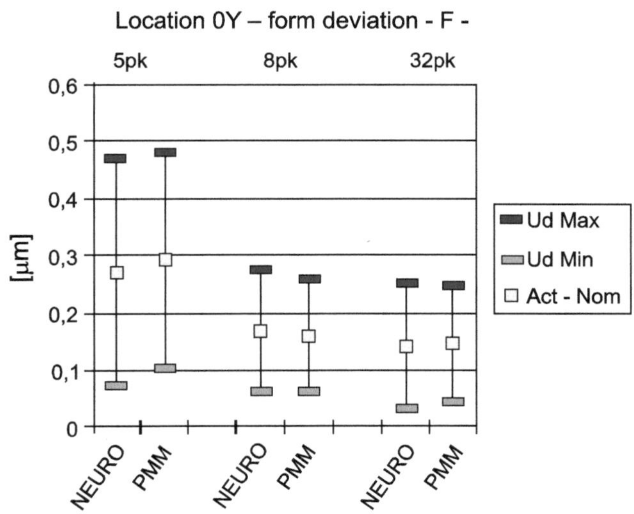
Location 0Y – form deviation - F -
5pk
8pk
32pk
0,6
0,5
0,4
0,3
0,2
0,1
0
[µm]
Ud Max
Ud Min
Act - Nom
NEURO
PMM
NEURO
PMM
NEURO
PMM

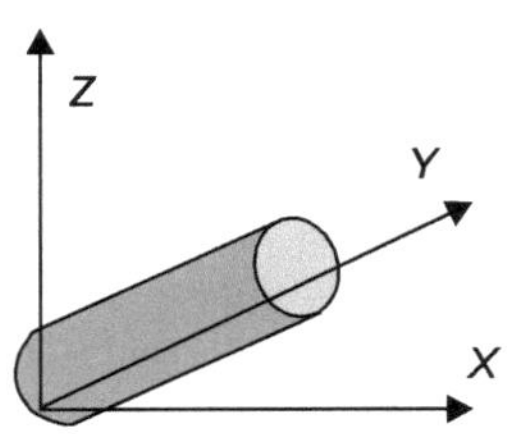
Z
Y
X

◀ **Fig. 5.42** Results of the model verification: Virtual Neuro CMM based on comparative measurements. The tested element type—the cylinder placed in the *Y*-axis direction, the circle measurement in the *ZX* plane: [89] *Neuro* tested Virtual Neuro CMM model, *PMM* the results of measurements on real HACMM; Points distributions: 5 pt—5-point strategy, 8 pt—8- point strategy, 32 pt—32- point strategy

on HA CMM was observed. Slightly worse results were obtained for MegaKal–PTB and Virtual CUT models, showing the superiority of the new model over compared solutions. The obtained differences between models are not yet understood, because the analytical models, due to a different way of error description, require an identification of particular components. This results in the necessity to estimate the uncertainty of particular error sources obtained from measurements with both the plate standard according to the PTB method described in Sect. 4.2.2, as well as with the laser interferometer shown in Sect. 4.2.3. The applicability of different methods of CMM error identification is an advantage of the Virtual CUT model. However, this results in inevitable errors coming from various methodologies of component determination. In the light of obtained results all models can be considered as properly modeling measurements on HA CMM, if we accept as a base the accuracy specified by the producer and also verified by the matrix method (Table 4.1).

The high quality of the developed model is also confirmed by a conducted compatibility test for different strategies in terms of tips changing during task performing. Results presented in Figs. 5.41, 5.42 and 5.43 confirm the high compatibility of the model and real measurements. The tests covered measurements performed both with the worst strategies (see results for 5-point strategy) and also with the strategy giving the most stable results (32-point strategy). Maximum observed differences are below 0.1 μm, which is below the machine resolution. The object of tests was also the surface curvature impact on measurement accuracy and the impact of a tip change during task realization. During 3-point strategy testing a good compatibility of results was demonstrated, similarly as in the case of favorable strategies. However, in this case, due to the accepted way of test cylinder placing (0*X* and 0*Y* locations) the influence of stylus change (measuring tip) was also tested (Figs. 5.41 and 5.42). But it is not as significant as the impact of point distribution, which indicates a stable probe head operation. It is also associated with the used stylus and with its balanced load (compare results for 8- and 32-point strategy, Figs. 4.35 and 4.36). In this case the model demonstrated its compliance with real measurements. On the basis of results of conducted tests it can be concluded that the adopted probe head operation model in this case has also proven its proper functioning in relation to real measurements. It was also shown that for the different accuracy characteristic of particular axes (compare the results shown in Figs. 5.37 and 5.38) the compliance of the model with real measurement was obtained. Similarly, for the measuring task with the use of the PRB1 tip for a curved surface, where the biggest errors were expected (compare results of coordinate *Z* of a sphere shown in Fig. 4.36 with results of the measurement with 5-point strategy for

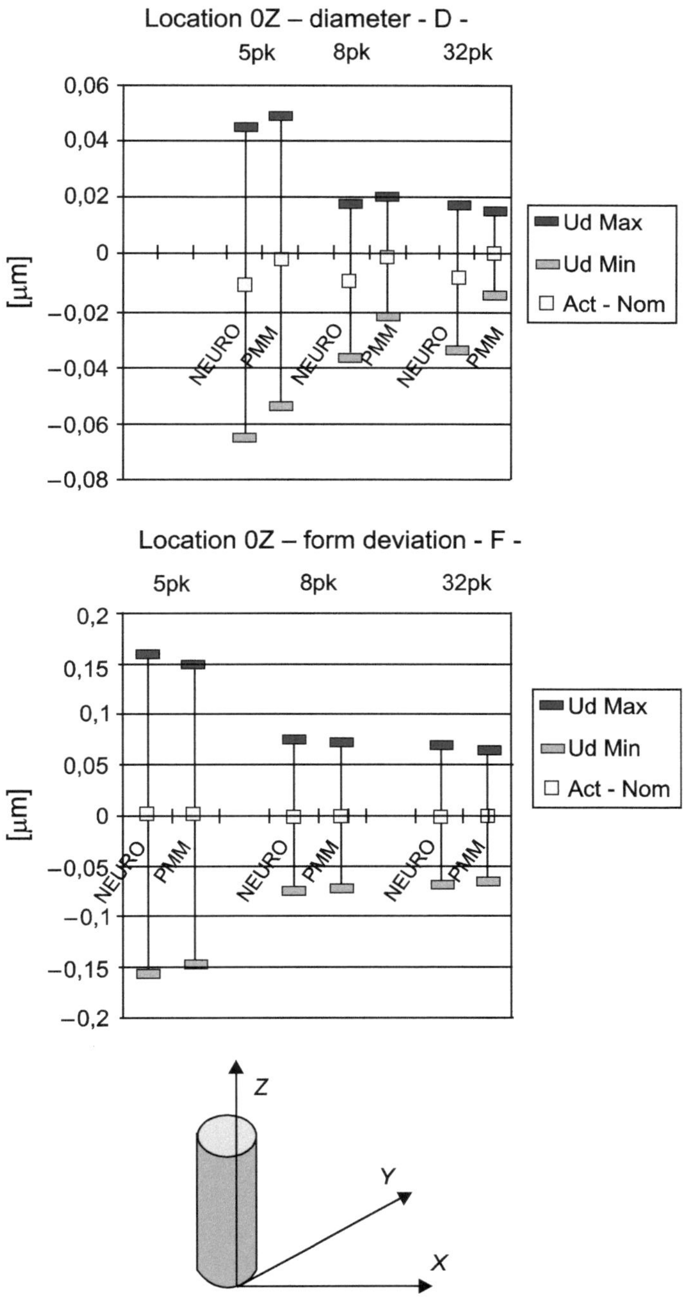
Location 0Z – diameter - D -
5pk
8pk
32pk
0,06
0,04
0,02
0
−0,02
−0,04
−0,06
−0,08
[μm]
NEURO
PMM
NEURO
PMM
NEURO
PMM
Ud Max
Ud Min
Act - Nom
Location 0Z – form deviation - F -
5pk
8pk
32pk
0,2
0,15
0,1
0,05
0
−0,05
−0,1
−0,15
−0,2
[μm]
NEURO
PMM
NEURO
PMM
NEURO
PMM
Ud Max
Ud Min
Act - Nom
Z
Y
X

◀ **Fig. 5.43** Results of the model verification: Virtual Neuro CMM based on comparative measurements. The tested element type—the cylinder placed in the *Z*-axis direction, the circle measurement in the *XY* plane: [89] *Neuro* tested Virtual Neuro CMM model, *PMM* the results of measurements on real HACMM; points distributions: 5 pt—5-point strategy, 8 pt—8-point strategy, 32 pt—32-point strategy

horizontal locations of the test cylinder, Figs. 5.41 and 5.42), the model functioned properly and re-created the real condition.

In summary, it can be said that the conducted analysis of model test results obtained in critical conditions—which in the case of metrological practice makes most problems with ensuring the measurement accuracy and shows the greatest variability of results—showed that the tested Virtual Neuro CMM model has been properly constructed and simulates the real CMM well. It also gives better results than models developed thus far and it is also much easier in construction and in practical use. Thanks to the popularity of the I++ protocol, this system can be used for any metrological software with this interface. In Fig. 5.44 the scheme of Virtual Neuro CMM use is presented.

5.2.3.6 Use of Neural Virtual Model for Measurement Accuracy Forecasting Task

The process of measurement planning and measurement program creating is more and more often realized in the system outside the machine (offline) with the use of

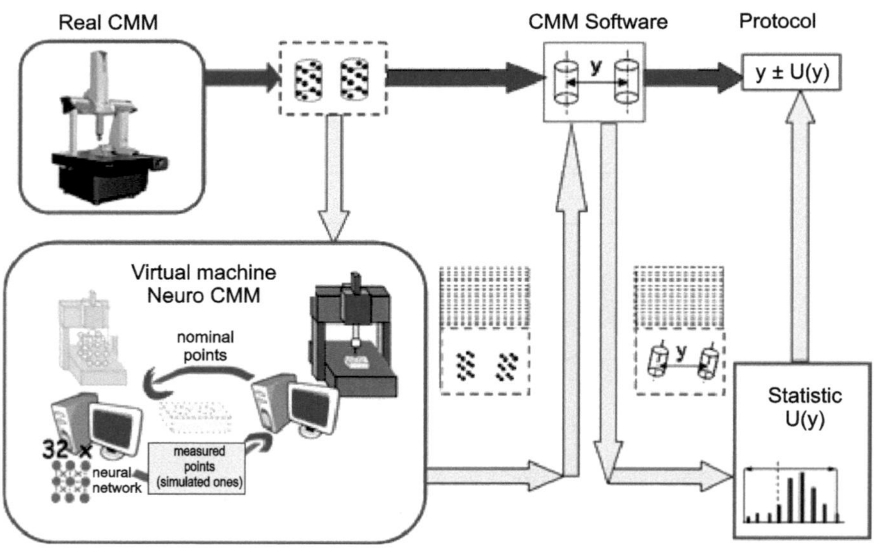

Fig. 5.44 Scheme of Virtual Neuro CMM use in conjunction with CMM metrological software

CAD files. Often complete measurement programs are supplied with the construction and technological documentation. This means that they are often developed in designers' stations based on visualization and CMM programming systems. There are already many such systems, allowing one to choose measuring styli, simulate the measurement process taking collision-free courses, and define clearance points finally to put together a complete program employing an optimal measurement strategy. All of the above-mentioned operations are done outside the CMM. For many years such systems have been available, and metrological software such as PCDMIS and Quindos Leitz, CALIPSO–Zeiss have these functions. However, there is a system developed specifically for a measurement simulation task, the Simulator I++ made by the Messtechnik Wetzlar Company. However, all of these systems lack the most important module from the metrological point of view, that is, the accuracy model (virtual machine). Therefore the connection of a CAD system, a system of measurement planning, and CMM programming with the issue of CMM accuracy modeling allows the realization of the measurement accuracy forecasting idea.

The idea of such a system was developed by the author and published in [82]. It is illustrated in Fig. 5.45. Here it is based on the simulation of a measuring task based on virtual CMM, including the object structure in the form of a CAD file together with the accepted tolerance of manufacturing.

The result of such simulation allows us to:

- Predict the measurement accuracy for both the partial measuring task and for the whole measured part.
- Assess the measurement abilities, and thus, answer the question of whether the given part can be measured and thus produced.
- Obtain the necessary information to make the proper decisions when taking the order for the parts production with the use of possessed machinery part.
- Significant shortening of the production preparation time by simulating the measurement process and the assessment of its accuracy without the necessity to create prototype components.
- Optimal planning of CMM use by calculating the measurement time and by planning the frequency of measurement realization for a given part and which set of measured characteristics has to be covered.
- Simulation of measuring tasks for selecting the optimal measurement strategy for the object in the project phase.

When forecasting the measurement accuracy for the object or its part in the project phase the possible measurement accuracy of a part can be determined, and possibilities of this accuracy change analyzed by the CMM user. The improvement of measurement accuracy can be obtained, for example, by changing the measurement strategy. Having in account tolerances set by the constructor, the measurement strategy can be chosen in such a way not only to specify all necessary parameters, but get the improvement of their accuracies, of course within certain limits. The measurement forecasting allows us to avoid costs related to wrong decisions.

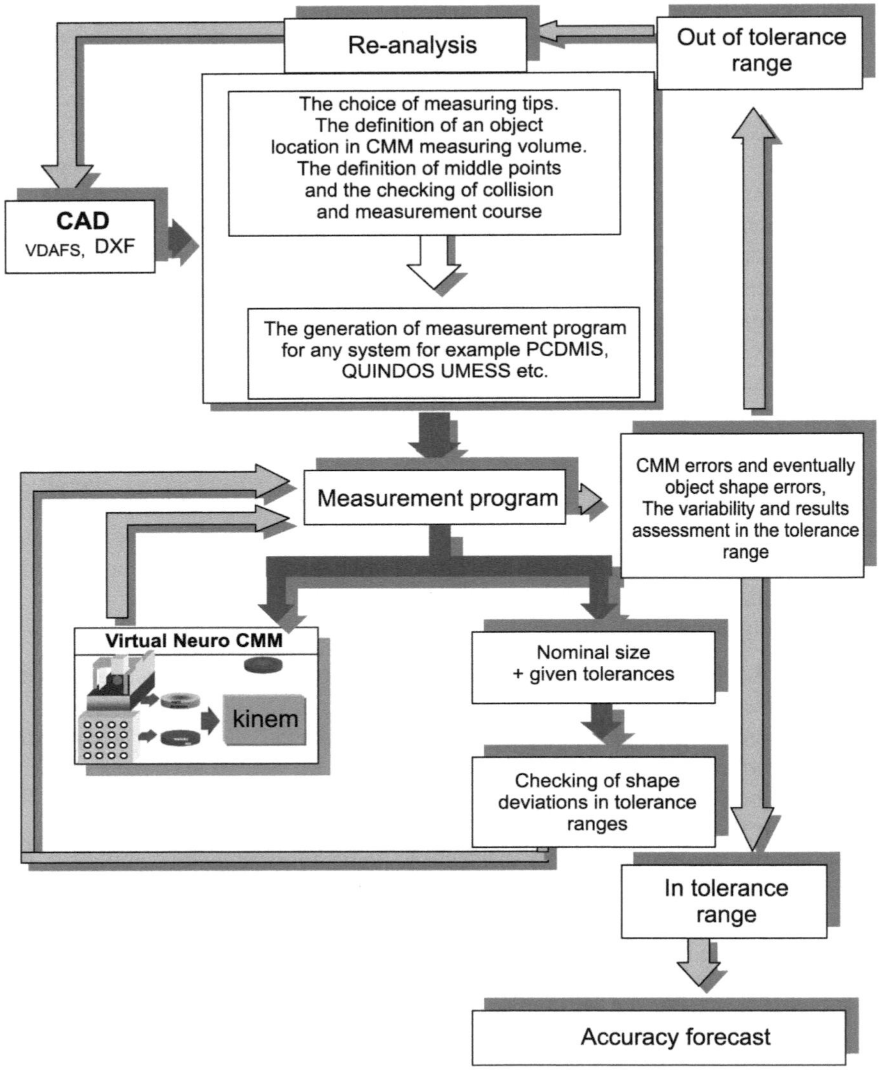

Fig. 5.45 Scheme of measurement accuracy forecasting in the offline system [82, 89]

In accordance with Berndt's principle (Golden Rule of Metrology) [4, 5] a tolerance reduction is not necessary if measurement uncertainty is not larger than a tenth ($U/T = 0.1$) or at the very most a fifth ($U/T = 0.2$) of tolerance.

- 4 × standard deviation of production variation $s_p = T$ (tolerance).
- Additional measuring process variation u is compensated through a reduced production variation s_{pV}.

$$T = 4s_{\mathrm{p}} = 4\sqrt{\left(s_{\mathrm{pV}}\right)^2 + u^2} = \text{const.} \tag{5.1}$$

Therefore the tolerance for production (T replaced by T_{P}) is reduced to keep the variation of production and measuring process constant:

$$T_{\mathrm{p}} = 4s_{\mathrm{pV}} = T\sqrt{1 - \left(2\frac{U}{T}\right)^2} \quad \text{when } 2u = U \tag{5.2}$$

The tolerance is not reduced if the uncertainty does not exceed 10 … 20 % of the tolerance.

Measuring and test equipment capability according to VDA5 and ISO 22514-7

Q—Relation between extended uncertainty $2U$ and tolerance T

$$Q = \frac{2U}{T} = \frac{2 \cdot 2u}{T} \tag{5.3}$$

Requirements:
For System: Q = 15 %
For Process: Q = 30 %
Comparison with "Golden Rule": U/T = 0.1 … 0.2 $\Longrightarrow$ Q = 0.2 … 0.4
From manufacturer specifications

$$u_{\mathrm{MS}} = \frac{\mathrm{MPE}}{\sqrt{3}} \tag{5.4}$$

$$u = u_{\mathrm{M}} \tag{5.5}$$

Calculation of the small measurable tolerance for a measuring system from MPE specification according to VDA 5 (Table 5.5)

$$Q_{\mathrm{MS}} = \frac{2U_{\mathrm{MS}}}{T} \quad \text{with} \quad U_{\mathrm{MS}} = 2u_{\mathrm{MS}} = 2\frac{\mathrm{MPE}}{\sqrt{3}} \tag{5.6}$$

Requirement for a measuring system

$$Q_{\mathrm{MS}} = 0.15 \quad \text{for } T_{\min} = \frac{2U_{\mathrm{MS}}}{Q_{\mathrm{MS}}} \tag{5.7}$$

$$\Rightarrow \quad T_{\min} = 4\frac{\mathrm{MPE}}{\sqrt{3}Q_{\mathrm{MS}}} = 15.4\mathrm{MPE} \tag{5.8}$$

Table 5.5 Example: CMM with specification: $MPE_{E0} = 1.7 + L/333$ μm

Length (mm)	MPE_{E0} (μm)	T_{min} (μm)
10	1.7	26.6
50	1.9	28.5
100	2.0	30.8
150	2.2	33.1
200	2.3	35.4
250	2.5	37.7

Requirement according to Q-DAS guideline (summary of the calculation) [48]

1. System point of view:
   c_{gk}: parameter, which describes the position (Bi) and variance (s) of the test in relationship to tolerance

$$c_{gk} = \frac{0.1 \cdot T - Bi}{2s} \tag{5.9}$$

   Requirement: $c_{gk} = 1.33$
   Comparison with "Golden Rule" $U/T = 0.1 \ldots 0.2$
   for: $Bi = 0$; $c_{gk} = c_g = 1$, $2s = 2u = U$: $U/T = 0.1$

2. Process point of view:
   $\%R \,\&\, R$: percentage of the variance (u_{EV} and u_{AV} from the analysis) of the test in the tolerance

$$\%R\&R = \frac{R\&R}{T} \quad \text{with } R\&R = 5.15 \cdot \sqrt{u_{EV}^2 - u_{AV}^2} \tag{5.10}$$

   Requirement: $\%R \,\&\, R = 20 \ldots 30\ \%$
   Comparison with "Golden Rule":
   $U/T = 0.1 \ldots 0.2 \rightarrow \%R \,\&\, R = 0.2$; $R \,\&\, R = 2U$: $U/T = 0.1$

Calculation of the small measurable tolerance for a measuring system from MPE specification using Q-DAS guideline (Table 5.6)

$$c_{gk} = \frac{0.1 \cdot T - Bi}{2s} \Rightarrow T = \frac{c_{gk} \cdot 2 \cdot s + Bi}{0.1} \tag{5.11}$$

Assumption:
MPE is composed 50 % from systematic and 50 % from random effects

$$s = 0.5\frac{\text{MPE}}{\sqrt{3}} \quad Bi = 0.5\text{MPE} \tag{5.12}$$

$$\Rightarrow \quad T_{\min} = \frac{c_{gk} \cdot 2 \cdot s + Bi}{0.1} = 9.4\mathrm{MPE} \tag{5.13}$$

For $c_{gk} = 1.33$
$T_{\min}$ = Smallest tolerance

Table 5.6 Example: CMM with specification: $\mathrm{MPE}_{E0} = 1.7 + L/333$ μm

Length (mm)	MPE_{E0} (μm)	$T_{\min}$ (μm)
10	1.7	16.3
50	1.9	17.4
100	2.0	18.8
150	2.2	20.2
200	2.3	21.6
250	2.5	23.0

As already described in previous chapters these relations concern simple measuring tasks in conditions of including length measurement and expanding the variation of the area of production errors. However, in the case of coordinate measurements with included impact of a given measurement strategy the calculation of measurement uncertainty should concern use of simulation methods, virtual CMM. Only in that case can we talk about determination of measurement uncertainty. Therefore the reference to production tolerances should be considered in two aspects: rational choice of the measuring tool with the measurement uncertainty imperative and also estimated share of measurement error in relation to the manufacturing process. This task was carried out for given measurement strategies. It was shown that estimated uncertainties had been dependent on the object and selected measurement strategy.

Using the virtual machine for measurement accuracy assessment in accordance with Fig. 5.46 it is possible to assess whether the obtained accuracy of realized or forecast measurement exhausts the terms of this principle. It also allows us to set the number and distribution of measuring points reasonably, which also affects the measurement costs. In selection of the measurement strategy it should be noted that there are connections between some of the components. For example, the choice of orientation imposes measuring styli configuration. After the measurement strategy selection follows measurement simulation. If the uncertainty obtained during the simulation is less than the expected uncertainty referred to the tolerance by the Berndt principle, the measurement strategy should be reselected.

While selecting the measurement strategy there should be defined:

- Number and distribution of measuring points
- Location and orientation of a given object in CMM volume
- Type of measuring stylus and its parameters
- CMM itself (if the database contains more than one machine)

Such a system was realized in the Laboratory of Coordinate Metrology at Cracow University of Technology, under the work [15], and its results were

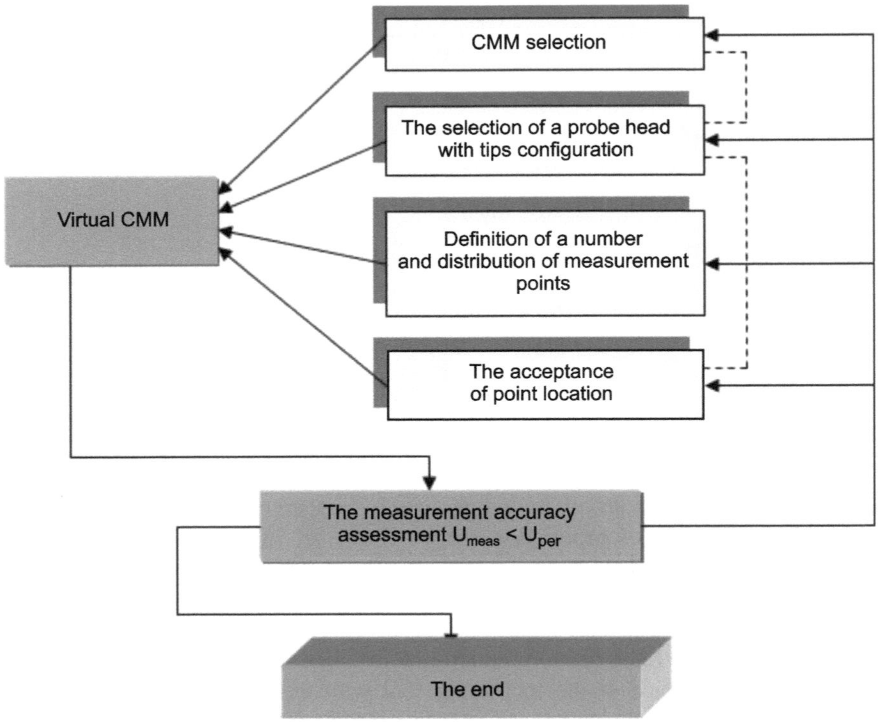

Fig. 5.46 Algorithm of steps during measurement accuracy assessment with the use of virtual CMM

published in [93, 94, 105]. Its operation is shown using a Virtual Neuro CMM model developed for MA CMM (Fig. 5.47). It is a machine built from light alloys and designed mainly to perform measurements in an industrial environment. Due to the accuracy ($MPE_E = 1.7 + L/333$ µm) and the construction of the machine, the Virtual Neuro CMM model was built with the use of a module including the effect of thermal conditions of its operation in the LCM CUT. The model was constructed for the temperature range of 20 ± 2 K. The modeled probe head was also different from the described model for HA CMM, that is, the contact probe head (Renishaw TP 200 probe with an articulating PH10 head). In the developed probe head model orientations of the articulating head according to the scheme shown in Fig. 5.48, were assumed to be sufficient.

According to the scheme in Fig. 5.45 the measuring task, which was the subject of simulation and of measurement accuracy forecasting, was realized based on the CAD model of the object shown in Fig. 5.49a. The simulated measurement was carried out in three different locations on the measuring table (Fig. 5.49b). This aimed to determine measurement accuracy depending on the object location on the table. The object orientation remained the same, so there was no need to change the

Fig. 5.47 Coordinate measuring machine global image 775 in the LCM CUT during the construction of the Virtual Neuro CMM model

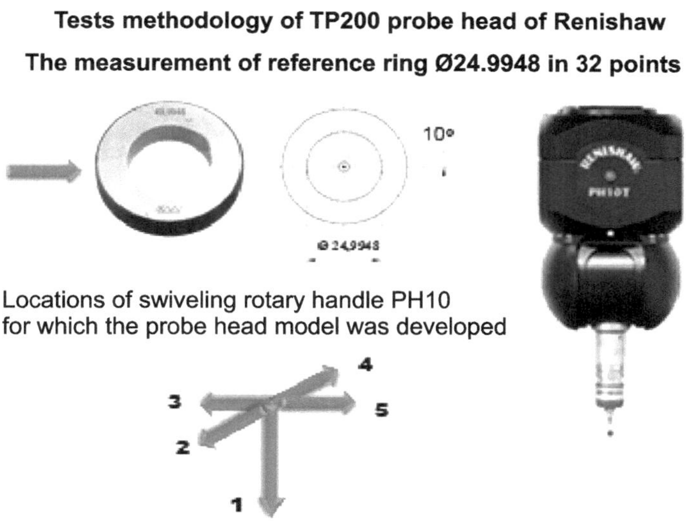

Fig. 5.48 Methodology of TP 200 probe head error identification with a PH10 head to build MA CMM model

configuration of the measuring styli. For tests a few representative geometric elements were selected:

(a) Plane of orientation (0, 0, 1)—PLN1
(b) Plane of orientation (0.5; 0; 0.866025)—PLN2
(c) Cylinder of diameter φ = 30 mm and length of 10 mm—CYL1
(d) Cylinder of diameter φ = 20 mm and length of 40 mm—CYL2

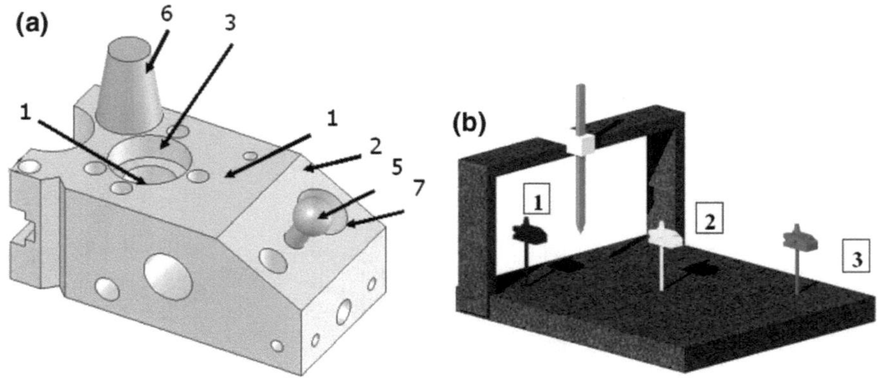

Fig. 5.49 Object with marked measuring tasks (**a**) and its distribution in the CMM measuring volume (**b**)

(e) Sphere of diameter $\varphi = 15$ mm—SPH1
(f) Cone—CON1
(g) Circle of diameter $\varphi = 20$ mm lying on a plane 2—CIR1

Two strategies of measuring point distribution were adopted, the first one of the regular distribution marked as **R**, and the second of the nonregular distribution marked as **N**. From the metrological practice it is known that regular distribution of measuring points gives better results, but very often we have to deal with the obligatory nonregular distribution of measuring points. It is caused by the shape of the object and technical possibilities of measurement realization, when it is not possible to reach the whole surface.

In Fig. 5.50, the strategies realized for planes of the cone and the sphere are presented. Based on simulation results, the most characteristic parameters of the measured geometric elements were tested. Each simulation was repeated 32 times. In Table 5.7 elements were set together with tested characteristics.

The number of measuring points was chosen as follows.

- Strategy 1—the lowest number of points
- Strategy 2—depending on the element from 6 to 12 points
- Strategy 3—depending on the element from 11 to 38 points

In order to facilitate the record symbols given in the table in Fig. 5.50 were used.

In Fig. 5.51 the extent to which the obtained simulated measurement uncertainties for determined flatness deviation depend on the adopted measured object placing in the modeled CMM volume is shown. In graphs shown in Figs. 5.52, 5.53 and 5.54 it is presented that the Berndt principle is met for all tested strategies.

1 R 1 str

Place on the table	Points distribution	Strategy type
1 – 1 place 2 – 2 place 3 – 3 place	**R – regular** N – non-regular	**1str – strategy 1** 2str – strategy 2 3str – strategy 3

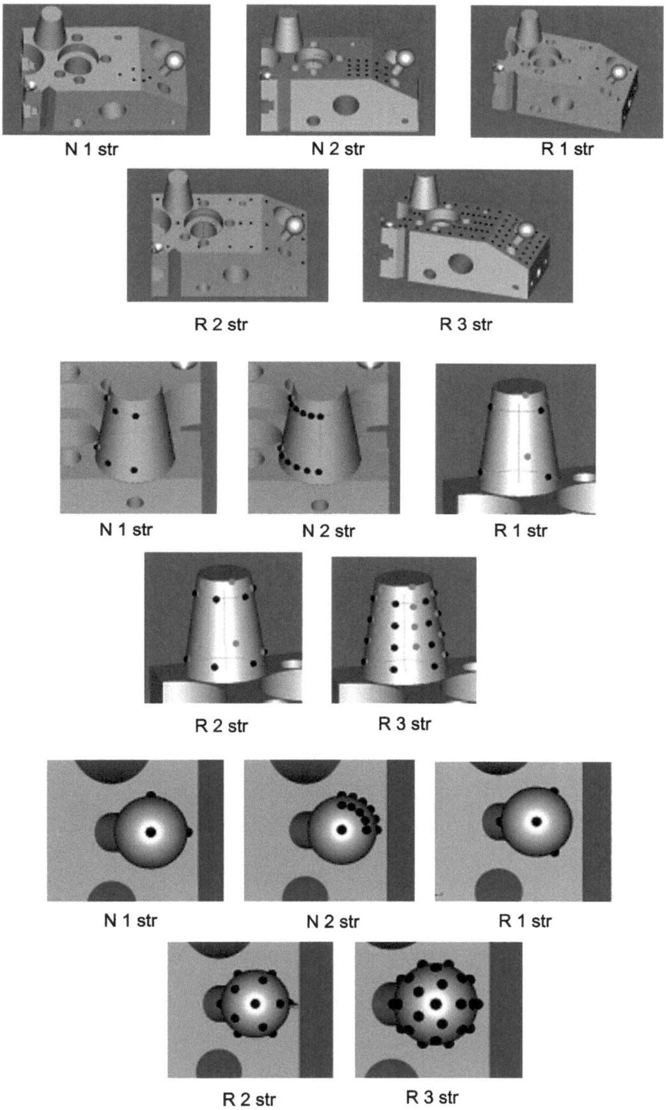

Fig. 5.50 Sampling strategies for the cone, sphere, and planes

Table 5.7 List of elements together with tested parameters using simulation

Element	Location			Direction vector units			Diameter D	Top angle	Shape deviation
	X	Y	Z	WX	WY	WZ			
PLN1	•	•	⊕	⊕	⊕	•	–	–	⊕
PLN2	⊕	⊕	⊕	⊕	⊕	⊕	–	–	×
CYL1	⊕	⊕	•	⊕	⊕	•	⊕	–	×
CYL2	⊕	⊕	•	⊕	⊕	•	⊕	–	×
SPH1	⊕	⊕	⊕	–	–	–	⊕	–	×
CIR1	⊕	⊕	⊕	–	–	–	⊕	–	×
CON1	⊕	⊕	–	×	×	×	×	⊕	×

⊕ parameter tested, × parameter not tested, – does not concern this element, • component of a negligible value or not examined because of the element degree of freedom in given direction

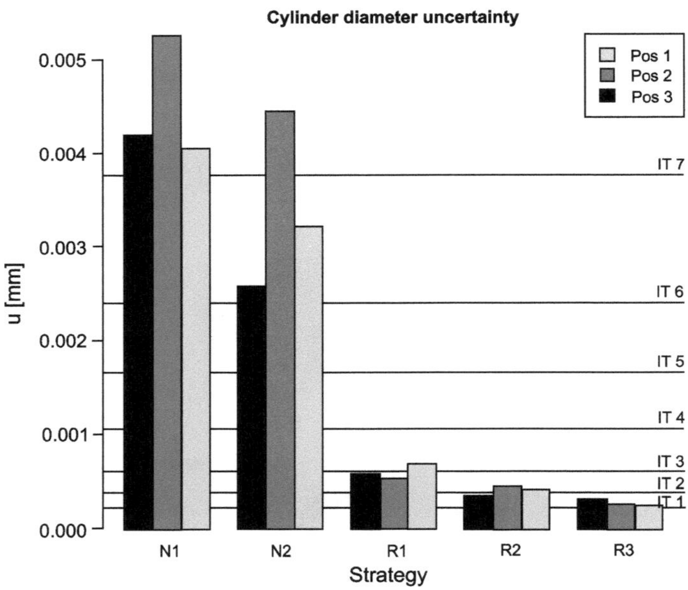

Fig. 5.51 Graphical presentation of the relation between the measurement uncertainty for cylinder diameter depending on the strategy and adopted location in CMM measuring volume

Verification Measurements

As a reference that allows for the verification of performed simulations the real measurements were carried out on a real MA CMM machine, for which a Virtual Neuro CMM model had been developed earlier. The real measurements of elements for which measurements had been simulated earlier with the use of their CAD

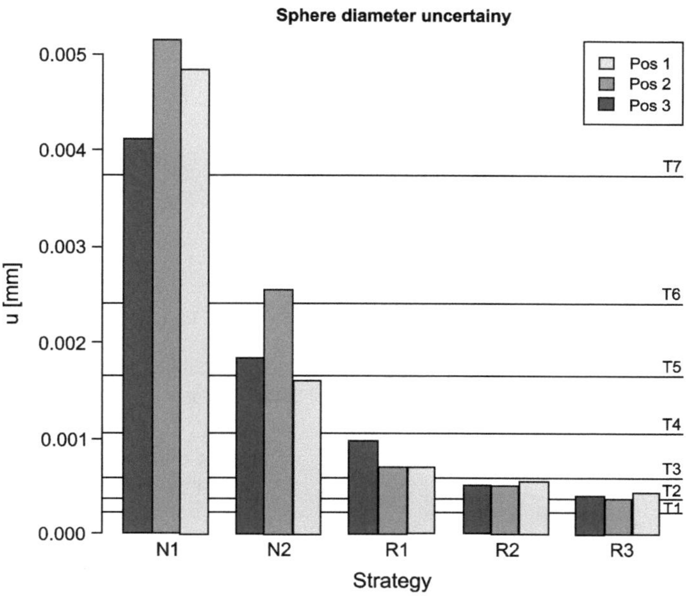

Fig. 5.52 Graphical presentation of the relation between the measurement uncertainty for sphere diameter depending on the strategy and adopted location in CMM measuring volume

models were performed. Measurements were carried out in one location on the measurement table (the center of measurement table, position 2 during simulation), and in regular distribution of points. The number of these points for each element is listed below, according to Fig. 5.49:

- Plane 1–22 points
- Plane 2–22 points
- Cylinder 1–8 points (2 circles)
- Cylinder 2–12 points (3 circles)
- Sphere—5 points
- Cone—8 points (2 circles)

Measurements were repeated 16 times for each element.

In Fig. 5.53 a comparison of the simulation results for R_2str and R_3str strategies is shown with real measurement results. Two elements were selected, the sphere and the cylinder. As can be seen in the attached charts, results are comparable. In some cases, the real measurement uncertainty is higher than the uncertainty calculated from the simulation results. The reason is that real measurements have not been carried out on standards but only for a part of relatively low manufacturing precision. Shape deviations, roughness, type of material, all of these factors affected the growth of real measurement uncertainty.

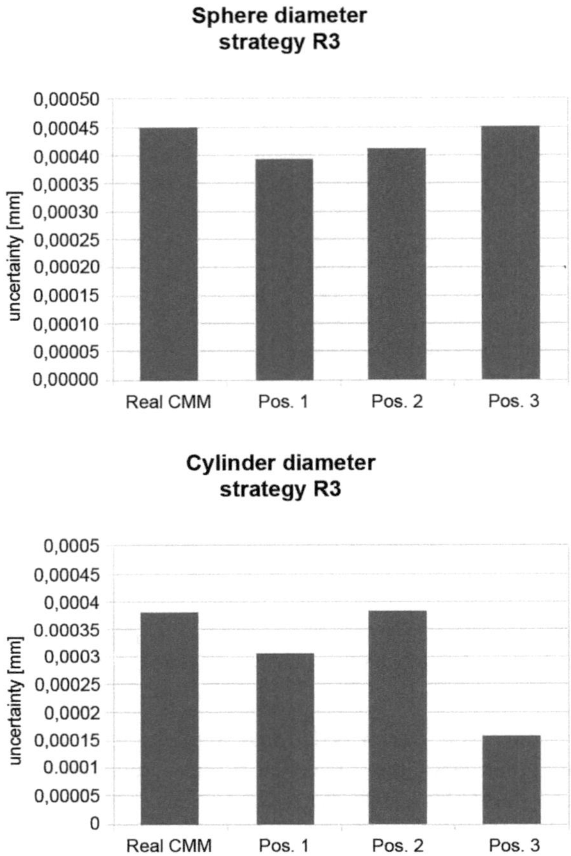

Fig. 5.53 Comparison of simulation results with real measurements for the sphere and cylinder 2. It is important here to compare the results of *D* diameter measurements

Another advantage of the measurement forecasting realization is earlier planning of appropriate process capacity *Cp* and thus reducing the cost of defects in the manufacturing process [7, 131]. This is the more important as along with documentation of the part and developed machining program more and more often the measurement program is supplied. The offline development of such a program should be carried out with a possibility of testing and with selection of the optimal measurement strategy. These options are given by the virtual machine system. Of course, the forecasting procedure should be consistent for the fully dimensioned part; then the dependencies of tolerance and results of measurement accuracy forecasting achieved for a given machine model allow us to make the decision both of the constructor and technologist leading to production realization with the use of a virtual prototype.

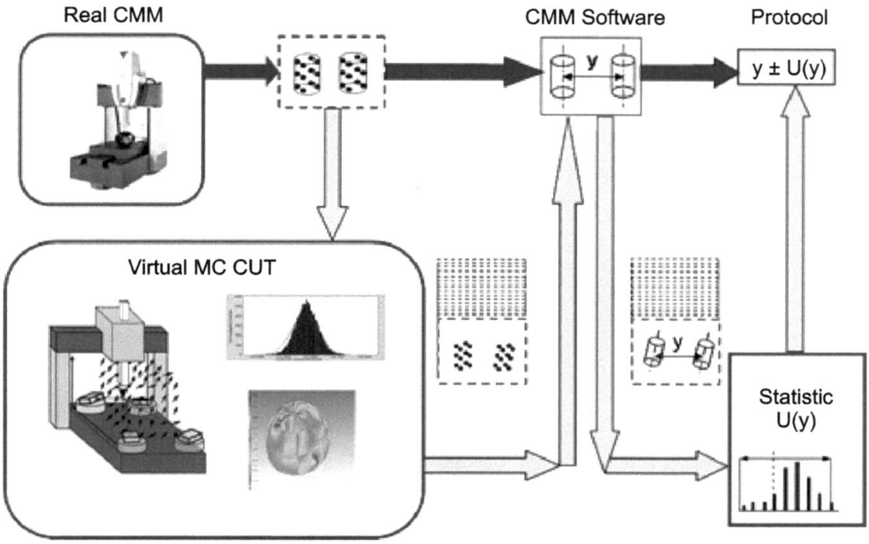

Fig. 5.54 Operation scheme of virtual MC CUT

5.2.4 *Virtual CMM Based on the Monte Carlo Method*

An increase in requirements for CMS accuracy can be observed as the development of better and better systems of software error correction, based on the matrix correction CAA (described in Sect. 4.2.1). This is related to the reduction of the participation of CMM systematic component error impact. Its consequence is the movement of the coordinate system modeling issue towards the area of residual and random errors described in Sect. 2.3, which is also confirmed by works [14, 16, 17, 20–22, 35, 40, 67, 74, 89, 96]. The assumption, which was adopted in the development of this model of virtual CMM, is an earlier correction of geometric, dynamic, and temperature errors by the use of the CAA correction matrix. Its effect is shown in Fig. 4.3 in the form of compensation of kinematic errors with residual errors remaining after correction. In Figs. 4.7 and 4.8 there is also shown how the correction matrix affects CMM accuracy. Therefore an idea to develop a new simulation system based on modeling of residual and random errors and on simulation with the use of the Monte Carlo method [20, 57, 96] was formed. This system was called the Virtual MC CMM CUT [132], and the operation scheme is presented in Fig. 5.54.

In the assumption the system operates similarly as described for virtual machines, but with another error model and simulation method. Furthermore, it is also a continuation of research on CMM error identification with the use of reference points—the matrix method (described in Sect. 4.4)—and presents a new approach to this method. The first research on the use of the Monte Carlo method

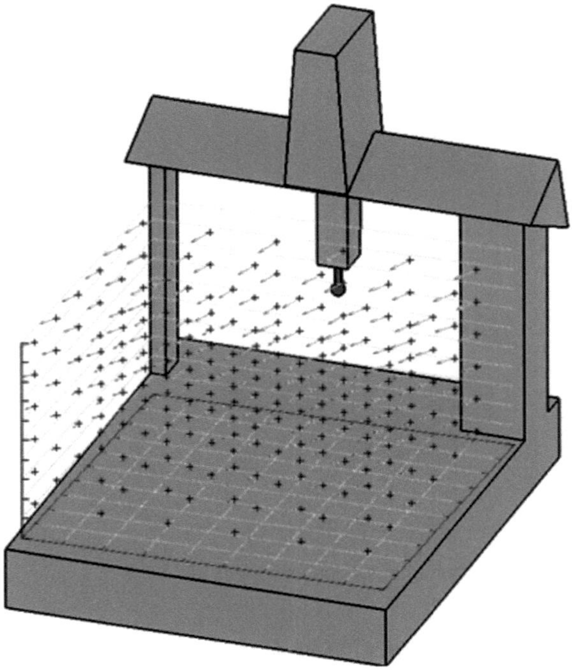

Fig. 5.55 Map of errors compensated by correction matrix CAA [18]

for measurement accuracy assessment was done in the LCM CUT in 2007 [106]. This model was dedicated to modern systems equipped with full, active error correction. According to the author, CAA can be rapidly implemented in an industrial environment, and it is even necessary due to the growth of requirements for necessary accuracy assessment of realized measurements. It was also accepted that it would consist of two basic modules:

- Module that simulates the residual and random errors dependent on kinematic system of CMM
- Module that simulates the workings of the contact system of CMM

Developing **the first module** assumed that the key task is to describe the measuring volume with a reference point grid (Fig. 5.55).

For machines equipped with a correction system it should be assumed that errors causing the inaccuracy of reproduction of the measuring point are residual and/or random errors. The identification of these errors is realized in accordance with the matrix method. In Fig. 5.56 residual errors and the method of their identification are schematically shown. It is also reasonable to assume that the uncorrected residual and random errors should be treated together and analyzed as random ones, which is also suggested by works [23, 49, 121]. This allows us to describe each point in the reference grid with a probability distribution, with which its reproduction by a machine occurs (the best in this application is the *t* distribution, according to [18]).

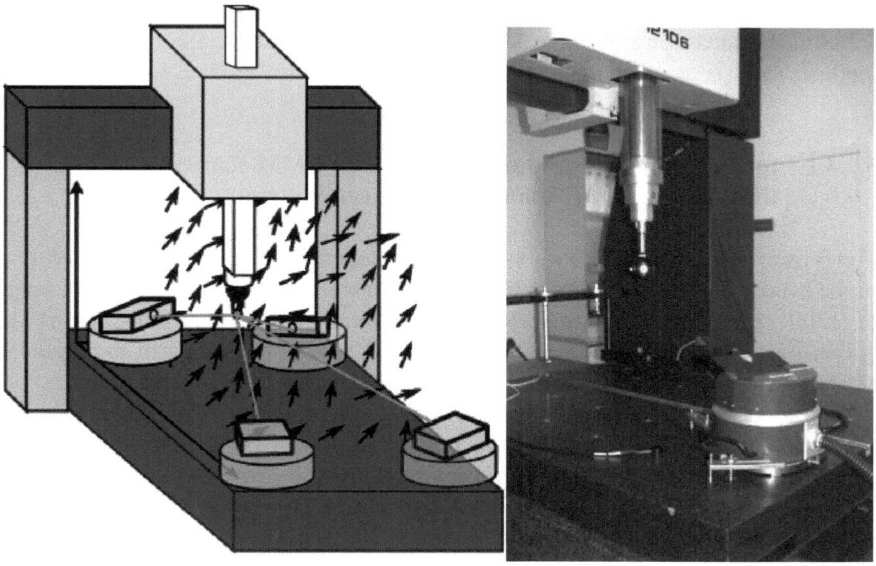

Fig. 5.56 Error vectors in reference points—on the *left*, and the method of their identification with laser tracer (LT)—on the *right*

The key issues here are the accuracy and the number of data, reference points that should be tested. The use of plate standards is not an option here due to the limited number of points and the time of model displacing, and also their accuracy. Currently, only the Etalon Company laser tracer allows the construction of a reference grid for measuring machines of average and high accuracy. The laser tracer is a device shown in Fig. 5.56 and described in detail in Sect. 4.2, with the uncertainty of $U = 0.2 \pm 0.3 * L/1000$ μm similar to the uncertainty of step gauges used for CMM calibration. It is also important to compact the reference point grid in order to obtain information about errors in any point in the CMM volume.

The second module is responsible for the simulation of CMM contact system operation. The probe head error function was used for the description of this system. This function may be described with formula (4.13). Similar to the case of the random and residual error modeling module it was decided to apply the Monte Carlo method for the simulation of contact probe head spatial operation. Data to build this module are collected by multiple measurement of the reference sphere. The requirement for a proper functioning model is the use of a standard with appropriately small form deviations (less than $0.2 \cdot P_{FTU,\ MPE}$ defined according to [31]). In addition, the modeled contact system should be checked in accordance with the [31] standard. The obvious limitation coming from the measurement nature is the lack of possibility to measure the standard on its whole surface. Therefore, it is measured at selected points. As a result of performed measurements the reference points grid is obtained, in which for known deflection angles a and β (Sect. 4.3) the value of the measuring probe head error is known. Similar to the case of the first module, there is an issue of the error grid concentration.

5.2.4.1 Identification of Residual and Random CMM Errors with the Use of MM and Laser Tracer System

The Virtual MC CMM CUT model was developed for HA CMM equipped in a probe head with an internal measuring system. The machine has a measuring range of 1200 × 1000 × 600 mm. The part of the measuring machine volume of dimensions 700 × 600 × 500 mm was modeled. As the beginning of the modeled area was adopted a point with coordinates 150, 50, and 50 mm was taken.

To model the **residual and random errors coming from the kinematic system, the first module**, the reference point grid should be created. It should be closely related to the grid at which machine correction matrix CAA is described. The grid consisting of 56 points in the measuring machine volume was built. According to [112] such a number of points are big enough to complete representation of the machine residual and random errors field. The assumed mutual overlapping of nodes of the reference grid and the grid for the correction matrix CAA is necessary due to the adopted model concept, that is, due to their active connection. This also allows us to modify the model in accordance with the matrices data of the same adopted interpolation nodes, obtained in the future.

The random error values in points being the nodes of the reference grid were determined by multiple approaches to the given point from different directions. In place of the measuring tip the "cateye" type retroreflector, was mounted, as shown in Fig. 5.57.

The retroreflector location was tracked in a dynamic mode by a laser tracer installed in the measuring machine volume and working under the control of Trac-Cal software described in Sect. 4.2.3 and **Annex D**. After a sequence of performed approaches, the machine moved to the next node point, and the cycle was repeated. The sequence consisted of six approaches in directions compatible with three machine axes in a positive and negative orientation, and of eight approaches in directions sloped at a 45° angle for each axis (with positive and negative orientation). This sequence is presented in Fig. 5.57.

In that way measurement paths are obtained. The paths are repeated five times, each time in a different laser tracer position. The reference point grid is presented in Fig. 5.58. The distances, and finally the coordinates, of particular points (14 sets of coordinates for a single point) are determined with use of the multilateration method, described with Eq. (5.14). Point coordinates are determined according to equation:

Fig. 5.57 Sequence of approaches to the measuring point (*yellow arrows* indicate the direction of the approach), *left* side; "cateye" retroreflector mounted on the measuring machine, *right* side (Color figure online)

$$w_{ij} = l_{ij} + l_{0j}\sqrt{\left(x_i - X_{0j}\right)^2\left(y_i - Y_{0j}\right)^2 + \left(z_i - Z_{0j}\right)^2} \quad (5.14)$$

where

i	number of the point
j	number of the position of laser tracer
w_{ij}	residue of a distance function
l_{ij}	distance to *i*th point measured by laser tracer being in the *j*th position
l_{0j}	length of "deathpath" for laser tracer in *j*th position
x_{ij}, y_{ij}, z_{ij}	particular coordinates of *i*th point
X_{0j}, Y_{0j}, Z_{0j}	particular coordinates of laser tracer location in *i*th position

Coordinates in the reference points are obtained by minimizing the sum of squares of distance function residuals w_{ij}.

$$\sum w_{ij}^2 \rightarrow \min \quad (5.15)$$

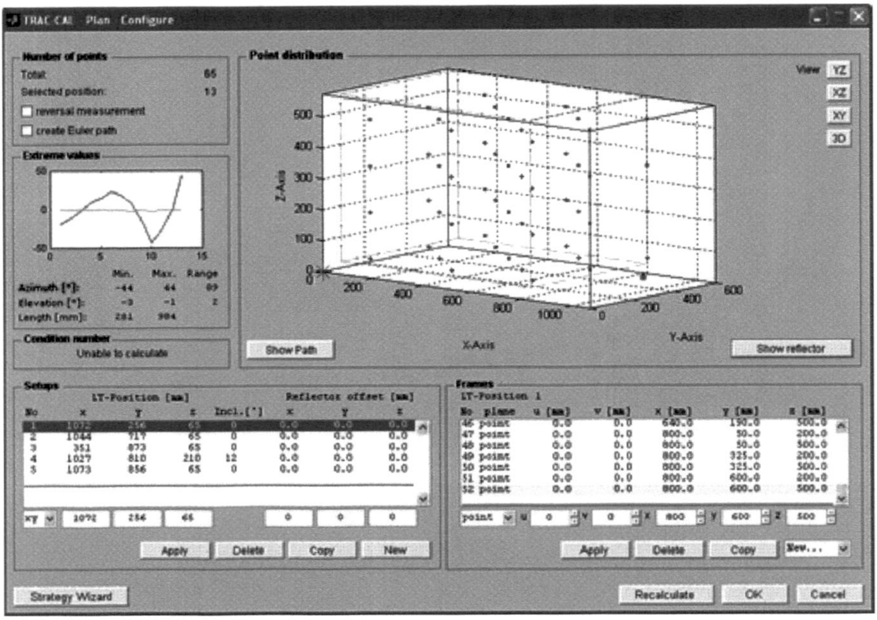

Fig. 5.58 Reference point grid together with an analysis of reference points, Trac-Cal program [122]

This method is performed using Trac-Cal software. In Fig. 5.58 the window of this program is shown with the reference point grid and with their analysis.

Identification results shown in Fig. 5.59, in the form of vectors, illustrate standard deviations (scale 100 000:1) of reproduction of particular coordinates of reference points (values on the axes given in mm).

Maximum values of deviation in each axis directions were:

$$d_{\max}(X) = 0.002595\,\text{mm}$$
$$d_{\max}(Y) = 0.001773\,\text{mm}$$
$$d_{\max}(Z) = 0.001473\,\text{mm}$$

And average values of deviations:

$$\bar{d}(X) = 0.000912\,\text{mm}$$
$$\bar{d}(Y) = 0.000791\,\text{mm}$$
$$\bar{d}(Z) = 0.000749\,\text{mm}$$

The simulation of measuring points is carried out with the use of the Monte Carlo method. The concentration of the grid for points that are not nodes of the grid is done using the interpolation method. Each point in the measuring machine

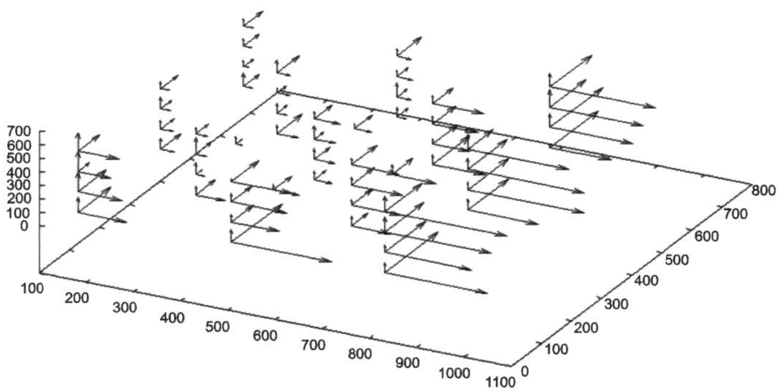

Fig. 5.59 Vectors in the figure illustrate standard deviations (scale 100,000:1) of reproduction of particular coordinates of reference points (values on the axes given in mm) [18]

volume is described with a specific probability distribution. For all points τ-distribution with parameters (x, o, υ) is assigned, where x is the mean value of the distribution, o is the standard deviation, and u is the number of degrees of freedom. As the mean value for each point its nominal coordinates are taken. The standard deviation is obtained through multiple simulations. The number of degrees of freedom is 13 for each point, because the number of degrees is equal to the number of measurements for each point minus one $14 - 1 = 13$ [18, 122].

Contact probe head errors, the second module, was also determined experimentally [18, 89]. To create the measuring probe head error model one should have an appropriate set of data that shows the overall picture of the probe head error function, in dependence on a and b angles (formula (4.20), Sect. 4.3.1). These data can be determined experimentally by reference sphere measurement. The diameter of this standard must be so small that during its measurement the errors derived from the kinematic system would be negligible. In addition, standard form deviations should also be negligibly small in relation to the designated values [89]. Measurements should be carried out regularly increasing a and b angles of the same value. It is assumed that the angle a should be increased 20° each time and the angle β, 10°, wherein α varies from 0° to 360°, and β from 0° to 90° (the 0°, 90° point is on the sphere pole). Therefore one measurement sequence consists of 163 points, which is presented in Fig. 5.60. Such a large number of measuring points is caused by the necessity of a true representation of the modeled probe head operation. The correctness of the model operation based on similar assumptions described in this chapter, and built on only 37 reference points, had also been initially proven in [110], and is very important for industrial implementation of this solution.

The measurement sequence should be repeated several times to obtain a statistically representative dataset describing measuring probe head errors. The model presented in this chapter was created on the basis of data obtained from 15 repetitions of the measurement sequence. After collecting the necessary data, the mean radius and its standard deviation are determined. These values are used next as a

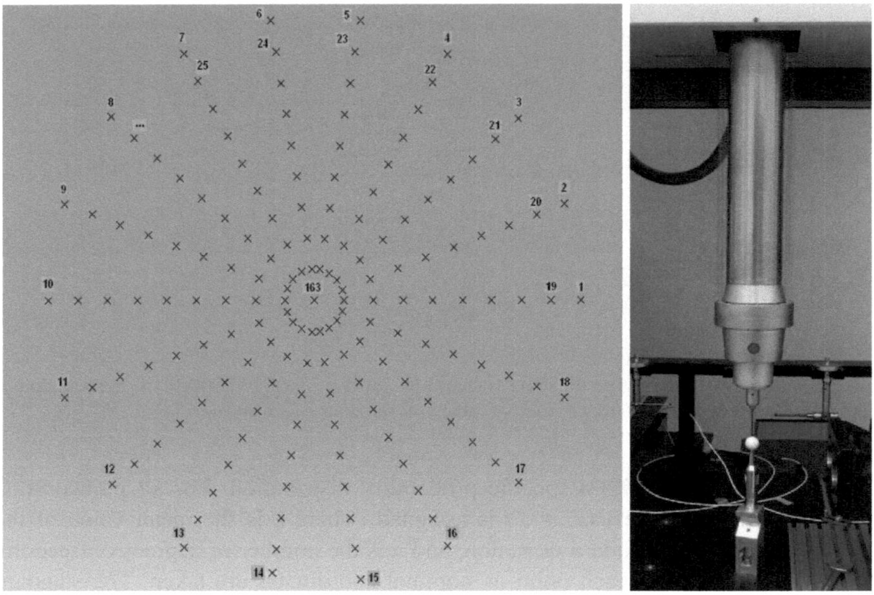

Fig. 5.60 Reference sphere measurement method: Leitz PMM machine and determined reference points being a measurement sequence (view from the top in the direction of *XY* plane) [18]

basis for the simulation of measuring probe head errors for deflection angles equal to (a, b), which are the reference grid nodes as a direct representation of PEF. The simulation is carried out accordingly with the Monte Carlo method. Probe head errors corresponding to each angle (α, β) pair are described similarly to the case of kinematic errors with appropriate probability distributions. To each pair of measuring stylus deflection angles a probability distribution T of parameters $(\bar{x}, \sigma, \upsilon)$ is assigned, where x is the mean value of the distribution, σ is the standard deviation, and υ is the number of degrees of freedom. As the mean value for each point in the reference grid a mean from values obtained from measurement of the reference sphere is taken. The standard deviation is also calculated from results of standard measurements. The number of degrees of freedom is 14 for each point because the number of degrees is equal to the number of measurements for each pair of angles minus one, $15 - 1 = 14$. For angle pairs, which do not determine one of the reference points, values are obtained by the use of two-dimensional linear interpolation adapted for use in a spherical system. The results of the performed standard measurements are presented below. As a standard the reference sphere with a diameter of $D = 24.9944$ mm was measured. Points were collected according to the sequence shown in Fig. 5.61.

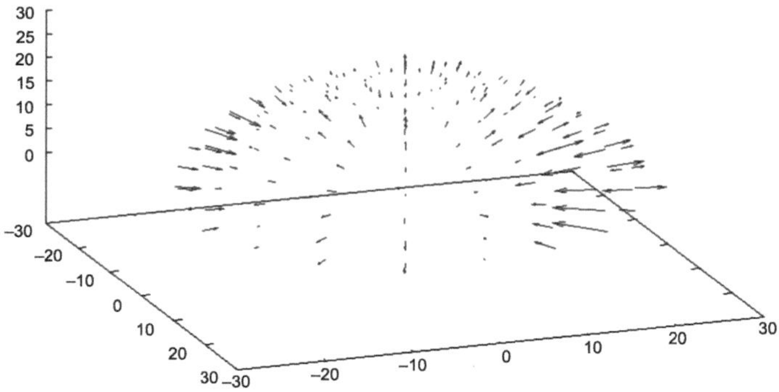

Fig. 5.61 Measuring probe head errors dependent on the measuring stylus deflection angles *a* and *b* [18, 110]

5.2.4.2 Virtual MC CUT Model Algorithm

The software realizing the operation of virtual measuring machine Virtual MC CUT is based on the algorithm shown in Fig. 5.62. The Virtual MC PCUT system is developed in the Python programming environment. For data processing the metrological program Quindos 7 was used. This program is also an object-oriented database programming environment.

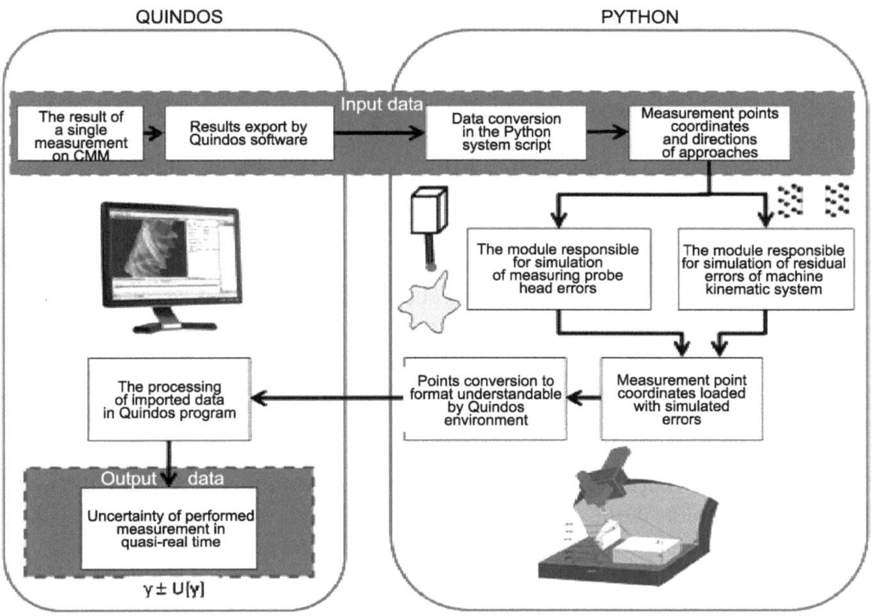

Fig. 5.62 Operation algorithm of virtual machine model Virtual MC CUT showing the relations between used programming environments [18, 110]

A description of the processes is presented below.

Export of Results by Quindos Software

After performing the measurement, the measured object data, which are coordinates of particular measuring points for measured characteristics are stored in the Quindos program as an object. There is a possibility to export it to the text file, which next can be read in the Python programming environment.

Data Conversion in Python Program Script

The use of the Python language allows us to read the exported results from a text file and to convert the file to the form of a matrix. The constructed matrix contains data of all measuring points. The recorded data are measuring point coordinates *X*, *Y*, *Z* in the machine coordinate system and three direction cosines of the approach vector on a given point. Thus prepared data are the input quantity to the simulation model.

Module Responsible for Simulation of Kinematic System Residual Errors

As an input value to this module a matrix containing values of measuring point coordinates returned by the previous process is taken. Its main task is to assign to each set of coordinates the value of variability of measuring point reproduction which for points being the nodes of the reference grid is expressed as the standard deviation from multiple approaches to the point. Then the reproduction of a given point is simulated by the use of Monte Carlo simulation.

For points lying in the reference grid nodes the variability of measuring point reproduction is known, because it has been determined experimentally. In this case, the simulation is performed for given reference points taken directly from the module input matrix, which contains (segregated in rows): *X*, *Y*, *Z* coordinates of the point and standard deviations of reproduction of particular coordinates $d(X)$, $d(Y)$, $d(Z)$. This case is a marginal percentage for all performed measurements, because most points of the real measurement will be between the reference grid nodes. To simulate the variability of reproduction of points lying between measurement grid reference points, aside from the Monte Carlo simulation, it is also necessary to use interpolation methods. The first step of this interpolation is the creation of three-dimensional spline functions of B-spline type, describing the relation that occurs between the reference points situated along the *Z*-axis direction. Therefore 13 B-spline curves should be created; each spread over 4 reference points. Values of particular coordinates of reference points are also simulated with the Monte Carlo method using *T*-distribution. An example spline curve of B-spline type spread over four reference points is shown in Fig. 5.63.

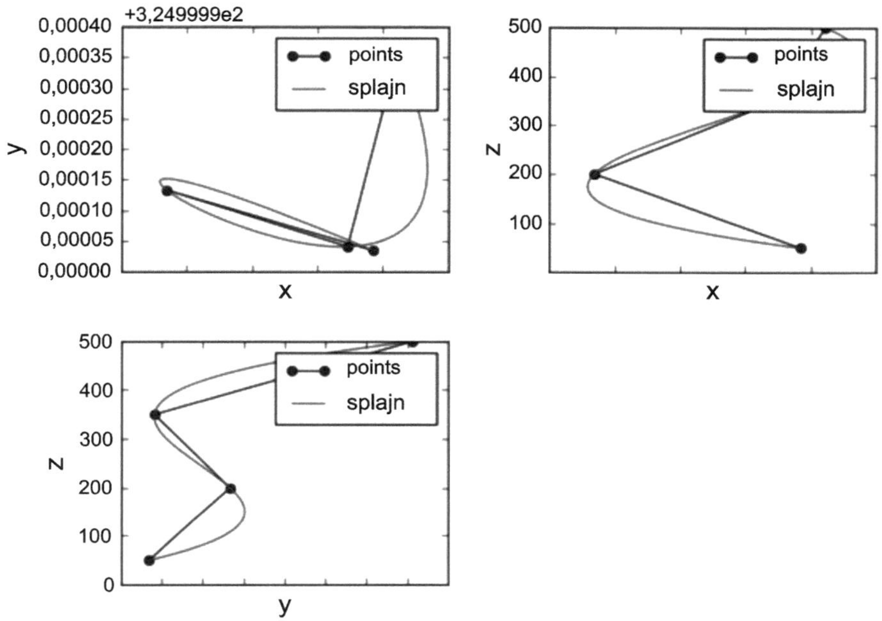

Fig. 5.63 Example B-spline curve spread over four reference points (view in projections from the *top*, *front*, and *side*)

The next step is the intersection of all curves with a plane parallel to the *XY* plane, located at a distance equal to Z_1 from this plane (where Z_1 is a simulated measuring point coordinate). The created plane intersects with each B-spline curve giving 13 points, in which values of coordinates (*x*, *y*, *z*) are known. The points obtained in this way are then the basis for the performance of interpolation according to the "nearest neighbor" principle [18] for two-dimensional functions. Assuming that the function *f* (*x*, *y*) is described and its values in ^-selected nodes are known, it is possible to interpolate the values of the function *F* in any place that belongs to the function domain and is located between selected nodes. The value interpolating given point is the value copied from the node that is closest to that point.

To illustrate the operation of the nearest neighbor method an example of interpolation with this function method is shown in Fig. 5.64,

$$f(x, y) = x \cdot (1 - x) \cdot \cos(4\pi x) \cdot \sin\left(4\pi y^2\right)^2 \tag{5.16}$$

implemented in the Python environment. The figure on the left shows a graph of the function *f*(*x*, *y*), and the figure on the right shows the reproduction of the function value with the use of interpolation with the nearest neighbor method, which is described in 1000 points (black dots in the graph on the left) randomly selected from the original function [18].

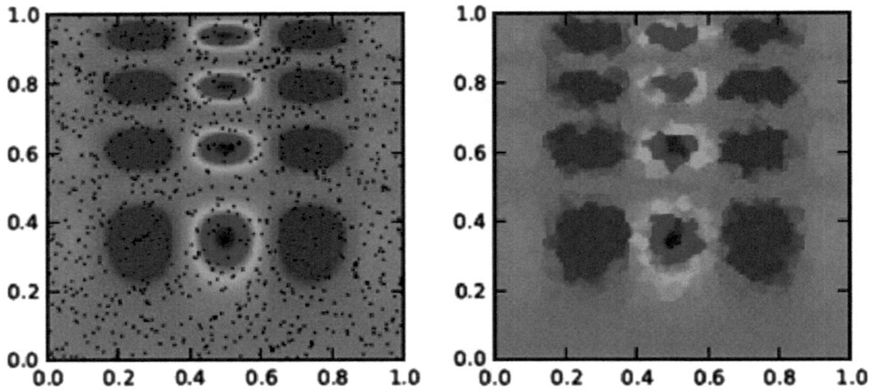

Fig. 5.64 Interpolation with the nearest neighbor method: value of the original function (**a**); the interpolated function value obtained for the interpolation based on 1000 nodes (**b**) [18]

Using this interpolation method it is possible to obtain values that describe the x coordinate, taking as input data the values of the (y, z) coordinates and y coordinate, taking as an input data the values of the (x, z) coordinates. As the value of the simulated z coordinate the value of this coordinate obtained from the Monte Carlo simulation in the nearest node of a three-dimensional grid is taken.

Thus obtained (x_1, y_1, z_1) coordinates of the considered point are then simulated N times, where N is the number of Monte Carlo trials [18]. The schematic algorithm of operation of the module simulating residual kinematic system errors for a single measuring point is presented in Fig. 5.65.

Module Responsible for Simulating Measuring Probe Head Errors

As an input value to this module a matrix containing direction cosines (nx, ny, nz) of the approach to the measuring point is taken. The basis is to assign to each direction set a value of the probe head error in that point. Similarly as in the case of the previously described module as a constant is given a matrix containing data of probe head errors for angle pairs of measuring stylus deflection α, β, determined experimentally, and described in Sect. 4.3.1. For angle pairs lying in nodes of the reference grid that models the measuring probe head operation, and having known parameters of the probability distribution assigned to them, the output value of the probe head error is obtained after N Monte Carlo simulations. For angle pairs that are not reference grid nodes in addition to simulation with the Monte Carlo method (similar as for kinematic errors) interpolation methods should also be used. In this case a two-dimensional method of linear approximation adapted for use in a spherical coordinate system is used. This method for one dimension is based on assigning to the interpolated point the value being the weighted mean of obtained values in nodes surrounding the interpolated point and distances to particular nodes.

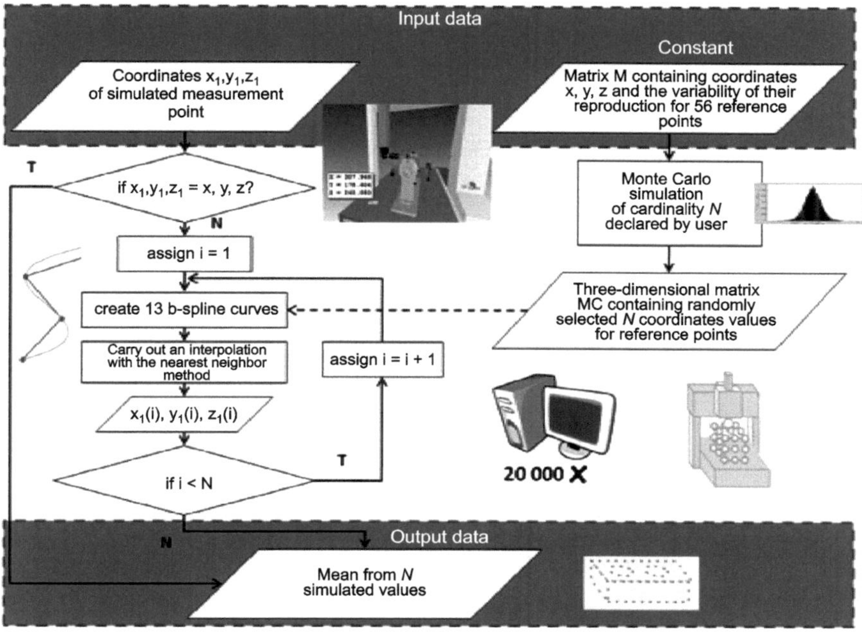

Fig. 5.65 Algorithm realized by the module simulating residual errors [18, 110]

The two-dimensional method, in turn, is based on estimation of the value in the first direction and then in a second one [18].

The value of the measuring probe head error (PE) for the simulated angle pair of measuring stylus deflection is simulated n-times in this way, where N is the number of Monte Carlo trials [53].

The determined probe head error (according to Sect. 4.3.1) is a vector compatible with the direction of the approach and with the orientation depending on the sign of this error value. Therefore at the end of the simulation each simulated error must be recalculated to specific changes in the location of the final simulated measuring point marked as BG_X, BG_Y, BG_Z. This is done according to the formula:

$$
\begin{aligned}
BG_x &= \mathrm{PE} * NX \\
BG_y &= \mathrm{PE} * NY \\
BG_z &= \mathrm{PE} * NZ
\end{aligned}
\tag{5.17}
$$

In Fig. 5.66 the algorithm realized by the described module (for one measuring point) is presented.

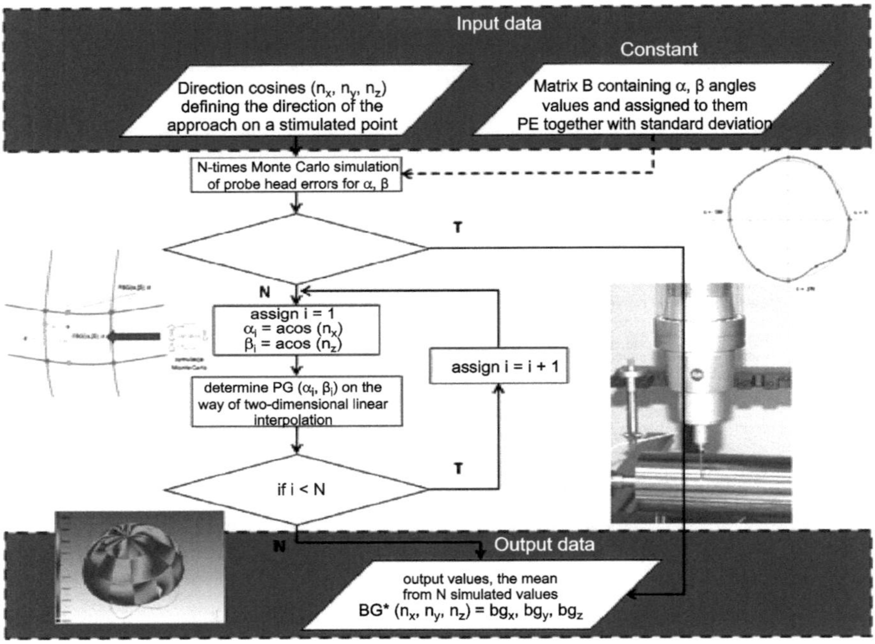

Fig. 5.66 Algorithm realized by the module simulating probe head errors [18, 110]

Coordinate Conversion to a Format Understood by Quindos Environment

After completing the work of the module simulating the residual errors of the CMM kinematic system and of the module simulating the work of the measuring probe head, the determined values x_1, y_1, z_1 and BG_X, BG_Y, BG_Z are added to each other. In this way the real coordinates of the simulated points are obtained. These coordinates, together with the corresponding approach directions NX, NY, NZ are then saved to a file in a format recognized by the Quindos metrological program environment.

Processing of Imported Data in Quindos Program

Converted actual coordinates are imported by the Quindos program and the geometric elements, corresponding to real measurements, are repeatedly determined from them. Independence of realized measuring task geometric parameters of the measured characteristic are determined and by using a script created in the Quindos program an uncertainty of these parameter determinations is calculated. This uncertainty is the output value of the described virtual measuring machine model.

5.2.4.3 Model Verification Tests

Verification of the developed model consists of two stages. In the first one the Virtual MC CUT model was tested according to VDI/VDE contained in [128]. The second stage involves the performance of common measuring tasks, such as measurements of a distance determined with the point–point method, the plane–plane distance, the sphere diameter, sphere form deviations, and the distance between the centers of two spheres. Then for each measurement a standard uncertainty was determined according to the methodology used in classical methods of measurement accuracy determination (according to Chap. 3). Uncertainty determined in the classical way was compared to the uncertainty obtained with the simulation performed by the Virtual MC CUT model.

Requirements contained in [128] recommend checking the functioning correctness of the simulation model by reference element measurements. A characteristic of an object (a cylindrical standard) is given in these recommendations and presented in Fig. 5.67 and in Table 5.8.

According to [128] measurements should be carried out with the use of different strategies and performed for different standard locations; thus it will be possible to check the functioning of the model in the whole measuring machine volume. Recommended object orientations are shown in Fig. 5.68. The simulation model is considered to be operating properly when the following inequality is met:

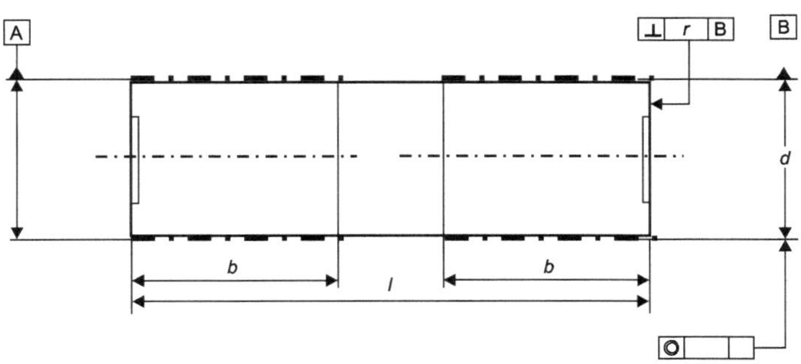

Fig. 5.67 Cylindrical standard to verify VCMM MC CUT according to [128]

Table 5.8 Measured characteristics of the cylindrical standard

Symbol	Measured/simulated feature
l	The distance of frontal surfaces
d	Cylinder diameter
r	Perpendicular deviations of frontal surfaces to the cylinder axis
c	Coaxial of two associated cylinders 1,2
b	The length of the reference for the coaxial and perpendicularity measurement in accordance with ISO 1101

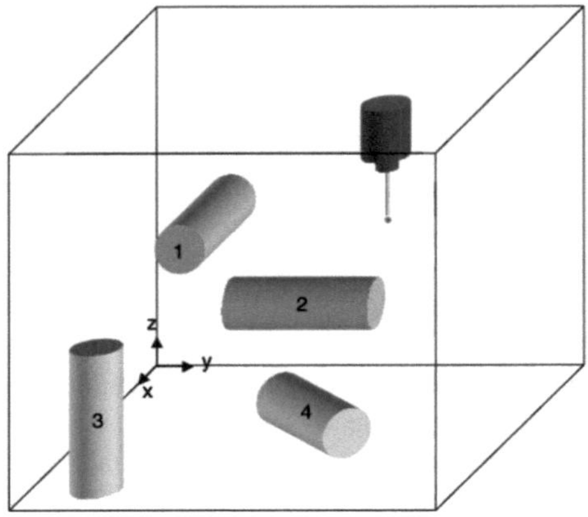

Fig. 5.68 Recommended positions of the cylindrical standard during checking of correctness of a simulation model [128]

Table 5.9 Dimensions of cylindrical square calibrated with the use of measuring machine PMM 12106

	r (mm)	*d* (mm)	*l* (mm)	*c* (mm)
U_k	0.052	115.2411	329.6112	0.0421
	0.0087	0.00069	0.00077	0.00088

$$|y_k - y| \leq U_k + U \tag{5.18}$$

where

y_k Value of standard feature obtained in its calibration process
y Measured value of standard feature
U_k Expanded uncertainty of standard calibration
U Expanded uncertainty of measurement (determined with the use of checked simulation model).

In tests a cylindrical standard of dimensions: $D = 115.241$; $L = 329.611$ was used (Table 5.9).

According to requirements [128] the tests of a standard were performed using the specially prepared holder that provides stability and ease of movement (Fig. 5.69). Positions 1–3 were selected for measurements, as shown in Fig. 5.68. Measurements were carried out using different measurement strategies. In each of the selected positions the cylinder was created from 9, 15, or 24 measuring points. The uncertainty determination of measurement of diameter, form deviations, and the distance between the cylinder front planes was assessed.

In the second model verification stage, results of its operation were compared with uncertainties determined in accordance with the classical method of type ***A*** uncertainty determination, comparative and multiposition methods.

For measurements a ball-bar standard developed and created in the Laboratory of Coordinate Metrology (which had not yet passed a calibration procedure and

Fig. 5.69 Measurements of cylindrical standard on the modeled coordinate machine

therefore was treated as a noncalibrated object) and a step gauge Koba Step 620 having a calibration certificate were used.

Measurements were realized on the same machine, Leitz PMM 12106, for which Virtual MC CUT was developed to achieve a common reference. The machine characteristic and laboratory conditions are specified in Sect. 3.5. For temperature correction of each measured object a system consisting of two temperature sensors was used, which allowed reduction of thermal expansion coefficient impact of the measured object. The mounting of sensors is shown in Fig. 5.70 in the example of measurement of the distance between two ball centers. The measuring machine is also equipped with a functioning system for temperature correction.

The distance measurements between two points and between two planes were realized on the Koba Step step gauge. Measurements were repeated 30 times in order to ensure a representative statistical sample to determine the uncertainty, which in this case was determined with the classical type *A* method. Obtained results were then compared with results obtained by using the Virtual MC CUT software; that is, a single simulated measurement of the distance between two points and a single measurement of the distance between planes was realized. For this measurement with the use of the Virtual MC CUT model a result together with the uncertainty determined using a simulation technique was given. The other measurements, that is, the measurement of the sphere diameter, sphere shape deviations, and the distance between the centers of two spheres was performed on the ball-bar standard. For measurements of sphere form deviations and the sphere

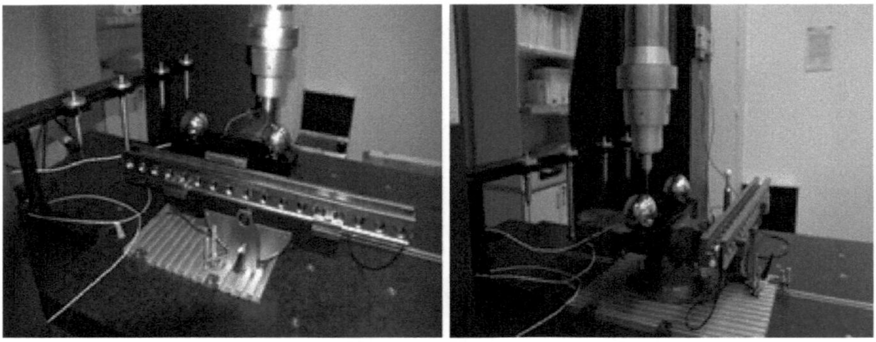

Fig. 5.70 Measurement of ball-bar standard in accordance with multiposition method guidelines

diameter the realized measurement uncertainty was determined in accordance with guidelines of [32, 128] and [34, 47, 111, 112] standards, and for measurement of the distance between centers of spheres with the use of multiposition (described in Sect. 3.1) and substitution (described in Sect. 3.2) methods. In Fig. 5.70 the measurements of a ball-bar standard performed in accordance with multiposition method guidelines are presented. Obtained results were compared with results of Virtual MC CUT software operation [18, 132]. As a length standard appropriate steps of a step gauge were used. In measurements realized in accordance with guidelines of the substitution method the measured length on the standard was chosen in a way that satisfied the comparative conditions. Also the measurement strategy for the object and standard should be consistent (Sect. 3.1). Therefore, step gauge measurements were performed as follows: first two planes on the same element were measured; then the plane of symmetry between measured planes was determined. These steps were repeated for two following planes on the other measured object. The value measured on a standard was represented by the distance between two planes of symmetry of two different measurement elements. In this way indirect measurements of the distance between centers of spheres corresponding to the measurement strategy were carried out.

Results of Performed Verification Tests and Simulations with the Use of Virtual MC CUT

Stage 1:

Measurements of cylindrical standard set in positions consistent with Fig. 5.70 [123]. Each time the measurements were performed using a different measurement strategy.

Results for position no. 1:

- Cylinder created by Gauss method from 9 points (symbols in Tables 5.10, 5.11, 5.12, 5.13, 5.14, 5.15, 5.16, 5.17, 5.18 and 5.19 are consistent with Eq. 5.18).
- Cylinder created by Gauss method from 15 points.
- Cylinder created by Gauss method from 24 points.

Table 5.10 Results of the cylindrical standard measurement: position number 1, the cylinder described at 9 points

Measured feature	y (mm)	y_k (mm)	U (mm)	U_k (mm)
Cylinder diameter	100.2454	100.2454	0.00095	0.00069
Form errors	0.0016	0.0015	0.00061	0.00029

Table 5.11 Results of the cylindrical standard measurement: position number 1, the cylinder described at 15 points

Measured feature	y (mm)	y_k (mm)	U (mm)	U_k (mm)
Cylinder diameter	100.2453	100.2454	0.00097	0.00069
Form errors	0.0018	0.0015	0.00056	0.00029

Table 5.12 Results of the cylindrical standard measurement: position number 1, the cylinder described at 24 points

Measured feature	y (mm)	y_k (mm)	U (mm)	U_k (mm)
Cylinder diameter	100.2454	100.2454	0.00096	0.00069
Form errors	0.0016	0.0015	0.00043	0.00029

Table 5.13 Results of the cylindrical standard measurement: position number 2, the cylinder described at 9 points

Measured feature	y (mm)	y_k (mm)	U (mm)	U_k (mm)
Cylinder diameter	100.2459	100.2454	0.00088	0.00069
Form errors	0.00155	0.00158	0.00063	0.00029

Table 5.14 Results of the cylindrical standard measurement: position number 2, the cylinder described at 15 points

Measured feature	y (mm)	y_k (mm)	U (mm)	U_k (mm)
Cylinder diameter	100.2459	100.2454	0.00075	0.00069
Form errors	0.00171	0.00158	0.00055	0.00029

Table 5.15 Results of the cylindrical standard measurement: position number 2, the cylinder described at 24 points

Measured feature	y (mm)	y_k (mm)	U (mm)	U_k (mm)
Cylinder diameter	100.2461	100.2454	0.00072	0.00069
Form errors	0.00158	0.00158	0.00046	0.00029

Table 5.16 Results of the cylindrical standard measurement: position number 3, the cylinder described at 9 points

Measured feature	y (mm)	y_k (mm)	U (mm)	U_k (mm)
Cylinder diameter	100.2459	100.2454	0.00082	0.00069
Form errors	0.0013	0.0015	0.00059	0.00029

Table 5.17 Results of the cylindrical standard measurement: position number 3, the cylinder described at 15 points

Measured feature	y (mm)	y_k (mm)	U (mm)	U_k (mm)
Cylinder diameter	100.2452	100.2454	0.00074	0.00069
Form errors	0.0014	0.0015	0.00051	0.00029

Table 5.18 Results of the cylindrical standard measurement: position number 3, the cylinder described at 24 points

Measured feature	y (mm)	y_k (mm)	U (mm)	U_k (mm)
Cylinder diameter	100.2451	100.2454	0.00071	0.00069
Form errors	0.0014	0.0015	0.00042	0.00029

Table 5.19 Results of the cylindrical standard frontal plane distance measurement

Frontal plane distance	y (mm)	y_k (mm)	U (mm)	U_k (mm)
Position 1	314.6595	314.6596	0.00091	0.00077
Position 2	314.6599	314.6596	0.00093	0.00077
Position 3	314.6595	314.6596	0.00086	0.00077

Results for position no. 2:

- Cylinder created by Gauss method from 9 points.
- Cylinder created by Gauss method from 15 points.
- Cylinder created by Gauss method from 24 points.

Results for position no. 3:

- Cylinder created by Gauss method from 9 points.
- Cylinder created by Gauss method from 15 points.
- Cylinder created by Gauss method from 24 points.

Measurements of the distance of standard frontal planes were also performed. Each plane was measured at six points. Results are presented in Table 5.17 (Fig. 5.74).

For all the results presented above a relation described by Eq. (5.18) is true, so the built model can be considered as consistent with recommendations of [128] and should be considered as a correctly functioning one.

Stage 2:

(a) Measurements of a distance between two points:
Two points were measured for which a nominal distance, consistent with the calibration certificate is $d = 420.9802$ mm. The results are as follows.

(b) Measurements of a distance between two planes:
Two planes were measured on measurement elements of the step gauge, for which a nominal distance read from the calibration certificate is $d = 259.9836$ mm. Each plane was measured in four places. The results are as follows.

(c) Measurements of a sphere diameter:
Sphere no. 1 diameter was measured on the standard of *ball-bar* type. The sphere was measured at 15 points. As previously mentioned, the standard does not have a calibration certificate, so the correct value of the measured length is not known. The results are as follows.

(d) Measurements of sphere form deviations:
Sphere no. 1 was measured on the *ball-bar* standard. The sphere was measured in 15 points. Form deviations of the measured sphere were determined. The results are as follows.

(e) Measurements of a distance between centers of two spheres:
Two spheres were measured on the *ball-bar* standard. Each sphere was measured in 15 points. Then the distance between centers of spheres was determined. The results are as follows.

The tests depicted above are also presented in Figs. 5.71, 5.72, 5.73, 5.74, 5.75 and 5.76. Results fulfill the condition from Eq. (5.16). Differences between results obtained with the use of the Virtual MC CUT model and classical methods (Tables 5.20, 5.21, 5.22, 5.23 and 5.24) are small enough and the developed model can be accepted as a correct one and the one that faithfully reproduces the uncertainty area of a real measurement carried out on the PMM machine. Therefore the method in the range of performed research may be considered a validated one. Of course, it is necessary to validate other machines every time, for which the Virtual MC CUT model is under development. In the case of a practical application and adaptation of the method, its validation for a machine with an average accuracy will be interesting. Especially for machines with modern floating stress-free construction (e.g., DEA Global or Zeiss Eclipse) operating in industrial air-conditioning conditions where the functioning of an active system of CAA correction matrixes can be fully seen.

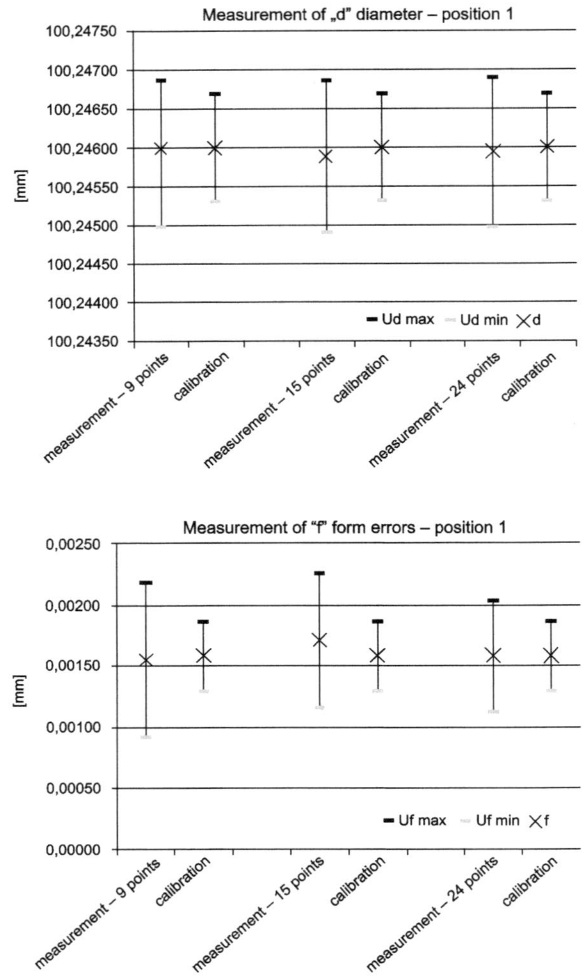

Fig. 5.71 Results of Virtual MC CUT model verification: results of diameter and shape deviation measurements related to results of the cylindrical standard calibration, position 1 [18, 111]

5.2.5 *CMM Simulator and Virtual Machine*

The concept of simulator use in teaching the use of complex and expensive systems already has a decades-long tradition. Simulators are used in military technology, the high-tech industries, and as a response to the complexity of scientific experiments. They make it possible to teach flying planes, driving vehicles, or operating technological machines. In simulators elements of so-called virtual reality application appear more often, "imitating" the real parameters of an object. Computer simulations are nowadays commonly used for manipulation of tested system parameters under variable conditions of its use and to test consequences of various decisions and their optimization. Simulators are also a huge games market and a way to fill

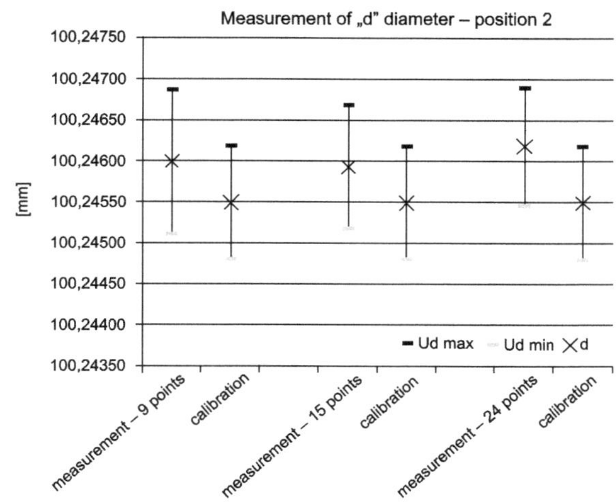

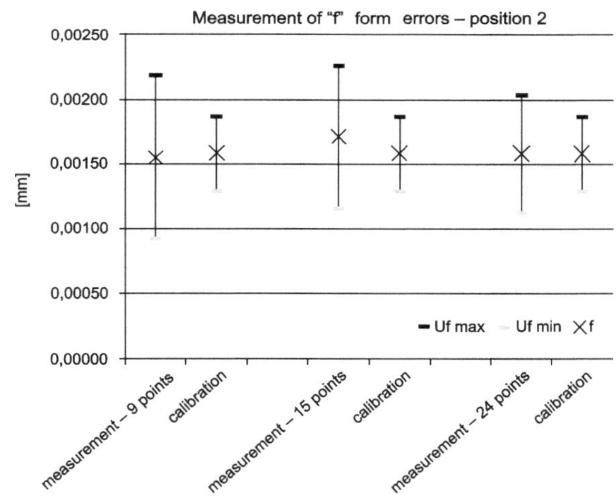

Fig. 5.72 Results of Virtual MC CUT model verification: results of diameter and shape deviation measurements related to results of the cylindrical standard calibration, position 2 [18, 111]

the free time of modern people. In the coordinate measuring technique metrological software often has the capability of simulating the program path, the object location, and the movement of the measuring tip. However, lack of simulation of machine work during measurement with simultaneous access to metrological software could be seen. Many simulation systems that repeat the measurement path—starting from the SILMA system through the concepts of Wilcox, Zeiss, or Metrologic—have appeared lately on the market. All systems are dedicated to machine systems and software of one manufacturer. This limited the wider use and flexibility of the training. Only the concept of the Messtechnik Wetzlar Company in connection with the use of the universal I++ protocol allowed the construction of such a system.

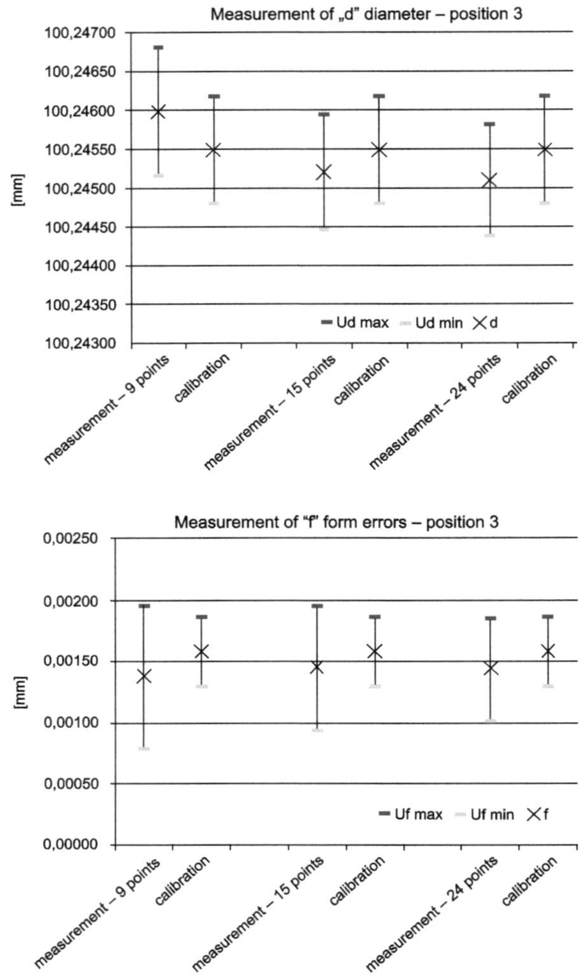

Fig. 5.73 Results of Virtual MC CUT model verification: results of diameter and shape deviations measurements related to results of the cylindrical standard calibration, position 3 [18, 111]

5.2.5.1 I++ DME Protocol, as a Common Platform for Metrological Software

The I++ Simulator of the Messtechnik Wetzlar Company [28] is an IT tool that uses two monitors, graphic engines, and game pads familiar from games technology. At the same time it allows spatial visualization of measuring machines with the ability to move its kinematic sets and connection of metrological software. The system not only allows the use of any metrological software, but also the construction of models of measuring systems already existing on the market. The name I++ Simulator is a kind of abbreviation that comes from the name "Inspection++", which means control in the area of quality assurance. It is based on a description of C++ programming language.

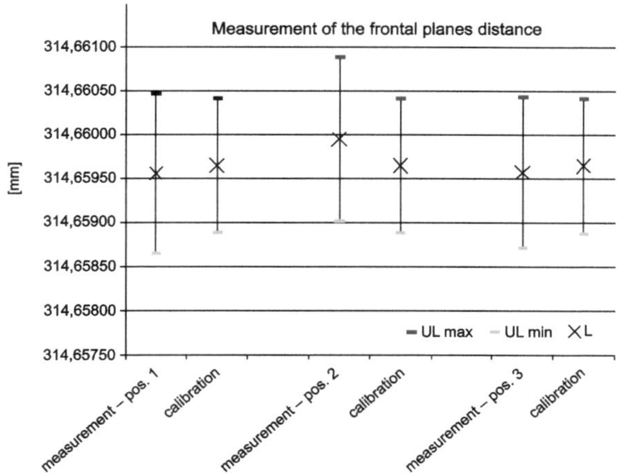

Fig. 5.74 Results of Virtual MC CUT model verification: results of frontal planes distance measurements related to results of the cylindrical standard calibration [18, 110]

Quindos Measuring Report

3D-Application Center

Description		Customer	
Drawing Number		Serial Number	
Remarks		Article Number	
Supplier		Delivery Date	
Delivery Note		Delivery Volume	
Lot Number		Lot Size	
Test Schedule		Sample Size	
Production Machine		Production Tool	
Production Date		Production Time	
Order		Department	
Inspector	AG	Inspection Date	23-MAR-2011, 19:14:56
Measuring Device	RMT 12 10 6 #4711	Measuring Program	Quindos7 - V 7.5.10317
User Name	Adam	WKP Name	BALLBAR250

Text	Eval.	Actual	Nominal	Up.Tol.	Low.Tol.	Act-Nom	Graphic
u(KULA1_DM) = 0.000560							
KULA1			SPH				
	DM	79.998529	80.000000	0.000500	-0.000500	-0.001471	

Fig. 5.75 Example report from the Quindos program presenting the result of a single measurement with the uncertainty determined with simulation technique

On the other hand, in 1999 a working group named I++ was set up. It aimed at unification of the system of interfacing the coordinate machines, allowing, mainly in the automotive industry, to exchange and unify metrological software regardless of the measuring machines used. The effect of their work is among others the development of a universal protocol called I++ DME (dimensional measuring

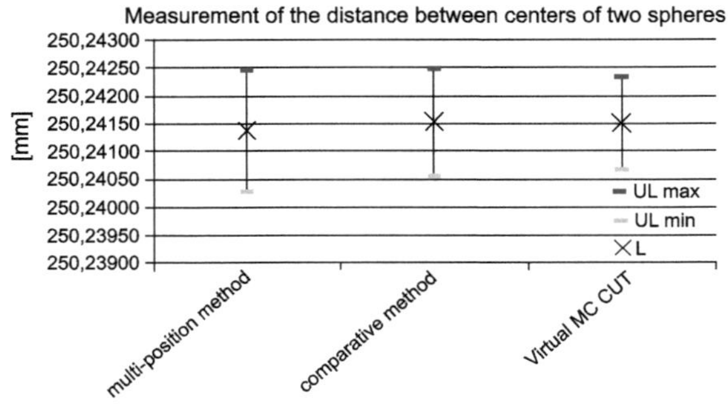

Fig. 5.76 Comparison of results of distance measurements between centers of two spheres [18, 110]

Table 5.20 Comparison of results of distance measurement between two points together with uncertainties determined with simulation and classical technique

	Distance (mm)	Standard uncertainty (mm)
Multiple measurement/uncertainty type A	419.98064	0.000106
Single measurement/uncertainty for Virtual MC CUT model	419.98047	0.000080

Table 5.21 Comparison of results of distance measurement between two planes together with uncertainties determined with simulation and classical technique

	Distance (mm)	Standard uncertainty (mm)
Multiple measurement/uncertainty type A	259.98343	0.000072
Single measurement/uncertainty for MC CUT model	259.98342	0.000051

Table 5.22 Comparison of results of diameter measurement together with uncertainties determined with simulation and classical technique

	Diameter (mm)	Standard uncertainty (mm)
Multiple measurement/uncertainty determined with multiposition method	79.99895	0.000759
Single measurement/uncertainty from MC CUT model	79.99852	0.000560

Table 5.23 Comparison of results of sphere shape deviation measurement together with uncertainties determined with simulation and classical technique

	Deviation (mm)	Standard uncertainty (mm)
Multiple measurement/uncertainty determined with multiposition method	0.00097	0.000736
Single measurement/uncertainty from MC CUT model	0.00139	0.000438

Table 5.24 Comparison of results of distance between centers of two spheres measurement together with uncertainties determined with simulation and classical technique

	Distance (mm)	Expanded uncertainty (mm)
Multiple measurement/uncertainty determined with multiposition method	250.24136	0.001103
Multiple measurement/uncertainty determined with comparative method	250.24153	0.000985
Single measurement/uncertainty from MC CUT model	250.24151	0.000838

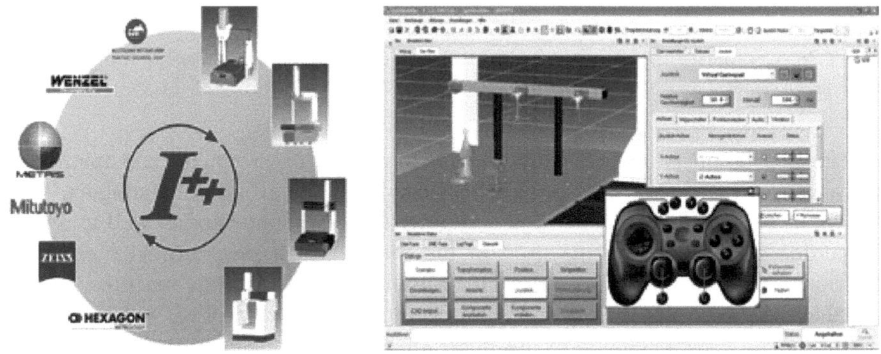

Fig. 5.77 I++ DME protocol and the I++ Simulator screen [19, 79, 100]

equipment). It allows the selection of CMM and its connection with any metrological software such as PCDMIS, METROLOG, Quindos, CALIPSO, COSMOS, and so on, so that companies gain an opportunity of optimal configuration of the measuring system according to their needs and its integration in a given facility or laboratory. Such a solution was adopted under the I++ Simulator, hence the other, broader meaning of its name, illustrated in Fig. 5.77.

5.2.5.2 Simulated Measurement with the Use of the I++ Simulator

The operation of the I++ Simulator is realistic because it is possible to use a very effective tool, which is the option of compilation of any real measuring machine model components with appropriate accessories, such as magazines or rotary tables. The option covers all machines of well-known producers. The simulator also allows the creation of very realistic measuring volumes. Using a Simulator I++ library in the LCM CUT several models of CMMs were built and activated. In Fig. 5.78 an example implementation for the PMM12106 machine of Leitz Messtechnik [100] is presented.

Fig. 5.78 Coordinate measuring machine with its visualization implemented using the I++ Simulator system

The measurement begins with the I++ Simulator environment and metrology software (for LCM Quindos7 or PCDMIS) start-up. The group of "Dialogues" tags allows us to define "Scenario" and computer system parameters as well as set all properties of the components used in the measurement program. The group of "Actions" tags allows us to define the simulator to measuring machine control together with metrology software. The machine can be controlled with a joystick connected to a computer or by its programming from the metrological software level. Both manual and automatic measurements can be simulated. The camera control option allows us to observe the machine and its measuring volume. It also enables an easy way to view a measured object location, thus control of the machine during measurements becomes much easier. There is also a possibility to define views fitted to the operator's needs and to measuring tasks that are carried out.

The simulation course is recorded, providing the user appropriate filtered information of errors and warnings with the possibility to create temporary indexes and connections exactly as in the real measurement. The type of machine and measuring tip together with their parameters are downloaded from applications and added to the metrological software (e.g., Quindos 7). It happens in exactly the same way as it is implemented in the real measurement. Items that are not available as default in the library can be created and imported in CAD file format and modified in IGP format as predefined modules (equipment, rotary tables, impulse and scanning probe heads, etc.) [19, 79, 100].

Figure 5.79 shows the import of the measured model to the I++ Simulator on the example of a standard for measuring the accuracy of surface projection for computed tomography. To make the import possible the model had to be created and saved in IGES format in the CAD environment; in this case the model was modeled in CATIA. The simulator allows us to write a program and run it to perform standard tests and optimization of measuring paths and correct selection of measuring tips to carry out complete calibration measurement. There is no need to use a

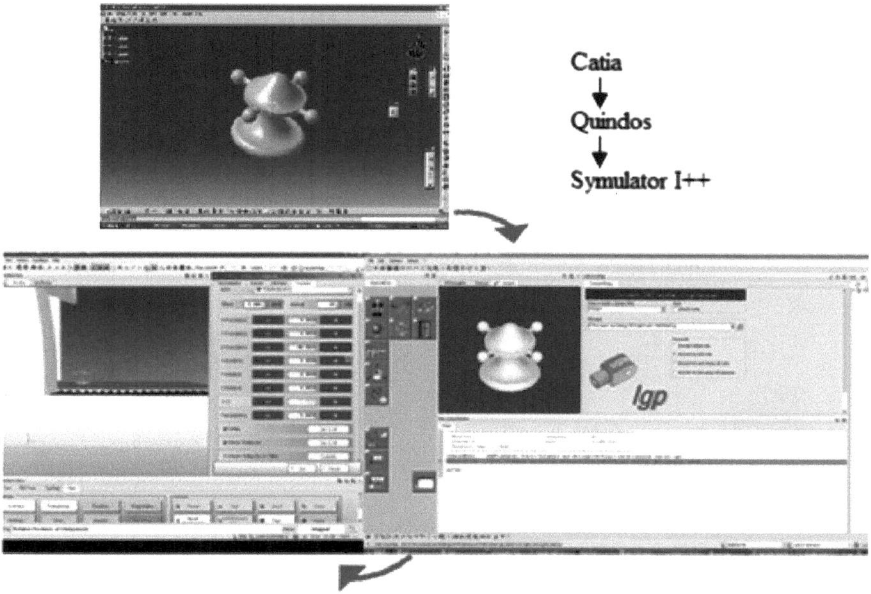

Fig. 5.79 Import of an example model saved in IGES format [19, 79]

huge number of measuring tips and determine the optimal coordinate system of an object. This may increase the accuracy of real measurements.

5.2.5.3 Virtual Coordinate Measuring Machine and I++ Simulator System

It is worth noting that the use of the I++ Simulator itself allows us to use nominal dimensions and does not provide information on such an important area as measurement errors. The concept of connecting Simulator I++ with the CMM model (Virtual CMM) results in operations in a virtual reality volume. The simulator has a possibility of standard installing the virtual machine options, from the metrology software level, based on options of Quindos, or the CALIPSO system, based on the PTB method.

In the LCM CUT such a connection for developed models of DEA machines and Quindos software was realized. At first, the Virtual Neuro CMM system in the I++ Simulator was installed, schematically shown in Fig. 5.80. Tasks solved due to this solution are: simulation forecast of measurement accuracy for objects in the CAD project phase, research and optimization of measurement strategies, and conclusive verification of whether it is possible to develop a measurement technology for a given constructional solution with the required manufacturing tolerances [15].

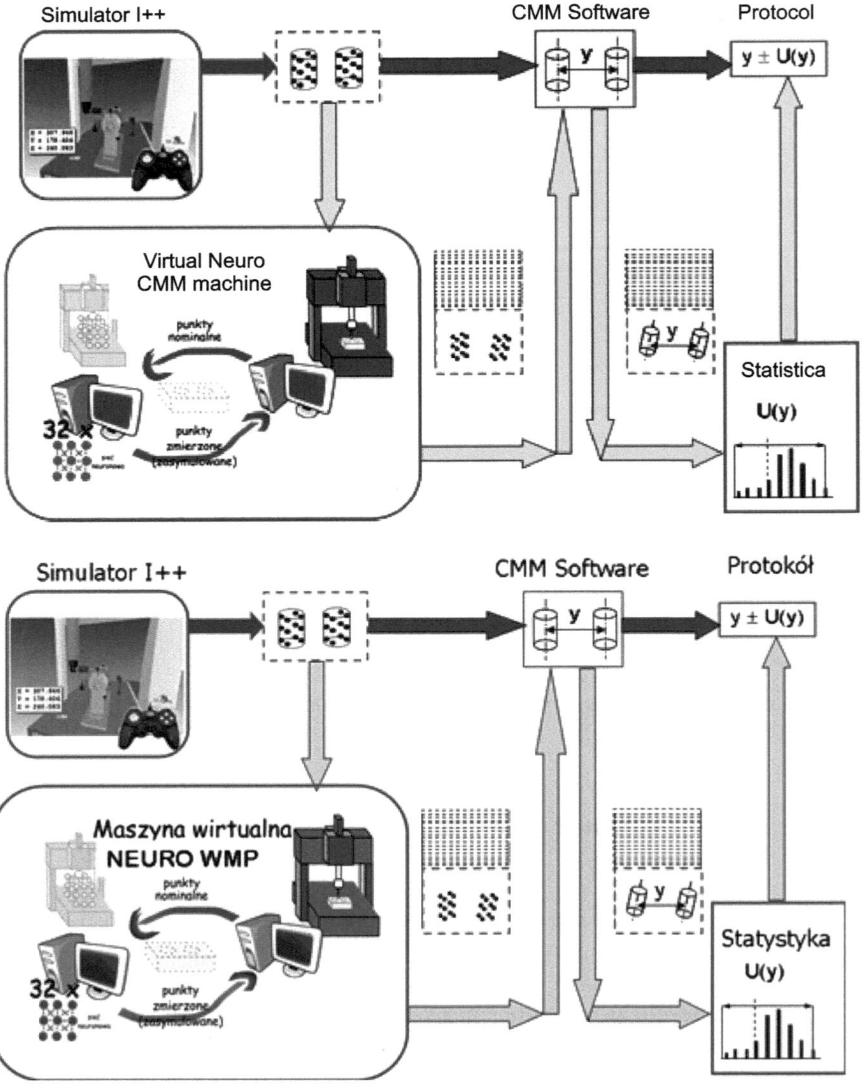

Fig. 5.80 Use of I++ Simulator and Virtual Neuro CMM system

Another virtual machine system connected with the I++ Simulator is Virtual MC CUT [132]. Because this system is based on cooperation of the Python programming environment [69] with the Quindos metrology software it can be used without major modifications in connection with the I++ Simulator, which also interacts with the Quindos program. Only the Python environment, which is an open-type software and has small hardware requirements must be installed. The Virtual MC CUT

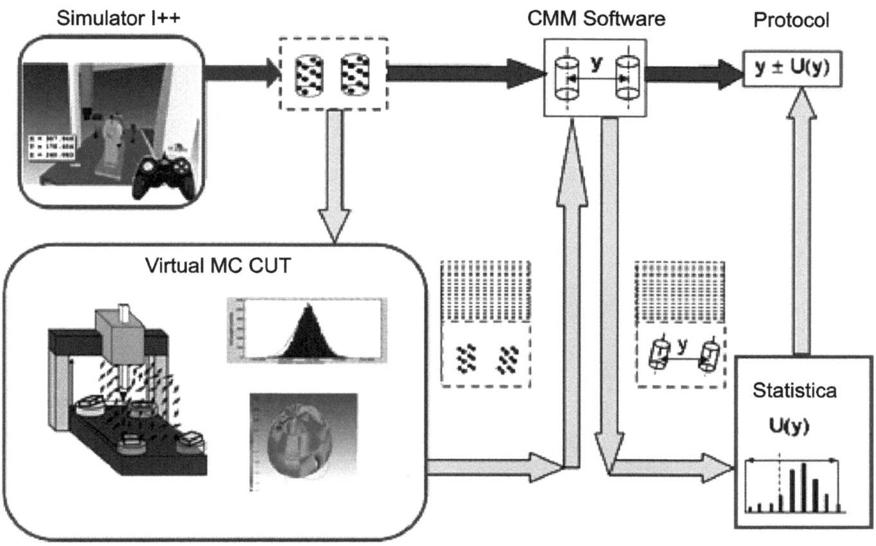

Fig. 5.81 Use of I++ Simulator and Virtual MC CUT system

system [132] was conceived to be used in cooperation with other metrological software based on the I++ metrological protocol. In Fig. 5.81 the scheme of the Virtual MC CUT connection with the I++ Simulator is shown.

The model operation together with the I++ Simulator has been tested for two basic measuring tasks, performed for the engine body, presented in Fig. 5.82. The hole diameter measurement in the engine body as well as two plane distance measurements and the uncertainty determination for them with the simulation technique using Virtual MC CUT were performed. Figure 5.83 shows the hole diameter measurement in the engine body. The hole was measured in two sections of 12 points, for a total number of 24 measuring points. Figure 5.84 shows the protocol from simulated measurements in the Quindos system. Figure 5.85 illustrates the plane distance measurement in the engine body. Each surface was measured at 8 points. The results are presented in Fig. 5.86.

5.2.6 *Virtual Articulated Arm Coordinate Measuring Machine (VAACMM)*

As noted at the end of Sect. 4.5.4 the most important application of the developed KmAACMM kinematic model is construction of a metrological model—virtual AACMM—which would be used for online assessment of measurement accuracy with a simulation method as schematically shown in Fig. 5.87. From a programming point of view this system was developed in the C++ language as V (virtual)

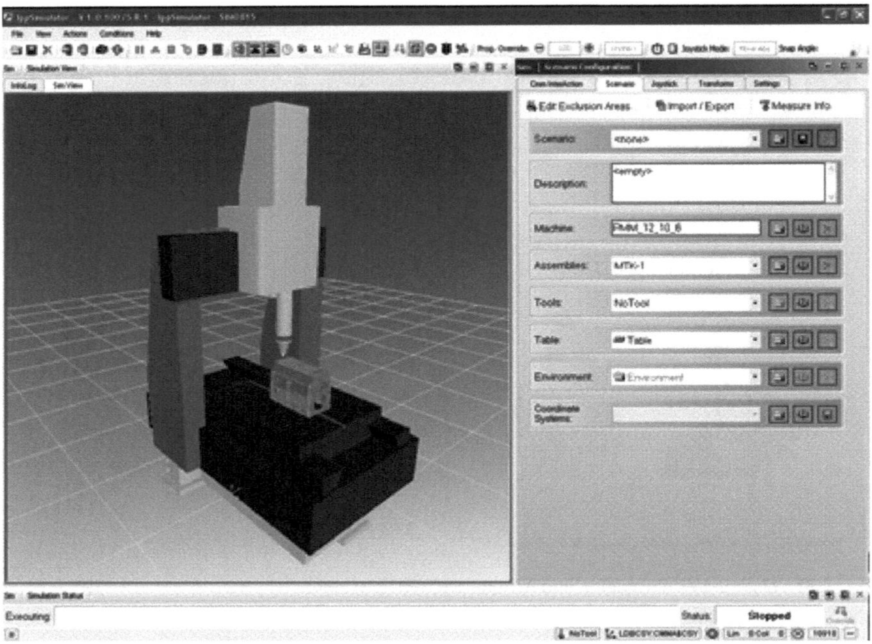

Fig. 5.82 I++ Simulator of Messtechnik Wetzlar Company PMM12106 Leitz machine model cooperating with Quindos software and with Virtual MC CUT system. Performing of measuring task for selected engine features

AACMM CUT [64, 65, 92, 107]. Thus far in the LCM CUT, three models have been built for articulated arms of ROMER: SIGMA 2000, Omega 2025, and Multi-Gage (Fig. 5.87).

5.2.6.1 Concept of Construction and Operation of Virtual AACMM

The VAACMM CUT operation can be described as follows: after each measurement (contact with the measured surface and acquisition of the measuring point coordinates) the VCMA software (using a module written with the use of the Monte Carlo method), simulates multiple measurements of the same point. VAACMM CUT is presented in Fig. 5.88. Its connection with the kinematic model KmAACMM is described in Sect. 4.5. There the simulation module is also shown (indicated by a dotted line), where MCM simulation plays an important role. In a special program module written for the purpose of VAACMM construction and based on the data transmitted from encoders, treated as input data with simultaneous analysis of their uncertainties, point coordinates for a given articulated arm configuration and tip orientation were collected. The function that reproduces its probability density (PDF) is assigned to each input value. Figure 5.89 shows the

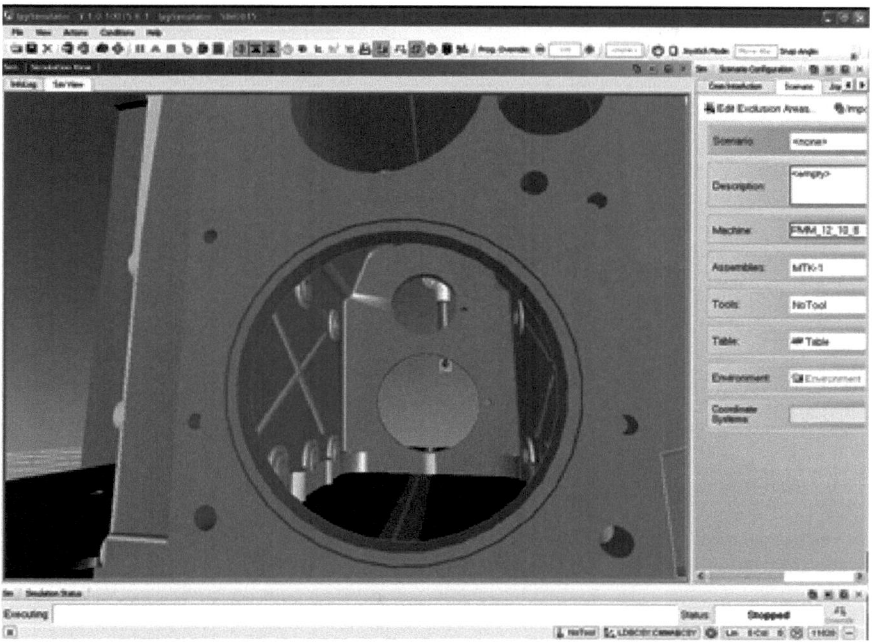

Fig. 5.83 Measurements of the hole diameter in the engine body realized with I++ Simulator

Quindos Measuring Report

3D-Application Center

Description		Customer	
Drawing Number		Serial Number	
Remarks		Article Number	
Supplier		Delivery Date	
Delivery Note		Delivery Volume	
Lot Number		Lot Size	
Test Schedule		Sample Size	
Production Machine		Production Tool	
Production Date		Production Time	
Order		Department	
Inspector	AG	Inspection Date	23-MAR-2011, 19:14:56
Measuring Device	RMT 12 10 6 #4711	Measuring Program	Quindos7 - V 7.5.10317
User Name	Adam	WKP Name	BALLBAR250

Text	Eval.	Actual	Nominal	Up.Tol.	Low.Tol.	Act-Nom	Graphic

U(CYL_KORPUS_DM) = 0.00098

CYL_KORPUS		CYL				
	DM	100.246033	0.000000	0.050000	-0.050000	100.246033

Fig. 5.84 Measurement results of the hole diameter in the engine body performed using the I++ Simulator together with measurement uncertainty, the Virtual MC CUT program

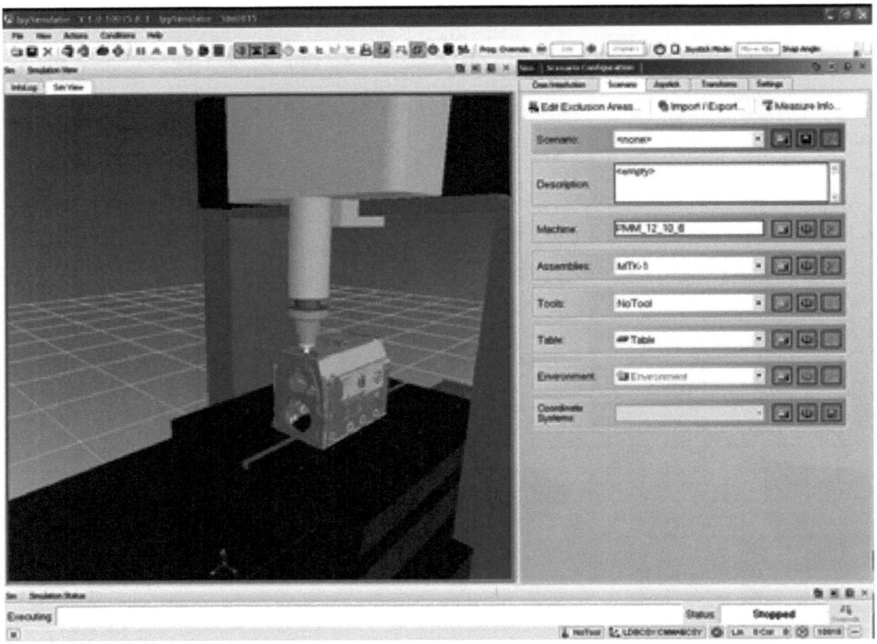

Fig. 5.85 Measurement of plane distance in the engine body realized with I++ Simulator

Quindos Measuring Report

3D-Application Center

Description		Customer	
Drawing Number		Serial Number	
Remarks		Article Number	
Supplier		Delivery Date	
Delivery Note		Delivery Volume	
Lot Number		Lot Size	
Test Schedule		Sample Size	
Production Machine		Production Tool	
Production Date		Production Time	
Order		Department	LMW PK
Inspector	AG	Inspection Date	29-MAR-2011, 18:10:54
Measuring Device	RMT 12 10 6 #4711	Measuring Program	Quindos7 - V 7.5.10317
User Name	Adam	WKP Name	KORPUS

Text	Eval.	Actual	Nominal	Up.Tol.	Low.Tol.	Act-Nom	Graphic

U(DIS_PLN)=0.00117

DIS_PLN **AXI**

DY	510.522842	0.000000	0.050000	-0.050000	510.522842

Fig. 5.86 Measurement results of plane distance performed with the use of the I++ Simulator together with measurement uncertainty, the Virtual MC CUT program

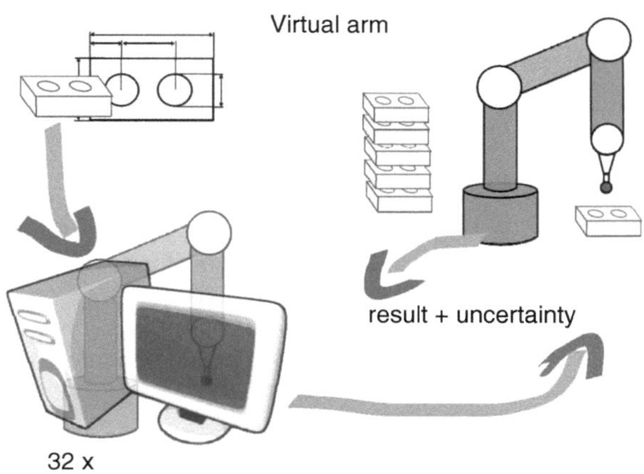

Fig. 5.87 Virtual articulated arm coordinate measuring machine

window of the program module that realizes simulation and loads the resulting data into the KmAACMM kinematic model. Finally the simulated value corresponding to that measured by real AACMM quantity value is obtained. Figure 5.90 presents the operation of Virtual AACMM CUT software. This program reads the data consisting of point coordinates together with the measuring stylus direction in order to calculate measured shape elements. Results from the simulation module are saved in a text file. Then they are loaded into the PC-DMIS program, which reads them as coordinates of the points. Then from these points geometric elements and relations between them are calculated. For example, if a circle is measured at four points, each point is simulated 30 times. Then obtained results are exported to the metrology software (e.g., PCDMIS). The software calculates circles from the obtained results, and next with the use of the *A* method estimates uncertainty of the realized measuring task. Virtual AACMM has a structure of metrology software. It runs in the background of realized measurement. Thanks to that obtained uncertainty is given with the measurement result, as presented in Fig. 5.90, and selected in the module protocol.

5.2.6.2 Test Verification of Virtual AACMM Simulation Model

The verification of the developed model was based on recommendations of VDI/VDE 2617 Part 7th [128] described in Sect. 5.2.4.3. A Virtual AACMM model developed for the Multi-Gage measuring arm of ROMER equipped with 15-mm steel contact tip was chosen for verification tests. The method of VAACMM model verification is presented in Fig. 5.91.

Tests were performed in an air-conditioned laboratory in temperature = 20 ± 0.5 K. Tests were at first performed with real AACMM by measuring the cylindrical standard (the same as described in Sect. 5.2.4.3) according to specifications given in Fig. 5.67, in four positions: parallel to *X*, *Y*,

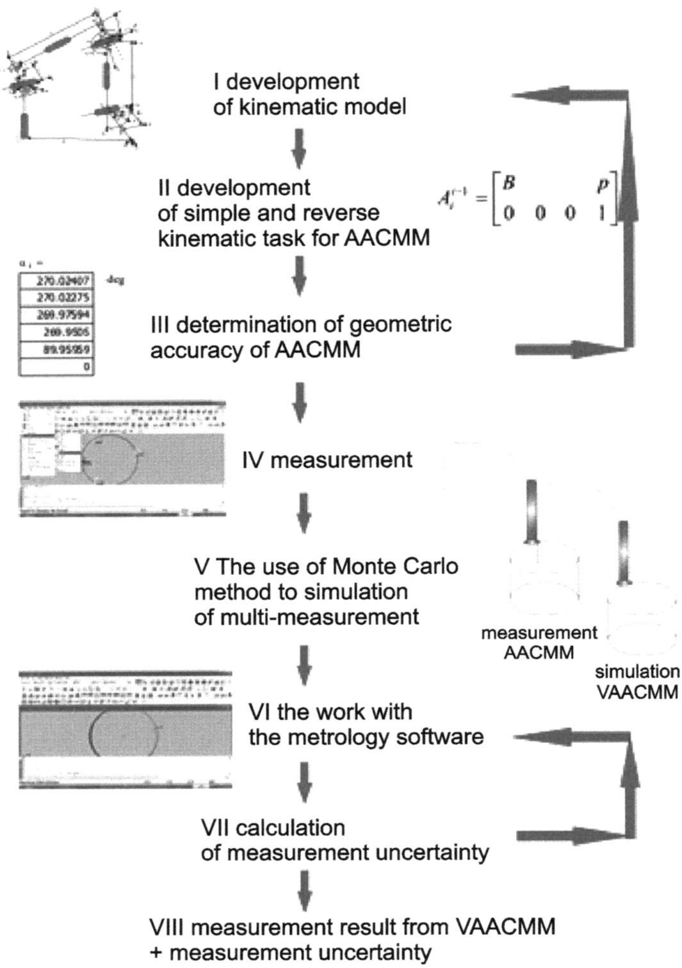

Fig. 5.88 Virtual AACMM construction and its operation, when frame simulation module is selected [64, 92, 107]

Z axes and at angle of 45° between X, Y axes. The cylinder was placed in a special prism facilitating position changes.

Measurements realization is shown in Fig. 5.92. The measurement strategy was as follows: on frontal surfaces four regularly distributed areas were determined, and in each 10 points were shown, giving 40 points in total. Such a strategy was necessary because AACMM is a manual device and it is not possible to reproduce the strict adopted measurement sequence (Fig. 5.93).

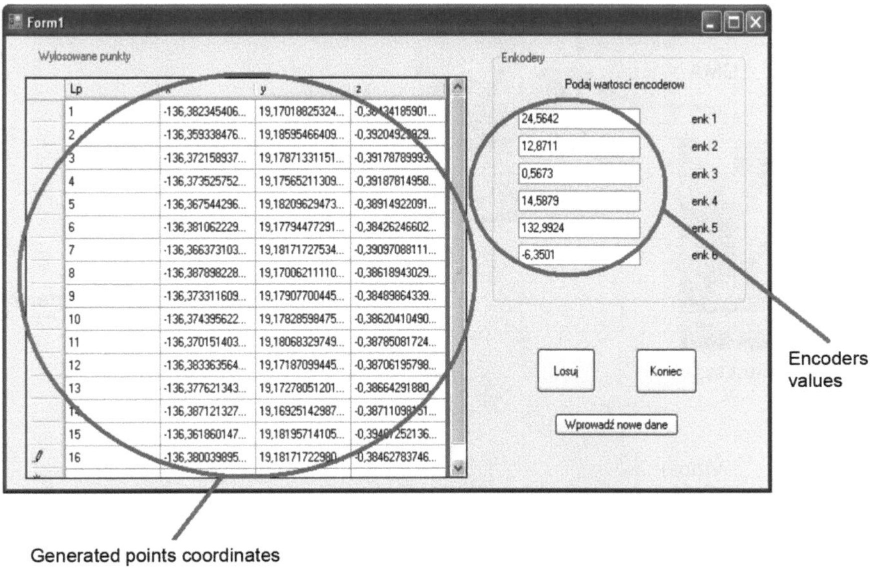

Fig. 5.89 Program that simulates measurement [106]

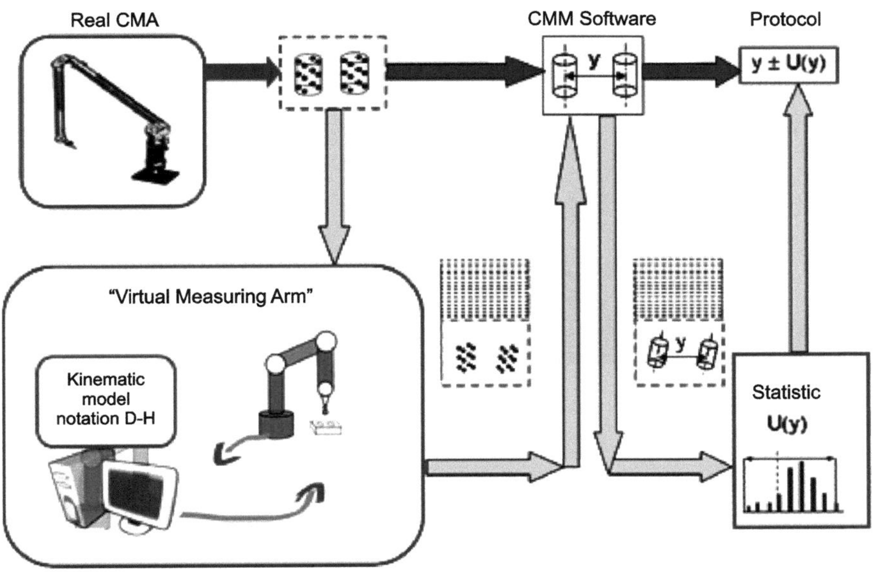

Fig. 5.90 Connection of real AACMM and Virtual AACMM CUT into an accuracy assessment system

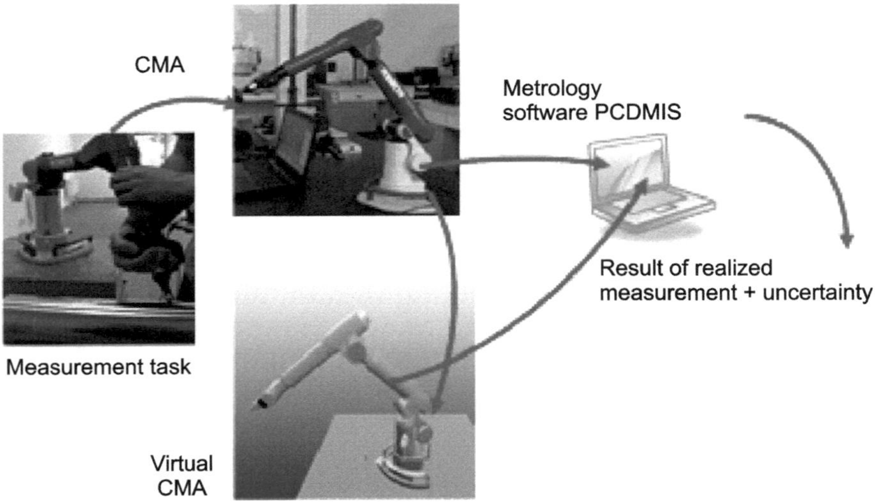

Fig. 5.91 Method of CMA model verification [64, 107]

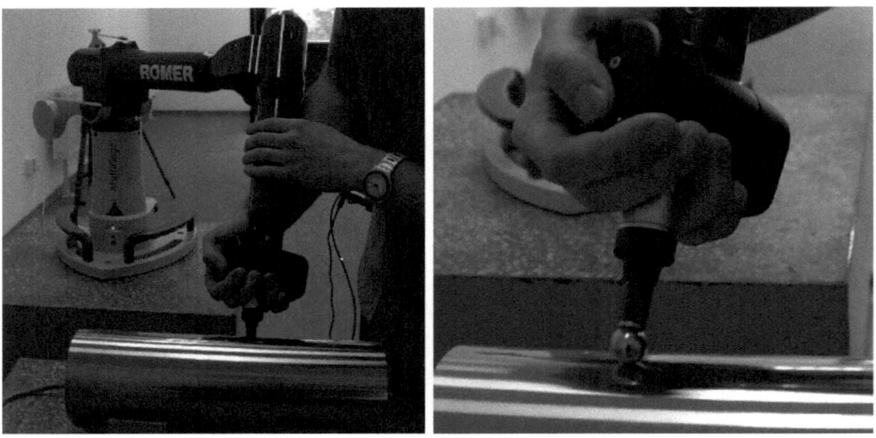

Fig. 5.92 Measurement of the reference cylinder, Romer Multi-Gage articulated arm with steel tip of 15 mm diameter

Then the two extreme cylinder surfaces were measured. Four areas that are regularly distributed over the cylinder were determined. Ten points were measured in each area, then measurements in a similar area, but moved by 100 mm were repeated. Therefore it is possible to determine two cylinders 1 and 2 (as shown in Fig. 5.67, Sect. 5.2.4.3). The complete measurement strategy is shown in Fig. 5.94. Simultaneously for each measuring point VAACMM system operation is initiated and simulation is realized (each point was simulated 32 times). According to the

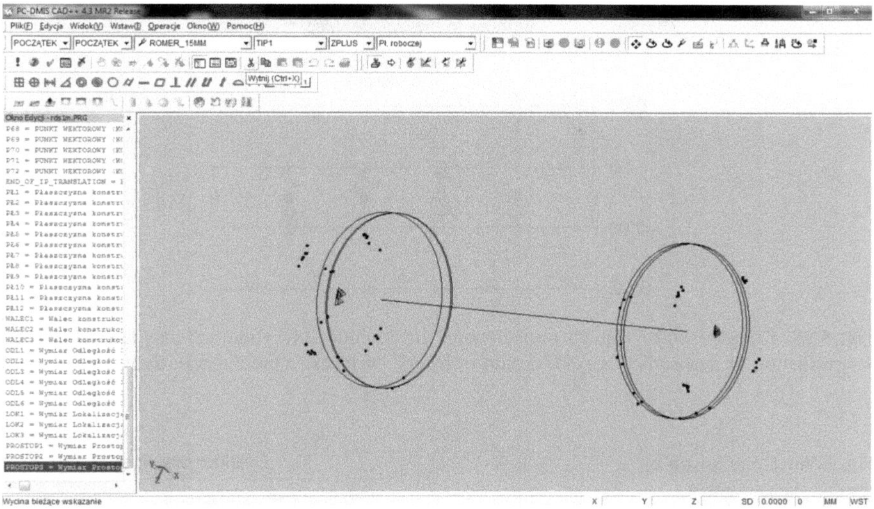

Fig. 5.93 PCDMIS system screen, measurement strategy for frontal surfaces of the cylindrical standard

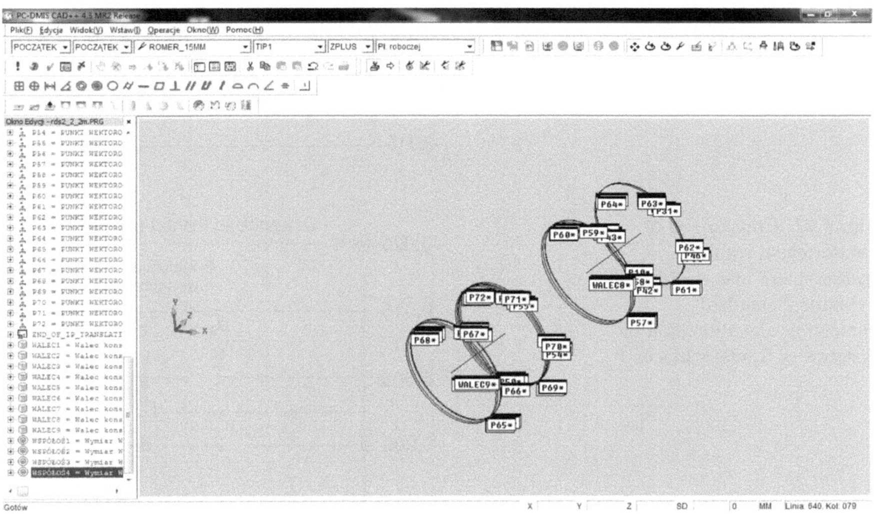

Fig. 5.94 PCDMIS system screen, measurement strategy for the cylindrical standard

VAACMM system concept and the operation of real AACMM, the metrology software is used to calculate both real and obtained as a result of simulation following features and characteristics of the cylinder together with their expanded uncertainties:

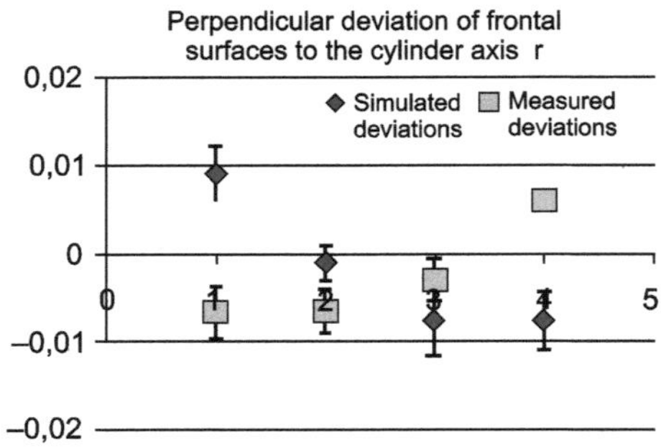

Fig. 5.95 Comparison of results, measurement uncertainties of cylindrical standard simulation: determination of perpendicularity deviation of frontal surfaces to the cylinder axis, *r*

Fig. 5.96 Comparison of results; measurement uncertainties of cylindrical standard simulation: determination of reference standard diameter

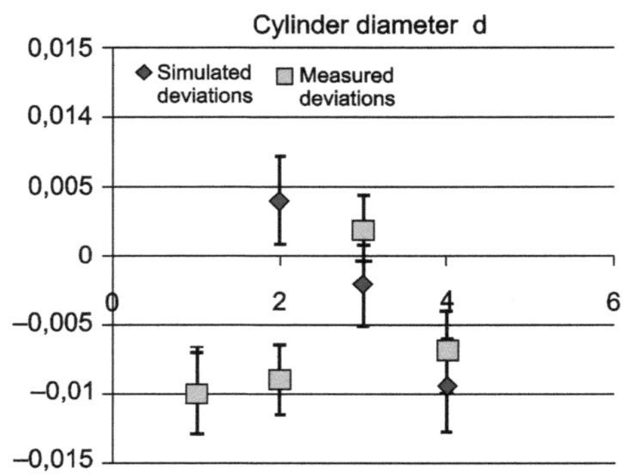

Fig. 5.97 Comparison of measurement results; measurement uncertainties of cylindrical standard simulation, determination of distance of frontal surfaces l

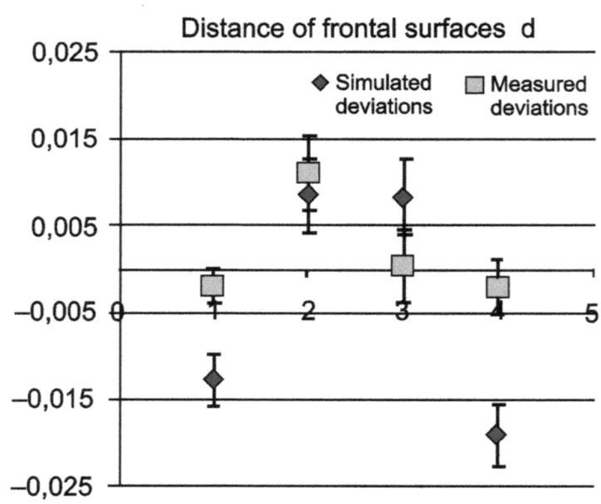

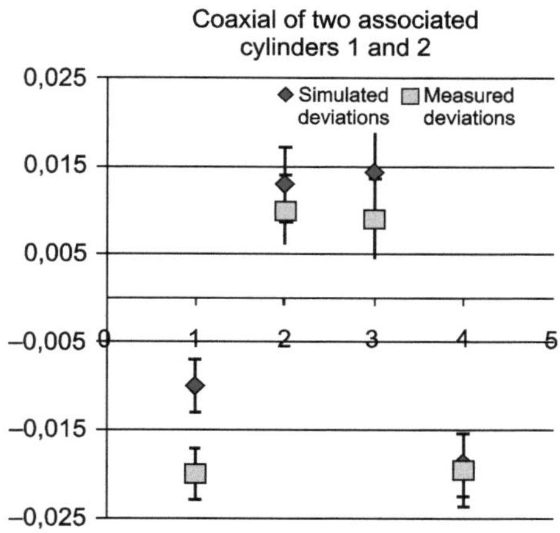

Fig. 5.98 Comparison of results; measurement uncertainty of cylindrical standard: determination of coaxiality of cylinder 1 and 2 axes

- Distance between frontal surfaces, *l*...
- Cylinder diameter, *d*...
- Perpendicular deviations of frontal surfaces to the cylinder axis, *r*...
- Coaxiality of two associated cylinders 1 and 2. Measurements and simulation results are given in Figs. 5.95, 5.96, 5.97 and 5.98.

It was confirmed simultaneously that the obtained results meet the inequality given by Eq. (5.18). That is why it can be concluded that the built VAACMM model meets recommendations contained in [128], so it should be accepted as correctly functioning.

Due to verification tests conducted in this way it was possible to state that developed method—measurement simulation with the use of the Virtual AACMM system achieves acceptable results, comparable with results from real measurements in ranges of obtained uncertainties.

Summarizing, the results of comparative and verifying tests of software operation, virtual systems built on the base of various models and methods, it can be clearly said that simulation methods are the most effective in the assessment of the accuracy of realized coordinate measurements. However, their use is determined by a number of factors. Therefore, their implementation in industrial environments requires further study. A solution based on the modeling of residual errors for systems equipped with CAA correction systems is closest to being introduced in industry, whereas, for articulated arms only, the model using Denavit–Hartenberg notation can be used as a basis for implementation and application.

References

1. Abacerli A.J., Pereira P.H., Calonego N.: A case study on testing CMM uncertainty simulation software (VCMM). J. Braz. Soc. Mech. Sci. Eng. **XXXII**, No.1/9 (2010)
2. Beaman, J., Morse, E.: Experimental evaluation of software estimates of task specific measurement uncertainty for CMMs. Precision Eng. **34**(1), 28–33 (2010)
3. Bendat, J.S., Piersol, A.G.: Random data: analysis and measurement procedures. Wiley, New York (1971)
4. Berndt, G.: Grudlagen und Geräte technischer Längenmessungen. Julius Springer, Berlin (1921)
5. Berndt, G., Hultzsch, E., Weinhold, H.: Funktionstoleranz und Meßunsicherheit. Wissenschaftliche Zeitschrift der Universität Dresden **17**, 465–471 (1968)
6. Breyer K.H.: Prüfen von Mass, Form und Lage mit Koordinatenmessgeräten. VDI-B 529/84
7. Dietrich, E., Schultze, A.: Pruefprozesseignung. Hanser Verlag, Munich (2007)
8. van Dorp, B., Haitjema, H., Delbressine, F., Bergmans, R., Schellekens, P.: Virtual CMM using Monte Carlo methods based on frequency content of the error signal. In: Decker, J.E., Brown, N. (eds.) Proceedings of SPIE 4401 Recent Developments in Traceable Dimensional Measurements, SPIE (2001)
9. EASYTRAC Project: Easier and cheaper traceability in industry by up-to-date methods of calibration. EU Contract No. G6RD-CT-2000-00188 Work package 6.4 "Reversal technique to calibrate gear and thread standards". Final Report from DTU—Calibration of Thread Gauges on CMMs
10. Eumetron GmbH. www.eumetron.de
11. FMT. www.qfm.uni-erlangen.de
12. FEINMESS GmbH & Co. KG. www.feinmess.com
13. Forbes A.B., Peggs G.N.: A large reference artefact for CMM verification. In: Ford, D.G., Postlethwaite, S.R. (eds.) Laser Metrology and Machine Performance III, pp. 394–400. Computational Mechanic Publications, Southampton-Boston (1997)
14. Franke, M., Hartig, F., Wendt, K.: Measuring large 3D structures using a portable 4-arm laser interferometer. In: Sładek, J., Jakubiec, W. (eds.) Advances in Coordinate Metrology, pp. 35–42. University of Bielsko-Biała, Bielsko-Biała, ISBN 978-83-62292-52-3 (2010)
15. Krzysztof, G.: Metoda symulacyjna prognozowania dokładności pomiaru współrzędnościowego (Simulation method for forecasting of the accuracy of coordinate measurement). Ph.D. thesis dissertation, Kielce University of Technology, Faculty of Mechatronics and Machine Design (2010) (Advisor, Prof. J.A. Sladek)
16. Gawlik, J., Kowalski, M., Ryniewicz, A.: Analysis of measurement of precision machines guides. In: 4th International Conference on Ultraprecision in Manu-facturing Engineering. Progress in Precision Engineering and Nanotechno-logy, Braunschweig, Germany, vol. 1, pp. 319–322 (1997)
17. Gąska, A.: Different uses of Monte Carlo Method in coordinate metrology. In: Xth International Scientific Conference Automation in Production Planning and Manufacturing, Zilina (2009)
18. Gąska, A.: Modeling of accuracy of coordinate measurement with use of Monte Carlo Method. Ph.D. thesis, Cracow University of Technology, Faculty of Mechanical Engineering (2011) (Advisor, Prof. J.A. Sladek)
19. Gąska, A., Szewczyk, D., Gąska, P., Gruza, M., Sładek, J.: Usage of I++ simulator to program coordinate measuring machines when common programming methods are difficult to apply. Meas. Sci. Rev. **14**(1) (2014). ISSN 1335-8871
20. Gąska, A., Krawczyk, M., Kupiec, R., Ostrowska, K., Gąska, P., Sładek, J.: Modeling of the residual kinematic errors of coordinate measuring machines using lasertracer system. Int. J. Adv. Manuf. Technol. (2014). doi:10.1007/s00170-014-5836-1
21. Giusca, C.L., Leach, R.K., Forbes, A.B.: A virtual machine-based uncertainty evaluation for a traceable areal surface texture measuring instrument. Measurement **44**(5), 988–993 (2011)

22. Grzelka, M., Gapiński, B., Marciniak, L., Wieczorowski, M., Matliński, K., Olszewska, I.: Pomiary cech geometrycznych przedmiotów obrotowo-symetrycznych (Measurements of geometrical features of rotationally symetrical elements). Pomiary Automatyka Kontrola (Measurement Automation and Monitoring), pp. 35–37 (2010)
23. Hartmann, W.W., Geise, G.: Displaydarstellungen als Entscheidungshilfe für Messauswertungsstrategien in der Koordinatenmesstechnik. Feingerätetech-nik (1984)
24. Haitjema, H., van Dorp, B., Morel, M., Schellekens, P.H.J.: Uncertainty estimation by the concept of virtual instruments. In: Decker, J.E., Brown, N. (eds.) Proceedings of SPIE 4401 Recent Developments in Traceable Dimensional Measurements, SPIE 2001
25. Herbst, C., Tutsch, R.: Simulation of micro coordinate measuring machine for parallel measurement of microstructures. In: Sładek, J., Jakubiec, W. (eds.) Advances in Coordinate Metrology, pp. 15–24. University of Bielsko-Biała (2010).
26. Hernla, M.: Aufgabenspezifische Meßunsicherheit bei Koordinatenmessung TM 64 7/8, pp. 286–293. Oldenburg Verlag (1997)
27. Hertz, J., Krogh, A., Palmer R.G.: Introduction to the Theory of Neural Computation, pol. edition II, Warsaw (1995)
28. Hexagon Metrology. www.mtwz.de
29. Hughes, E.B., Wilson, A, Peggs, G.N.: Design of a high-accuracy CMM based on multi-lateration techniques. CIRP Ann. Manufact. Technol. **49**(1), 391–394 (2000)
30. Immarco P.: Neural Nets for the Rest of Us. NeuroShell 2 Opens a Window, PC-AJ, pp. 34–36 (1993)
31. ISO 10360 part 5—CMMs using single and multiple stylus contacting probing systems, 2nd edn, 15 Sept 2010
32. ISO/CD TS 15530-2—Use of multiple measurement strategies
33. ISO/TS 15530-4:2008—Evaluating tasc-specific measurement uncertainty using simulation
34. ISO/TS 23165:2006—Geometrical product specifications (GPS)—Guidelines for the evaluation of coordinate measuring machine (CMM) test uncertainty
35. Jaeger, G.: Limitations of precision length measurements based on interferometers. Measurement **43**(5), 652–658 (2010)
36. Jakubiec, W., Starczak, M.: Identification of parameters of virtual coordinate measuring machines model. In: IVth International Scientific Conference—Coordinate Measuring Technique, Lodz University of Technology Scientific Books, Branch in Bielsko-Biała, no. 53, pp. 115–118, Bielsko-Biała (2000)
37. Joskowicz, L., Ostrovsky-Berman, Y., Myers, Y.: Efficient representation and computation of geometric uncertainty: The linear parametric model. Precision Eng. **34**(1), 2–6 (2010)
38. Juras, B.: Metoda pomiaru zarysu o zmiennej krzywiźnie na współrzędnościowej maszynie pomiarowej (Method for measuring the freeform profile with the use of coordinate measuring machine). Ph.D. thesis, Cracow University of Technology (1997)
39. Korbicz, J., Obuchowski, A., Uciński, D.: Sztuczne sieci neuronowe. Podstawy i zastosowania (Artificial neural networks. Bases and applications), Warsaw (1994)
40. Kowalski, M.: Analiza pomiarów na współrzędnościowych maszynach pomiarowych obarczonych błędami kształtu (Analisys of measurements done with the use of coordinate measuring machines affected by form errors). Ph.D. thesis, Cracow University of Technology (1993)
41. Kowalski, M.: Model wirtualnej wielowspółrzędnościowej maszyny pomiarowej i jej zastosowanie (Model of virtual multi- coordinate measuring machine and its application). In: Proceedings of VIth Scientific-Technical Conference Metrology in Production Engineering, pp. 218–223. Rzeszow University of Technology (1995)
42. Kowalski, M.: Wybrane aspekty realizacji modelu wirtualnej współrzędnościo-wej maszyny pomiarowej (Chosen aspects of realization of the virtual coordinate measuring machine model). Polish Academy of Sciences—KBM, Scientific Book no. 63, Works of Institute of Machine Technology and Production Automation, Cracow University of Technology, pp. 117–126 (1999)

43. Kowalski, M.: Zastosowanie wirtualnej WMP do wyznaczania i minimalizacji błędów pomiarowych (The use of virtual CMM for determintaion and minimization of measurement errors). In: Proceedings of VIIIth Scientific-technical Conference Metrologia w technikach wytwarzania maszyn (Metrology in Machine Production), vol. II, pp. 401–408. Szczecin–Międzyzdroje (1999)
44. Kowalski, M., Sładek, J., et al.: Opracowanie wirtualnej wielowspółrzędnościowej maszyny pomiarowej z zastosowaniem do badań i korekcji błędów obiektów rzeczywistych i optymalizacji pomiarów (Development of virtual multi-coordinate measuring machine with the application for researches and correction of errors of real objects and for measurements optimization). Final Report of KBN Research Project PB 1367/T07/95/08, Cracow University of Technology 1998, not published
45. Kowalski, M., Sładek, J., Rakoczy, R.: Opis wirtualnej współrzędnościowej maszyny pomiarowej (Decription of virtual coordinate measuring machine). In: IIIrd International Scientific Conference—Coordinate Measuring Technique, Lodz University of Technology Scientific Books, branch in Bielsko-Biała, no. 44, pp. 71–80 (1998)
46. Kruth, J.-P., Van Gestel, N., Bleys, P., Welkenhuyzen, F.: Uncertainty determination for CMMs by Monte Carlo simulation integrating feature form deviations. CIRP Ann. Manufact. Technol. **58**(1), 463–466 (2009)
47. Kupiec, M.: Optyczno-stykowa metoda pomiarów współrzędnościowych (Optical-contact method for coordinate measurements). Ph.D thesis, Cracow University of Technology (Advisor, Prof. J.A. Sladek Faculty of Mechanical Engineering 2008 work realized under research grant with Department of Photonics Engineering of Institute of Micromechanics and Photonics—Faculty of Mechatronics at Warsaw University of Technology no.3 T10C 010 29 nt: Optonumeryczny system do pomiaru elementów geometrycznych zintegrowany z Współrzędnościową Maszyną Pomiarową (Opto-numerical system for measuring the geometrical elements integrated with Coordinate Measuring Machine)—2005–2008, Ministry of Science and Higher Education)
48. Leitfaden zum Fähigkeitsnachweis von Messsystemen/Measurement System Capability. Reference Manual, Q-DAS GmbH, Weinheim (2002). www.qdas.com/fileadmin/files2/qdasguidelinemsa/Leitfaden_v21_me.pdf
49. Lotze, W., Teichmann, U.: Einfluss von Gestalt—und Lageabweichungen auf die Unsicherheit der Rechnergestützehn Koordinatenmessung. Feingerätetechnik **8**, 139–343 (1976)
50. Lotze, W.: Form testing by means of the universal CMM ScanMax. In: IIIrd International Scientific Conference—Coordinate Measuring Technique, Lodz University of Technology Scientific Books, branch in Bielsko-Biała, no. 44, pp. 82–90 (1998)
51. Mailhe, J., Aranda, S., Linares, J.-M., Sprauel, J.-M.: Impact of datum feature uncertainties on tolerance zone. In: 10th CIRP Conference on Computer Aided Tolerancing, Erlangen, Germany, March 2007
52. Mailhe, J., Linares, J.M., Sprauel, J.M., Bourdet, P.: Geometrical checking by virtual gauge, including measurement uncertainties. CIRP Ann. Manufact. Technol. **57**, pp. 513–516 (2008)
53. Maihle, J., Linares, J.M., Sprauel, J.M.: The statistical gauge in geometrical verification: Part I. Field of probability of the presence of matter. Precision Eng. **33**(4), 333–341 (2009)
54. Maihle, J., Linares, J.M., Sprauel, J.M.: The statistical gauge in geometrical verification. Part II. The virtual gauge and verification process. Precision Eng. **33**(4), 342–352 (2009)
55. Mandziuk, J.: Sieci neuronowe typu Hopfielda. Teoria i przykłady zastosowań (Hopfield-type neural networks. Theory and application examples) EXIT, Warsaw (2000)
56. MegaKal—Virtual Machine developed in PTB Reference manual (1995)
57. Metropolis, N., Ulam, S.: The Monte Carlo Metod. J. Am. Stat. Assoc. **44**(247), pp. 335–341 (1949)
58. Metrosage. www.metrosage.com
59. Nałęcz, M.: In: Duch, W., Korbicz, J., Rutkowski, L., Tadeusiewicz, R. (eds.) Biocybernetyka i inżynieria biomedyczna (Biocybernetics and biomedical engineering),

vol. 6: Sieci neuronowe (Neural networks). Polish Academy of Sciences, EXIT, Warsaw (2000)
60. Nawara, L., Sładek, J.: Influence of Surface State on the Measuring Accuracy Measuring Machine. In: 7th International Conference on Production Engineering, Tokyo Japan, July 1984, pp. 696–671
61. Nawara, L., Kowalski, M., Sładek, J.: Badania metrologiczne maszyn pomiarowych i urządzeń ze szczególnym uwzględnieniem dokładności geometrycznej, pozycjonowania i strategii pomiarów (Metrology tests of measuring machines and devices with special include of geometrical accuracy, positioning and measurement strategy), Subtopic 1, Works of ZMiKT PK under CPBP no. 02.20, 1985–1989
62. NeuroShell2 4.0—Software for creating neural networks of Ward System INC Ref. Man
63. Oberkampf, W.L., DeLand, S.M., Rutherford, B.M., Diegert, K.V., Alvin, K.F.: Error and uncertainty in modeling and simulation. Reliab. Eng. Syst. Saf. **75**(3), 333–357 (2002)
64. Ostrowska, K.: Accuracy assessment method for measurements done with the use of articulated arm coordinate measuring machines. Ph.D thesis, Cracow University of Technology, Faculty of Mechanical Engineering, 2010 (Thesis Advisor, Prof. J.A. Sladek)
65. Ostrowska, K., Gąska, A., Sładek, J.: Determining the uncertainty of measurement with the use of a Virtual Coordinate Measuring Arm. Int. J. Adv. Manufact. Technol. **71**(1–4), 529–537(9) (2014). doi:10.1007/s00170-013-5486-8
66. PCDMIS Hexagon Metrology—User Manual
67. Pedone, P., Romano, D.: Designing small samples for form error estimation with coordinate measuring machines. Precision Eng. **35**(2), 262–270 (2011)
68. Phillips, S.D., et al.: The calculation of CMM measurement uncertainty via the method of simulation by constraints. Am. Soc. Precision Eng. **16**, 443–446 (1997)
69. Phyton Open Software
70. PN/EN/ISO 14253-2—Przewodnik do oceny niepewności pomiarów podczas wzorcowania sprzętu pomiarowego i kontroli wyrobów (Guidance for the estimation of uncertainty in GPS measurement, in calibration of measuring equipment and in product verification)
71. Pressel, H.G., Hageney, T.: Messunsicherheit von Pruefmerkmalen iin der koordinatenmesstechnik. Expert Verlag (2008)
72. QUINDOS Programming System for Dimension Metrology, Rreference Manual
73. Renishaw Sp. z.o.o., www.renishaw.com.pl
74. Ryniewicz, A., Kowalski, M., Rewilak, J., Sładek, J.: Laser measurement of guides and travelling units of CMMs and CNC machine tools. In: 7th International DAAAM Symposium: Product and Manufacturing, Flexibility, Integration, Intelligence, Technical University of Vienna, 17–19 Oct 1996
75. Schwenke, H.: Bedienanleitung MegaKal, KalKom, TKal, PKal, KalKmess, PTB, Braunschweig (1996–1999)
76. Schwenke, H.: Abschätzung von Messunsicherheiten durch Simulation an Beispielen aus der Fertigungsmesstechnik. Dissertation der Technischen Universität Chemnitz—PTB-F 36 Braunschweig, Juli 1999
77. Schwenke, H., Trapet, E., Wäldele, F.: Calibration of coordinate measuring machines to improve and to know the measurement uncertainty. In: IIIrd International Scientific Conference—Coordinate Measuring Technique, pp. 213–232. Lodz University of Technology Scientific Books, branch in Bielsko-Biała, no. 44, Bielsko-Biała (1998)
78. Schwenke, H., Siebert, B.R.L., Wäldele, F., Kunzmann, H.: Assessment of Uncertainties in Dimensional Metrology by Monte Carlo simulation: proposal of a modular and visual software. CIRP Ann. Manufact. Technol. **49**(1), 395–398 (2000)
79. Simulator I++ Messtechnik Wetzlar—software description
80. Sładek, J.: Assessment of the accuracy of contact probe heads used in coordinate measuring machines. Ph.D. thesis, Cracow University of Technology, Faculty of Mechanical Engineering (1990)

81. Sładek, J.: The examination of multicoordinate measuring devices (MMD) by means of circular standards. In: Proceedings of International Conference MATE—"Automation 92", Budapest, Hungary, 18–19 Feb 1992, pp. 333–339
82. Sładek, J.: Metody oceny dokładności WMP a prognozowanie dokładności pomiarów pomiarów (Methods for accuracy assessment of CMM vs forecasting of measurement accuracy). In: IInd National Scientific Conference (with international participation)—Coordinate Measuring Technique, Lodz University of Technology Scientific Books, branch in Bielsko-Biała, Bielsko-Biała, pp. 211–221 (1996)
83. Sładek, J.: Metrological assessment of CMMs and the possibilities of forecasting a measurement accuracy. In: 13 Conference, BIAM'96, Zagreb, Croatia, 18–21 June 1996, pp. D15–D183
84. Sładek, J.: Strategy of applying coordinate metrology in quality assurances systems with regards to a measuring accuracy. In: D.G. Ford, S.R. Postlethwaite (eds.) Laser Metrology and Machine Performance, 3rd edn. Computational Mechanic Publications, Southampton-Boston (1997)
85. Sładek, J.: Modelowanie współrzędnościowych maszyn pomiarowych—macierzowa metoda identyfikacji błędów (Modeling of coordinate measuring machines—Matrix method for errors identification). In: Proceedings of VIIIth Scientific-Technical Conference Metrologia w technikach wytwarzania maszyn (Metrology in Machine Production), Szczecin, pp. 437–444 (1999)
86. Sładek, J.: Errors identification and measurement accuracy assessment of coordinate measuring machines (CMM). Adv. Technol. Mach. Equip., PAN **23**(4), 113–136 (1999)
87. Sładek, J.: Model wirtualny WMP z zastosowaniem do oceny dokładności realizowanych pomiarów w warunkach czasu quasi-rzeczywistego (CMM virtual model with the application to accuracy assessment of measurements realized in equations of quasi- real time). In: IVth International Scientific Conference—Coordinate Measuring Technique, pp. 193–208. Lodz University of Technology Scientific Books, branch in Bielsko-Biała, no. 53, Bielsko-Biała (2000)
88. Sładek, J.: A neural network model of CMM applied for measurement accuracy assessment. In: Proceedings IMECO World Congress, vol. 11, Vienna 23–26 Sept 2000, pp. 281–286
89. Sładek, J.: Modelowanie i ocena dokładności maszyn oraz pomiarów współrzędnościowych (Modeling and accuracy assessment of coordinate machines and measurements). Cracow University of Technology Scientific Books, Mechanika, no. 87, Cracow (2001)
90. Sładek, J.: The relationship between measuring machines performance specifications and their real capability to solve given measurement tasks. In: METROMEET 2005 International Conference on Industrial Dimensional Metrology Bilbao (Spain), Euskalduna Jauregia Invited lecture, 7–8 Apr 2005
91. Sładek, J.: Opracowanie systemu oceny dokładności pomiarów oraz metody identyfikacji własnego pola błędów współrzędnościowych maszyn pomiarowych (Developing of a system for assessing the accuracy of measurements and of a method for identification coordinate measuring machine's errors field). Final Report of the Research Project funded by the KBN under contract no. PB-0921/T07/99/1
92. Sładek, J.: Metoda oceny dokładności pomiarów realizowanych redundantnymi systemami współrzędnościowymi (RSW) [Method for accuracy assessment of measurements done with the use of redundant coordinate systems (RCS)]. Research Project No. N505 255935, Ministry of Science and Higher Education (2008–2010)
93. Sładek, J., Gawlik, K.: Symulacja pomiarów współrzędnościowych jako narzędzie oceny i prognozy ich dokładności. Inżynieria Maszyn (Coordinate measurements simulation as a tool for assessment and prediction of their accuracy. Machine Engineering). Scientific Book No. 4, pp. 74–89 (2007)
94. Sładek, J., Gawlik, K.: Looking for uncertainty of measurement—Virtual Machines based on the Matrix Method using artificial neural networks. In: IV International Congress on Precision Machining 2007, Sandomierz-Kielce (2007)

95. Sładek, J, Gąska, A.: Modeling of CMM probe head errors. In: Proceedings of 10th ISMQC —TC14 IMEKO, Osaka, Japan, 5–9 Sep 2010
96. Sładek, J., Gąska, A.: Evaluation of coordinate measurement uncertainty with use of virtual machine model based on Monte Carlo method. Measurement **45**, 1564–1575 (2012)
97. Sładek, J., Kowalski, M.: Anwendung des Virtuellen normales zur Schätzung der Genauigkeit bei Mehrkoordinaten-Messungen. In: 6 Internationales Symposium von Donau Adria Assoziation für Automation und Messtechnik, DAAAM, Automation und Metrologie: Mensch, Technik, Umwelt, TU Krakow, Fakultät Maschinenbau, Kraków 24–26 Okt 1995, pp. 187–188
98. Sładek, J., Rakoczy, R.: CMM virtual modelling applied for measuring accuracy assessment. In: Proceedings of 1st International Workshop on CMM Calibration. The Czech Metrological Institute, Prague Czech Republic, 1–2 June 1999
99. Sładek, J., Rakoczy, R.: Wykorzystanie koncepcji sztucznych sieci neuronowych do modelowania współrzędnościowej maszyny pomiarowej (WMP) przy zastosowaniu wzorca przestrzennego (The use of the concept of artificial neural networks for modeling the coordinate measuring machine (CMM) with the use of spatial standard) Works of Institute of Machine Technology and Production Automation, Committee on Machine Building of the Polish Academy of Sciences, Technology Bases Section, Scientific Book no. 63, Cracow University of Technology, pp. 193–206 (1999)
100. Sładek, J., Szewczyk, D.: Wykorzystanie Simulatora I++ w nauczaniu metrologii współrzędnościowej i pracach badawczych (The application of the I++ simulator to coordinate metrology teaching and to research work). In: Proccedings of XIV National and V International scientific-technical conference "Metrology in Production Engineering". Warsaw University of Technology (2011)
101. Sładek, J., Rakoczy, R., Szwajkowski, A.: Application of neural networks for modelling coordinate measuring machines. In: Kunzmann, H., Waeldele, F., Wilkening, G., Corbet, J., MacKeow, P., Weck, M., Huemmler, J. (eds.) Progress in Precision Engineering and Nanotechnology, vol. 1, pp. 323–326. PTB-Braunschweig u. Berlin P.u.Oe–Verlag (1997)
102. Sładek, J., Rakoczy, R., Majdosz, R.: Realizacja i badanie powiązań informatycznych CAD WMP CAD z uwzględnieniem możliwości prognozowania dokładności pomiarów (Realization and testing of the CAD CMM CAD IT couplings, considering the aspect of possibilities of measurement accuracy prognostication). In: IVth International Scientific Conference—Coordinate Measuring Technique, pp. 209–221. Lodz University of Technology Scientific Books, branch in Bielsko-Biała, no. 53, Bielsko-Biała (2000)
103. Sładek, J., Rakoczy, R., Majdosz, R.: Research and realization of the CAD-CMM-CAD couplings, considering the aspect of possibilities of measurement accuracy prognostication. In: Proceedings of the 2nd International Conference on Surface Machining and Measurements of Sculptured Surfaces, Cracow 20–23 Sept 2000, pp. 71–84
104. Sładek, J., et al.: System metrologicznego nadzoru nad dokładnością maszyn pomiarowych robotów i obrabiarek z wykorzystaniem wzorców interferometrycznych laserowych systemów śledzących jako podst. wzrostu jakości produkcji przemysłu maszynowego (Metrological monitoring system for accuracy of measuring machines, robots and tools using interferometric standards of laser tracking systems as a base for a growth of production quality of machinery industry). Research and Development Project no. R03 029 01 Ministry of Science and Higher Education (2006–2009)
105. Sładek, J., Krawczyk, M., Gawlik, K.: Muzyka-Żmudzki M.: the assessment of the coordinate measurement accuracy based on Matrix Method with use of artificial neural networks. In: X CIRP Conference on Computer Aided Tolerancing, Germany, p. 33 (2007)
106. Sładek, J., Krawczyk, M., Ostrowska, K., Gąska, A.: Zastosowanie metody Monte Carlo do wyrażania niepewności pomiarów współrzędnościowych na przykładzie metody wielopozycyjnej (The use of the Monte Carlo method for estimation of coordinate measurements uncertainty on the example of multi-position method). Bielsko Biała Scientific books no. 81-2008

107. Sładek, J., Ostrowska, K., Gacek, K.: Kinematic metrological model of the coordinate measuring arm (MCMA). In: XIX IMEKO World Congress Fundamental and Applied Metrology, Lisbon, Portugal, 6–11 Sept 2009
108. Sładek, J., Kupiec, R., Gąska, A., Kmita, A.: Modelowanie czynników temperaturowych, wpływających na niepewność pomiaru, przy pomocy metody Monte Carlo (Temperature factors affecting the measurement uncertainty modeling with the use of Monte Carlo method). Pomiary Automatyka Kontrola (Measurement Automation and Monitoring) 01/2010
109. Sładek, J., Ostrowska, K., Gąska, A.: Wirtualne współrzędnościowe ramie pomiarowe (WWRP) [Virtual articulated arm coordinate measuring machine (VAACMM)]. Pomiary Automatyka Kontrola (Measurement Automation and Monitoring) 01/2010
110. Sładek, J., Juras, B., Krawczyk, M., Gąska, A.: Modelling of the probing system errors by use of Monte Carlo Method. In: Sładek, J., Jakubiec, W. (eds.) Advances in Coordinate Metrology, Monograph. University of Bielsko-Biała (2010)
111. Sładek, J., Sitnik, R., Kupiec, M., Błaszczyk, P.: The hybrid coordinate measurement system as a response to industrial requirements. Metrol. Meas. Syst. **XVII**(1), 109–118 (2010)
112. Sładek, J., Błaszczyk, P.M., Kupiec, M., Sitnik, R.: The hybrid contact–optical coordinate measuring system—Elservier. Measurement **44**, 503–510 (2011)
113. Sładek, J., Gąska, A., Olszewska, M., Kupiec, R., Krawczyk, M.: Virtual coordinate measuring machine built using laser tracer system and spherical standard. Metrol. Meas. Syst. M&M **XX**(1), pp. 77–86, doi:10.2478/mms-2013-007.JCR, ISSN 0860-8229 (2013)
114. Sommer, K.D., Siebert, B.R.L.: Systematic approach to the modelling of measurements for uncertainty evaluation. Metrologia **43**, 200–210 (2006)
115. Summerhays, K.D., Baldwin, J.M., Campbell, D.A., Henke R.P.: Application of simulation software to coordinate measurement uncertainty evaluation. In: Proceedings of Aspe, Uncertainty Analysis in Measurement and Design, June–July 2004
116. Suplement 1 to the Guide to the expression of uncertainty measurement .Propagation of distribution using Monte Carlo method, JCGM 2006
117. Sweeney, J.: NeuroShell2 Technical Analysis, materials of Ward System INC
118. Sweeney, J.: NeuroShell 4.1. Technical Analysis of STOCK&COMMODITIES 1992–98
119. Tadeusiewicz, R.: Sieci neuronowe,(Neural networks) PWN, Warsaw 1993
120. Tadeusiewicz, R.: Elementarne wprowadzenie do techniki sieci neuronowych z przykładowymi programami (Elementary introduction to neural network technique with examplary programs). Akademicka Oficyna Wydawnicza, Warsaw (1998)
121. Teichmann, U.: Metrologische Probleme bei der Pruefung von koordinatenmessgeraeten. Feingeraetentechnik (1978)
122. Trac Cal—software for identification of CMS component errors with the use of laser tracer or laser tracker
123. Trapet, E.: Introduction to traceability with the virtual CMM. In: Proceedings of Workshop on Traceability of CMM. PTB-Braunschweig, 9–10 Oct 1997
124. Trapet, E., Waeldele, F.: The virtual CMM concept. In: Ciarlini, P., Cox, M.G., Pavese, F., Richter, D. (eds.) Advanced Mathematical Tools in Metrology. II World Conference Scientific, Singapore, pp. 239–247 (1996)
125. Trapet, E., Franke, M., Härtig, H., Schwenke, H., Wäldele, F., Cox, M., Forbes, A., Delbressine, F., Schellekens, P., Trenk, M., Meyer, H., Moritz, G., Guth, Th., Wanner, E.: Traceability of coordinate measurements according to method of virtual measuring machine. PTB F-35, Braunschweig (1999)
126. Trapet, E., et al.: Traceability of coordinate measurements according to the method of the virtual measuring machine. European Commission SMT- EDG XII/C/5, Project No. Mat.1 CT94-0076, final report
127. Trenk, M., Franke, M., Schwenke, H.I.: The "Virtual CMM", a Software Tool for Uncertainty Evaluation—Practical Application in an Accredited Calibration Lab. In: ASPE Proceedings: Uncertainty Analysis in Measurement and Design, July 2004

128. VDI/VDE 2617—Blatt7.Ermittlung der Unsicherheit von Messungen auf Koordinatenmessgeräten durch Simulation VDI/VDE- Richlinien ver. 2008
129. Wäldele, F., Drieschner, R., Elligsen, R.: Testing of coordinate measuring machines algorithmus. Phase 2, BCR-Information EUR 13 417 EN Brussels, Luxemburg (1991)
130. Wäldele, F., Bittner, B., Busch, K., Drieschner, R., Elligsen, R.: Testing of coordinate measuring machine software. Precision Eng. **15**, 121–123 (1993)
131. Weckenmann, A., Knauer, M.: Causes and consequences of measurement uncertainty in production metrology. In: Ossana, P.H., Postrednik, D., Durakbasa, N.M. (eds.) Proceedings of 6th ISMQC IMEKO Symposium, Wien, Austria, 8–10 Sept 98, pp. 693–698
132. Wirtual MMC /Virtual MMC—simulation software for determining uncertainty of CMM measurements with the use of Monte Carlo method Python Software Application developed at Cracow University of Technology Lab of Coordinate Metrology
133. Virtual Neuro WMP/Virtual Neuro CMM—Neural CMM model (Application C++) developed at Cracow University of Technology
134. Wirtual PK/Virtual CUT—CMM model (Application C++) developed at Cracow University of Technology under the development project no. PB1367/T07/95/08
135. Wyrażanie niepewności pomiaru—przewodnik. Główny Urząd Miar [Guide to the Expression of Uncertainty in Measurement (GUM)], ISO 1993/1995 with the addition to polish edition of J.M. Jaworski Warsaw (1999)

Chapter 6
Accuracy of Modern Coordinate Measuring Systems

Abstract This chapter is dedicated to monitoring CMS accuracy, described here as an important task that together with a system of accredited laboratories maintains measurement traceability. The system of ISO 10360 standards plays a key role here. In this system, determination of the indication error is based on the measurement of known length. This chapter describes the requirements for the assessment of accuracy of contact and noncontact systems together with optical tomography. It describes standards and laser systems, including those developed in the LCM CUT and used in applications of procedures developed on the basis of standards requirements. Moreover, it reports on the importance of thermal influence correction, including a correction system originally developed in the LCM CUT that uses Internet data transmission between corrected CMM and the monitoring laboratory. It describes the key role played by an accredited calibration laboratory in monitoring the accuracy of coordinate measuring systems. It shows how important it is for each user of a measuring system to ensure appropriate calibration methods and to separate the calibration procedure from other adjustments including repair and correction of measuring system accuracy. Although the latter can be performed by a producer's authorized service, the calibration of the measuring system should be carried out only by an independent accredited laboratory.

The accuracy of coordinate measurements is closely related to the issue of the measuring systems and machine accuracy monitoring. The required and specified accuracy of the machine or articulated arm can be maintained using calibration, acceptance tests, monitoring, and finally, reverification tests. From the normative point of view a key role belongs to the ISO 10360 standard [31], which is constantly expanded to cover new measuring systems and updated when new calibration methods and standards appear. Normative requirements for management systems, described in Chap. 2, obligate users to regular calibration and monitoring of the measuring equipment accuracy and these activities should be done only by accredited laboratories independent of the measuring equipment producers. However, in the case of coordinate measuring systems usage, users must deal with many difficulties in the implementation of effective monitoring of measuring system accuracy. It has to be started from an appropriate selection of the measuring system

J.A. Sładek, *Coordinate Metrology*, Springer Tracts in Mechanical Engineering,
DOI 10.1007/978-3-662-48465-4_6

through the acceptance test, ending on determination of the length of a period between reverification tests, depending on the manner and conditions of the equipment usage. The real problem, however, is protection from the use of measuring equipment that does not meet established requirements. This problem discovered during the reverification test is not correctly analyzed. It is reduced to receiving the calibration certificate, often without investigating the reasons in the case when the equipment exceeds the established values of maximum permissible errors. Sometimes it is even not tested in the conditions in which the overrun occurred, or how it affects production quality. It would be sufficient to perform a cursory analysis of the sources and causes of noncompliance. However, the problem of interim monitoring appears then, together with problems of temporary tests, the ability to use standards, and the monitoring of thermal conditions during measurement realization. There is also the problem of insufficient knowledge of the ISO 10360 standard [31] or of very important and helpful German recommendations from the VDI/VDE 2617 series [67].

6.1 Coordinate Systems Accuracy Test in Accordance with ISO Standards

The ISO 10360 series of standards plays an important role in the calibration and monitoring of accuracy of coordinate measuring machines. Determination of the indication error is based on the measurement of known length. This standard, as stated earlier, provides the basis for the development of procedures for keeping the maximum permissible errors (MPEs) of used coordinate systems at a constant level.

ISO 10360 consists of the following parts [31].

Part 1: Basic terms and definitions
Part 2: Tests of CMMs used for linear dimension measurements
Part 3: Tests of CMMs with a rotary table
Part 4: Tests of CMMs used in scanning mode
Part 5: Test of CMM probing systems equipped with one tip or set of tips
Part 6: Tests of CMM's software operation (error estimation of Gauss best fitting elements calculations)
Part 7: Test of CMMs equipped with imaging probing systems
Part 9: Test of CMMs equipped with multisensory probing systems

The following parts are under preparation.

Part 8: Test of CMMs equipped with optical distance sensors
Part 10: Tests of laser trackers
Part 12: Tests of articulated arm CMMs

6.1.1 Accuracy Testing and Calibration of Contact Systems

Parts 2 and 5 may be considered as the most important from the implementation point of view. Recently these parts of the standard, that is, ISO 10360-2: 2009 [18] and ISO 10360-5:2010 [19], were completely revised by changing not only the requirements, but also by dividing the point coordinate acquisition system calibration procedures (e.g., of contact probe head by moving them to the fifth part of the standard, i.e., ISO 10360-5: 2010). The previous version of the ISO 10360-2:2003 standard [32] recommended in CMM tests (reverification and acceptance tests) usage of a length step gauge or a set of gauge blocks (Fig. 6.1). The new version also allows the use of other standards, but only if they enable bidirectional measurements, ball-bar type (Fig. 6.13). However, if the standard enabling only one-direction measurements is applied then additional measurements of a reference plate or a reference sphere [18] should proceed. In addition, the novelty introduced in this part of ISO 10360-2:2009 [18] is the possibility of laser interferometer usage, including laser tracker systems and especially the laser tracer system, Fig. 4.26 of Sect. 4.2.3, with the TRAC-CHECK software [15, 61], Figs. 6.5a, b and 6.6.

Three main technical objectives to be tested according to the ISO 10360-2:2009 standard are [18]:

- The indication error of the stylus tip oriented parallel to the ram axis
- The indication error of the stylus tip oriented orthogonally to the standard measuring axis and compatible with the orientation of one of the machine axes, where the stylus tip center is located at a specified distance from the ram axis
- The repeatability of length measurement

The requirements, terms, and definitions that have been substantially changed are specified below.

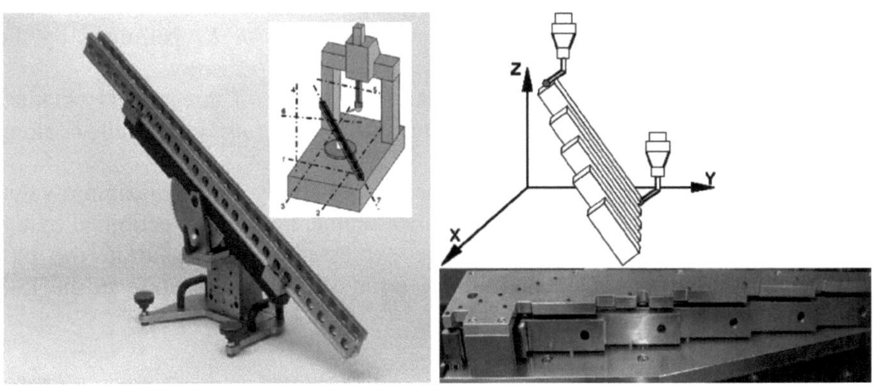

Fig. 6.1 Step gauge: *left side* and gauge blocks; *right side* used for CMMs calibration in the LCM CUT

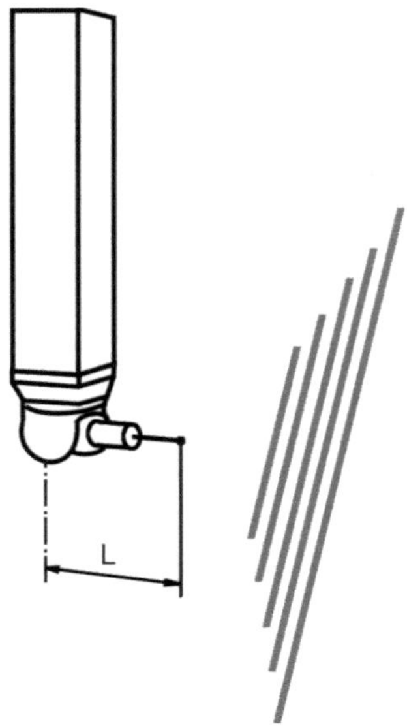

Fig. 6.2 Example orientation of measuring tip in determination of E_{150}

The ISO 10360-2 standard introduces new terms and definitions. The most important are [18]:

L Distance from the center of tip ball and ram axis: the distance between the stylus tip and a reference point measured orthogonally to the ram axis. The manufacturer has to define the reference point. If there is no such point defined by the manufacturer then the user has to choose a reference point close to the probe head mounting system

E_L Indication error: CMM indication error determined by standard length measurement, including an offset (*L*) from ram to tip ball. The default values for *L* are 0 mm (or other near zero) and 150 mm

E_{lmpe} Maximum permissible error of length measurement, defined as a maximum value of the length measurement error E_L permitted by the specification (possible parameters: E_{0MPE} and E_{150MPE})

R_0 Repeatability of the length measurement; range of three errors obtained during measurements of the same length with the use of tip ball set along the ram axis

$R_{0\ \mathrm{MPL}}$ Maximum permissible measurement repeatability value; maximum value of the repeatability of the length measurement error (R_0) permitted by the specification. The MPL may be expressed as presented in the plot illustrating maximum permissible error of CMM indication in the ISO 10360-1:2000 standard.

The principle of the method assumes usage of a length standard to establish whether the CMM does not exceed maximum permissible indication error *E* and maximum permissible limit for the repeatability range R_{0MPL} [18]. The new version

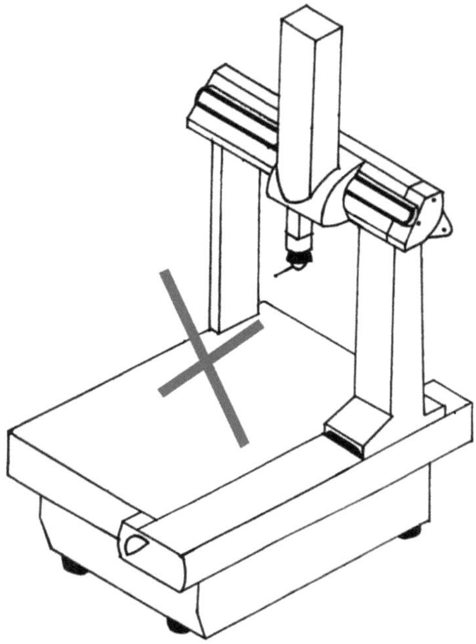

Fig. 6.3 Example positions of the calibrated test length with appropriate measuring tip orientations

Fig. 6.4 Example of the measuring tip configuration in the determination of E_{150} error: measurement of the step gauge

of the ISO 10360-2 standard systematized artifact orientations in which indication error tests should be performed. Orientations are shown in Tables 6.1 and 6.2. It should be noted that the idea itself, contained in earlier versions of the standard, was kept here in case of determination of the indication error E_0. Five different lengths to be measured have to be placed in seven different positions (which differ in

Table 6.1 Four required orientations in the measuring volume for E_0

Orientations in measuring volume	Position number
Along the diagonal in space from point (1, 0, 0) to (0, 1, 1)	1
Along the diagonal in space from point (1, 1, 0) to (0, 0, 1)	2
Along the diagonal in space from point (0, 1, 0) to (1, 0, 1)	3
Along the diagonal in space from point (0, 0, 0) to (1, 1, 1)	4
Coordinates of the points are given in Cartesian coordinates (*X*, *Y*, *Z*)	

locations and orientations) in the CMM's measuring volume. Each length has to be measured three times, giving a total of 105 measurements. Four of the seven positions have to be the space diagonals (Table 6.1; Fig. 6.5a) and another three positions are parallel to each of the CMM axes (Table 6.2; Fig. 6.5a) [58]. The longest tested length in each position must be not less than 66 % of the length of the axis or diagonal in which the measurement is made. Additionally lengths should be regularly distributed over the measurement line (e.g., for tested length of 1 m: 100, 200, 400, 600, 800 mm).

It can be noticed that there are no recommendations for the minimum length that was previously required. It is associated with the transfer of the probe head test to the fifth part of the standard where error related to determination of the sphere diameter is designated.

The requirements for the coefficient of thermal expansion (called CTE) for the used standard were also clarified. By default it should be in the range from 8×10^{-6}

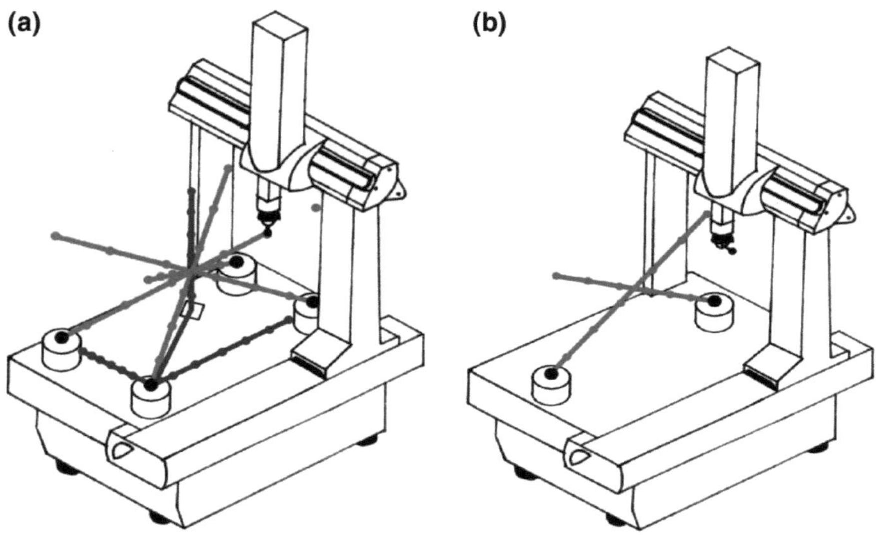

Fig. 6.5 Laser tracer positions and measurement lines during determining E_0 error (**a**); laser tracer positions and measurement lines during determining E_{150} error (**b**)

Table 6.2 Three default orientations in the measuring volume for E_0

Orientation in measuring volume	Position number
Parallel to the machine scales from point (0, 1/2, 1/2) to (1, 1/2, 1/2)	1
Parallel to the machine scales from point (1/2, 0, 1/2) to (1/2, 1, 1/2)	2
Parallel to the machine scales from point (1/2, 1/2, 0) to (1/2, 1/2, 1)	3
Coordinates of the points are given in Cartesian coordinates (*X*, *Y*, *Z*)	

Table 6.3 Orientations in the measuring volume for E_{150}

Orientations in measuring volume
Along the *YZ* plane diagonal from point (1/2, 0, 0) to (1/2, 1, 1)—1A
Along the *YZ* plane diagonal from point (1/2, 0, 1) to (1/2, 1, 0)—1B
Along the *XZ* plane diagonal from point (0, 1/2, 0) to (1, 1/2, 1)—2A
Along the *XZ* plane diagonal from point (0, 1/2, 1) to (1, 1/2, 0)—2B
Coordinates of the points are given in Cartesian coordinates (*X*, *Y*, *Z*)

[1/°C] to 13×10^{-6} [1/°C] (it is called normal CTE), unless the manufacturer's specification clearly provides otherwise. If the coefficient of thermal expansion of the artifact is equal to or less than 2×10^{-6} [1/°C], the standard requires three additional length measurements using the standard with a normal expansion coefficient.

For all 105 measurements (if they were performed) and for three additional errors E_0 should be calculated as a difference between the indicated value and the nominal (calibrated) value. All values of length measurement errors are put on the plot as shown in the plot that presents the maximum permissible error of CMM indication in the ISO 10360-1:2000 standard.

The CMM used to measure the dimensions works correctly if length measurement errors, measured with the zero offset of the center of the measuring tip (E_0), do not exceed the maximum permissible length measurement error E_{0MPE}. This error is set by the manufacturer and it includes the measurement uncertainty according to ISO 14253-1 [33] and ISO/TS 23165 [21]. In this edition of the standard additional tests for error of indication with established offset of a measuring tip center E_{150} are introduced. This allows the assessment of the ram rotation error, which is proportional to the center of a measuring tip offset [58].

By default a test is performed for the center of a measuring tip offset of L = 150 mm (Fig. 6.2). The direction of the measuring stylus has to be perpendicular to the measurement line defined by the tested length and directed along the CMM axis (positive or negative) (Figs. 6.3 and 6.4).

The user may specify one or two positions of calibrated test length from four possible (shown in Table 6.3) and one or two orientations of the measuring probe head (e.g., +X or −X for position 1A or 1B, as well as +Y or −Y for position 2A or 2B).

In total two from among eight possible combinations of tested length location and measuring probe head orientation are required. In each selected combination five different calibrated test lengths should be measured, each three times, which together give 30 measurements [58]. Examples of possible positions are shown in Fig. 6.5b.

The most unfavorable tested CMM situation, because of the greatest impact of rotational errors, appears when only one position is chosen and the test is performed with two mutually opposite orientations of the measuring probe head with the use of the center of measuring tip offset of $L = 150$ mm.

Also in this test the longest calibrated test length for each position has to be at least 66 % of the maximum CMM displacement along a line determined by this length. Each calibrated test length has to be significantly different from the others. Lengths have to be evenly distributed over the line. Five calibrated test lengths used in one position may be different from those used in other positions, for example, due to different size of CMM axes.

For all 30 measurements, the length measurement error E_{150} should be calculated as a difference between the indicated value and the calibrated value of each test length. All length measurement error values are put on the appropriate plot.

CMM works correctly if length measurement errors measured with the center of the measuring tip offset equal to 150 mm (E_{150}) do not exceed the maximum

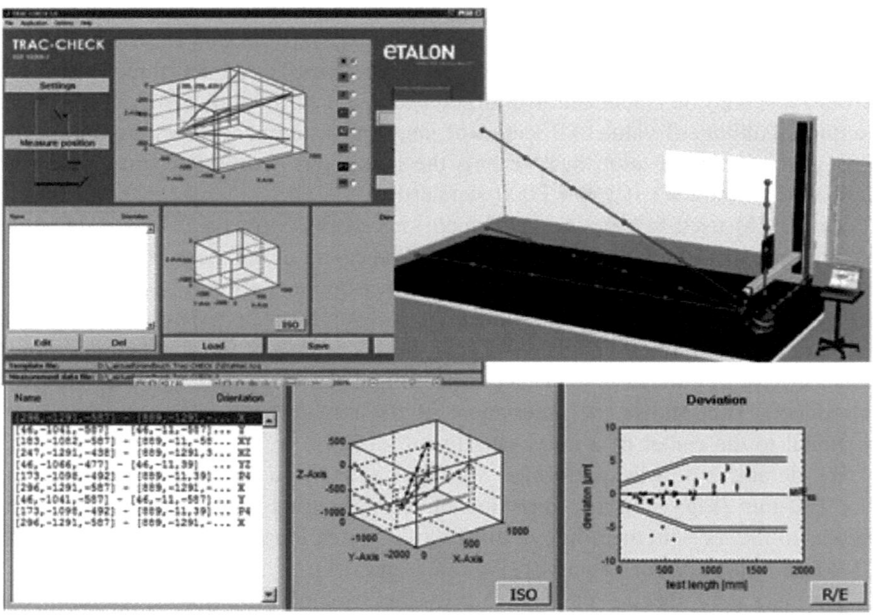

Fig. 6.6 TRAC-CHECK software for CMM accuracy testing according to ISO 10360–2:2009 with the use of laser tracer [15]

permissible length measurement error $E_{150,MPE}$ set by the manufacturer, including the measurement uncertainty according to ISO 14253–1 [33] and ISO/TS 23165 [21].

For each set of three repeated measurements repeatability of length measurement is calculated (R_0) by assessing the range of three repeated measurements of the same length. The CMM accuracy is correct, if repeatability of the length measurement does not exceed the maximum permissible limit of the repeatability range $R_{0,MPL}$ set by the manufacturer, including the measurement uncertainty according to ISO 14253-1 [33] and ISO/TS 23165 [21].

The new version of the standard also includes the examination of horizontal arm CMMs working in duplex mode and the accuracy tests for the whole machine set.

The fifth part of the ISO 10360 standard presents the testing procedure of contact probe heads of coordinate measuring machines. In the current edition of the ISO 10360-5:2010 standard [19] the assessment of measuring probe heads is separated from CMM error checking. However, in Sect. 6.3.1 of ISO 10360-2:2009 standard it is recommended to carry out the tests provided by Part 5 of the standard first, and then perform the tests required in Sect. 2. In this way the integrity of activities concerning the probe head and machine tests is methodologically maintained. The ISO 10360-5 standard gives the procedure for examination of the probe heads equipped with a single measuring tip and with a set of measuring tips arranged, for example, in a star combination, or for an articulating probe head set. It also includes the extensions and magazines. For a single measuring tip the verification of the contacting system is based on measurement of the test sphere of a diameter not less than 10 mm and not greater than 50 mm, in 25 regularly distributed points. Points should be distributed in accordance with Fig. 6.7.

The standard states that results of performed tests can significantly differ depending on the length of the measuring stylus. Regarding this fact it is recommended to carry out measurements for the lengths of the stylus suggested by the manufacturer (suggested lengths: 20, 30, 50, and 100 mm). The direction of the measuring stylus should be parallel to the axis of movement of the measuring probe head. The associated element should be determined according to the Gauss method from all the measured 25 points. For each measuring point the radius should be determined from the center of the associated element. On the basis of obtained results $R_{max} - R_{min}$ should be calculated; it determines P_{FTU}. The compliance with the requirements of the probing system equipped with a single measuring tip is confirmed if the determined P_{FTU} value (together with the uncertainty of this parameter determination in accordance with the ISO 14253-1 standard) is less than the maximum permissible error limit of the contacting system MPE_{PFTU} [19]. If measurements are carried out with the use of several measuring tips, then in order to determine the contact system error the measurements should be carried out according to a certain scheme. Therefore it is assumed that for the measurements the measuring probe head equipped with a set of tips arranged in a "star" is used; that is, one tip is oriented towards the vertical direction and the other four in a horizontal direction at an angle of 90° to each other. The spherical standard (the requirements for the standard are consistent with the ones stated for a single

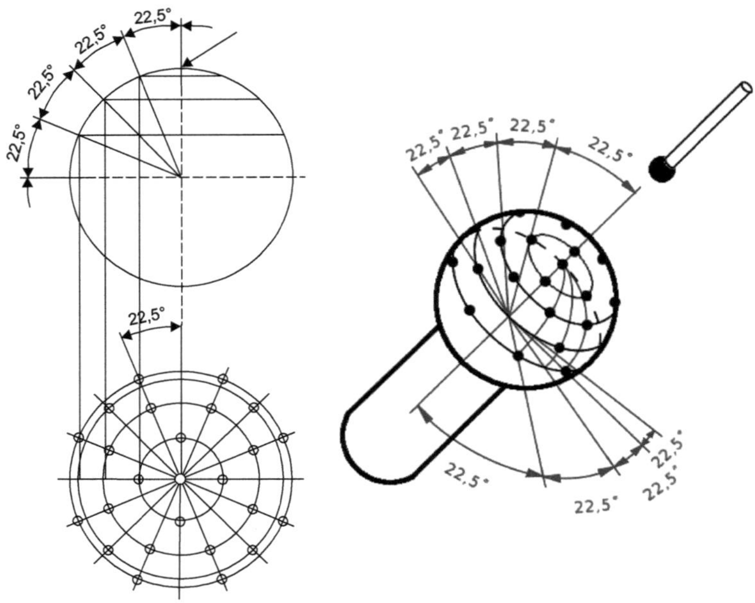

Fig. 6.7 Distribution of measuring points on the test sphere when determining the measuring probe head error P [19]

measuring tip) set in one location should be measured with the use of each measuring tip. Measurement with each tip should consist of 25 regularly distributed measuring points.

Then, for each tip the associated sphere should be determined from the measured points in accordance with the Gauss method. In this way five spheres will be determined. Then ranges of the values of measured sphere centers should be determined for each coordinate. The maximum value from the three ranges obtained in that way constitutes the P_{LTE} parameter that corresponds to location errors. The next two parameters are determined from all 125 points measured on a spherical standard (using all tips). From these points the associated element should be created with the Gauss method and subsequently the diameter and shape errors for this element should be set. The difference between the measured value of the diameter and the nominal value given by the calibration certificate of the spherical standard is treated as a parameter P_{STE} related to errors of dimension representation of the contact system. Shape errors of a sphere determined at 125 points correspond to the parameter P_{FTE}. The indication of the point acquisition system equipped with several measuring tips is confirmed when determined values of parameters P_{LTE}, P_{STE}, and P_{FTE} (together with the uncertainties for each parameter determined in accordance with the ISO 14253-1 standard) are smaller than the maximum permissible errors limits $P_{\mathrm{LTE,MPE}}$, $P_{\mathrm{STE,MPE}}$, and $P_{\mathrm{FTE,MPE}}$ corresponding to them [19].

In the case of the articulating probe head testing, analogously to the case of the probe head with a set of tips of a "star" type, the measurement of shape, size, and location of the examined sphere is carried out in five different angular locations of the probe tip. In each angular location 25 points on the test sphere are measured, which gives a total of 125 points for all five positions. If a set of changeable measuring styli or changeable measuring probe heads were supplied together with CMM, five exchanges should be made, one before the use of each angular position. For each group of 25 points collected from different angular locations, the sphere should be determined by the least squares method, giving a total of five replacement spheres. For all five spheres ranges of the center coordinates *(X, Y,* and *Z)* are calculated. The largest of the three ranges is the location error, P_{LTE}. Furthermore, in order to determine the indication error from all 125 points the sphere is determined by the least squares method. This analysis gives an indication error P_{STE} and P_{FTE}. Because results of these tests are strongly dependent on the length of measuring probe head extensions, different lengths of measuring styli are taken into account. However, only the lengths described by the CMM manufacturer as applicable for the articulating probe head are tested. The chosen location of the test sphere in a measuring volume of the machine can significantly affect the test results. The size of the material standard (i.e., the test sphere) has to have a diameter not less than 10 mm and not greater than 50 mm. The sphere has to be calibrated, before tests, both for size and shape. The sphere used for CMM qualification should not be used as a test sphere. Before carrying out the test, the articulating probe head set should be equipped with a short simple measuring stylus (20 mm, if not stated otherwise) and an extension of the measuring probe head. The length of the measuring probe head extension l_{pE} should be chosen from the following values: 100, 200, and 300 mm. If not stated otherwise, the measuring stylus and elements of the measuring probe head extension provided by the CMM producer should be used. Then, for each of the five angular locations of the articulating probe head set the qualification according to the standard procedure specified by the CMM manufacturer should be carried out. These five angular locations should consist of one vertical and four horizontal locations, oriented to each other at 90° [19]. It is recommended to move the test sphere in relation to the location of the sphere used for qualification of the measuring probe head set both in the *X* as well as in the *Y* direction at a distance at least equal to the length of the longest extension used during calibration. Then, the measurement of the test sphere using 25 points for each angular location is carried out (in total 125 points).

In order to calculate results, for each group of 25 points collected in a given angular location the associated sphere should be determined by the least squares method (five spheres in total). For all five spheres ranges of the center coordinates (*X*, *Y*, and *Z*) are calculated, where the largest of these three ranges is the indication error P_{LTE}. Next, a sphere is determined by the least squares method for all 125 points collected with the use of all five angular locations. The deviation of the associated sphere diameter, in relation to the calibrated value of the physical standard of size is the size error P_{STE} of the articulating probe head set. Form deviation P_{FTE} of the articulating probe head is determined for a sphere calculated for 125 measuring points. Similar calculations are carried out for every extension length. Requirements for CMM equipped with an articulating probe head are met, if for each provided value of measuring probe extension length, l_{PE}, the following relations occur: $P_{FTE} \leq MPE$, $P_{STE} \leq MPE$, and $P_{LTE} \leq MPE$, (with consideration to the measurement uncertainty) [19].

6.1.2 Accuracy Testing and Calibration of Optical Systems and Computed Tomography

Coordinate optical systems (COS) and computed tomography (CT) are the combination of a coordinate technique with a noncontact method for identifying points. The noncontact system in their operation uses various phenomena among others: structured light, photogrammetry, and laser triangulation. Some systems are based on the measurement of the detector beam return time (TOF, time of flight), there are also systems of computed tomography or magnetic resonance imaging (MRI).

Currently, there are three documents related to the subject of optical and tomography system accuracy that give requirements for their calibration: ISO 10360-8 [20] and the recommendations of VDI/VDE 2634 [73] and VDI/VDE2617 Blatt 12/VDI/VDE 2630 Blatt 1.3 [72].

Recommendations of German PTB for optical system testing collected in VDI/VDE 2634 “Optical systems based on area scanning” describe examination methods and standards for optical 3-D measuring systems. These methods include acceptance test and reverification of 3-D optical measuring systems.

The first specified quality parameter, that is, probing error, describes testing of optical 3-D measuring systems based on area scanning within the small measuring range. To determine it an operating mode should be specified. Spheres made of materials with a diffusely reflecting surface, for example, ceramic or steel with the diameter of $Dp = (0.1 \ldots 0.2) \cdot L0$ are used as a standard where: L_0 is the body diagonal of the smallest rectangular parallelepiped that encloses the measuring volume. Testing procedure, presented in Fig. 6.8, consists of measurement of the sphere in at least 10 positions within the whole measuring volume. From all measured points (for each position) the best-fit element should be calculated according to the Gaussian method. Then for each measuring point the radius should

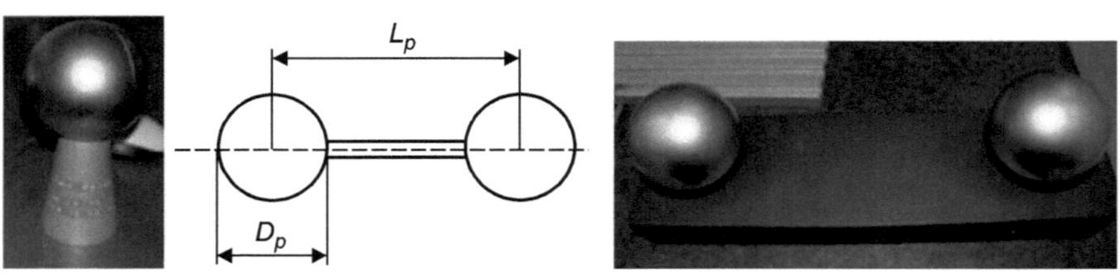

Fig. 6.8 **a** Test sphere for the calibration of COS; **b** standard scheme; **c** ball-bar type standard

be determined from the center of the best-fit element. Parameter R defined as $R_{max} - R_{min}$ is calculated on the basis of obtained radii. Compliance with the stated requirement is confirmed when the determined value R (together with the uncertainty of this parameter designation according to ISO 14253-1 [33]) is less than the maximum permissible errors of probing MPE_R.

Another quality parameter called the *sphere-spacing error* is used for verification of the length-measuring capability of the measuring system. This parameter is defined as the difference between the measured and calibrated values of distance between spheres of the one-dimensional standard. To determine the sphere-spacing error the standard with two spheres should be used. As before it should be made of materials with a diffusely reflecting surface (low reflection), for example, ceramic or steel. The dimensions of standards for tested coordinate measuring systems result from the machine measuring volume: it should have the diameter of $(0.1 \ldots 0.2) \cdot L_0$, where L_0 is the body diagonal of the smallest rectangular parallelepiped that encloses the measuring volume of the tested system. The distance between spheres L_p should meet following requirement $L_p \geq 0.3 \cdot L_0$ (Fig. 6.8). For sphere-spacing error determination the measurement of the standard has to be done in seven positions. Recommended orientations are presented in Fig. 6.9. The standard should be set parallel to axes (positions 1, 2, and 3), and then in accordance with diagonals of planes forming a measuring space cube (4, 5, and 6) and along one of the spatial diagonals of the mentioned cube (7). Analogously to the previously described determination of probing error the spheres should be determined in accordance with the least squares method.

To determine the sphere-spacing error the sphere centers are used. The principle is to determine whether the indication of a given system does not exceed its maximum permissible error of SD_{MPE} indication.

The last (third) quality parameter concerns the verification of flatness measurement error as the distance of measured points from the best-fit plane constructed according to the least squares method. The standards used here are in a form of

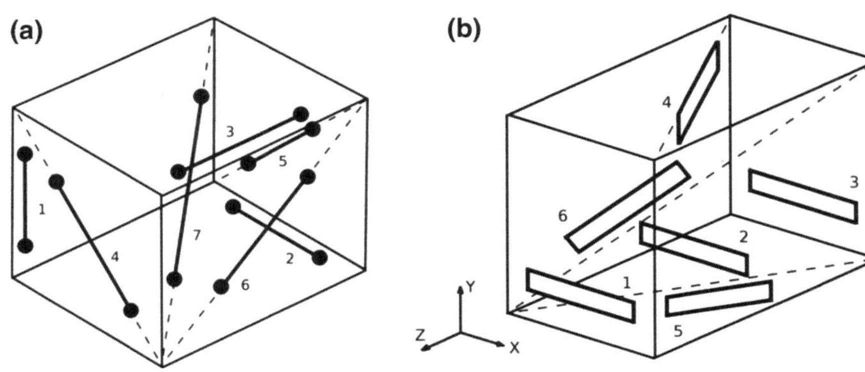

Fig. 6.9 **a** Recommended arrangement in determination of the sphere-spacing error; **b** recommended arrangement in determination of the flatness measurement error [73]

rectangular parallelepipeds. They are made of steel, ceramic, aluminum, or other suitable materials with a diffusely scattering surface, the widths of which cannot be less than 50 mm, and lengths not smaller than $0.5 \cdot L_0$. Flatness error of the standard should not be bigger than 1/5 of the measured characteristic. In order to determine the flatness measurement error the measurement should be carried out in at least six positions. Figure 6.9 presents a recommended arrangement of the standard: parallel to the *XY* plane in three distances (1, 2, and 3), in accordance with diagonals of the cube that determines measuring volume bases (4 and 5), along the diagonal of the mentioned cube (6). The distance of measuring points from the plane, defined by the least squares method, is a flatness measurement error R_E. Correct operation of the device is confirmed if the maximum error specified by the manufacturer $R_{E,MPE}$ is not exceeded. The reverification involves examination cycle repetition in order to confirm (or not confirm) the quality of the optical system. Multiple verification allows us to assess the trend of system characteristic changes. The methodology used in ISO 10360–8 for testing is analogous to the ISO 10360-2 standard. Two general objectives for tests, accordingly to ISO 10360-8 [20], are: "to test the error of indication of a calibrated test length using an optical distance sensor and to test the errors in the optical distance sensor." The acceptance and reverification tests given in ISO 10360-8 are applicable only to Cartesian systems with optical distance sensors. The calibration of optical 3-D measuring systems involves determination of the error in the optical distance sensor in the form of probing form error P_{FX}, probing dispersion value $P_{95\ \%\ X}$, probing size error P_{SX}, probing size error all $P_{S\text{-}ALL\ X}$, and determination of the error of indication of a calibrated test length, length measurement error $E_{O\ X}$. The performance of a system equipped with optical distance sensors is confirmed, when determined values of parameters P_{FX}, $P_{95\ \%\ X}$, P_{SX}, $P_{S\text{-}ALL\ X}$, and $E_{O\ X}$ (together with uncertainties for each parameter determined in accordance with the ISO 14253-1 standard) are less than maximum permissible errors $P_{FX,MPE}$, $P_{95\ \%\ XMPE}$, $P_{S,X,MPE}$, $P_{S\text{-}ALL,\ X,MPE}$, and $E_{O,\ X,MPE}$ corresponding to them.

When determining the error in the optical distance sensor, probing form error $P_{F,X}$, the measurement of the sphere has to be carried out in one position. To perform calculations the points taken on the sphere should be divided in 25 areas, and then each area should be reduced to one representative point. From the obtained 25 points the sphere radii have to be calculated with the least squares method. The difference of obtained radii is the indication error marked as $P_{F,X}$. When determining the error in the optical distance sensor, probing dispersion value $P_{95\ \%\ X}$, the procedure is similar to the previous measurement. Then the range of variation of radii belonging to 95 % of obtained results is to be found. When determining the error in the optical distance sensor, probing size error $P_{S,X}$, with the least squares method the sphere diameter is calculated from 25 points mentioned above. The difference between the obtained and reference diameter is the error $P_{S,X}$. When determining the error in the optical distance sensor, probing size error all $P_{S\text{-}ALL,X}$ the sphere diameter is calculated from all points with the least squares method. The difference between the obtained and reference diameter is the error $P_{S\text{-}ALL,X}$.

When calibrating the length measurement error $E_{O,\,X}$ the user can select seven different locations and orientations to measure five calibrated test lengths within the limits specified by the manufacturer. It is recommended to use the arrangement shown in Fig. 6.10. Every test length is measured three times, which gives a total of 105 measurements. Analogously as above the spheres are created (in accordance with the least squares method). To determine length, parameter sphere centers are used. For all 105 measurements (if performed) the length measurement error E_0 should be calculated as a difference between the indicated and the correct value. All the values of the length measurement errors are put on the graph as shown, describing the maximum permissible error of CMM indication in the ISO 10360-1 standard. The operation of COS used for dimension measurement is correct if length measurement indication errors do not exceed the maximum permissible length measurement error E_{0MPE} specified by the manufacturer, including the measurement uncertainty according to ISO 14253-1 [33] and ISO/TS 23165 [21].

The recommended arrangement is: three positions along the axes of the coordinate system and four diagonal positions [4].

The ISO 10360–8 standard also includes surface mapping tests which give information about system operation in the case of more complex surfaces.

When performing flat form error tests the standard plane has to be placed on the *x*–*y* plane and as close as possible along a diagonal in measuring the volume of the tested coordinate system. The arrangement shown in Fig. 6.10b is recommended. The flatness standard measurement is performed in two positions: one along the diagonal in the reference *x*–*y* plate and the other along the diagonal in space. Using all measured points the distribution in accordance with Gauss is determined. For each of the *nF* measurements the signed normal distance *N* should be calculated. Then the flatness measurement error (F_x) can be calculated as the range n_F:

$$F_x = N_{max} - N_{min} \tag{6.1}$$

The second test, called a depth measurement test, is the comparison of the depth measured with the use of a probe head and the calibrated depth. The result is used to determine the structural resolution. Another test is the test of edge slope

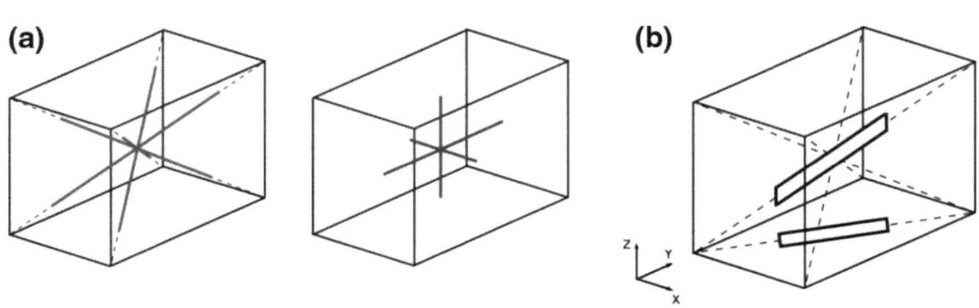

Fig. 6.10 Recommended standards arrangement when determining **a** length measurement error; **b** flatness measurement error

measurement. Finally there is a test of wavy shape measurement, where the sinusoidal area of the standard is tested. The wavelength is characterized by the measured amplitude ratio over the calibrated amplitude (greater than 0.7) and it is compared to the specified resolution.

In determination of flatness error standards in the form of parallelepipeds, steel, ceramic, aluminum, or other material with low reflection (Fig. 6.11) is used. Their widths cannot be less than 50 mm, and lengths not smaller than $0.5 \cdot L_0$. Standard flatness error should not be greater than 1/5 of the measured characteristic.

In addition, the ISO standard gives a procedure for the so-called structural resolution test *Rs*. This resolution should be clearly distinguished from the resolution understood in the current sense, that is, the minimum quantity for which it is possible to increase the output data of the instrument. Structural resolution characterizes the smallest structure size possible to measure with the use of scanner.

Quantities influencing structural resolution in accordance to ISO 10360-8 [20] are: quantization of A/D converters, measuring noises, pixel grid of the image sensor, resolution of the optical system, spot size of laser/beam diameter, and size of the image-processing window (sensors with autofocus function).

The large number of quantities influencing the structural resolution shows that the popular formula, "pixel grid = resolution," is not true. The real structural resolution is much lower. It is therefore important to specify and test this characteristic, with the use of standards presented in Fig. 6.12.

The material used for standard construction should not reflect wavelength emitted by the given COS (different for laser systems, and different for systems that operate on the basis of structured light) to avoid reflections that are seen as noise during measurement. Despite VDI/VDE recommendations and the ISO 10360-8 standard dealing with the same issue, they differ from each other on several points. It should be noted that Sect. 8 of ISO 10360 is just entered whereas VDI/VDE has operated for a considerable time. Even a summary review of compared recommendations allows us to draw attention to the comprehensiveness of ISO 10360-8 in comparison to VDI/VDE. In VDI/VDE there are recommendations for the determination of the three basic types of errors: probing error, sphere-spacing error, and flatness measurement error. The ISO standard also includes specified quality parameters, but also contains additional tests dedicated to the optical probe head.

Fig. 6.11 Plate standard

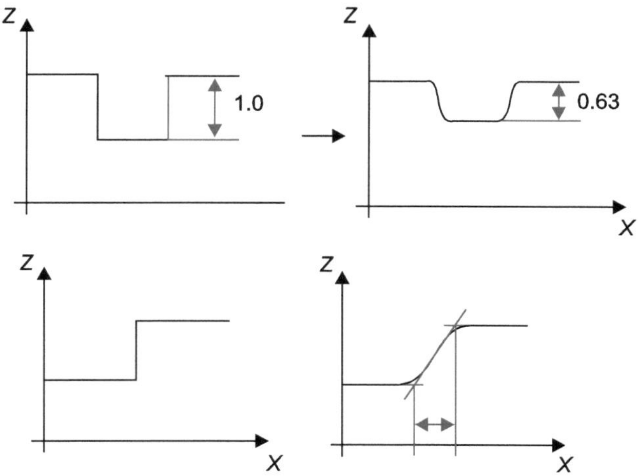

Fig. 6.12 Recommendations for standards testing the structural resolution

Differences arise when four different final results are obtained, which significantly extend the measurement duration. It is caused by exporting obtained points to other software, distributing them in 25 areas, and then finding one point that would be representative for each area. Additionally, in the ISO standard there are three types of tests for mapping: depth, edge slope, and wave shapes. Requirements for standards used in both the ISO standard and recommendations VDI/VDE for determination of probing errors and flatness measurement errors are the same. Differences can be observed in the case of the requirement for determining the length measurement error: in the ISO standard ball-bar is used, which should cover at least 66 % of the measuring volume, which is not required by VDI/VDE. According to the ISO standard additional artifacts should be used, such as the sinusoidal standard or the standard for determination of slot depth.

VDI/VDE 2630 recommendations [72] listed in the literature as VDI/VDE 2617 [Blatt 13] characterize computed tomography accuracy by following four groups of parameters: parameter linked to the length measurement error, ***E***; parameters associated with the scanning error, ***PF***, ***PS***; parameters describing size-material dependences, ***GS***, ***GF***, ***GG***; and parameters describing the resolution, ***Dg***. The parameters from the first three groups are determined for two tomography systems: for 2D, flat radiation beam, *TS* is added to the parameter such as *PF* (*TS*), and for 3D, cone radiation beam, *TT* is added to the parameter such as *PF* (*TT*); the measurement is referred to as "on the image." Similarly as for other coordinate systems, according to PN EN ISO 10360, length measurement error ***E*** is determined by physical length standards, such as gauge blocks, sphere standards, step cylindrical standards, and additionally standards of special construction, such as a set of styli ended with spherical elements (Fig. 6.29). The determined value of error *E* is related to the maximum permissible error of length measurement in the form of $MPE_e = \pm (A + L/K)$ μm, where *L* is a measured length given in millimeters. For example, for the C. Zeiss METROTOM 800 tomograph this dependency has the form of $MPE_e = \pm (4.5 + L/100)$ µm, and for the Werth TomoScope HV tomograph, $MPE_e = (4.5 + L/75)$ µm (for 20°C ± 1°C). As mentioned before, the length measurement error ***E*** is determined using physical length standards. In the case of X-ray, the beam passes through the standard and the measurement consists of the determination of the distance between two opposite maximum surfaces (plate standard). For ball-bar type standards the measurement consists of coordinates of sphere centers determined as for optical systems from sphere approximation based on a certain number of point coordinates collected on the sphere surface (Fig. 6.13). Thus the error E is determined from the relation $E = L_a - L_r$, where L_a is determined from the measurement of the plate length, L_r, the reference length of the plate.

In the case of the use of sphere standards the determination of the sphere center position, different from the determination of the surface location on the end standard, is based on averaging and matching the shape and dimensions. Therefore, it must include the averaging effect correction and the compensation of systematic errors. VDI/VDE 2630 guidelines suggest the two following procedures: Procedure A, which consists of referring to given values of scanning errors (*PS* and *PF*). It should be noted that standard properties such as its material, size, shape, and surface, as well as measurement strategy and measuring point distribution should be relatively uniform in determining the length measurement error and scanning error. Length measurement error E consists of the difference between the measured length L_{ka} and the standard length L_{kr}, assigned to the measured diameter of *PS* error and the scanning error *PF*. Relations between them can be described with one of the three following equations:

$$(L_{ka} - L_{kr} + PS) > 0 \rightarrow E = L_{ka} - L_{kr} + PS + PF$$
$$(L_{ka} - L_{kr} + PS) = 0 \rightarrow E = +PS \text{ or } E = -PS$$
$$(L_{ka} - L_{kr} + PS) < 0 \rightarrow E = L_{ka} - L_{kr} + PS - PF$$

Procedure *A* is recommended for sphere standards, when an error connected with the averaging of the sphere center location from many measuring points occurs, so that the impact of individual points is negligible. Procedure B is more accurate, but more complex. It consists of measurement of the additional (in addition to the sphere standard) short end standard for each length. Length measurement error results from the difference between the measured distance L_{ka} and the distance

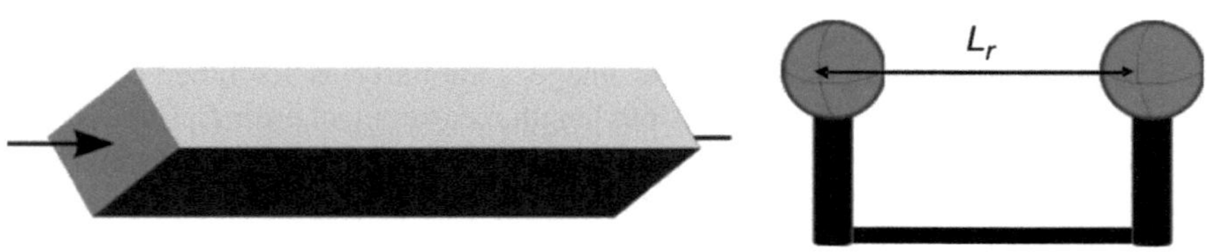

Fig. 6.13 Standards for the determination of length measurement error E: **a** end standard; **b** ball-bar type standard

between spheres *Lkr* and from the additional error E_e estimated on the basis of measurements of short end standard $E = L_a - L_r + E_E$. The parameters associated with scanning error, *PF*, *PS*, are also determined. *PF* is the form scanning error whereas the *PS* is the size scanning error. In both cases reference spheres are used. Their roughness and shape deviations in relation to scanning error should be negligible. Standards can also be used that contain balls in their structure, such as the ball-bar type as well as plate ones with ball and spherical elements. However, standards in the form of individual spheres are most popular. In parameters *PF*(*TS*) and *PS*(*TS*) the determination device should work in the measurement mode "in the image." Also the diameter of each sphere should be in the range of 0.1 to 0.2 of the diagonal of the measurement area at the magnification used. However, when determining parameters *PF*(*TT*) and *PS*(*TT*) in the measurement mode "on the image" the diameters of the spheres should not be less than in the mode "in the image." Form scanning error *PF* is determined from the difference of radii R_{max} and R_{min}, reflecting the maximum and minimum distance from the sphere center point which is designated with the Gaussian method from 25 points (Fig. 6.14a). In this test it is not required to know the certified diameter of the sphere. Size scanning error *PS* is determined from the difference between the sphere diameter determined from D_a measurement and the diameter of known standard size D_r (Fig. 6.14b).

Parameters that describe size-material dependences are *GS*, *GF*, *GG*. In computed tomography measurements the impact of the material and geometry of the measured object is especially visible. That is why the material of the standard should be as similar as possible to the material of the most commonly measured elements. Only in this way is it possible to determine the impact of the material and the geometry on the measurement of a real typical element. All the parameters described below are made on the basis of the Gaussian best-fit of relevant shape elements: the *GS* parameter is given for each degree of internal and external diameter. The *GS* parameter value is a measure of the deviation of measured internal or external diameter D_a from a certain reference diameter D_r. Therefore *GS* is equal to: $GS = D_a - D_r$. The *GG* parameter specifies the straightness of the inner hole axis. When determining the uncertainty, the real deviation of inner hole straightness as well as deviation of the shape of the

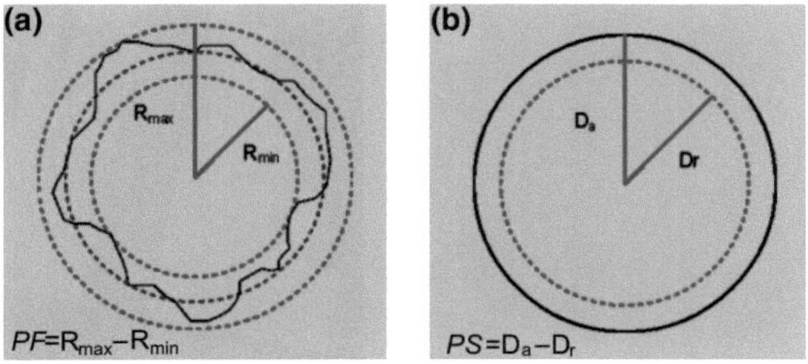

Fig. 6.14 Graphic illustration of scanning error: **a** form *PF*; **b** size *PS* [37]

standard cross have to be taken into account. The *GF* parameter of the measured curve is defined as the extension of the radial deviation from the fitted shape. As a standard step cylinder it can be used here, with minimum five and maximum ten significantly different external diameters with a coaxial pass-through hole. In Fig. 6.15 a standard with five diameters and with the equal-sized hole in the middle is shown. Some requirements concerning the standard are formulated: the ratio of external radii to internal radii must be at least 5. The step heights should be equal and the ratio between the standard height and its largest diameter should be between 0.8 and 1.2. Diameter graduation has to be linear.

The value of standard calibration uncertainty with respect to sizes and form must be documented. Form deviation should be much smaller than the length measurement error given by the manufacturer. The measurement uncertainty for the validation is set by ISO/TS 23165 [21].

Parameters Defining the Resolution

There are two distinct types of resolutions: structural resolution and spatial resolution. *Structural resolution* for spatial measurements D_g is defined as the diameter (maximum diameter) of the smallest sphere with which can be determined the maximum value MPE of form scanning error *PF* and size scanning error *PS* given by the manufacturer. Structural resolution describes the size of the smallest structure, which is measurable with given tolerance. The smallest measurable sphere (specified by the machine manufacturer) is the normal structure. It is used to check the structural resolution. Declaration of structural resolution for coordinate measurements is closely related to other accuracy parameters and has a big impact on them. In principle the resolution is independent of coordinate machine detectors. Therefore the tomograph manufacturer is required to specify the value of the resolution parameter. Procedures for determining the structural resolution are not currently specified in standards. However, in requirements of VDI/VDE 2617 Sheet

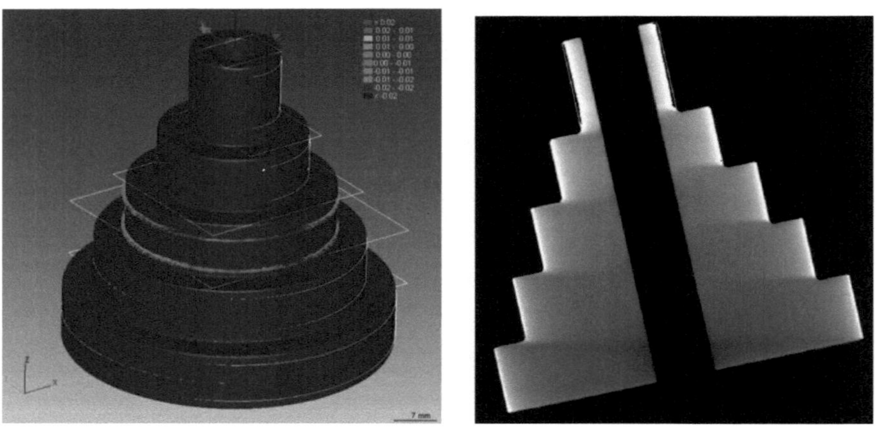

Fig. 6.15 Step cylindrical standard for determination of *GS*, *GG*, and *GF* parameters [37]

6.1 and Sheet 6.1 [67, 71] for optical coordinate machines procedures for determining the resolution are described, and therefore they may be suitable for CT scanners. An example procedure based on the mentioned requirements can be found in VDI/VDE 2617 Sheet 1.3 [67].

For spatial measurements, when data are derived from measurements with a CT scanner for a given maximum value of a process, the structural resolution in a Voxel's gray-scale does not cover the whole measurement chain for spatial measurements. To obtain certain information of the geometry the structural resolution for spatial measurements should generally be distinguished from Voxel's gray-scale resolution. Structural resolution in Voxel's gray-scale is described (ISO 15708) by a modulation transfer function (*MTF*). The *MTF* parameter at 10 % of modulations in units of line pairs per mm (lp/mm) estimates the maximum visibility of the structure for measurement data from CT. High resolution of the structure in the Voxel's gray-scale may favorably affect spatial measurements of CT, but it is not a sufficient condition for good metrological properties of the measurement. Structural resolution is affected by the following factors: the size and shape of the focus (focus spot) of the X-ray tube, that depends in turn on the voltage and current, prefiltration of X-rays, zoom, detector pixel size, number of sections (projections) when downloading data, filtering operations, averaging, calculation functions, and thresholds for the designation of the surface.

Spatial resolution describes the smallest measurable change (movement) in a given measuring direction (x, y, and z). Spatial resolution is included both in length and in scanning measurement. Factors affecting the spatial resolution are zoom, quantization of the A/C system noise, and software algorithms such as Subvoxeling.

6.2 Standards Used for Reverification and Interim Tests for CMM

The choice of a standard for reverification and interim tests depends on the requirements of the quality system of the institution. They should incorporate the ISO 10360 standard or VDI/VDE recommendations and be carried out in an optimal way both in terms of uncertainty and measurement time, as well as in terms of the information about the tested system that the user wants to obtain [3, 10, 26–29, 40, 60, 62]. In the case of standard uncertainty, the ISO 10360-2:2009 standard recommends that uncertainty of reproduction of the given length by a standard should not exceed 20 % of the maximum permissible error of the tested length.

One-dimensional standards, including the end standards, are characterized by the highest accuracy. Thus far, for interim tests it was recommended to use ball-bar type standards. These standards are made mainly of carbon fiber with many ceramic balls (Figs. 6.16 and 6.17). Recent changes in the current ISO 103602:2009 standard [18] create the possibility of using them in reverification tests in the normal calibration procedure. They are mainly used for large machine examination but also

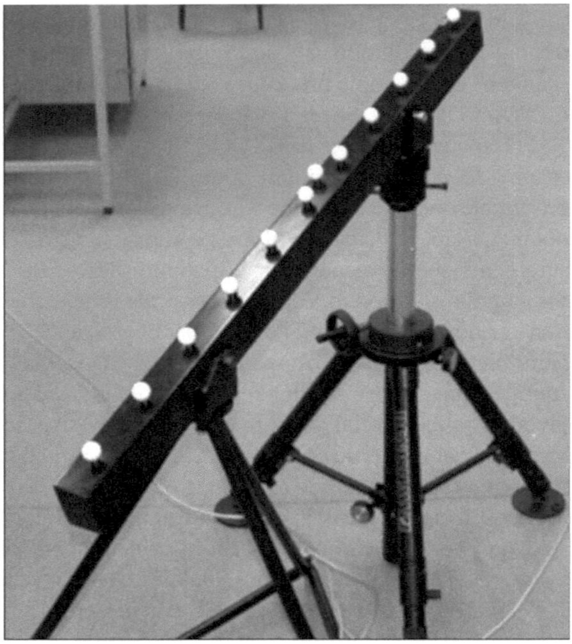

Fig. 6.16 Ball-bar type standard of a length of 1 m used in the LCM CUT

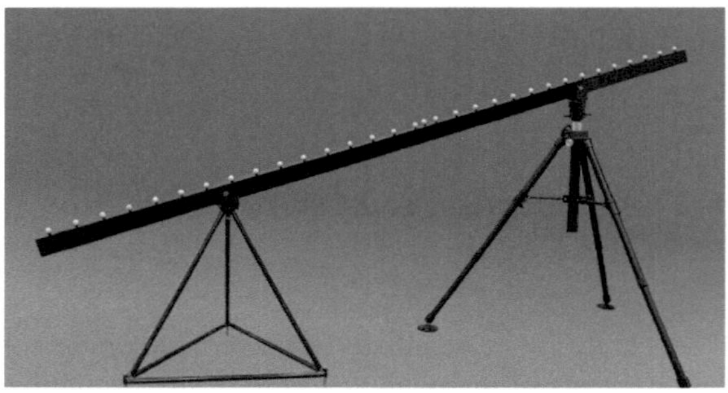

Fig. 6.17 Ball-bar type standard of a length of 3 m used in the LCM CUT

for redundant systems or less accurate machine testing. They are easy to transport, and the dimension of 3 m is sufficient to calibrate most of the CMMs, fulfilling the requirements of the ISO standard through two displacements of the standard. Standards made of carbon fiber are resistant to thermal changes; however, following the requirements of the new version of the standard the results of their measurements should be calculated for the value of the expansion coefficient for steel. They are used in conjunction with the necessary tests of step gauges made of steel of the range above 500 mm; that meets the requirements of the standard (in LCM KOBA a step standard of the range of 620 mm is used). The use of KOBA ball-bar rod standards (Figs. 6.18 and 6.19) has practically been abandoned due to the wearing

Fig. 6.18 Rod ball-bar standard, KOBA [22]

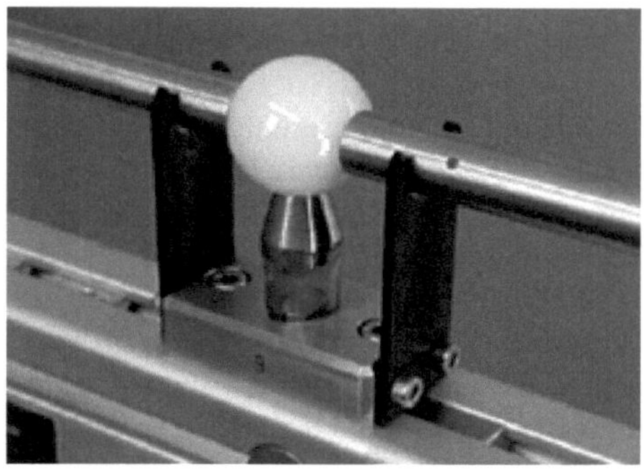

Fig. 6.19 Mounting of the reference sphere rod standard, Retter [38]

of mounting sockets. The simple tube and L-shaped tube standards, developed in NPL and shown in Figs. 6.20 and 6.21, are also rarely used.

If the measuring volume of a machine requires longer standards, kinematic standards with translator pairs are used [13, 14, 42–44]. Their name comes from the fact that movement is indispensable for length determination. The standard of the length here is the laser light wave. The probe head of the laser and the

Fig. 6.20 L-shaped standard developed in NPL under [28]

Fig. 6.21 One-dimensional ball standard, developed in NPL under [28]

interferometer are mounted on the fixed part. The retroreflector and reference element (sphere or reference plate) are placed on the moving trolley traveling on a linear guide. The representation of the length by a standard of this type is done by measuring the reference element in the zero and target position. An example is the laser step gauge (LSG) [23, 47, 51–53] developed at Cracow University of Technology by M. Krawczyk, presented in Fig. 6.22.

When using this type of standard, because of their construction they must be calibrated with consideration to a software correction of geometry errors. Therefore, it should be possible to identify geometric errors of coordinate machines also during their use [23, 47]. LGS has such capabilities, so it meets the requirements of ISO 10360–2:2009 standard [18].

Another group consists of two-dimensional standards shown in Figs. 6.23, 6.24, 6.25 and 6.26. They allow us to obtain results of a similar character only in two different locations [11]. It should be noted, however, that the uncertainty of the calibration of such standards is higher than the calibration of one-dimensional standards. It is related both to the reference elements used, usually sphere-shaped, as well as to the need to determine two coordinates of each reference element. Dimensions of these standards predispose them to machines of small and medium ranges of measuring volume (up to 2 m). High rigidity of plate standards is an important advantage that makes the mounting system much simpler than in the case of one-dimensional standards, where the two-point support in Aire's points is recommended. In reverification and interim tests the procedures with usage of this type of standard are described in VDI/VDE 2617 recommendations Part 5 [69]. The best results are obtained for thermostable standards made of Zerodur (6.24), Robax (Fig. 4.10), or carbon fiber (Fig. 6.23) when the change in coefficient of expansion of the material in relation to steel is included.

The last group of standards consists of three-dimensional standards. In interim tests of CMS accuracy it is enough to carry out the measurement of all reference elements in one orientation. In this case it is possible to obtain two kinds of information related to the error of the tested system. One of them is the test of errors associated with the measurement of the length between the various reference elements. Due to their spatial distribution their number may be much smaller than for

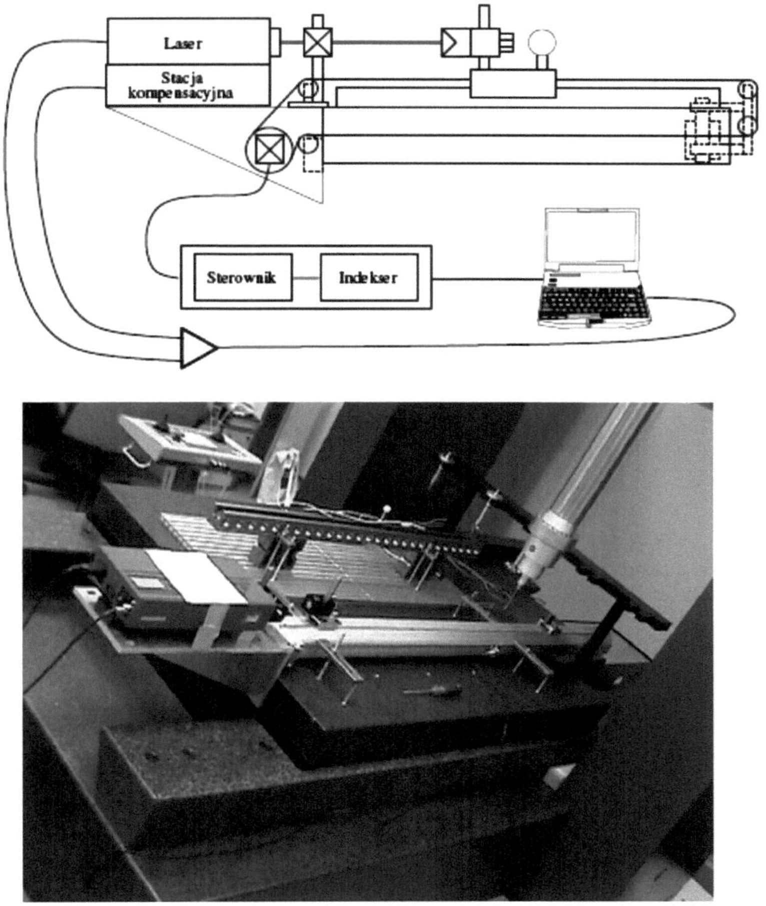

Fig. 6.22 Laser step gauge, the operation scheme (*above*), and during calibrations on PMM12106 (*below*) [23, 47]

two- or one-dimensional standards but still it gives same amount of information of the machine state. The second one is the information on the coordinates of centers of spherical elements.

This allows us to determine errors associated with the reproduction of the point in measuring volume by the tested CMM. An example of a three-dimensional standard is a *thetraedr* type standard shown in Fig. 6.27. As the name indicates this is a regular tetrahedron with spherical elements on vertices. The standard data are given as six lengths representing the distances between every four reference elements. The obtained lengths allow us to check if the equation of maximum permissible errors is met in the CMM measuring volume. On the other hand, as presented above, the determined lengths give us the opportunity to determine the

Fig. 6.23 Plate standard 960 × 960 mm Spider Plate (PTB) [30]

Fig. 6.24 Plate standard, hole plate made of Zerodur [76]

coordinates of centers of all measured spheres. Making a comparison between these coordinates and the coordinates obtained on the basis of calibration data it is possible to obtain the position deviations of particular spheres. This gives the basis for the determination of six basic geometric errors, which are usually corrected. These are the three position errors *(xtx*, *yty*, and *ztz)* and three errors of mutual axes orientation (*xwy*, *ywz*, *zwx*). The similar application has the Quader type standard shown in Fig. 6.28 and used in interim tests of arm machines developed in accordance with [25, 48, 50, 55–57].

Fig. 6.25 Retter sphere standard used in the LCM CUT and ATH Bielsko-Biala [38]

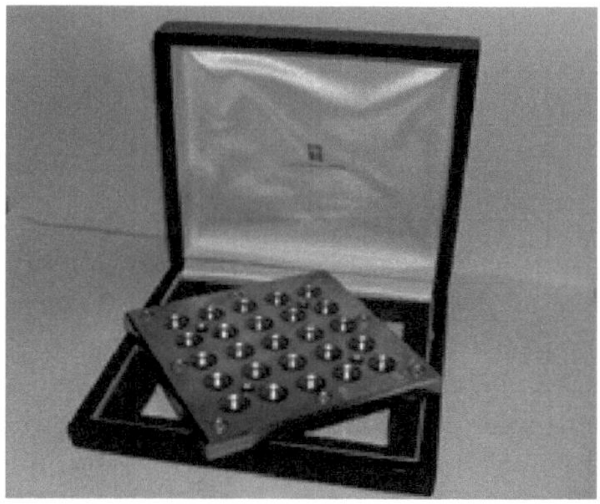

Fig. 6.26 Danish IPU plate standard for calibration and monitoring of coordinate optical systems and certain multisensory systems used in the LCM CUT [17]

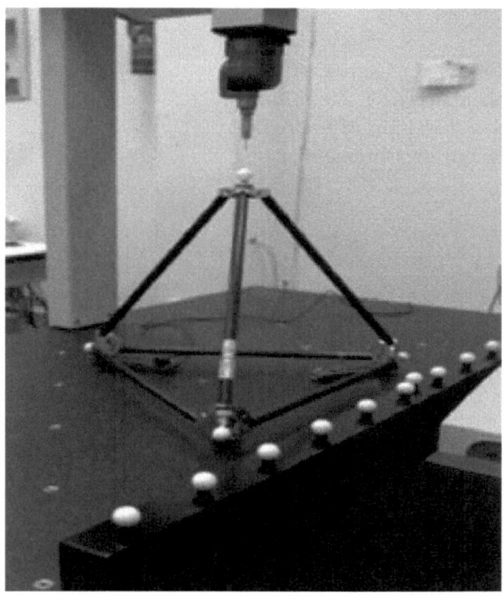

Fig. 6.27 Unimetrik standard of *thetraedr* type [66] used in the LCM CUT

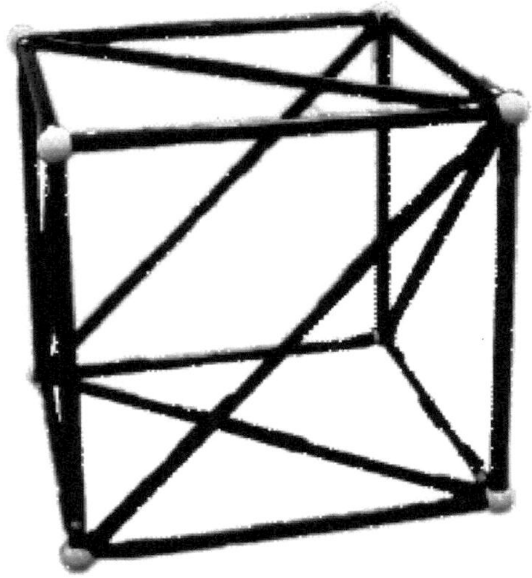

Fig. 6.28 Unimetrik standard of Quader type [66]

To determine length measurement error E for tomography systems more complex standards can be used as shown in Fig. 6.13, the standard with sphere elements that allows us to determine more dimensions (Fig. 6.29).

To determine length measurement error E step plate or spatial standards are also used. When selecting a standard, attention should be drawn to its radiation absorption coefficient. The standard's roughness should be negligibly small. For different materials and different absorption coefficients of the standard the determined error value of length measurement may be different. In Fig. 6.30 the plate standard developed by PTB (Physikalisch-Technische Bundesanstalt) and BAM

Fig. 6.29 MetrotomCheck standard (IMT Zeiss) for the determination of tomography length measurement error E

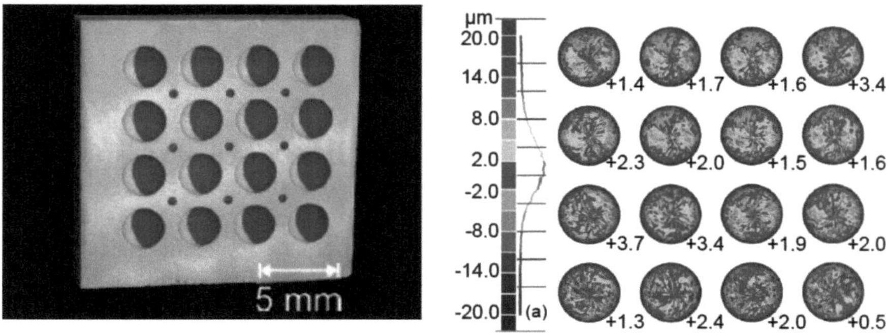

Fig. 6.30 Plate standard with spherical reference elements, tomographic measurement results with the determination of sphere shape errors [8]

(Federal Institute for Materials Research and Testing) is shown. It has spherical reference elements in the form of internal spheres. This standard, of the size of 20 × 20 mm, is made of Zerodur and has a very low susceptibility to temperature influences, because its expansion coefficient is $\alpha = 0 \pm 0.1 \times 10^{-6}\ K^{-1}$. It contains 16 standard elements, and sphere form errors not exceeding 2.5 μm, positioned with an accuracy of 1.5÷2 μm. In this figure in accordance with [8] results of these standard measurements are also shown as areas of form errors inside spheres. The numbers next to them present values of their center distances. Sphere form deviation values are shown in the vertical diagram.

Figure 6.31 presents the spatial standard in the form of a cube for testing the accuracy of Feinmess GmbH & Co. KG industrial tomographs, made of titanium alloy, with dimensions of 10 × 10 × 10 mm^3, inner spheres in three areas, and grid 5 × 5 (25 spheres for each surface); the sphere radius is 0.4 mm.

Fig. 6.31 Feinmess GmbH & Co. KG spatial standard with spherical reference elements [16]

6.3 Monitoring of Coordinate Measuring System Accuracy

Monitoring of the accuracy of a coordinate measuring system (CMS) in industrial environments is often reduced to a reverification test, realized every year or even every two years. Such a test is often carried out by the manufacturer's service after a periodic inspection. Often during the test for certain maximum permissible errors the service makes CAA correction of CMM accuracy of recalibration. This is always carried out by the same service, which realized the accuracy correction without the necessary evaluation of uncertainty of the calibration. All these activities should be considered inappropriate actions. Of course the manufacturer is a party in the accuracy assessment, and does not guarantee the independence of the carried-out assessment. Therefore, in accordance to standard requirements previously specified in Sect. 2.4, such tests should be carried out by independent laboratories (not related to the manufacturer or the user of the system) that have management systems accredited by, for example, PCA, for compliance with the requirements of PN/EN/ISO 17025:2005 [34]. The CMS manufacturer or its specialized service can and should provide periodic technical service, ensuring the maintenance of system accuracy. It should not, however, perform the reverification test and calibration related to it, because it is not an objective party in this action.

During analysis of the reverification test results the question often arises as to when the system lost the specified accuracy and for how long a period of time the measurement results were incompatible. These are key questions for the industrial plant because in extreme cases they can be a cause of calling into question the conformity of parts produced with their specification. The problem of manufacturer responsibility for measurement equipment, the accuracy of which can be questioned, is undisputed and can lead to obvious, in this case, economic losses. If this situation repeats, the length of the period of time between tests should be changed in such a way that the time in which the machine can remain outside the specification would be as short as possible.

The designation of the period of time between calibrations is associated with frequent CMS accuracy testing in the regular cycle of interim tests. The length of the period of time after which a more precise test is performed should be designated individually for each CMS. In this case the following parameters should be taken into account: device workload, number of operators, failure frequency, and environmental conditions. When the machine is more charged and more operators use it, then the period of time between tests should be shorter. If proceeding in accordance with ISO 10360-2, it is possible to use to this aim the shorter procedure described in the standard as an interim test. In this case, the standard allows the use of artifacts of a different construction than the standards required during the reverification or acceptance test. For this purpose such standards can be used: plate standards and ball-bar type standards (Figs. 6.19 and 6.20), as well as the kinematic length standards. The procedure itself is less time consuming than the reverification test. Depending on the standard type and the number of its orientations the number of repeats may be from one, when the standard is three-dimensional, to four different orientations, when the standard is one-dimensional. The ISO 10360-2 standard itself recommends the interim test as the requirement of a proper monitoring of system accuracy. It should be carried out in between reverification tests.

The use of three-dimensional standards for monitoring the CMS condition drastically reduces the time interval necessary to carry out measurements, so that the interim checks can be implemented under industrial conditions. According to ISO 10360-2 measurements should be carried out several times during the time period between reverification tests. The use of these types of standards allows using the same acceptance procedure for a variety of CMS. That makes it possible to compare the accuracy of each other. This is especially important if in the plant's various measuring systems the coordinate measuring principle is used for a similar or the same measuring tasks. This enables the comprehensive monitoring of coordinate measuring system accuracy.

It should be noted that the results obtained from the interim check cannot be treated in the same way as the results from the acceptance or reverification tests. This is generally due to the shorter form of such a test, and in many cases also due to increased uncertainty of the length representation by the standard used. If obtained results show that the tested device exceeds the maximum permissible error, additional tests should be performed to confirm the results. Preferably in this case the reverification test procedure is to be conducted.

An interesting issue that has been recently noted is the need for monitoring of the installed virtual machine system. Generally this monitoring is performed by the same laboratory that installed the virtual machine system, such as Feinmess for Leitz Messtechnik machines or Eumetron for Zeiss machines. A plate standard is generally used for monitoring but measurements are carried out according to the simplified procedure or with the use of *thetraedr* type standard. This is schematically presented in Fig. 6.32.

Set test deadlines must be especially strictly observed and the use conditions monitored, which is also confirmed by experience of the LCM CUT in maintenance of their own virtual models for PMM 12106 and DEA machines.

6.3.1 Monitoring and Correction of Thermal Effects

Increasing the accuracy of CMM measurement and technical difficulties associated with the provision of reference conditions in production halls caused the great interest in systems of thermal effect corrections on CMM accuracy. Although these systems are more and more complex and more often applied, the problem has not been completely solved yet [5–7, 9, 12, 24, 25, 35, 36, 39, 41, 45, 50, 55–57, 59, 63–65, 74, 75]. The causes must be searched in a poorly designed air-conditioning

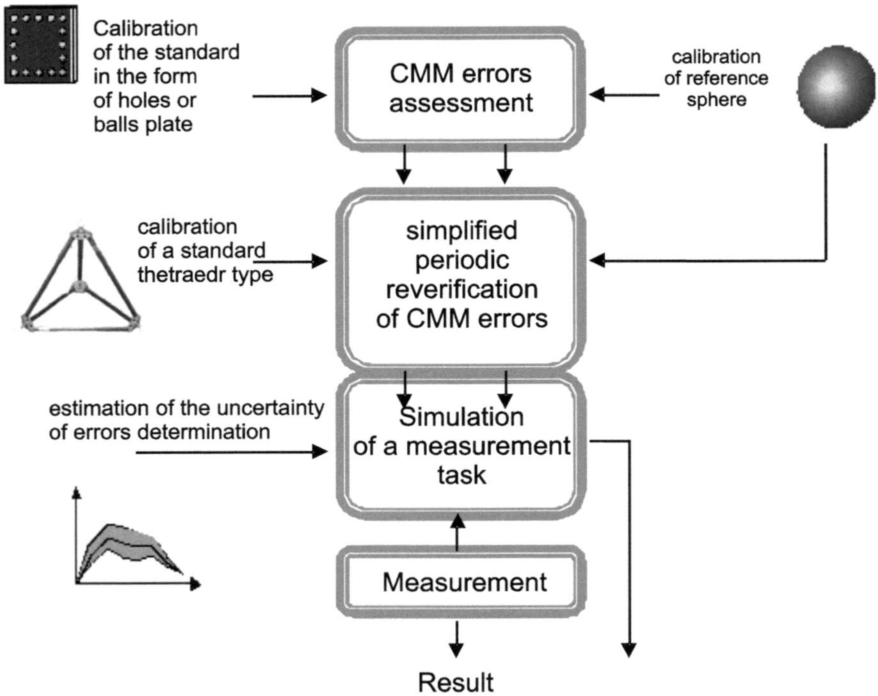

Fig. 6.32 The scheme of monitoring of the virtual machine system [46]

system or in its unstable operation, often associated with its turning off or on—for example, the lighting—in an air-conditioned room. Another problem is the use of machines of older construction or machines purchased without a system of active compensation of thermal effects, while expecting higher measurement accuracy.

Under the work in [25, 48] a system was designed that allows the external correction of thermal effects on the measurement accuracy of any mid-size CMM without interfering with the machine construction. The idea of the system is to develop a model of the error field of the coordinate measuring machine under varying thermal conditions, using a thermostable standard and external temperature measurement sensors. To build this system, first the CMM error field should be identified, and next the thermal effects on the construction deformation should be modeled in order to develop an effective system of their correction under exploitation conditions. It is assumed to get the fixed maximum errors MPE of the given CMM. It is also possible to use this system to reduce the thermal effects similarly as was solved in the case of so-called active correction systems mounted in more accurate machines in their construction nodes.

It is known that the thermal effects result in CMM element deformation, causing an increase of its geometric errors (Sect. 3.5, Figs. 3.25–3.27). It is obviously the cause of temporary changes of its measurement accuracy. The effect of the functioning of the thermal effects correction system and its impact on CMM indication errors is presented in Figs. 6.33 and 6.34. These studies were carried out for

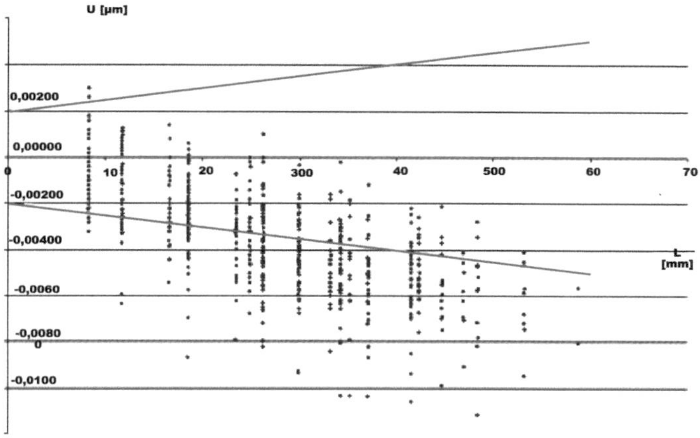

Fig. 6.33 Chart of indication errors before the use of the temperature effect correction system ACTIV for machines of the DEA Global series; the area of maximum permissible errors (MPE) is marked [48]

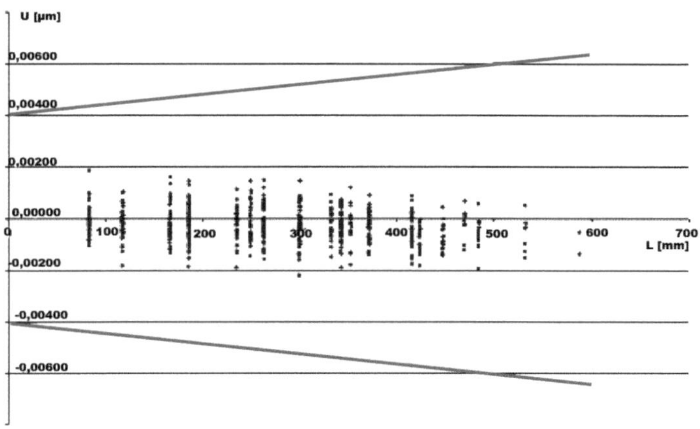

Fig. 6.34 Chart of indication errors after the use of the temperature effect correction system ACTIV for machines of the DEA Global series; the area of maximum permissible errors (MPE) is marked [48]

machines of the Global series equipped with the ACTIV system. The graphs present indication errors obtained for the machine with the temperature effect correction system switched off and on. Also marked is the area of maximum permissible errors to enable observation of how the used correction system affects the machine accuracy.

The analysis of reported results demonstrates how important the discussed problem of thermal effects is to CMM accuracy, and simultaneously to the accuracy of realized measurements.

The basis of the CMM error field model is a database founded on the long-term monitoring of thermal conditions using a specially designed placement method of a set of sensors for a given CMM and its workspace. The analysis of temperature changes in the CMM environment allows the development of a model of CMM thermal effects and the determination of maximum temperatures, for which the CMM accuracy tests are planned using the thermostable standard shown in Fig. 4.10.

It is very important to determine the maximum temperatures for which it is necessary to use the correction of CAA matrix, that is, to determine CMM accuracy changes in relation to its maximum permissible errors. The system was designed so that the information on the current CMM temperature would be transferred by the Internet to the monitoring laboratory server. Next the system installed on the server would send back correction data of a new CAA matrix to the computer controlling the accuracy of coordinate measuring machine. This is illustrated in Fig. 6.35.

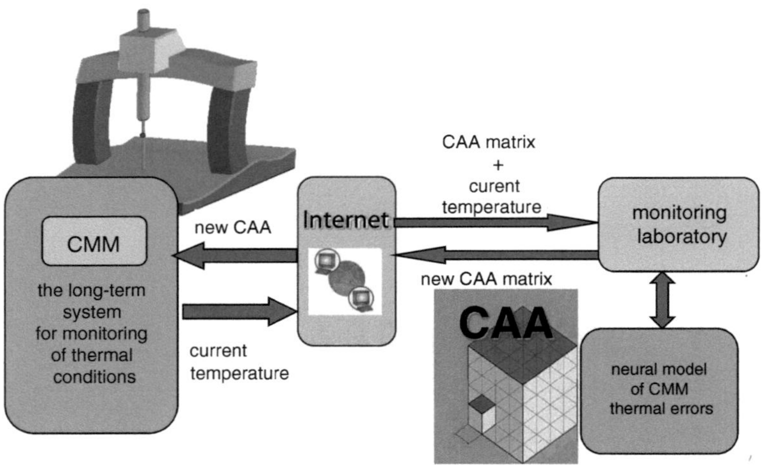

Fig. 6.35 Scheme showing the concept of the correction system of thermal effects using the Internet data flow between the corrected CMM and monitoring laboratory

The novelty is a model of thermal effects based on the use of artificial neural networks calculating correction factors of the CAA matrix for measured temperature.

6.3.2 *The Internet Platform Tracesys.eu.—A System of Metrological Monitoring of Coordinate Measurement Systems*

The importance of monitoring of coordinate system accuracy for widely understood production quality was confirmed by long-term studies of PTB and other European metrology institutes and key laboratories. The result was the development of the European project: TRACESYS [60].

The project had the task of implementation of the interim test on the European market. Under the project an Internet platform was developed that was dedicated to help machines and other coordinate systems users in implementation of an interim test by a simplified measurement procedure and by use of standards that do not require advanced metrological knowledge from the user. The measurement itself is carried out by an operator who usually operates a given measurement system. The Internet platform is designed additionally to support the user. Thanks to it the service provider—accredited laboratory—does not have to be present at every interim check. On the other hand the user is not left alone during tests and interpretation of the results. The user also receives support in the case of obtaining a negative result of interim tests concerning the determination of the nonconformity causes. If the user implements the TRACESYS monitoring system at his or her site, appropriate training on the Internet platform and standard operation is guaranteed. During the training the measurement program for the supplied standard testing is also installed at the site. Then, for the realization of interim tests the user receives on the specified date the standard for measurement. Obtained results are introduced to the platform TRACESYS through a Web browser. For the user the platform is a combination of a database of his measuring devices and an application for analysis of measurement results. The user himself and the calibration laboratory have access to the user's devices. In the case of the service supplier the platform is used for systems monitoring and management of standards and data (test results) of particular users. The user also receives the shortened information on interim test results. Example data for the device entered into the system are shown in Fig. 6.36.

The obtained measurement results allow us to forecast the measurement accuracy of the tested device and changes in a longer period of time, and makes it possible to determine the length of the period of time at which the reverification test should be carried out. The exemplary data obtained from the interim test with use of the standard type *thetraedr* are presented in Fig. 6.37.

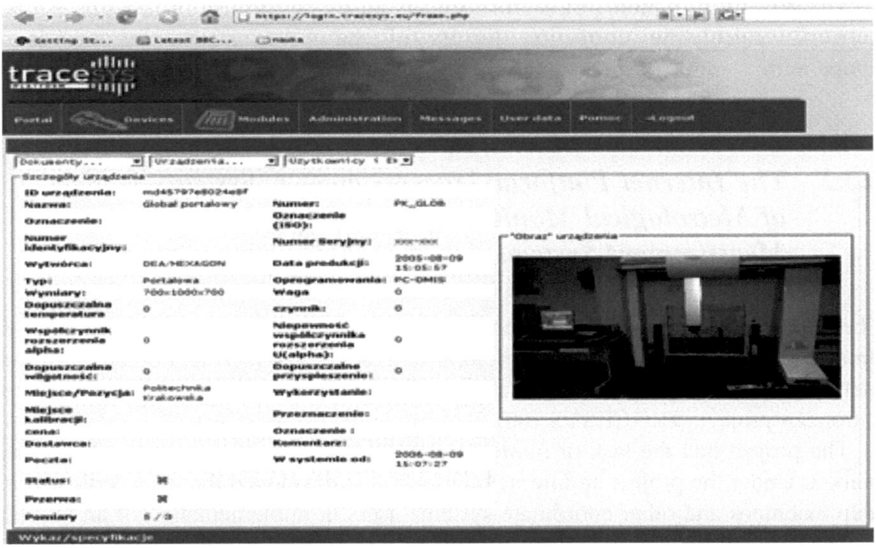

Fig. 6.36 www.tracesys.eu portal: The page with the characteristic of the monitored machine

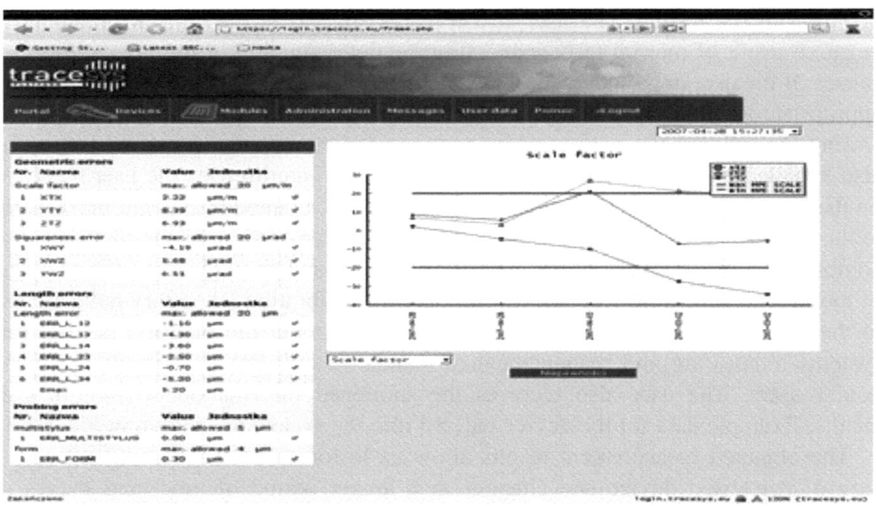

Fig. 6.37 Monitoring results for a standard of *thetraedr* type

In addition to PTB and CMI in the project TRACESYS [54, 60] these accredited calibration laboratories have also participated:

1. Physikalish-Technishe Bundesanstalt—As a coordinator
2. Laboratory of Coordinate Metrology at Cracow University of Technology
3. Unimetrik from Spain
4. IPU—Spin-off company of Technical University of Denmark
5. Cermet—Italian calibration laboratory
6. CMI—Czech Republic National Metrology Institute
7. Feinmess—German calibration laboratory
8. Etalon AG company
9. VUO—Czech research laboratory

The Internet thermal effects correction system described in Sect. 6.2.1 is the essence of the TRACESYS[1] monitoring system.

In Poland work on implementation of the metrological monitoring system of the accuracy has been done under the development project R03 029 01 [49], which also covered the development of a Web portal tracesys.eu interface.

The long-term aim of the monitoring is to keep CMS accuracy in established limits under the whole period of its use. The implementation of the metrological monitoring system of CMS accuracy, MMS-CMS, with the use of the platform Traces allows better use of the interim test to monitor the coordinate measuring system. It also reduces the danger of the use of devices, the accuracy of which is not compatible with their specification, and therefore manufacturer responsibility for the product. This allows accuracy control over a longer time, permitting the appropriate selection of the period of time between reverification checks. In the case of having a larger number of CMS of different construction it is possible through the use of appropriate standards to compare their accuracies, and thereby to establish their mutual measurement consistency. The implementation of a MMS-CMS system by a given laboratory stabilizes the accuracy of carried-out measurements and simplifies the development of an effective procedure for estimating measurement uncertainties. Finally, the importance of close cooperation of the company's service with coordinate system manufacturers should be distinguished. It is a key element in maintaining the accuracy of the coordinate system by maintaining good technical condition and quick response to failures, and above all, accuracy correction in the case of a negative result of the reverification test, which is connected with the redetermination of the matrix of geometric errors.

[1]TRACESYS system is available in the Internet platform on the address www.tracesys.eu.

6.3.3 Role of Accredited Calibration Laboratory in the Coordinate Measuring System Accuracy Monitoring

In accordance with the modern concept of the management system through quality, the monitoring of CMS accuracy is performed by accredited laboratories. Their role and importance and normative regulations are described in Sect. 2.4 and also in Sect. 6.2. The scheme of the monitoring of CMS accuracy with the participation of a calibration laboratory, the companies' service, manufacturers, and the Internet monitoring system are shown in Fig. 6.38. To provide an objective complex monitoring of the coordinate system's accuracy, it is necessary to build a network of accredited calibration laboratories in our country. Currently (year 2014) in Poland there are three laboratories with accreditation scope of calibration of coordinate systems and one laboratory calibrating objects with coordinate methods. Only the Laboratory of Coordinate Metrology at Cracow University of Technology is accredited in both areas, and its accreditation scope covers:

1. Calibration and verification of coordinate measuring systems
2. Calibration and verification measurements with the use of coordinate technique

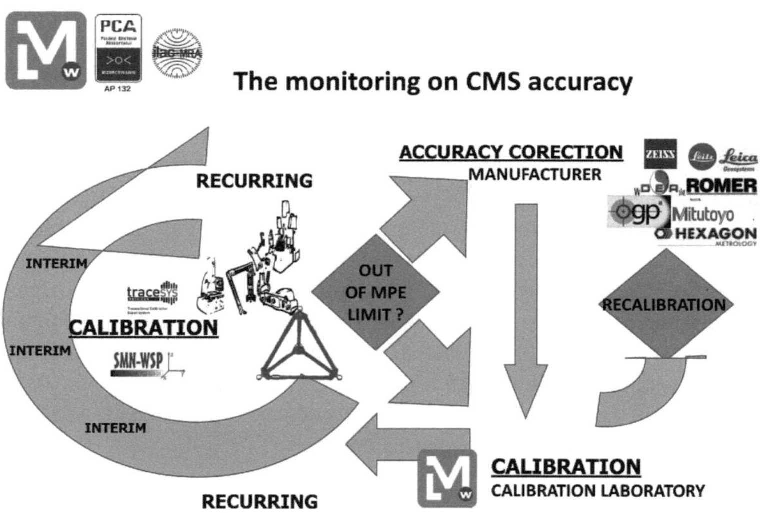

Fig. 6.38 Monitoring system of CMS accuracy with the participation of calibration laboratory, TRACESYS and its national implementation SMN-SYS [49]

The key objectives for the accredited laboratory are staff training and necessary equipment that must be systematically completed with new calibration standards and calibration technologies meeting changing requirements. Currently the LCM CUT is equipped with a set of standards that allows the calibration of almost all coordinate systems. For interim checks the *thetraedr* type standard is used, in ranges of 0.5 m and 1 m. In tests a thermostable plate made of Robax with a coefficient of expansion 0.5 μ m/m/K of Feinmess (Fig. 4.16 and Annex C) is used. It makes it possible to conduct the procedure consistent with VDI/VDE 2617 Part. 5, but mainly it is used to monitor the accuracy of virtual models and interim control of PMM12106 machine accuracy. In order to perform reverification checks the laboratory has step gauge type standards of ranges of 620 and 1024 mm of KOBA and standards of a ball-bar type of a range of 1 m and 3 m. The 3-m standard is used for the verification and calibration of large machines and laser trackers, Figs. 6.39, 6.40, 6.41, 6.42 and 6.43.

Etalon AG's laser tracer has an uncertainty of $U = 0.2 + 0.3$ L and measurement range 0–15 m also used in the LCM; it is described in Sects. 4.2.3 and 6.1. For the calibration of optical scanners of a structural light system, reference spheres (ceramic and steel balls) of ball-bar type are also used. They have a nonreflective surface and they are presented in Fig. 6.44. Such standards are also used to test the CMS measuring point identification system with an optical distance sensor (laser triangulation probe heads) [71].

Measurement consistency is provided by the use of KOBA step standards (Step Gauge type) of sizes 1024 and 620 and the uncertainty $U = 0.1 + 0.5$ L μ m. Measuring machine PMM12106 from Leitz Messtechnik is used for the maintenance of the calibration chain for element calibration with usage of coordinate methods, and for calibration of large elements, the Laser Tracker Leica LTD 840 equipped with T-Probe. The maintenance of the constant measurement capability and CMC verification provides an interlaboratory comparison system.

Fig. 6.39 Standards for testing LCMM ball-bar of Unimetrik and KOBA step standard used by the LCM CUT and meeting the requirements of the standard

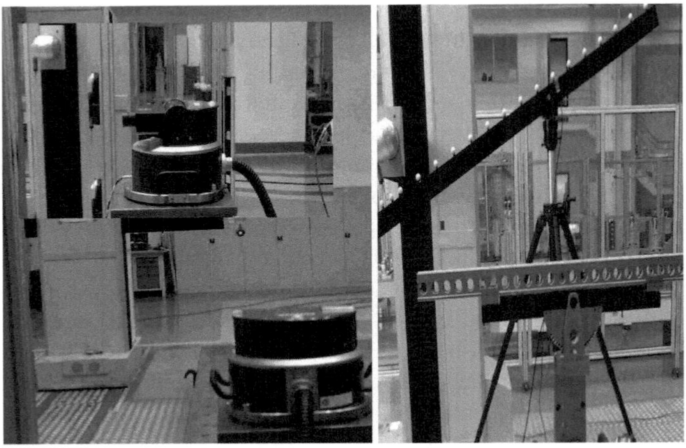

Fig. 6.40 Etalon laser tracer [15] and standards for testing Unimetrik's LCMM ball-bar and KOBA step standard used by LCM CUT and meeting the requirements of the standard

Fig. 6.41 Standards for testing Unimetrik's LCMM ball-bar and KOBA step standard used by LCM CUT and meeting the requirements of the standard dual ram system calibration

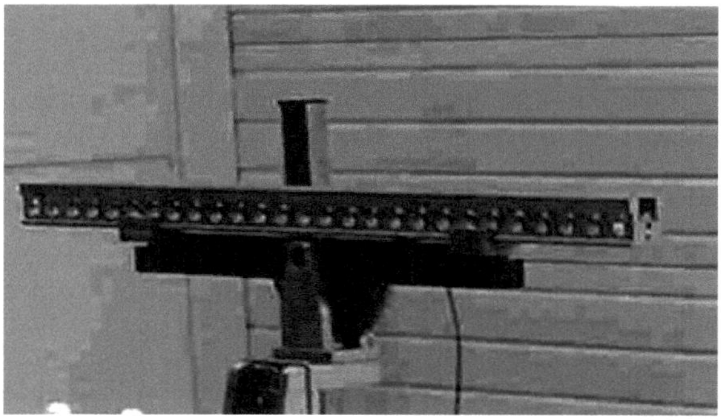

Fig. 6.42 KOBA step standard used by the LCM CUT

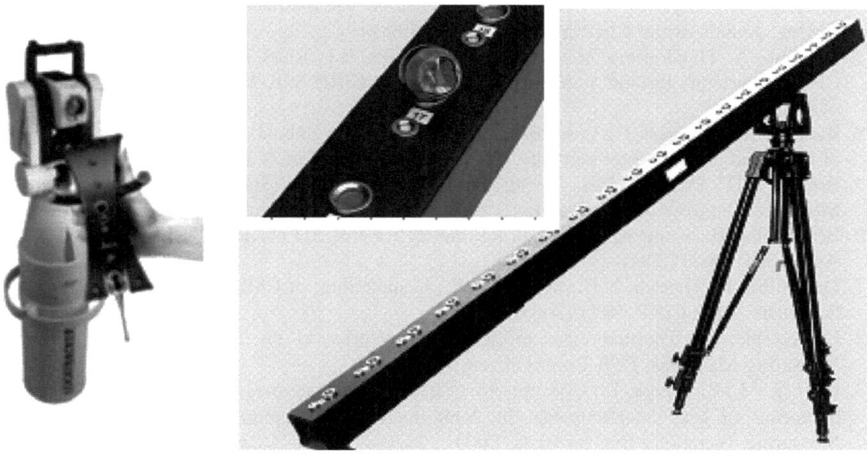

Fig. 6.43 Laser trackers Leica LTD 840 with T-Probe and 3-m standard for calibration of this type of Unimetrik system, used by the LCM CUT

Fig. 6.44 Standards for checking the Unimetrik optical ball-bar system, used by LCM CUT

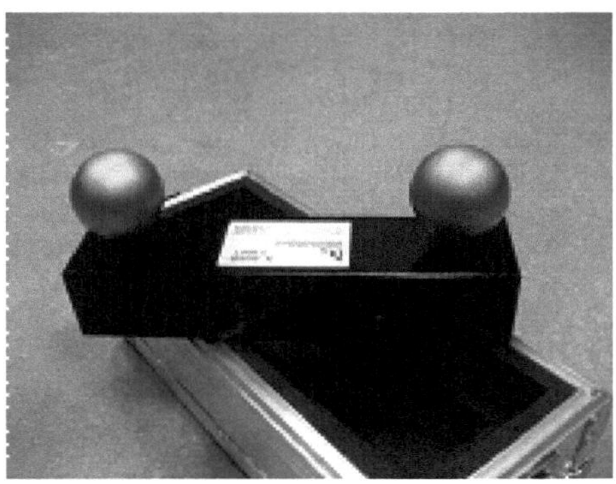

References

1. Abacerli, A.J, Pereira, P.H., Calonego, N.: A case study on testing CMM uncertainty simulation software (VCMM). J, Braz. Soc. Mech. Sci. Eng. **XXXII**(1/9), January–March 2010
2. Abbe, M., Verdi, M., Wanner, J.: Revision of CMM performance specifications. Paper of the International Association of Coordinate Measuring Machine Manufacturers (ia.cmm)
3. Arriba, L., Trapet, E., Bartscher, M., Frankem M., Balsamo, A., Costelli, G., Torre, S., Kitzsteiner, F., San Martín, F.: Methods and artifacts to calibrate large CMMs. In: Proceedings of the 1st International EUSPEN Conference, vol. 2, pp. 391–394. Bremen, ISBN 3-8265-6085-X, 1999
4. ASME B89.4.22 2004: Methods for performance evaluation of articulated arm coordinate measuring machines
5. Attia, M.H., Fraser, S.: A generalized modeling methodology for optimized real-time compensation of thermal deformation of machine tool and CMM structures. Pergamon Int. J. Mach. Tools Manuf. **39**, 1001–1016 (1999)
6. Balsamo, A., Marques, D., Sartori, S.: A method for thermal-deformation corrections of CMMs. Annals of the CIRP **39**(1), 557–560 (1990)
7. Balsamo, A., Di Ciommo, M., Mugno, R., Rebaglia, B.I., Ricci, E., Grella, R.: Evaluation of CMM uncertainty through monte carlo simulations. CIRP Ann. Manuf. Technol. **48**(1), 425–428 (1999)
8. Bartscher, M., Hilpert, U., Goebbles, J., Weidemann, G.: Enhancement and proof of accuracy of industrial computed tomography (CT) measurements. Ann. CIRP. **56**(1), 495–498 (2007)
9. Breyer, K.H., Pressel, H.G.: Paving the Way to Thermally Stabile Coordinate Measuring Machines. Progress in Precision Engineering, pp. 56–76. Springer, Berlin (1991)
10. Bringmann, B., Kung A.: A measuring artefact for true 3D machine testing and calibration. Ann. CIRP **54**(1) (2005)
11. De Chiffre, L., Hansen, N.H., Morace, R.E.: Comparation of CMM using an optomechanical hole plate. Ann. CIRP **54**(1) (2005)
12. Delbressine, F.: Thermal error modeling. In: Workshop on Traceability of Coordinate Measuring Machines, PTB Braunschweig
13. Dobosz, M., Ratajczyk, E.: Conceptions of CMM's distance measurement inaccuracy testing by means of laser interferometer. In: VIth International Scientific Conference Coordinate Measuring Technique Bielsko-Biała, 2004
14. Dobosz, M., Ratajczyk, E.: Conception of accuracy test for CMM's during measuring distance error checking by means of laser interferometer. In: VIIth International Scientific Conference Coordinate Measuring Technique Bielsko-Biała, 2006
15. ETALON AG, www.etalon.de
16. FEINMESS GmbH & Co. KG, www.feinmess.com
17. IPU, www.ipu.dk
18. ISO 10360 part 2—CMMs used for measuring linear dimensions, Third edition 2009
19. ISO 10360 part 5—CMMs using single and multiple stylus contacting probingsystems, Second edition 2010-09-15
20. ISO 10360 part 8—CMMs with optical distance sensors
21. ISO/TS 23165:2006—Geometrical product specifications (GPS) — Guidelines for the evaluation of coordinate measuring machine (CMM) test uncertainty
22. Kolb & Baumann GmbH & Co. KG, www.koba.de
23. Krawczyk. M.: Metoda oceny dokładności Współrzędnościowych Maszyn Pomiarowych z zastosowaniem Laserowego Wzorca Stopniowego (Coordinate measuring machines accuracy assessment method with the use of laser step standard) Ph.D Thesis Dissertation, Cracow University of Technology, Faculty of Mechanical Engineering 2005 Thesis Advisor—Prof. J. A. Sladek

24. Kunzmann, H., Schuster, H.J.: Temperature measurements in dimensional metrology. PTB, Bericht F-17 Braunschweig, Nov 1994
25. Kupiec, R.: Korekcja oddziaływań termicznych we współrzędnościowych maszynach pomiarowych (Correction of thermal interactions in coordinate measuring machines) Ph.D Thesis, Cracow University of Technology, Faculty of Mechanical Engineering 2005, Thesis Advisor—Prof. J.A. Sladek Giving a doctoral degree by the Council of Cracow University of Technology, October 2007
26. Lee, S.D., Chang, D.Y.,1D: Space ball array calibration methodology for CMM. In: 7th International Symposium on Measurement Technology and Intelligent Instruments. Journal of Physics: Conference Series, vol. 13, pp. 194–197 (2005)
27. Lim, C.K., Rapid volumetric calibration of coordinate measuring machines using a hole bar artefact. In: Proceedings of the Institution of Mechanical Engineers Part B, vol. 216, pp. 1083–1093 (2002)
28. MESTRAL Project Artefacts and methods to establish traceability of large CMMs EU PROJECT (SMT4-PL97-2330) Final Report
29. Osawa, S., Takatsuji, T., Kurosawa, T.: Uncertainty analysis of calibration of geometrical gauges. Precis. Eng. J. Int. Soc. Precis. Eng. Nanotechnol. **26**, 24–29 (2002)
30. Physikalisch-Technische Bundesanstalt, www.ptb.de
31. PN EN ISO 10360 Specyfikacje Geometrii Wyrobów (GPS) Badania odbiorcze i okresowe współrzędnościowych maszyn pomiarowych (CMM) (Geometrical Product Specifications (GPS)—Acceptance and reverification tests for coordinate measuring machines (CMM))
32. PN EN ISO 10360–2 Specyfikacje Geometrii Wyrobów (GPS) Badania odbiorcze i okresowe współrzędnościowych maszyn pomiarowych (CMM). Cześć 2—CMM Stosowane do pomiaru długości wyd. 2 (Geometrical Product Specifications (GPS) –Acceptance and reverification tests for coordinate measuring machines (CMM). Part 2—CMMs used for measuring linear dimensions)—Second edition 2003
33. PN/EN/ISO 14253—Kontrola wyrobów i sprzętu za pomocą pomiarów (Inspection by measurement of workpieces and measuring equipment)
34. PN-EN ISO/IEC 17025:2005—Ogólne wymagania dotyczące kompetencji laboratoriów badawczych i wzorcujących (General requirements for the competence of testing and calibration laboratories)
35. Ratajczyk. E., Skuteczność kompensacji wpływu temperatury na proces kalibracji współrzędnościowych maszyn pomiarowych (The effectiveness of compensation of temperature impact on the coordinate measuring machines calibration process). In: Proceedings of KKM 2001, Warsaw, vol. 1, pp. 71–74
36. Ratajczyk, E. The influence of temperature changes and effectiveness of their compensation in investigations of CMM accuracy. In: VI International Scientific Conference Coordinate Measuring Technique, vol. 10, pp. 259–268. University of Bielsko-Biała, Scientific books, 21–23 Apr 2004
37. Ratajczyk, E.: Tomografia komputerowa CT w zastosowaniach przemysłowych, cz. II Tomografy i ich parametry, przykład zastosowań (Computed Tomography CT in industrial applications, part. II tomographs and their parameters, example of applications) Mechanik, no. 3/2011, pp. 226–231
38. RETTER—automation +Messtechnik GmbH, www.retter.de
39. Ricci, E., Verdi, M., Siemieniako, W.: Testing of temperature influence on a large CMM based on real installation case. In: Proceeding of scientific-technical conference Metrology in Production Engineering, Cracow 25–27 IX 2003
40. Rupik, J.; Jendrollik, J.; Wieczorowski, M.: Wykorzystanie techniki współrzędnościowej do pomiaru precyzyjnych elementów wielkogabarytowych (The use of coordinate technique to measuring of precise large parts). In: Advances in Coordinate Metrology, pp. 115–124. University of Bielsko-Biała, Bielsko-Biała, 2010
41. Ryniewicz, A., Kowalski, M., Gawlik, J.: Theoretical and experimental aspects of temperature analyses of coordinate measuring machines. Postępy Technologii Maszyn i Urządzeń, Kwartalnik Komitetu Budowy Maszy, Polish Academy of Sciences **23**(4), 99–112 (1999)

42. Schüssler, H.H.: Einsatz von Prüfkörpern zur Überwachung von Mehrkoordi-naten-Messgräten. VDI-B 378, pp. 29–38 (1980)
43. Schüssler H.H.: Absolute Kalibrierung von Stufenendmassen und Endmassen mit einem zweiachsigen Laserinterferonmeter-Komparator. Technische Messen **50**(5), 191–195 (1983)
44. Schüssler H.: Prüfkörper für Koordinatenmessgeräte Werkzeugmaschinen und Meßroboter. Techn. Mess 51 J. 84 H.3
45. Sładek, J.: Współrzędnościowa technika pomiarowa w nadzorowaniu jakości wyrobów (Coordinate measuring technique in the supervision of product quality). In: Tabor, A., Zając, A., Rączka, M. (eds.) Zarządzanie jakością – metody oceny jakości wyrobów technicznych (Quality management—methods for assessing the quality of technical products), textbook for university students, vol. IV, pp. 249–290, Carcow university of Technology, 2000
46. Sładek, J., Modelowanie i ocena dokładności maszyn oraz pomiarów współrzędnościowych (Modeling and accuracy assessment of coordinate machines and measurements) Cracow University of Technology Scientific Books—Mechanika, no. 87, Cracow 2001
47. Sładek, J.: Metoda oceny dokładności współrzędnościowych maszyn pomiarowych z zastosowaniem laserowego wzorca stopniowego (Coordinate measuring machines accuracy assessment method with the use of laser step standard)—PhD grant no. 4 T07D 015 27 Ministry of Science and Higher Education (2005)
48. Sładek, J.: Korekcja oddziaływań termicznych we współrzędnościowych maszynach pomiarowych (Correction of thermal interactions in coordinate measuring machines) – PhD grant no. 4, T07D 015 30 Ministry of Science and Higher Education (2007)
49. Sładek, J., et al.: System metrologicznego nadzoru nad dokładnością maszyn pomiarowych robotów i obrabiarek z wykorzystaniem wzorców interferometrycznych laserowych systemów śledzących jako podst. wzrostu jakości produkcji przemysłu maszynowego (Metrological monitoring system for accuracy of measuring machines, robots and tools using interferometric standards of laser tracking systems as a base for a growth of production quality of machinery industry) Research and Development Project no. (R03 029 01): implemented in 2006–2009
50. Sładek, J., Kowalski, M., Kupiec, R.: Monitorowanie warunków termicznych przy ścisłej klimatyzacji pomieszczeń do kalibracji maszyn pomiarowych. Problemy jakościowe, energetyczne i eksploatacyjne w maszynach cieplnych (Monitoring of thermal conditions in high air-conditioning of spaces to calibrate measuring machines. Quality, energy and operating problems of thermal machines). In: Proceeding Bydgoszcz—Duszniki Zdrój 2–5, pp. 129–136 (2004)
51. Sładek, J., Krawczyk, M.: Identyfikacja błędów wielkogabarytowych WMP z uwzględnieniem laserowych systemów nadążnych (Identification of errors of large CMM including laser tracker systems) IX scientific-technical conference Metrology in Production Engineering, pp. 511–524. Czestochowa University of Technology, 17–19 IX 2001
52. Sładek, J., Krawczyk, M., Modeling and assessment of large CMM's accuracy. In: XIII IMEKO World Congres—Metrology in the 3rd Millenium, Cavat—Dubrovnik, Croatia, Jun 2003
53. Sładek, J, Krawczyk, M.: Model of uncertainty for laser Gauge as an endgauge. Adv. Manuf. Sci. Technol. **29**, 59–70 (2005) (Quartely of Committee on Machine Building of the Polish Academy of Science)
54. Sładek, J, Krawczyk, M: Nadzorowanie dokładności pomiaru współrzędnościowego- Rozdział w Współrzędnościowa Technika Pomiarowa – Problemy i Zastosowania (Supervision of the accuracy of coordinate measurement- chapter in Coordinate Measuring Technique - Problems and Applications). In: Ratajczyk, E., Jakubiec W. (eds.) University of Bielsko-Biała 2008, pp. 29–38. ISBN 978-83-60714-41
55. Sładek, J., Kupiec, R.: Active system correction of thermal interactions designed for CMM. In: Adamczak, S. Ossana, I.H. (eds.) Modern Metrology in Quality Management Systems, CEEPUS Science Report, PSw-2006, pp. 237– 248
56. Sładek, J., Kupiec R.: Correction system of the CMM thermal interactions based on neural networks IV International Congress on Precision Machining 2007. In: ICMP 2007 Proceedings, Sandomierz-Kielce, 2007

57. Sładek, J., Kupiec, R., Muzyka-Żmudzki, M.: The concept of a system for the correction of CMM errors resulting from temperature changes. In: Measuring Technology in Advanced Machine Manufacturing Systems, Science Report, Project PL-127, pp. 147–158. CEEPUS, Kielce University of Technology, 2004
58. Sładek, J, Kupiec R, Gaska A, Krawczyk M: Badanie dokładności współrzędnościowej maszyny pomiarowej z wykorzystaniem laser tracera - zgodnie z ISO 10360–2:2009 (The accuracy of the coordinate measuring machine testing with the use of laser tracer—accordingly to ISO 10360-2: 2009). In: Proceedings of XIV national and V International Scientific—Technical Conference Coordinate Measuring Technique, Warsaw University of Technology, 2011
59. Sładek, J., Kupiec, R., Gąska, A., Kmita, A.: Modelowanie czynników temperaturowych, wpływających na niepewność pomiaru, przy pomocy metody Monte Carlo (Temperature factors affecting the measurement uncertainty modeling with the use of Monte Carlo method) Pomiary Automatyka Kontrola (Measurement Automation and Monitoring), 01/2010
60. TRACES—Transnational Calibration Expert System (2005–2007) C517456 TRACES Project —European Reasearch and Development Project implemented under e-Ten to develop new type of service targeted to the industry using the most advanced measurement technologies in the production of machines and tools. Its paricipants beside Laboratory of Coordinate Metrology were: leading German scientific institute Physikalisch Technischen Bundesanstalt in Braunscheig (German Office of Measures) as the project coordinator, Laboratory of Coordinate Metrology in Cracow University of Technology—Poland Danish Technical University in Copenhagen, CMI—Czech Metrology Institute, Spanish research center UNIMETRIK, Italian CERMET, German calibration center DKD-1-FEINMESS, Czech VUOS and ETALON AG Germany. 2005–2007
61. Trac Chec—software for CMM check and calibration in accordance to ISO 10360–2:2009 with the use of laser tracer
62. Trapet, E.: Large coordinate measuring machines - state of art and future for geometry errorcorrection, performace, verification and calibration. Proceedings of 1st International Workshop on CMM Calibration, Published by The Czech Metrological Institute, Prague Czech Republic, 1–2 June 1999
63. Trapet, E., Wäldele, F.: Coordinate measuring machines in the production line influence of temperature on measuring uncertainties. In: Proceedings of the IV Congress International de Metrologia i Industrial, Zaragoza (Spain), pp. 229–242 (1989)
64. Trapet, E., Wäldele, F.: Koordinatenmessgeräte in der Fertigung – Temperatureneinflüsse und erreichbare Messunsichercheit. VDI Berichte Nr **751**, 209–227 (1989)
65. Trapet, E., Wäldele F.: Temperature measurement in coordinate metrology. In: Proceedings of Euronet Workshop, Braunschweig, 5–6 Oct 1993
66. UNIMETRIK Metrology and Calibration, www.unimetrik.es
67. VDI/VDE 2617 Genauigkeit von Koordinatenmessgeräten Kenngrößen und deren Prüfung
68. VDI/VDE 2617 Genauigkeit von Koordinatenmessgeräten Kenngrößen und deren Prüfung. Blat 1 - Grundlagen
69. VDI/VDE 2617 Genauigkeit von Koordinatenmessgeräten Kenngrößen und deren Prüfung. Blatt 5.1 Überwachung mit Kugelplatten 2009
70. VDI/VDE 2617 Genauigkeit von Koordinatenmessgeräten Kenngrößen und deren Prüfung. Blatt 6.1 - Genauigkeit von Koordinatenmessgeräten Kenngrößen und deren Prüfung Koordinatenmessgeräte mit optischer Antastung Leitfaden zur Anwendung von DIN EN ISO 10360 für Koordinatenmessgeräte mit optischen Sensoren für laterale Strukturen VDI/VDE-Richlinien 2007
71. VDI/VDE 2617 Genauigkeit von Koordinatenmessgeräten Kenngrößen und deren Prüfung. Blatt 6.2- Genauigkeit von Koordinatenmessgeräten Kenngrößen und deren Prüfung Leitfaden zur Anwendung von DIN EN ISO 10 360 für Koordinatenmessgeräte mit optischen Abstandssensoren VDI/VDE- Richlinien 2005
72. VDI/VDE 2617 Genauigkeit von Koordinatenmessgeräten Kenngrößen und deren Prüfung. Blatt 13 - Genauigkeit von Koordinatenmessgeräten Kenngrößen und deren Prüfung

Leitfaden zur Anwendung von DIN EN ISO 10360 für Koordinatenmessgeräte mit CT-Sensoren—VDI/VDE-R Richlinien 2009
73. VDI/VDE 2634 Optische 3-D-Messsysteme Systeme mit flächenhafter Antastung Blatt 2
74. Wieczorowski, M., Jackson, G.: Coordinate measurements of small plastic workpieces. In: VIth International Scientific Conference Coordinate Measuring Technique, Bielsko-Biała. University of Bielsku-Białej. Scientific books 10, 2004
75. Wieczorowski, M.; Chajda, J.; Gapiński, B.; Matliński, K.: Coordinate measuring machine application for machine tool correction. In: VIIth International Scientific Conference Coordinate Measuring Technique, Bielsko-Biała. University of Bielsku-Białej. Scientific books 22, 2006
76. Zeiss Europa, www.zeiss.pl/imt

Chapter 7
Summary and Directions for Future Works on Coordinate Measurements Accuracy

Abstract This chapter summarizes the described issues, and provides directions for future developments in the field of coordinate metrology. The main problem, identified in early 2014, is concerned with the existing limits to the accuracy of coordinate systems. We don't have in fact standards that allow the verification and assurance of the traceability with the meter unit for coordinate measuring systems of accuracies below 100 nm. Another important issue regards the correct accuracy assessment of measurements in production conditions, which may significantly contribute to the quality of the final products. A series of actions must be implemented, among them raising awareness towards metrology throughout factories by means of: training programs dedicated to methods of estimation of coordinate measurement uncertainty; implementation of the methods described in this book; promotion of skills needed to deal with budget errors and uncertainty budget; and implementation of simulation methods, initially in calibration laboratories and afterwards in industrial conditions. This chapter also highlights the key role of calibration laboratories in ensuring traceability, as well as the importance of undertaking research tasks on coordinate systems including those oriented towards the development of multisensor systems.

In the present state of knowledge, the identification of coordinate system errors and their connection with the assessment of measurement accuracy is still a difficult task. As author of this book, I think that all coordinate systems should be based on a single common feature, namely the determination of the reproducibility error of the measurement point (REMP) for certain reference points, and in contact or noncontact mode. Measurement traceability should be provided in this case by the standard set of reference points, not as currently, the length standard. This concept is described in detail in Chap. 2. The original matrix method developed here (Sect. 4.4) is a verified solution. It can be used for both contact and noncontact systems including computed tomography. Currently its development by, for example, the use of laser tracer and the multilateration method allows for calibration of almost all coordinate systems with accuracy limited by the uncertainty of the laser tracer ($U = 0.0002 + 0.0003 \quad L/1000$ mm). Therefore the current task is research and methodological mastering of the tracer laser system, and its implementation in industrial conditions [6].

J.A. Sładek, *Coordinate Metrology*, Springer Tracts in Mechanical Engineering,
DOI 10.1007/978-3-662-48465-4_7

Another problem that should be undertaken (at the beginning of 2014) is the task of developing the technology for calibration of coordinate systems with accuracy less than 0.1 μm in the measuring volume described by a cube with a side of 1 m. The need of developing an appropriate standard is connected with this task. In this case, we should also move away from the concept of length standards and go to standards forming a grid of reference points. The solution of this task is necessary because of coordinate measurements in micro- and nano-dimension scope.

The issue of coordinate measurement accuracy covered in this book was presented by descriptions of methods, models, and results of our own tests. The uncertainty estimation methods and their implementation in the form of the substitution and multiposition method and *B* type method that uses the so-called uncertainty budget were also discussed. The calculation methods and the analytical method were described as well.

The calibrated object or standard method is clear and the best documented and validated, but in measurement practice sometimes it is technically more difficult because its use obliges the user to have a base of differentiated standards. On the other hand, its use raises fewer doubts relating to the connection with the unit of length, thus it was described in the normative development in the form of technical specification ISO/TS 15530-3:2004, [1]. Based on this specification, a procedure used for many years in LCM CUT was developed and because of the uniqueness of obtained uncertainties used for particularly difficult tasks. It can be recommended wherever it is possible to prepare so-called production standards, as shown in Fig. 3.6 in Sect. 3.1. This method is also being continuously developed, for example, by the concept of uncertainty determination for freeform contours, the MFG method developed by E. Savio. Additionally, for the calibrated object method we usually obtain lower uncertainties in comparison to other methods (the example is the results shown in Fig. 3.23 in Sect. 3.4), which is also very important when there is a critical problem with determination of compliance with the specification.

The multiposition method can be regarded from a practical point of view as a comprehensive method. It is associated with the lack of restrictions to ensure the similarity of used standards (as in the case of the calibrated object method). In addition, it includes the uncertainties associated not only with the standards, but also with the components coming from the coordinate machine (as shown in Table 3.3) It consists of the errors associated with the mounting or switching of the measurement objects and with the individual fit of a machine coordinate system to the local coordinate system of an object. Another group of components should be connected to technical problems caused by a lack of possibility of placing the noncalibrated object in different orientations. Then some locations may be skipped, but it requires a detailed justification. Indirectly, this procedure is verified for validity because wrongly selected elements of the measurement strategy may cause geometric errors from particular axes to propagate, which can cause a large dispersion of results. Therefore special precision is required while applying it. It is also the main reason why it has not been previously agreed upon by ISO (there is its technical specification as Draft ISO TS 15530-2) [1]. LCM CUT has accredited this procedure and applies it successfully. Some application examples are given in Annex A.

The identification of error sources and the development of their budget is a method that directly evolved from the B method requirements for uncertainty estimation. It is an accurate and comprehensive method, although very time consuming. It requires a deep examination of the coordinate system using laser interferometers, electronic levels, and identification of error sources. This examination should be done in reference to the actual conditions in which the measurement was performed and to the measurement strategy that was used. Also the uncertainty caused by this source should be estimated. The uncertainty budget requires a strict mathematical function defining the measurement accuracy in relation to the influential quantities. Often even simple measuring tasks exceed the capabilities of many laboratories. For complex measuring devices this functional relationship becomes difficult to determine even for specialists in that field. Using the numerical calculations often brings falsifications of the result, because it is not always possible to present the process as a single function. Taking into account the correlation of some input quantities is often a difficult task, sometimes even impossible to solve. An especially difficult case of including the correlation occurs when the measurement result is the result of a large number of particular measured values whose input quantities often correlate with each other. The uncertainty budget according to GUM [5] allows for including the known systematic input quantities only in special cases, such as a *priori* distributions. That procedure is too accurate and leads to overestimation of the uncertainty. The uncertainty budget is used especially during calibration of standards, as shown in the example described in Sect. 3.5. It is also recommended always to use the uncertainty budget when we cannot construct a coherent standard.

7.1 Analytical and Calculation Methods Based on Dependencies Resulting from the Model of Maximum Permissible Errors of Coordinate Measuring System

These methods, in spite of the seeming simplicity and universality, are difficult to implement and use, especially when it concerns the intuitive uncertainty estimation requiring the qualifications of an expert, which limits the use of these methods to the accredited laboratories, and each time requires a precise description of the realized procedure. The analytical calculation methods are the development of the measurement uncertainty calculation based on the maximum permissible error equation with the whole problem of a large underestimation of such calculated measurement uncertainty. Although the authors of these concepts try to correct it by taking the maximum errors, resulting from the geometric relationships, this is possible only for simple cases (mainly length measurements), thus the propagation of measurement uncertainty, typical for coordinate measurements—resulting from the variable accuracy in different areas of measuring volume, from the impact of the calculation algorithm, and especially from the impact of the measuring point identification system—is not included in this concept. This is indirectly confirmed by the example of comparison of the uncertainty determined for a simple task by direct calculation of measuring machine maximum errors and the error budget (Figs. 3.42–3.44). Only the analytical method of W. Jakubiec [4] includes the characteristic of CMM accuracy resulting from its geometric errors, which can be considered as its important advantage over Hernla's and Pressel's methods. Moreover, the use of EMU software in this case creates for this method a compact system for coordinate measurement uncertainty assessment.

However, in all methods we find significant problems, which examined from the measurement practice point of view negate the possibility of using them, and we tend to look for universal solutions, especially for solutions that can be used in industrial environments. Such a solution is certainly *the use of simulative methods* based primarily on the development of coordinate measuring system models and on the construction of virtual systems (ISO 15530-4:2008 standard) [3]. Summarizing the results of comparative and verifying tests of operation of software implementing virtual system algorithms that were built on different models and methods, it can be clearly stated that simulation methods are the most effective in assessing the accuracy of performed coordinate measurements. This situation was described in detail in Chap. 5. However, simulative methods require tests and at the current stage they can be used under conditions accessible in good laboratories and accredited laboratories.

The implementation of virtual systems for direct use in industrial conditions requires further tests. The closest to it is a solution based on the modeling of residual errors for systems equipped with CAA correction systems (Monte Carlo method and artificial neural networks). For articulated arm coordinate measuring machines, the model using Denavit–Hartenberg notation may be a base for building the virtual articulated arm coordinate measuring machine and its implementation in industrial environments to assess online measurement accuracy.

One important research direction currently is the use of a virtual machine model together with the machine working simulator for forecasting measurement accuracy for objects in the project phase. This combination creates an effective tool for evaluation of the suitability of using the given measuring machine for the chosen measuring task, regarding the specified manufacturing tolerance. Another task that is possible to be solved with the use of coordinate system simulators is the establishment of the optimal measurement strategy due to a technically economic indicator (accuracy/measurement time). Thanks to this combination, a system is created that can be fully operational in virtual reality. It is also possible to undertake research on the impact of the measured object, including its shape deviations and influence of the chosen coordinate system on the obtained measurement accuracy. Experiences (described in Sect. 5.2.5) gathered during the work with the I++

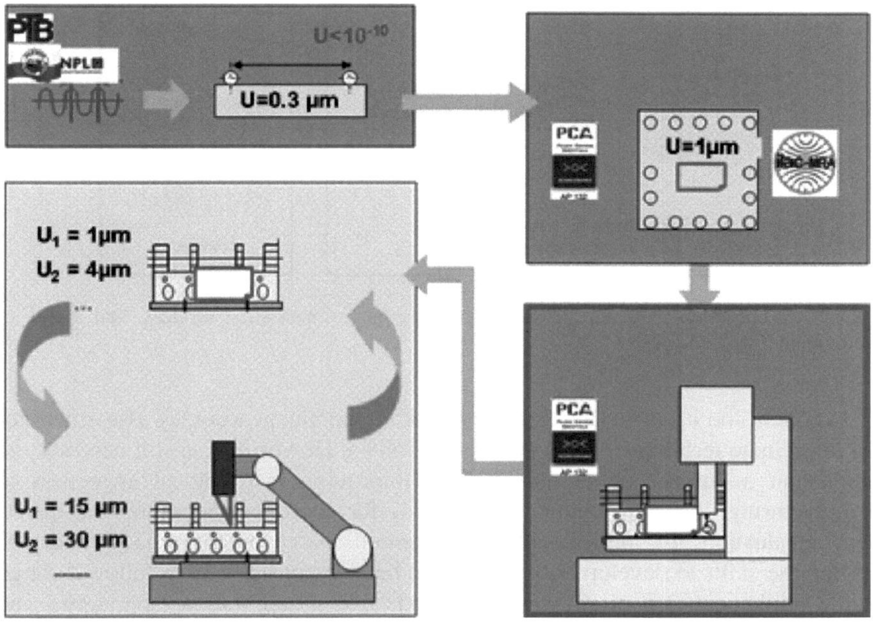

Fig. 7.1 Connection with the unit of length and the calibration chain, together with the assessment of the coordinate measurement accuracy [6, 7]

Simulator of the Messtechnik Wetzlar Company with solutions of virtual machine models implemented in it create such opportunities.

An example of the application of the calibrated object method under industrial conditions described in ISO/TS 15530-3 [2] in conjunction with the simulation method in the form of an installed virtual machine system in the accredited laboratory is shown in Fig. 7.1. The figure also presents the method for linkage with the unit of length by calibration of the standard in the form of a ball plate or a hole plate used to build the CMM model, virtual machine. It is currently the most effective system for accuracy assessment of coordinate measurements which is recommended mainly in German and British industry [7].

In this case it should be remembered that virtual systems running online require continuous cooperation between not only the CMM user and the calibration laboratory, but also with the coordinate system manufacturer. Cooperation with the manufacturer is required to provide the correction matrix formula (CAA) for CMM or correction data for AACMM encoders with the application for current accuracy correction. Cooperation with the calibration laboratory is based on periodic inspections of used calibration standards and of temperature measuring systems. It must be continuous and satisfied by appropriate monitoring of operation conditions for a given CMM and by establishment of the period of testing and monitoring the accuracy.

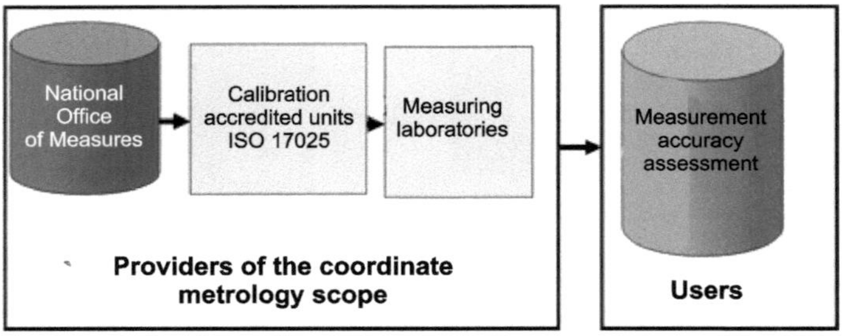

Fig. 7.2 Connection between providers of coordinate metrology services and users of measurement technologies

In conclusion it can be said that the task of measurement accuracy assessment for the coordinate technique can be effectively solved. However, it is still necessary to implement a number of actions such as improving metrological awareness in manufacturing plants by training of methods for coordinate measurement uncertainty evaluation, the implementation of methods based on ISO 15530, and promoting the skills to develop error budgets to be able consistently to follow through to the implementation of simulation methods first in calibration laboratories and then in industry conditions.

The realization of the necessary linkages between the service providers in the field of coordinate metrology and the users of coordinate systems is a crucial step in the improvement of metrological knowledge. Calibration laboratories accredited according to PN/EN/ISO 17025:2005 should be considered key partners in this cooperation and service laboratories should provide users with the necessary knowledge for determining measurement accuracy. These relations, shown in Fig. 7.2, are closely linked to quality requirements, a very important aspect of production engineering in modern industry.

References

1. ISO/CD TS 15530-2 – Geometrical Product Specifications (GPS)—Coordinate measuring machines (CMM): Techniques for evaluation of uncertainty of measurement Part 2: Use of multiple measurement strategies in measurements of artefacts (under development)
2. ISO/TS 15530-3:2004—Geometrical Product Specifications (GPS)—Coordinate measuring machines (CMM): Techniques for evaluation of uncertainty of measurement Part 3: Use of calibrated workpieces or standards
3. ISO/TS 15530-4—Geometrical Product Specifications (GPS)—Coordinate measuring machines (CMM): Techniques for evaluation of uncertainty of measurement Part 4: Evaluating tasc-specific measurement uncertainty using simulation
4. Jakubiec W. Analityczne wyznaczanie niepewności pomiaru we współrzędnościowej technice pomiarowej.(Analilitical evaluation of the measurement uncertainty in coordinate measurement technique) Bielsko-Biała 2008. University of Bielsko-Biała. Scientific Books 22

5. Central Office of Measures (Warsaw 1999) Wyrażanie niepewności pomiaru—przewodnik. Guide to the Expression of Uncertainty in Measurement (GUM) ISO 1993/1995 with the addition to polish edition of J.M. Jaworski
6. Sładek J.: Badania rozwojowe nad wzorcowaniem (z wykorzystaniem Laser Tracera) i wyznaczanie niepewności pomiarów dla systemów współrzędnościowych (Development researches on calibration (with the use of Laser Tracer) and determination of measurements uncertainty for coordinate systems) Grant no. 0869/R/T022010/10 National Centre of Research and Development 2010–2013
7. Trenk, M., Franke, M., Schwenke, H.I.: The virtual CMM, a software tool for uncertainty evaluation—practical application in an Accredited Calibration Lab. In: Proccedings of ASPE: Uncertainty Analysis in Measurement and Design, July 2004

Annex A
Procedure for Determining the Uncertainty of Coordinate Measurement Using Multiple Method

A.1 Procedure Rules

Equipment

1. Measuring tips needed for the measurement of the feature. Measuring tips necessary for characteristic measurement.
2. Standards that meet characteristics of similarity:
   - Length standards: in the case where the uncertainty is estimated for a characteristic that is expressed by length (width, height, position of holes from bases, etc.). The standard used should reflect the nature of the dimension (external, internal, etc.).
   - Diameter standards: in the case where the measuring tip is not qualified (in the case of used articulated arm CMMs or laser trackers).
3. Accessories for mounting the detail object and standard mounting.
4. Thermometer for measurement of standard temperature.

The standards used should have calibration certificates that are up to date in terms of applied management system and with specified uncertainty of dimension calibration. They should also fulfill similarity conditions according to Table 3.1 included in this book (Table A.1).

Procedure

The procedure involves multiple measurements of the same characteristic with the use of different orientations and various distributions of measurement points. The final measurement result is determined according to the equation:

J.A. Sładek, *Coordinate Metrology*, Springer Tracts in Mechanical Engineering,
DOI 10.1007/978-3-662-48465-4

Table A.1 Results of distance measurements on measuring machine PMM 12106

	Orientation 1	Orientation 2	Orientation 3	Orientation 4
Measurement cycle 1	133.9575	133.9571	133.9573	133.9573
Measurement cycle 2	133.9573	133.9567	133.9572	133.9576
Measurement cycle 3	133.9574	133.9570	133.9572	133.9577
Measurement cycle 4	133.9571	133.9565	133.9572	133.9575
Measurement cycle 5	133.9573	133.9568	133.9571	133.9574
Arithmetic mean for orientation ${}^{j}y$	133.9573	133.9568	133.9572	133.9575
Standard deviation ${}^{j}s$	0.00016	0.00023	0.00005	0.00014
Number of degrees of freedom for u_{rep}	16	16	16	16
Degrees of freedom for u_{geo}	3	3	3	3

All results (except degrees of freedom) are given in mm

$$y_{corr} = y - E_L - E_D \tag{A.1}$$

where

- y average value of given characteristic obtained from all measurements and all repetitions in certain orientations
- E_L length systematic error determined when the measured characteristic is a length
- E_D systematic error that occurs when the measuring tip is not subjected to qualification, but is only defined in the system

2.1. Perform five measurements of the given characteristic maintaining the following rules:

  2.1.1. The number of repetitions should exceed the number of orientations by at least one repetition.

  2.1.2. Each repetition should have a different distribution of measurement points on the measured surface.

2.2. Further orientations should be adopted as follows:

  2.2.1. Base location.

  2.2.2. Location twisted 90° in relation to the base one, around the *Z*-axis.

  2.2.3. Location twisted 90° in relation to the base one, around the *Y*-axis.

  2.2.4. Location twisted 90° in relation to the base one, around the *X*-axis. Determine the mean y value from all measurements carried out in all orientations.

2.3. Perform three measure standards three times; measurements in the same planes as measurements in the point 2.2 and determine systematic errors.

2.3.1. In case of the length standard:

$$E_L = \frac{L - L_{calstd}}{L_{calstd}} \cdot y \tag{A.2}$$

where

L result of length measurement considered as the mean value from all repetitions in all orientations

L_{calstd} correct value of the length standard (given in the calibration certificate)

y feature characteristic mean value determined at the point 2.1 of this procedure

2.3.2. In case of the diameter standard:

$$E_D = D - D_{calstd} \tag{A.3}$$

where

D diameter determined as the mean value from all repetitions in all orientations

D_{calstd} correct value of the diameter standard (given in the calibration certificate)

Determination of the Measurement Uncertainty

The measurement uncertainty is determined as given:

$$U = k \cdot \sqrt{u_{rep}^2 + u_{geo}^2 + u_D^2 + u_{corrL}^2} \tag{A.4}$$

where

k coverage factor adopted in accordance with the *t*-Student distribution for the effective number of degrees of freedom ν_{eff} and confidence level of 95 %

u_{rep} uncertainty component associated with the repeatability, determined as

$$u_{rep} = \frac{1}{\sqrt{n_1}} \cdot \frac{1}{\sqrt{n_2}} \sum \left({}^{j}s\right) \tag{A.5}$$

where

n_1 number of measurement repetitions in one orientation

n_2 number of orientations

${}^{J}S$ standard deviation of the mean value for the *j*th orientation

$$ {}^{j}S = \sqrt{\frac{1}{n_1 - 1} \cdot \left({}^{j}y_i - {}^{j}y\right)^2} \tag{A.6} $$

${}^{j}y$ *i*th measurement in the *j*th orientation

${}^{J}y$ mean value for the measured characteristic for the *j*th orientation

u_{geo} uncertainty component related to the reproducibility of the characteristic of given measurement

$$ u_{\mathrm{geo}} = \frac{1}{\sqrt{n_2}} \cdot \sqrt{\frac{1}{n_2 - 1} \sum \left({}^{j}y - y\right)} \tag{A.7} $$

u_{D} component associated with the uncertainty of measuring tip radius correction error determination

$$ u_{\mathrm{D}} = \sqrt{\left(\frac{U_{\mathrm{calstd}D}}{k}\right)^2 - u^2_{\mathrm{messtdD}}} \tag{A.8} $$

U_{calstdD} uncertainty of standard calibration read from the calibration certificate

k coverage factor of calibration given in the calibration certificate (generally $k = 2$)

u_{messtdD} uncertainty component associated with the repeatability of the standard measurement determined as the standard deviation of measurement results from all repetitions and orientations

$$ u_{\mathrm{corrL}} = \frac{L}{L_{\mathrm{calstd}}} \sqrt{\left(\frac{U_{\mathrm{calstdL}}}{k}\right)^2 + \frac{u^2_{\mathrm{messtdL}}}{3 \cdot n_3}} \tag{A.9} $$

U_{calstdL} uncertainty of standard calibration read from the calibration certificate k, coverage factor of calibration given in the calibration certificate (generally $k = 2$)

u_{messtdL} uncertainty component associated with the repeatability of the standard measurement, determined as the standard deviation of measurement results from all repetitions and orientations.

A.2. Measurement Performed on the High Accuracy Coordinate Measuring Machine

High accuracy CMM installed and operated used in a laboratory specially prepared for this purpose at Cracow University of Technology.

The machine has a portal type construction; one of the measurement axes is a sliding table (machine *X*-axis). The table is made of granite, and its slides and the slides of the other two measurement axes are mounted aerostatically. Dimensions of the machine measuring volume are: 1200 × 1000 × 600 mm. The machine is equipped with incremental optoelectronic measuring systems with glass standard, working in transmitting light. Together with PMM machine work, a PC computer with Quindos software works together with the PMM machine. The room is air-conditioned and a specially designed air exchange system provides so-called long-term stabilization of environmental conditions (20.0 ± 0.5 °C). The picture in Fig. A.1 shows the measurement station with the described machine as its main element.

Fig. A.1 Leitz Messtechnik PMM 12106 coordinate measuring machine

The station for the estimation of the coordinate measurement uncertainty is equipped with the following devices and elements:

(a) CMM PMM 12106
(b) Measured object
(c) Set of standards: gauge blocks, rings
(d) Temperature correction system for the machine and the measured object (temperature sensors, the reading system)
(e) Quindos software
(f) Elements and brackets used for mounting and manipulating of the measured object

Distance Measurement

The maximum permissible error HA of the Leitz CMM PMM 12106 is defined according to ISO 10360–2 and described by the equation:

$$\mathrm{MPE}_E = 0.0012 + 0.0025/1000 \cdot L\,\mathrm{mm},$$

and the value of the accuracy obtained while maintaining special measurement conditions[1] is:

$$\mathrm{MPE}_E = 0.0008 + 0.0025/1000 \cdot L\,\mathrm{mm},$$

where

L measured length in mm.

This means that for the HA CMM PMM machine the maximum permissible error of indications during length measurement in accordance with the manufacturer data does not exceed the value of MPE_E = 0.0008 + 0.00025/1000 · L mm for expanded accuracy and MPE_E = 0.0012 + 0.0025/1000 · L mm for standard accuracy. In further discussions as the value of the maximum permissible error the value equal to 1.2 µm was taken, due to standard conditions of the measurement performance.

The measured object was the pump body: the casting after mechanical treatment was done and then it was mechanically processed with very carefully made cylindrical surfaces that were made very precisely (manufacturing tolerance of

[1] Custom, higher measurement process stability conditions: temperature 20 °C ± 0.2) and when special tips are used as well as probe head correction and with maintaining the temperature regime at 20 ± 0.1 °C MPE_E = 0.5 + L/400 µm.

±0.007 mm and diameters of H7 and G7). Among others they served as base elements for determining the local coordinate system (datum) determining and measuring characteristic features. The measured characteristic was the distance between the collar's plane (symbolically marked with a triangle) and the axis of the hole; both elements are visible in Fig. A.2. All the surfaces with which the machine measuring tip has physical contact were carefully cleaned to eliminate the effect coming from their contamination.

For object mounting, elements of an "alufix" type mounting system made of aluminum alloys, were used for fixing, which in combination with the conventional object instrumentation (prisms, pressure pawclamps) precisely place the measured detail object in the machine measuring volume.

The measuring machine is equipped with a temperature correction system.

Temperature sensors are placed at: machine ruler guides, measured elements, and in its surroundings. In Fig. A.2 one of these sensors is visible, attached to the measured element. The space between the detail and the sensor is filled with a heat-conducting paste to provide better heat conduction.

The procedure requires that the measured characteristic be reproduced as accurately as possible by a certain standard. A gauge block and a standard ring that reproduces the measured hole diameter were used for this purpose. They were selected based on principles set out in Table 3.1, Sect. 3.1.

The nominal measured length was $134^{+0.05}_{-0.10}$ →. A standard plate gauge block of 150 mm made in IT 0 class was selected.

The nominal size of the measured diameter was $\varphi 34\,\mathrm{G7}$ →. A standard ring $\varphi\ 28_{-0002}$ was selected.

Measuring tips with a diameter of 8 mm were used for measurements; the length of the measuring stylus was 100 mm for each stylus. The measuring tip ball was made of synthetic ruby. To measure the object in given orientations appropriate constructions of measuring tips were made; they are visible in Fig. A.3.

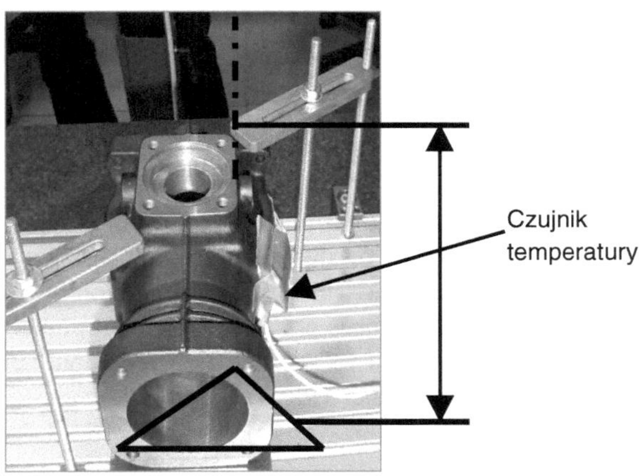

Fig. A.2 Measured object. Visible mounting system and the temperature sensor are visible. The measured distance was marked with *black*

Fig. A.3 Measuring tip configurations

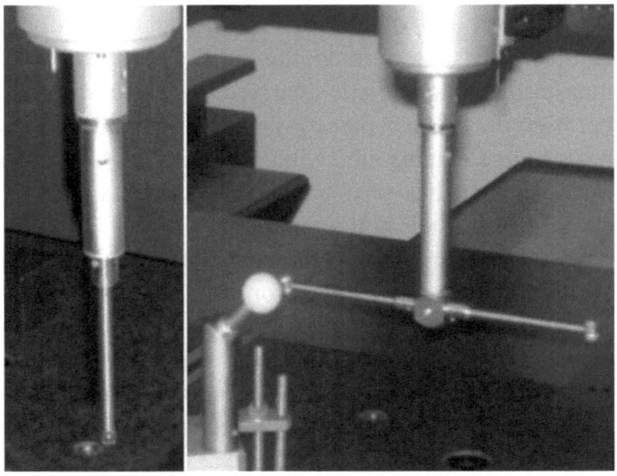

The measurement strategy was based on the procedure described in Chap. 3 of this book. It meets all the procedure requirements where possible.

Before performing measurements the qualifications of measuring tips were carried out. The next step was bringing the local machine coordinate system to a local system of the object. The base was "anchored" at the hole plane, and the beginning of the *X*- and *Y*-axes was placed in the hole axis.

The distance measurement was performed in five measuring cycles. The term "one cycle" is understood as a single measurement of the object (distance) and also measurement of the standards, all in the same orientation as the measured object, with one certain point distribution. The hole was measured at eight points in two sections, and measurement points of the collar plane were generated on the circle section, from 0° to 180°.

Position distribution of measurement points was changed in each cycle, if possible, without, if possible, changing the area of measured surface and the number of these points.

In order to measure the distance and the diameter standards, it was necessary to individually construct the coordinate system for each of the standards.

The measurement of the gauge block was carried out in three identical orientations as in the case of the measured object. It was carried out by measuring two mutually parallel standard planes, and by measuring five points on each of them.

The measurement of the standard ring was performed in three mutually perpendicular orientations, (the fourth orientation is a repetition of one of the other orientations). The number of 25 measurement points was assumed here for the ring diameter measurement on each of two measured sections. The ring measurement was repeated three times (Fig. A.4).

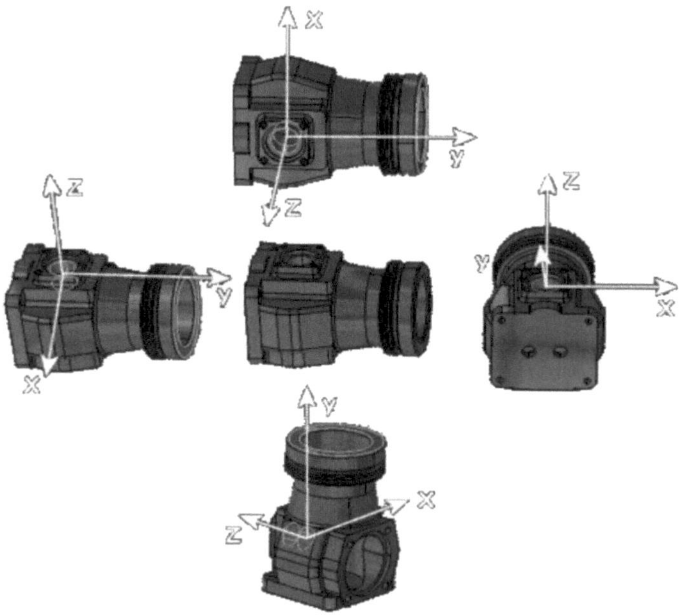

Fig. A.4 Adopted orientations of the measured object to calibrate

As a result of carrying out these measurements, following the calculation course in accordance with the calculation procedure described in Sect. A.3, the results posted in Table A.1 were obtained.

Uncertainty arising from machine geometric errors is

$$u_{\mathrm{geo}} = 0.00014\,\mathrm{mm}.$$

Table A.2 Results of the gauge block 150 mm measurement on the PMM 12106 machine

	Orientation 1	Orientation 2	Orientation 3
Measurement cycle 1	150.0002	150.0005	150.0012
Measurement cycle 2	150.0003	150.0005	150.0012
Measurement cycle 3	150.0002	150.0005	150.0012
Standard uncertainty	0.00009	0.00005	0.00004
Standard nominal size	149.99996	149.99996	149.99996
Ruler indications errors	0.00000	0.00000	0.00001
Uncertainty of standard calibration u_{calstd}	0.0005	0.0005	0.0005
Degrees of freedom for u_{geo}	6	?	?

All results (except degrees of freedom) are given in mm

Table A.3 Results of ring standard of a diameter 28-mm measurement on the PMM 12106 machine

	Orientation 1	Orientation 2	Orientation 3
Measurement cycle 1	27.9997	27.9986	27.9980
Measurement cycle 2	27.9997	27.9986	27.9979
Measurement cycle 3	27.9997	27.9986	27.9980
Correct value φ D_{cal}	27.99800	27.99800	27.99800
Standard uncertainty of the standard $u_{mes\ std}$	0.00030	0.00030	0.00030
Average diameter measurement error E_D	0.00064	0.00064	0.00064
Degrees of freedom v_e	8	8	8

All results (except degrees of freedom) are given in mm

Uncertainty related to machine repeatability is

$$u_{rep} = 0.00007\ \text{mm}.$$

The next step was the measurement of the standards, length, and internal diameter. Standards measurement results are presented in Tables A.2 and A.3.

The following errors were calculated based on the length standard measurements results:

- Systematic error of length measurement EL = 0.00050 mm.
- Standard uncertainty of length standard measurement caused by repeatability (u_{rep}) $u_{measstd}$ = 0.00006 mm.
- Standard uncertainty of the correction, which was introduced applied to length measurements (u_{rep}) $u_{measstd}$ = 0.00018 mm.
- Standard uncertainty of making the biggest mistake error in the diameter measurement is equal to u_D = 0.00025 mm.
- Systematic error for the diameter is E_D = 0.00064 mm.

Finally the results of the distance measurement uncertainty are:

- Measurements were performed with the assumed coverage factor k = 2.04; measured characteristic y = 133.9567 mm.
- The value of the expanded uncertainty is U = 0.00167 mm.
- The value of the standard uncertainty for the length measurement (expressed as a standard deviation) is $u_{measstd}$ = 0.00030 mm.
- The uncertainty of the coordinate measuring machine in accordance with the limit maximum permissible errors equation CMM MPE_E = 0.0012 + 0.0025/1000 · L mm is E = 1.134 μm.

A graph showing the estimated measurement uncertainty and the equation of the maximum permissible error is presented in Fig. A.5.

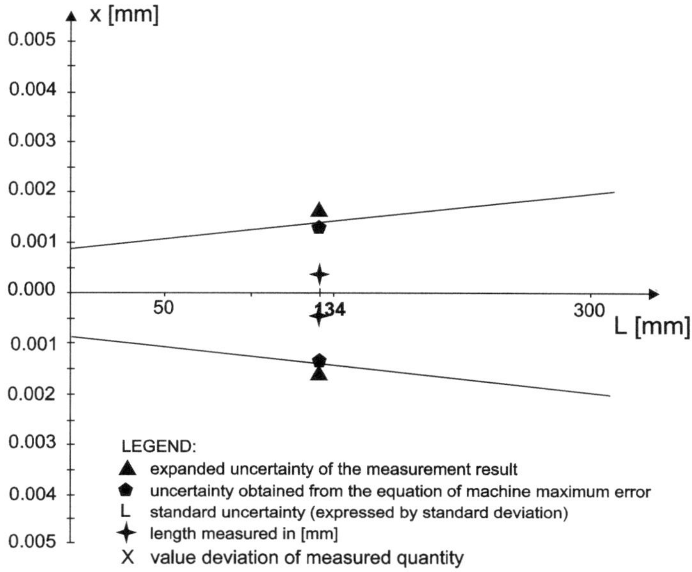

Fig. A.5 Presentation of the uncertainty of 134 mm length measurement in relation to MPE of PMM 12106 machine

Angle Measurement

The angle measurement was also carried out complying with the procedure described in Chap. 3 requirements. All conditions that had been created to carry out the measurement, discussed in Sect. 3.1.1, have also been kept for angle measurement. Thus, the angular characteristic measurement procedure was carried out analogous to the length measurement. The measurement object was the same element—water pump body, for which the measurement feature was carried out, was of the angle between the axes—determined from two nominally concentric cylindrical holes and the collar plane (Fig. A.6). For a perfectly made part this angle should be 0°. Similarly, as in the case of distance measurement on the PMM machine, the same orientations of the measured object were selected.

For this measurement task an appropriate configuration of measuring tips was used so as to allow free access to all measured surfaces. Best for this purpose was the tip with a diameter of 8 mm mounted on the stylus of 200 mm, oriented vertically and horizontally in relation to the measuring machine table. In Fig. A.7 the configuration of tips used in the measurement was presented.

For the angle measurement no standards were used because of the lack of a standard that would reproduce the measured quantity.

The measurement was carried out in five repetitions. In each repetition the object orientation of the measuring machine axes was cyclically changed. The rules for the

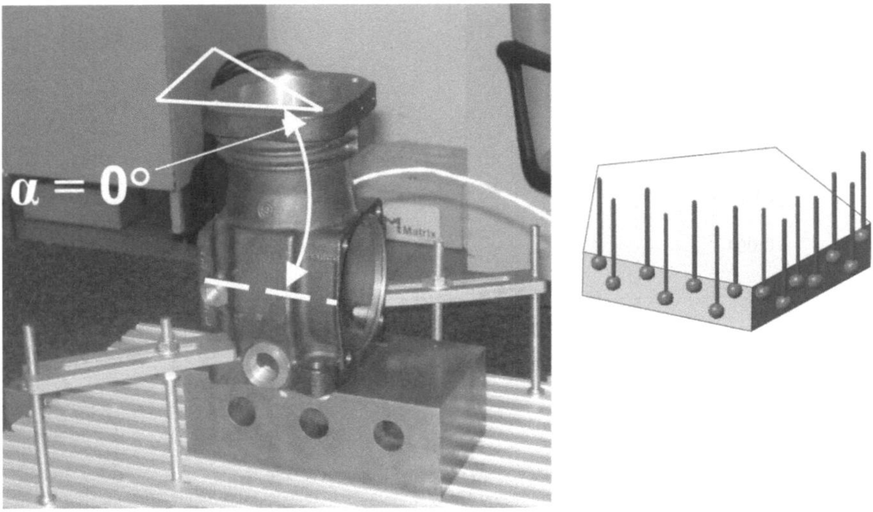

Fig. A.6 Measured characteristic of the object. With white marking the angle, the mounting in the sample orientation is visible

Fig. A.7 Standard ring measurement

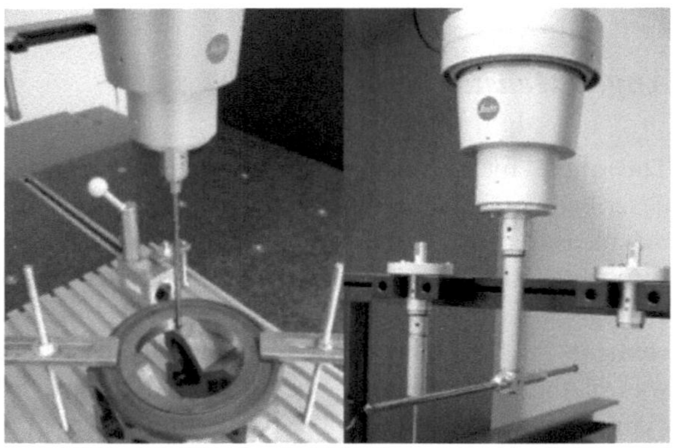

orientation selection are described in Sect. 3.1. Measurement results are shown in Table A.4.

The u_{rep} and u_{geo} components were calculated on the basis of data from Table A.12. For the angle measurement they are (Table A.5):

$$u_{\text{rep}} = 0.00190^{\circ}$$
$$u_{\text{geo}} = 0.00419^{\circ}$$

Table A.4 Angle measurement on Leitz PMM 12106 machine

	Orientation 1 base	Orientation 2 rotation of 90° around *X*-Axis	Orientation 3 rotation of 90° around *Y*-Axis	Orientation 4 rotation of 90° around *Z*-Axis
Cycle 1	0.01416°	0.03111°	0.00888°	0.01500°
Cycle 2	0.01416°	0.03083°	0.00888°	0.01472°
Cycle 3	0.01416°	0.03138°	0.00805°	0.01500°
Cycle 4	0.01361°	0.01666°	0.00861°	0.01555°
Cycle 5	0.01388°	0.03194°	0.00916°	0.01444°
Mean per orientation $^j y$	0.01399°	0.02838°	0.00872°	0.01494°
Standard deviation $^j s$	0.00025°	0.00657°	0.00042°	0.00041°
Degrees of freedom for u_{rep}	4			
Degrees of freedom for u_{geo}	3			

Angle values in the table are expressed in tenths of degree

Table A.5 Distance measurement results on DEA image clima measuring machine with temperature correction turned on

	Orientation 1	Orientation 2	Orientation 3	Orientation 4
Measurement cycle 1	133.9621	133.9613	133.9612	133.9629
Measurement cycle 2	133.9624	133.9613	133.9605	133.9627
Measurement cycle 3	133.9622	133.9612	133.9601	133.9626
Measurement cycle 4	133.9620	133.9613	133.9602	133.9627
Measurement cycle 5	133.9619	133.9613	133.9600	133.9626
Arithmetic mean for orientation $^j y$	133.96212	133.96128	133.96040	133.96270
Standard deviation $^j s$	0.00019	0.00004	0.00048	0.00012
Number of degrees of freedom for u_{rep}	16	16	16	16
Degrees of freedom for u_{geo}	3	3	3	3

All results (except dimensionless value of degrees of freedom) are given in mm

Finally, the uncertainties of the angle measurement between the axes of two concentric holes and the collar plane are:

The average result of all measurements $\alpha = 0.01651°$.
Expanded result uncertainty $U = 0.01279°$.
The standard uncertainty $u = 0.00803°$.
Calculations were made for the assumed coverage factor $k = 2.78$.

Table A.6 Distance measurements results on DEA image clima measuring machine with temperature correction turned off

	Orientation 1	Orientation 2	Orientation 3	Orientation 4
Measurement cycle 1	133.9575	133.9630	133.9664	133.9690
Measurement cycle 2	133.9578	133.9630	133.9656	133.9677
Measurement cycle 3	133.9578	133.9630	133.9655	133.9669
Measurement cycle 4	133.9578	133.9629	133.9654	133.9674
Measurement cycle 5	133.9576	133.9630	133.9653	133.9667
Arithmetic mean for orientation ${}^{j}y$	133.9577	133.96298	133.96564	133.96754
Standard deviation ${}^{j}S$	0.00014	0.00004	0.00044	0.00091
Number of degrees of freedom for u_{rep}	16	16	16	16
Degrees of freedom for u_{geo}	3	3	3	3

All results (except dimensionless value of degrees of freedom) are given in mm

Table A.7 Results of 150 mm gauge block measurement on DEA image clima machine with temperature correction turned on

	Orientation 1	Orientation 2	Orientation 3
Measurement cycle 1	149.9971	150.0023	149.9982
Measurement cycle 2	149.9970	150.0023	149.9982
Measurement cycle 3	149.9969	150.0022	149.9982
Standard uncertainty	0.00010	0.00006	0.00000
Correct value of standard	150.00041	150.00041	150.00041
Systematic errors	−0.00002	0.00001	−0.00001
Uncertainty of standard calibration u_{calstd}	0.000050	0.000050	0.000050
Nominal value of L standard	150	150	150
Degrees of freedom for u_{rep}	6	6	6

All results (except dimensionless value of degrees of freedom) are given in mm

Table A.8 Results of 150 mm gauge block measurement on DEA image clima machine with temperature correction turned off

	Orientation 1	Orientation 2	Orientation 3
Measurement cycle 1	149.9997	150.0061	150.0031
Measurement cycle 2	149.9997	150.0039	150.0027
Measurement cycle 3	150.0000	150.0039	150.0032
Standard uncertainty	0.00017	0.00127	0.00026
Nominal value of standard	150.00041	150.00041	150.00041
Ruler indications errors	0.00000	0.00003	0.00002
Uncertainty of standard calibration u_{calstd}	0.00050	0.00050	0.00050
Nominal value of L standard	150	150	150
Degrees of freedom for u_{rep}	6	6	6

All results (except dimensionless value of degrees of freedom) are given in mm

Table A.9 Results of ring φ 28 measurement on DEA image clima machine with temperature correction turned on

	Orientation 1	Orientation 2	Orientation 3
Measurement cycle 1	27.9997	27.9986	27.9980
Measurement cycle 2	27.9997	27.9986	27.9979
Measurement cycle 3	27.9997	27.9986	27.9980
Calibrated value $\varphi\ D_{cali}$	27.99800	27.99800	27.99800
Standard uncertainty of calibrated standard u_{std}	0.00030	0.00030	0.00030
Mean error of diameter measurement E_{int}	0.00064	0.00064	0.00064
Degrees of freedom v_e	8	8	8

All results (except degrees of freedom) are given in mm

In this case, the determination of a given parameter on the basis of the maximum permissible error equation is not possible (Fig. A.7).

A.3 Measurement on Medium Accuracy Coordinate Measuring Machine with Temperature Correction System DEA Image Clima

The DEA Image Clima coordinate measuring machine is made of lightweight construction materials, mainly aluminum, from which a sliding movable portal is made. This allows the machine to achieve significant acceleration of movement. The DEA machine was installed in the Laboratory of Coordinate Metrology at Cracow University of Technology, in a room that was not air-conditioned but had a stable temperature. The temperature correction was provided by a special system of automatic temperature correction, equipped with temperature sensors, cooperating with the machine software PC-DMIS.

The equation of maximum permissible errors of the machine is (Fig. A.8):

$$\mathrm{MPE}_E = 0.0017 + 0.003/1000 \cdot L\,\mathrm{mm}$$

The DEA machine is equipped with a Renishaw PH10MQ articulating probe head, which can change its angular location in two planes, and therefore it allows the orientation of the probe head measuring tip (here the TP200 head probe is installed) in 720 repeatable locations. Thus, the measuring tip mounted on the probe head of this type is provided with free access to all measured surfaces, without the necessity to manual "construct" the tip system, appropriate for a given measuring task. This allows more efficient and more complete machine use for inspection tasks of almost any shape of object detail.[2]

[2]Each measuring machine has established maximum weight, with which the measuring table may be charged and with which the accuracy of its indications does not change.

Fig. A.8 DEA Image Clima measuring machine

In order to compare measurement results obtained using the reference Leitz CMM with the DEA machine results, the same measurement tasks were carried out for the same part.

Distance Measurement

In order to compare uncertainty results for length measurement for a detail, the local coordinate system (associated with the measured object) was built in the same way (associated with the measured object). Also the same standards reproducing measured characteristics were selected. Measured objects were mounted on the measuring table using the alufix system in the same way as on the PMM 12 106 machine. Measuring styli configuration was chosen so as to reproduce conditions of measurement on the PMM machine as faithfully as possible. Selected tips of the same diameter were selected and mounted on styli of the same length. The measurement method presented in Fig. A.9 refers to one orientation for each measured object.

Each measuring probe head orientation is defined and changed with appropriate commands of the PC-DMIS system.[3]

[3]The change of probe head location is in two planes in relation to the WMP table, with a resolution of 7.5°.

Fig. A.9 Measurement objects mounted on the measuring table

Fig. A.10 Distance measurement (DEA Image Clima machine) in a set orientation

The measurement strategy had only one difference: in each cycle the same location distribution of measurement points (in the case of PMM, it was changed without changing units of the distribution area) was applied.

All distance measurements were carried out with machine temperature correction turned on and off.

Measurement results are shown in tabular form in Figs. A.9 and A.10.

At this measurement stage the uncertainty components u_{rep} and u_{geo} were calculated using appropriate dependences (A.5) and (A.7).

Measurement uncertainty component coming from repeatability:

$$u_{rep} = 0.00012\,\text{mm}.$$

Measurement uncertainty component coming from geometric errors:

$$u_{geo} = 0.00050\,\text{mm}.$$

Measurement uncertainty component coming from repeatability:

$$u_{rep} = 0.00023\,\text{mm}.$$

Measurement uncertainty component coming from geometric errors:

$$u_{geo} = 0.00214\,\text{mm}.$$

The second step was the measurement of standards that reproduce measured quantities. Measurements were carried out in three measurement cycles in three orientations (perpendicular to first base orientation).

Based on the length standard measurement results the following errors were calculated (Fig. A.11):

Fig. A.11 Measurement of the 150-mm gauge block. In the background the pressure pawclamp and temperature sensor are visible. The measurement is carried out by the contact of the measuring tip with plate planes at five points

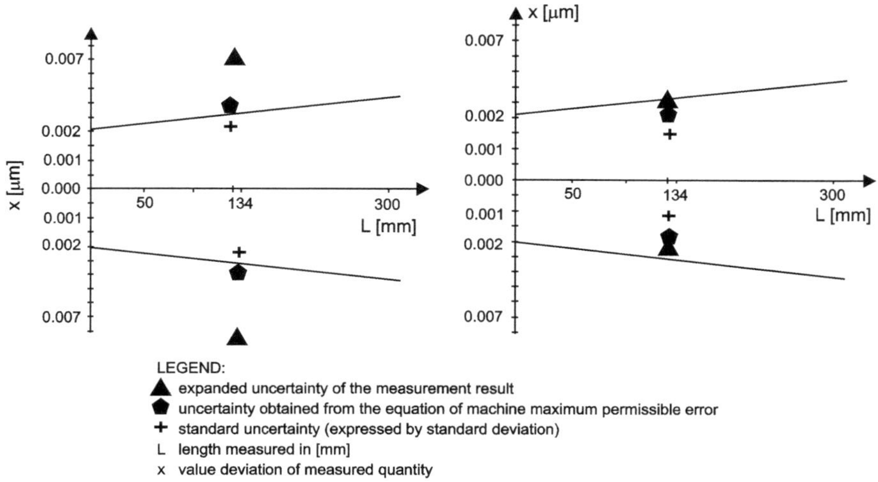

Fig. A.12 Presentation of the uncertainty: on the *left* measurement without temperature correction and on the *right*, with temperature correction turned on

Systematic error of length measurement $E_L = 0.00125$ mm.

Standard uncertainty of length standard measurement caused by repeatability (u_{rep}) $u_{measstd} = 0.00007$ mm.

Standard uncertainty of the correction $u_{corrL} = 0.00025$ mm.

Systematic error of length measurement $E_L = 0.00207$ mm.

Standard uncertainty of length standard measurement caused by repeatability (u_{rep}) $u_{measstd} = 0.00076$ mm.

Standard uncertainty of the correction, which was introduced to length measurements $u_{corrL} = 0.00035$ mm.

Standard uncertainty of diameter measurement is $u_D = 0.00025$ mm. Systematic error for the diameter measurement is $E_D = 0.00064$ mm. Standard uncertainty of

Table A.10 Results of ring φ 28 measurement on DEA image clima machine with temperature correction turned off

	Orientation 1	Orientation 2	Orientation 3
Measurement cycle 1	27.9998	27.9985	27.9983
Measurement cycle 2	27.9998	27.9985	27.9979
Measurement cycle 3	27.9997	27.9985	27.9982
Calibrated value φ D_{cali}	27.99800	27.99800	27.99800
Standard uncertainty of calibrated standard u_{std}	0.00030	0.00030	0.00030
Mean error of diameter measurement E_{int}	0.00066	0.00066	0.00066
Degrees of freedom ν_e	8	8	8

All results (except degrees of freedom) are given in mm

diameter measurement is u_D = 0.00025 mm. Absolute error for the diameter measurement is E_D = 0.00066 mm (Fig. A.12).

Finally, the results of the uncertainty of the distance measurement are shown in Table A.10.

Angle Measurement

The angle measurement was carried out analogously to its measurements done on the PMM machine. The measurement object was the same element—water pump body, for which were carried out measured feature elements of the angle between the axes, determined from two theoretically concentric holes—and the collar plane.

The nominal value of the measured angle should be equal to 0°. For the measurement the same orientations of the measured object were chosen as in the distance measurement.

For this measurement task an appropriate construction of measuring tips was used, expanding the measuring probe head with appropriate extensions so that the measurement strategies would not change too much on both machines. Best for this purpose was the tip with a ball diameter of 4 mm mounted on a stylus of 200 mm, oriented in relation to the angle A settings (rotation of the probe head in a plane perpendicular to the machine table) and the angle B (rotation in a plane parallel to its table) analogous to tip settings used on the Leitz CMM.

Measurements were carried out with temperature correction turned on. Results of the experiment performed are presented in Table A.11.

The components u_{rep} and u_{geo} were calculated based on data from Table A.11.

Table A.11 Angle measurement on DEA image clima machine

	Orientation 1 base	Orientation 2 rotation of 90° around *X*-Axis	Orientation 3 rotation of 90° around *Y*-Axis	Orientation 4 rotation of 90° Around *Z*-Axis
Cycle 1	0.01810°	0.03611°	0.01888°	0.01300°
Cycle 2	0.01826°	0.02222°	0.01888°	0.01372°
Cycle 3	0.01816°	0.03333°	0.01805°	0.01400°
Cycle 4	0.01861°	0.03055°	0.01861°	0.01355°
Cycle 5	0.01888°	0.02777°	0.01916°	0.01444°
Mean ${}^j y$	0.01840°	0.03000°	0.01872°	0.01374°
Standard deviation ${}^j s$	0.00033°	0.00534°	0.00042°	0.00053°
Degrees of freedom for u_{rep}	4			
Degrees of freedom for u_{geo}	3			

Angle values in the table are expressed in tenths of degree

For the angle measurement they are:

$$u_{\text{rep}} = 0.00156^\circ,$$
$$u_{\text{geo}} = 0.00345^\circ.$$

Finally, uncertainty results of angle measurement between the axes of two concentric holes and the collar plane are:

- The average result of all measurements $\alpha = 0.02021^\circ$.
- Expanded result uncertainty $U = 0.01053^\circ$.
- The standard uncertainty $u = 0.00662^\circ$.
- Calculations were made for the assumed coverage factor $k = 2.78$.

A.4. Measurement on Medium Accuracy Coordinate Measuring Machine Without Temperature Correction System DEA Global Status

The DEA Global Status 09.12.08 measuring machine was installed in an air-conditioned room, where the temperature was constantly maintained in the range of 19.5–21 °C. The machine was not, however, equipped with the system of

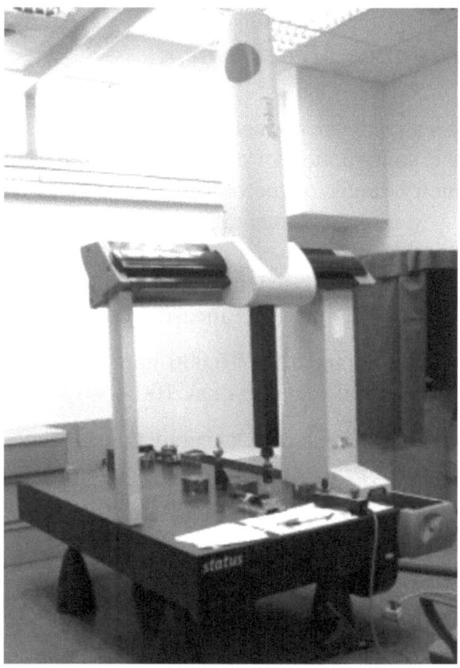

Fig. A.13 Coordinate measuring machine DEA global status 09.12.08

automatic temperature correction, as in the case of the DEA Clima machine. Its measuring range was 900 × 1200 (on the *Y*-axis) × 800 mm on the *Z*-axis. The machine was equipped with a Renishaw PH10MQ automatic articulating probe head, identical to the probe head installed in the DEA Image Clima machine and with a Renishaw SP600M probe head with the possibility that allowed scanning any shape surfaces (Fig. A.13).

Measurements in accordance with the same previously described procedure were also carried out on the DEA Global Status machine. They were carried out in order to check whether this method is suitable for the estimation of coordinate measurement uncertainty for machines also used in production conditions.

The equation of maximum permissible errors for this machine is: $MPE_E = 0.0027 + 0.0025\ L/1000$ mm

The measured object was the resistance plate of extractors, the element used in the production of pressure casting dies. The object was made of tool steel NC 10.

In the case of such a plate a high precision of manufacturing of drilled (and milled) holes and of their mutual location is required and expected. Axes of holes should be parallel. If, for example, axes of holes are made slantwise, it can lead to permanent damage or destruction of the tool.[4] Thus it appears to be technologically reasonable to choose measured feature values characteristic for a plate, such as:

- Diameters of drilled holes
- Distance between these holes
- Angle designated between hole axes

Distance Measurement

As already mentioned, the distance measurement was carried out on the element of the construction of an injection mold. This element was symmetrical to the symmetry center of the object. Figure A.14 shows the measured object.

Measured element dimensions were 246 × 112 × 17 mm. Measured distance was the distance between centers of two holes 16 H7 (in the photo four such holes are visible, marked with red arrows). The distance between these holes was 140 ± 0.01 mm. The measurement was carried out using the standards:

- Ring standard φ 16.006 mm
- Stack of gauge blocks 100 + 40 mm of nominal length 140_{-00009}

Therefore it was ensured that in general the standards reproduce the nominal measured quantity of the object.

To carry out the measurement the following sizes and measuring tip configurations were used.

- Tip diameter of ϕ 8 mm.

[4]As a “tool” a pressure form mold is named here.

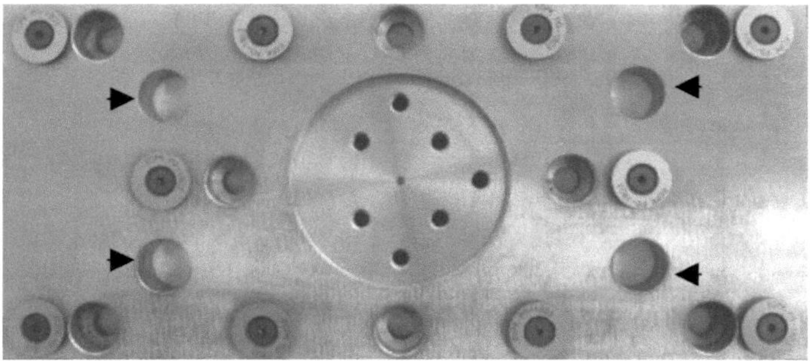

Fig. A.14 Resistance plate of extractors

- The same orientations as in the distance measurement on DEA Image Clima machine.
- The qualification of the measuring tip was carried out before measurement with the same instrument artifact (calibration sphere) as in the case of qualification of DEA Image machine.

The following equipment was also used.

- Magnetic holder
- Prisms
- Glue for mounting of objects

The measurement strategy:

- For each orientation the machine coordinate system was individually matched to the object system.
- Orientations of the object relative to the machine were chosen in accordance with the procedure from Sect. A.2.
- Within the one orientation, each characteristic was measured with a different location of equal point distribution (hole measurement at different sections).
- Each hole was measured using 16 measurement points.
- Each cylinder was constructed from two measured concentric holes.
- Measurement result was a distance between axes of cylinders.

Through the whole time of measurement test duration, the environment temperature fluctuated in the range of 20.1 ± 0.3 °C.

The obtained results are presented in the Table A.12.

Uncertainty coming from machine geometric errors is:

$$u_{\text{geo}} = 0.00078\,\text{mm}.$$

Uncertainty coming from machine repeatability is:

Table A.12 Distance measurement results on DEA global measuring machine

	Orientation 1	Orientation 2	Orientation 3	Orientation 4
Measurement cycle 1	139.9872	139.9895	139.9881	139.9824
Measurement cycle 2	139.9864	139.9891	139.9894	139.9845
Measurement cycle 3	139.9878	139.9809	139.9897	139.9853
Measurement cycle 4	139.9874	139.9881	139.9879	139.9855
Measurement cycle 5	139.9863	139.9890	139.9884	139.9863
Arithmetic mean for orientation ${}^{j}y$	139.98701	139.98732	139.98871	139.98481
Standard deviation ${}^{j}S$	0.00064	0.00364	0.00080	0.00149
Degrees of freedom for u_{rep}	16	16	16	16
Degrees of freedom for u_{geo}	3	3	3	3

All results (except degrees of freedom) are given in mm

Fig. A.15 Gauge blocks holder of KOBA company

$$u_{\mathrm{rep}} = 0.00091\ \mathrm{mm}.$$

The second step was the measurement of standards, of the length, and internal diameter.

The φ 16.006 mm ring standard has a second accuracy class of manufacturing. It was measured in 25 points in two section planes in each cycle at different heights. After construction of the cylinder from each pair of measured holes the cylinder was dimensioned. The fact that each hole had been measured at different heights provided a variable distribution of points in each orientation.

The distance standard measurement was carried out using the handle manufactured by KOBA, visible in Fig. A.15.

Standards measurement results are presented in Tables A.13 and A.14.

Based on the length standard measurement results the following errors were calculated (Fig. A.16).

- Systematic error of length measurement $\boldsymbol{E}_{\mathbf{L}}$ = 0.0010 mm.

Table A.13 Results of gauge blocks 140 mm measurement on DEA global status machine

	Orientation 1	Orientation 2	Orientation 3
Measurement cycle 1	139.9942	139.9984	140.00095
Measurement cycle 2	139.9949	139.9987	140.00104
Measurement cycle 3	139.9949	139.9989	140.00101
Standard uncertainty	0.00041	0.00025	0.00005
Nominal value of standard	139.99910	139.99910	139.99910
Ruler indication errors	−0.00003	0.00000	0.00001
Uncertainty of standard calibration u_{calstd}	0.00050	0.00050	0.00050
Degree of freedom for u_{rep}	6	6	6

All results (except degrees of freedom) are given in mm

Table A.14 Results of φ 16 mm ring standard measurement on DEA global status machine

	Orientation 1	Orientation 2	Orientation 3
Measurement cycle 1	15.9940	15.9970	15.9948
Measurement cycle 2	15.9954	15.9970	15.9953
Measurement cycle 3	15.9961	15.9971	15.9955
Calibrated value φ D_{cali}	16.00600	16.00600	16.00600
Standard uncertainty of calibrated standard u_{std}	0.00030	0.00030	0.00030
Mean error of diameter measurement E_{int}	−0.00662	−0.00662	−0.00662
Degrees of freedom v_e	8	8	8

All results (except degrees of freedom) are given in mm

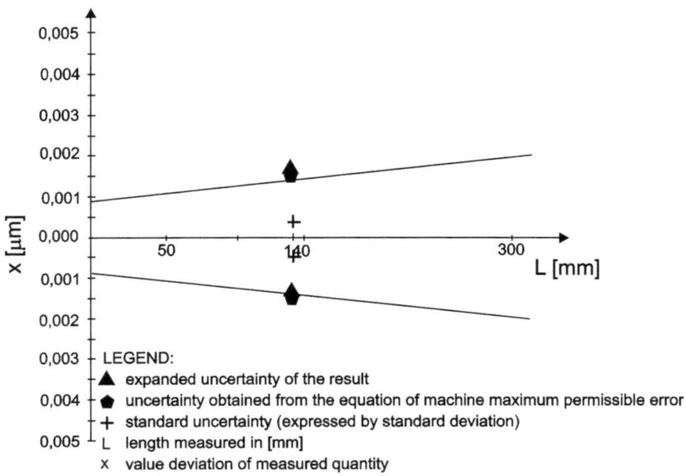

Fig. A.16 Presentation of the uncertainty of 140 mm length measurement

- Standard uncertainty of length standard measurement caused by repeatability (u_{rep}) $u_{measstd}$ = 0.00028 mm.
- Standard uncertainty of the correction, which was introduced applied to length measurements: $\boldsymbol{u}_{corrL}$ = 0.00027 mm.
- Standard uncertainty of making the biggest mistake in the diameter measurement is u_D = 0.00037 mm.
- Systematic error for the diameter measurement is $\boldsymbol{E}_{\mathbf{D}}$ = 0.00662 mm.

Finally, the results of the distance measurement uncertainty are as follows: measurements were performed with the assumed coverage factor k = 2.10. The corrected calibrated mean from all measurements y = 139.9880 mm.

The value of the expanded uncertainty is U = 0.00286 mm.

The value of the standard uncertainty is u = 0.00235 mm.

The value of CMM uncertainty is E = 2.260 μm.

Diameter Measurement

Measured diameters, according to assumed established technology, theoretically were made at φ 16 H7. According to DIN 11012 the tolerance of 16 H7 dimension can be written as the tolerance (for hole) $\varphi 16^{+0.018}_{0.0}$. In Fig. A.11 four such holes are visible. All these diameters were measured in each of three set orientations but only one of them was used for measurement (randomly selected hole) was used.

For this measurement task the ring standard of a nominal diameter of 16.006 mm was chosen, therefore the measured diameter and the standard reproducing it had almost equal nominal values.

Three orientations of the measured object were used, perpendicular to certain base orientation.

Uncertainty coming from machine geometric errors is u_{geo} = 0.00140 mm. Uncertainty coming from machine repeatability is u_{rep} = 0.00076 mm.

Results of diameter φ 16 H7 measurement were corrected with φ 16.006 mm standard measurement results (Table A.14). Finally the following uncertainties of diameter coordinate measurement were obtained.

- Measurements were carried out for assumed coverage factor k = 2.57.
- Measurement result y = 15.99158 mm.
- The value of the expanded uncertainty is U = 0.00422 mm.
- The value of the standard uncertainty is u = 0.00330 mm.
- The value of CMM uncertainty in accordance with (8.4) is E = 2.764 μm.

Obtained uncertainties of diameter measurement are presented in the graph in Fig. A.17.

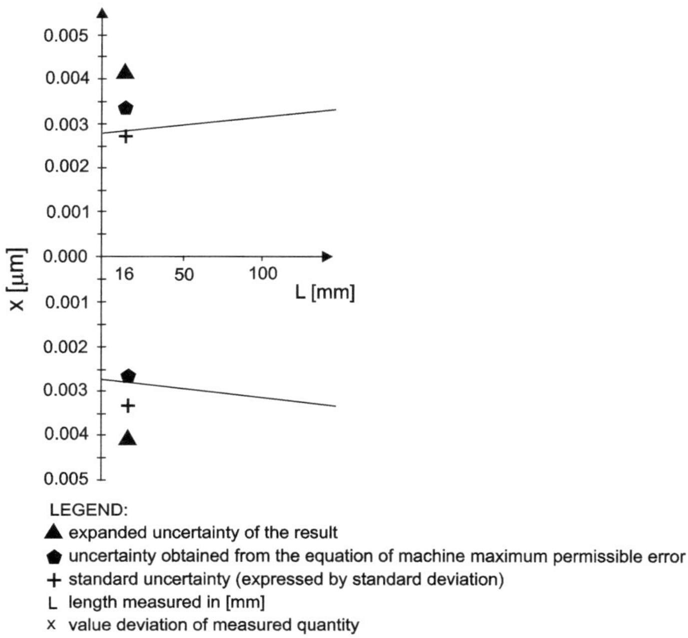

Fig. A.17 Presentation of the uncertainty of diameter measurement on DEA Status machine in relation to its MPE

Angle Measurement

At this stage of the experiment the angle measurement between hole axes was carried out, previously measured during the measurement of the distance between them. Thus the angle measurements boiled down to select appropriate relations with the use of PCDMIS software options.

The nominal value of the measured angle should be equal to 0°, because it is assumed that axes of cylinders made from measured holes are parallel.

Table A.15 Results of φ 16 mm ring standard measurement on DEA global status machine

	Orientation 1	Orientation 2	Orientation 3
Measurement cycle 1	15.9911	15.9894	15.9941
Measurement cycle 2	15.9933	15.9858	15.9952
Measurement cycle 3	15.9922	15.9881	15.9953
Calibrated standard value φ D_{calint}	15.9923	15.9894	15.9955
Standard uncertainty of calibrated standard u_{std}	15.9919	15.9887	15.9953
Mean error of diameter measurement E_{int}	15.9922	15.9883	15.9951
Degrees of freedom ν_e	8	8	8

All results (except degrees of freedom) are given in mm

Table A.16 The angle measurement on DEA global status machine

	Orientation 1 base	Orientation 2 rotation of 90° around X-Axis	Orientation 3 rotation of 90° around Y-Axis	Orientation 4 rotation of 90° around Z-Axis
Cycle 1	0.09790°	0.04038°	0.01164°	0.03251°
Cycle 2	0.02461°	0.04038°	0.01650°	0.02612°
Cycle 3	0.06826°	0.02219°	0.00619°	0.02219°
Cycle 4	0.07655°	0.02016°	0.00523°	0.02016°
Cycle 5	0.01505°	0.03126°	0.00430°	0.03126°
Mean per orientation ${}^{j}y$	0.05647°	0.03087°	0.00877°	0.02645°
Standard deviation ${}^{j}s$	0.03532°	0.00963°	0.00518°	0.00542°
Degrees of freedom for u_{rep}	16			
Degrees of freedom for u_{geo}	3			

Results of the experiment that was carried out are in Table A.15.

Based on data from Table A.16 components u_{rep} and u_{geo} were calculated. For the angle measurement they are:

$$u_{\text{rep}} = 0.00985^\circ,$$
$$u_{\text{geo}} = 0.01079^\circ.$$

Finally, uncertainty results of angle measurement between the axes of two concentric cylindrical holes and the collar plane are (Tables A.6, A.7, A.8, A.9 and A.10):

Average result of all measurements $\alpha = 0.03064^\circ$.

Expanded result uncertainty $U = 0.03257^\circ$.

Standard uncertainty $u = 0.0244^\circ$.

Calculations were made for the assumed coverage factor $k = 2.23$.

Annex B
Standard Accuracy CMM Geometric Error Identification Using Laser Interferometer and Correction Matrix CAA Determination

Introduction

Test research was carried out for a standard accuracy machine shown in Fig. B.1, with the maximum permissible error described by the equation: MPE = 0.0017 + 0.003/1000 *L! mm.

Measuring volume of the tested machine is 700 × 500 × 500 mm.

This machine cooperates with PC-DMIS and Quindos software systems. Tests were carried out with the use of Renishaw's ML10 laser interferometer.

Component rotation errors, translation errors (slide straightness), and machine positioning errors (as described in Sect. 4.2.1) were determined.

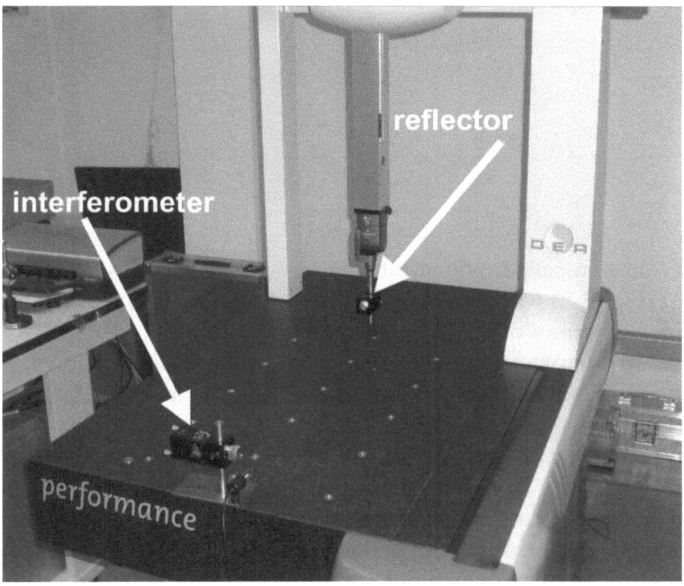

Fig. B.1 Measurement of positioning errors *xtx*

J.A. Sładek, *Coordinate Metrology*, Springer Tracts in Mechanical Engineering,
DOI 10.1007/978-3-662-48465-4

Fig. B.2 TESA's Microbevel level

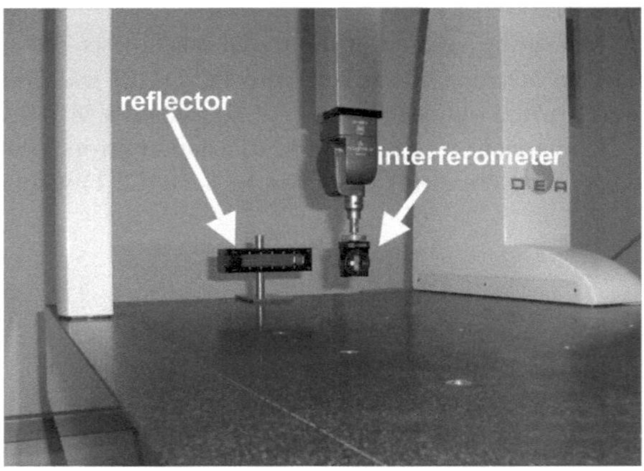

Fig. B.3 Measurement of the *X*-axis translation error in a horizontal plane *xty*

To determine so-called mutual rotation errors *xrx*, *yry* TESA's electronic level Microbevel (Fig. B.2) was used (Figs. B.2 and 4.4b from Chap. 4) in application to examine *xrx*, *yry* errors of the High accuracy machine.

The laser interferometer used was equipped with packages allowing the determination of positioning, straightness, perpendicularity, and angular location deviations (Figs. B.3, B.4 and B.5).

Changes in laser wavelength may be affected by temperature, humidity, and atmospheric pressure. To obtain the highest accuracy the correction of environmental conditions is required. Therefore, for measuring the temperature of air and material, atmospheric pressure, and humidity the EC10 station is used that measures and corrects environmental conditions and, on the basis of measured parameters, corrects the length of laser light wave.

Fig. B.4 Measurement of the X-axis translation error in a vertical plane xtz

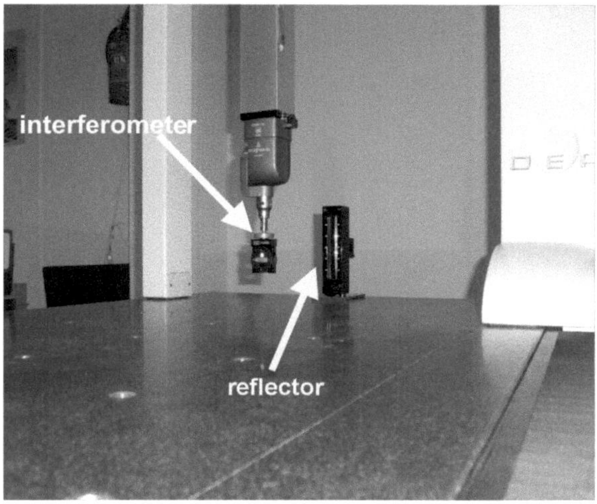

Fig. B.5 Angle measurement: mounting of optic elements on the measuring machine table

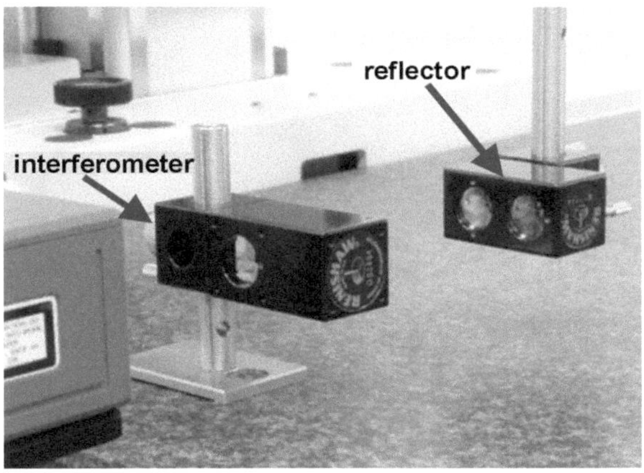

Determination of Machine Geometric Error Components—Rotation Error

At the beginning of the tests the rotation errors in the X-axis and later in the Y-axis were determined. Tests were conducted for the vertical pinole ram offset in relation to the machine system Z-axis by the value −100 mm and −400 mm, and the measuring step for all offsets was 50 mm. Test realization assumed the use of an interferometer optic package for measuring small angles, which was mounted on the machine table, and a mirror that was mounted inside the handle of the machine probe head. The calculations were done with the use of Renishaw software application, "Angular Measurement" (Figs. B.6, B.7, B.8 and B.9).

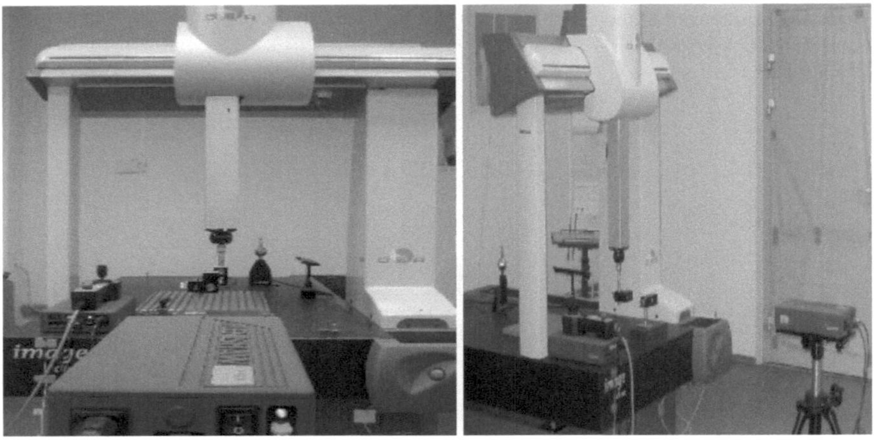

Fig. B.6 Measurement of rotation errors of Y-axis in relation to X-axis and $Zxrz$-axis, both lower settings $Z = -400$ mm

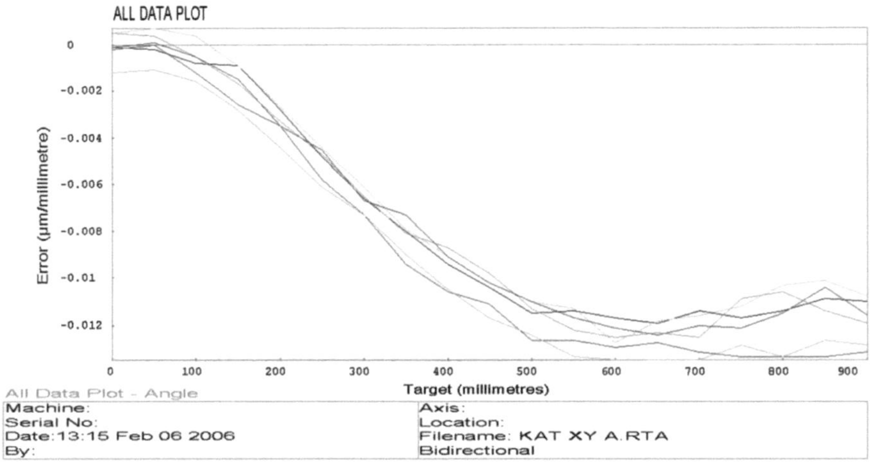

Fig. B.7 Measurement of Y-axis torsion in relation to X-axis; lower setting $Z = -400$ mm

Determination of Machine Geometric Error Components—Translation Error (of Slide Displacement/Shift Translational Straightness)

Slide shift straightness errors were identified along the main directions of machine axes X, Y, Z. Tests for each of the X- and Y-axes were carried out for the vertical pinole ram deflection in relation to the machine system Z-axis by the values

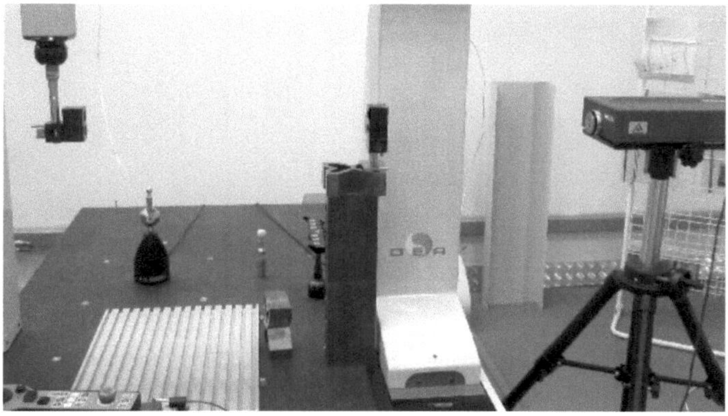

Fig. B.8 Measurement of *X*-axis torsion in relation to *Y*7-axis; upper setting $Z = -100$ mm

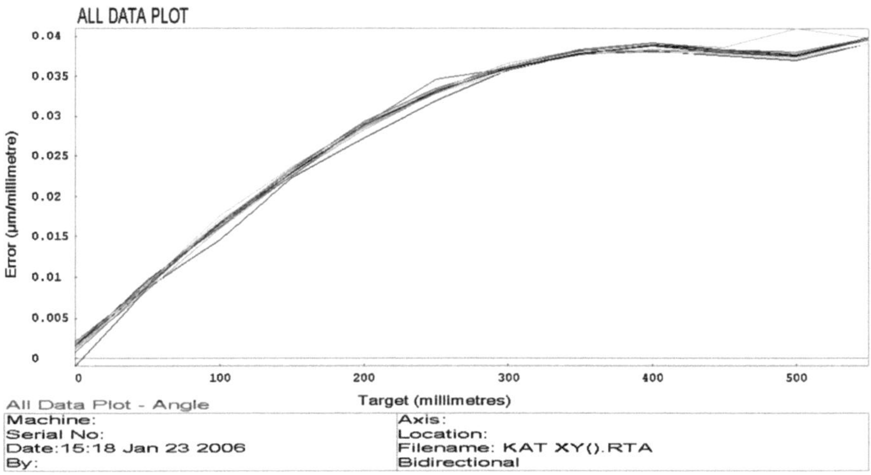

Fig. B.9 Measurement of *X*-axis torsion in relation to *Y*-axis; upper setting $Z = -100$ mm

−100 and −400 mm, and the measuring step for each measurement was 50 mm. As a result of such division, for the *X*-axis (the longer one) 8 measurement points were obtained on the length of 700 mm, and for the *Y*-axis (shorter) 11 measurement points on the length of 500 mm. A separating prism was mounted in the place of the main module of the PH10MQ articulating probe main module measuring tip and Wollastone's prism was lying on the machine measuring table.

In order to reduce random error impact the tests were repeated six times and carried out in two opposite directions, to determine the hysteresis error. To determine machine straightness errors Renishaw's "Short Range Straightness Measurement" application was used, with the long-term averaging of measurement results (Figs. B.10, B.11, B.12, B.13, B.14, B.15 and B.16). As a result the guide

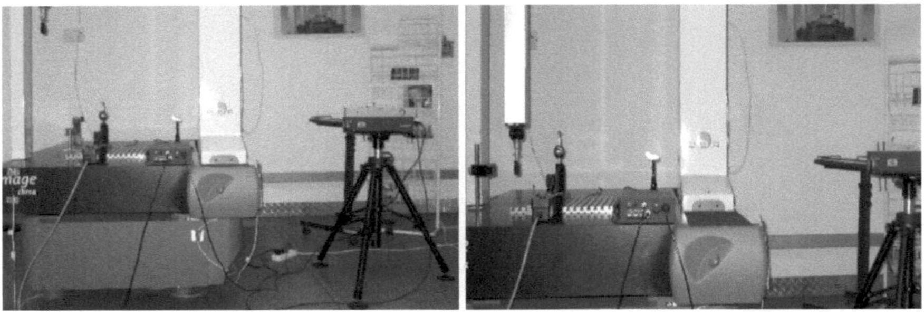

Fig. B.10 Measurements of *X*-axis straightness in relation to *Z*-axis; the vertical position and in relation to *X*-axis horizontal; for the lower bottom location *Z* = −604 mm

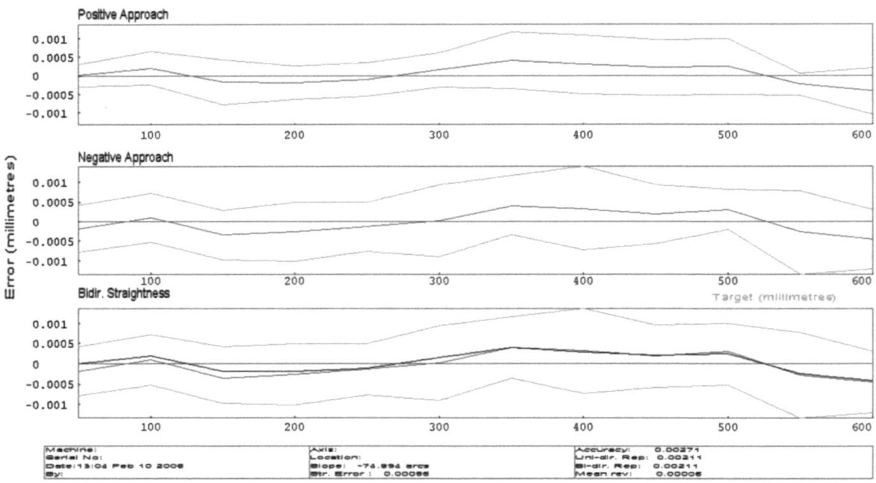

Fig. B.11 Measurement of *X*-axis straightness in relation to *Z*-axis of lower location *Z* = −400 mm

characteristic was obtained, determined as a mean value for particular positions in which the straightness measurement was carried out.

Determination of Machine Geometric Error Components—Positioning Error

Positioning errors were determined for all, *X*-, *Y*-, and Z-axes. Errors were determined in the lower and upper position for the *X*-axis. For the *Y*-axis positioning errors were tested twice in the lower location (right and left) and for one upper location. In the end *Z*-axis positioning errors were determined. Exemplary graphs of measurement results are shown in the figures. Measurements were carried out at

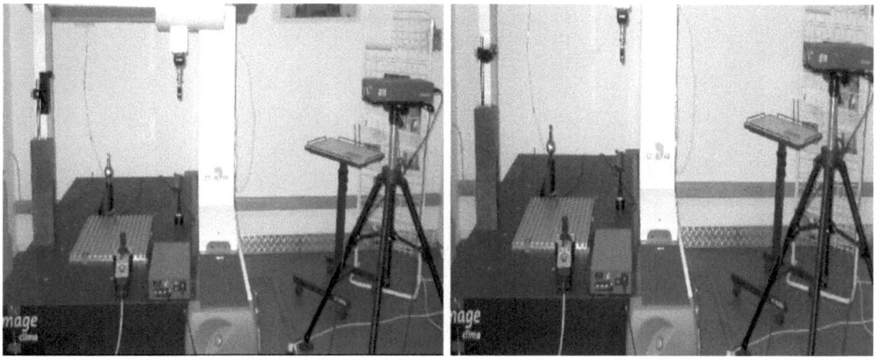

Fig. B.12 Measurement of *X*-axis straightness in relation to *Z*-axis, the vertical position, and to *X*-axis horizontal for the upper location *Z* = −100 mm

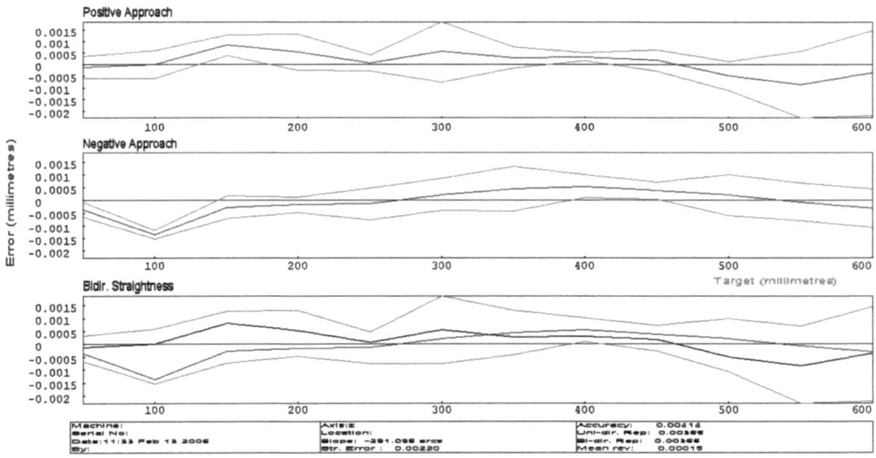

Fig. B.13 Measurement of *X*-axis straightness in relation to *Z*-axis of upper location *Z* = −100 mm

two heights of the *Z*-axis: −100 and −400 mm, and the measuring step for each measurement was 50 mm as well. Both measurements along the *X*-, *Y*-,and *Z*-axes were carried out maintaining the same number of measurement points as during straightness measurements. Tests were carried out with the use of the prism that divides the beam of the laser which was permanently mounted on the machine table, and using a retroreflector which was mounted instead of the main module of the articulating probe head. Tests were carried out bidirectionally in six repetitions. For the positioning measurements Renishaw's "Linear Measurement" application (Figs. B.17 and B.18) was used, and approximated median characteristics for all slides were obtained. Table B.1 presents results of the identification of CMM errors with use of the laser interferometer.

Fig. B.14 Measurement of Y-axis straightness in relation to X-axis, the horizontal position, and in relation to Z-axis vertical for both lower locations left and right $Z = -400$ mm

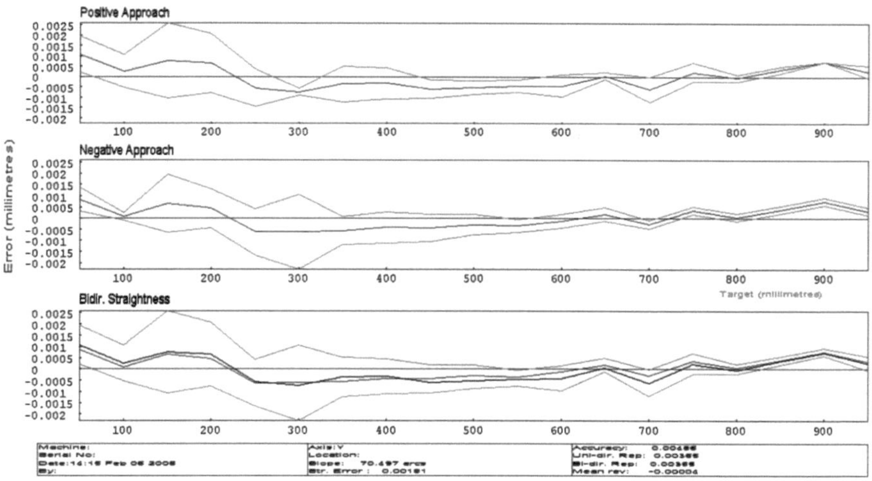

Fig. B.15 Measurement of *Y*-axis straightness in relation to *X*-axis horizontal, for lower location, *Z*

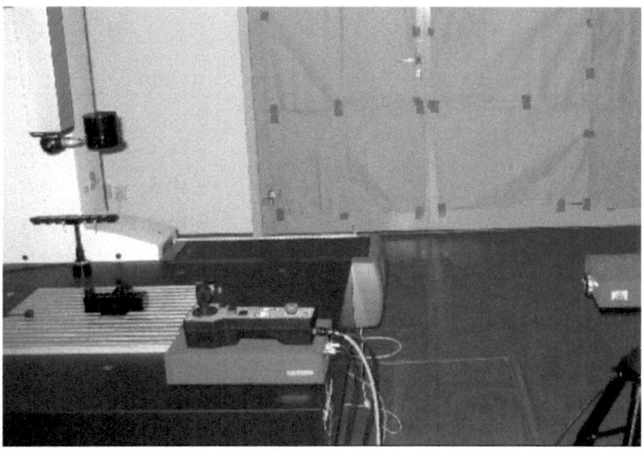

Fig. B.16 Measurement of *Z*-axis straightness in relation to *X*-axis

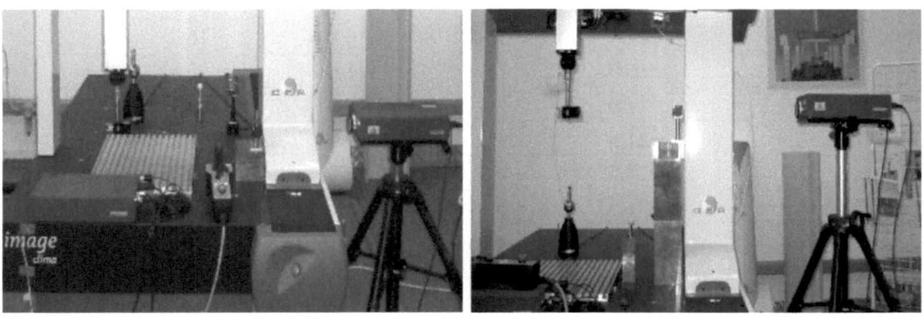

Fig. B.17 Measurements of *X*-axis positioning error in relation to *X*-axis, lower location $Z = -400$ mm and in relation to *X*-axis for upper location $Z = -100$

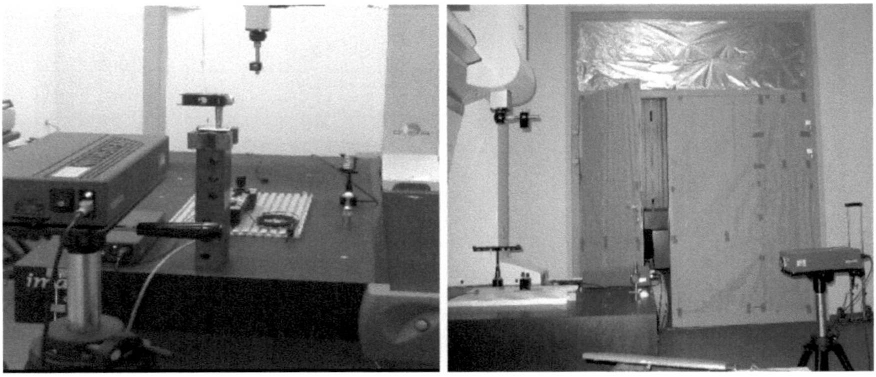

Fig. B.18 Measurements of *Y*-axis positioning, upper position $z = -220$ mm and in relation to *Z*-axis

Table B.1 Correction matrix containing geometric errors of: *X*-axis (**a**), *Y*-axis (**b**), *Z*-axis (**c**), and errors of axes mutual perpendicularity (**d**)

(a)						
Step	*xrx* μm/m	*xry* μm/m	*xrz* μm/m	*xtx* μm	*xty* μm	*xtz* μm
0	0	0	0	0	0	0
100	−0.5	3.7	5.1	−14.1	−0.8	0.1
200	−0.5	7.6	7.6	−30.5	−0.6	−0.2
300	−2.5	10.2	8.7	−46.9	−0.9	−0.3
400	−4.0	9.8	8.0	−62.5	0.1	1.3
500	−4.5	8.0	7.6	−77.4	−0.5	−0.1
600	−5.0	8.1	8.7	−90.0	−0.1	0.4
700	−6.0	7.2	10.4	−101.8	0	0
(b)						
Step	*yrx* μm/m	*yry* μm/m	*yrz* μm/m	*ytx* μm	*yty* μm	*ytz* μm
0	0	0	0	0	0	0
50	14.1	−2.8	8.6	1.3	−4.9	−1.3
100	26.6	−5.0	13.5	1.1	−10.4	−2.7
150	34.7	−6.0	13.2	−0.6	−16.2	−2.2
200	38.2	−10.0	11.4	0.1	−22.6	−1.8
250	38.7	−17.0	9.4	−0.1	−27.9	−2.2
300	36.7	−21.5	9.0	−0.1	−34.4	−1.2
350	32.7	−23.5	8.3	−0.3	−40.4	−0.6
400	31.3	−27.5	6.4	−0.1	−46.4	0.5

(continued)

Table B.1 (continued)

(b)

Step	*yrx* μm/m	*yry* μm/m	*yrz* μm/m	*ytx* μm	*yty* μm	*ytz* μm
450	34.8	−31.5	6.5	−0.7	−50.7	−1.0
500	35.1	−34.0	10.1	0	−53.8	0

(c)

Step	*zrx* μm/m	*zry* μm/m	*zrz* μm/m	*ztx* μm	*zty* μm	*ztz* μm
0	0	0	0	0	0	0
−50	10.0	6.3	−22.5	2.2	2.3	5.8
−100	18.5	16.5	−25.6	4.4	3.9	13.8
−150	27.5	23.5	−21.3	7.6	8.7	22.0
−200	34.5	28.5	−20.3	10.3	11.4	31.0
−250	37.5	30.5	−26.1	11.0	13.9	39.5
−300	37.0	31.0	−32.2	11.4	13.4	45.9
−350	36.0	25.0	−33.7	10.0	14.2	53.1
−400	30.0	17.0	−44.1	6.7	11.5	60.9
−450	21.5	7.0	−44.3	2.2	7.4	69.6
−500	12.5	−0.5	−50.3	0	0	78.3

(d)

xwy μm/m	*xwz* μm/m	*ywz* μm/m
49.83	42.37	53.24

Components of geometric errors and correction matrix CAA are presented in Table B.1.

Annex C
Identification of Geometric Error Components Using Plate Standard and KalKom 4.0 Program According to PTB Method

Measurements were carried out for the medium accuracy measuring machine (Figure C.1), maximum permissible error: MPE = 0.0017 + 0.003/1000 · L mm, and measuring volume: 700 × 500 × 500 mm. This machine cooperates with PC-DMIS and Quindos software. In geometric error tests the Feinmess GmbH thermostable plate standard made of Robax with expansion coefficient of $0.5 \times 10^{-6}\ \mathrm{K}^{-1}$ (Fig. C.1) was used during research.

This standard had 36 evenly placed spherical holes. During the measurements the part segment of the inner sphere was measured.

To estimate the CMM geometry errors the plate standard was measured in four set positions (Fig. 4.13 in Sect. 4.2.2). In order to eliminate "drift" effect and hysteresis the plate standard should be measured clockwise and counterclockwise.

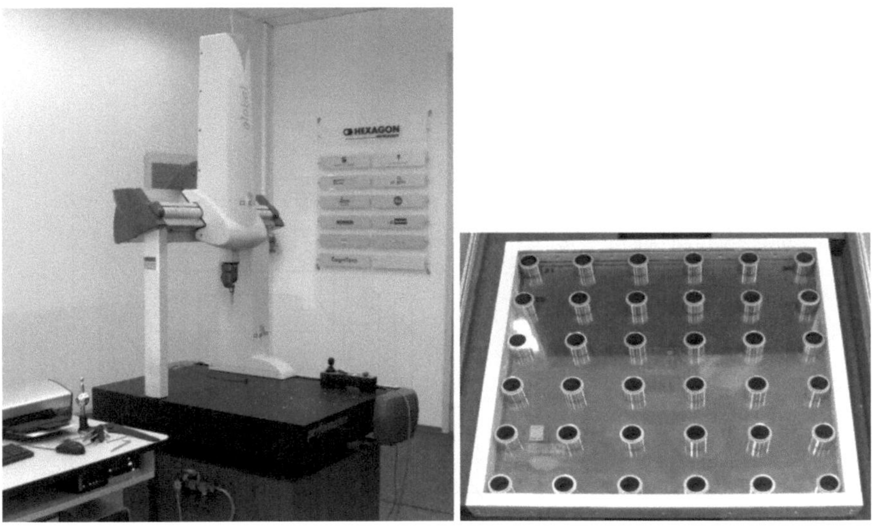

Fig. C.1 DEA's global image coordinate measuring machine and Feinmess's plate standard

J.A. Sładek, *Coordinate Metrology*, Springer Tracts in Mechanical Engineering,
DOI 10.1007/978-3-662-48465-4

Fig. C.2 Measurement order in one position of the plate standard

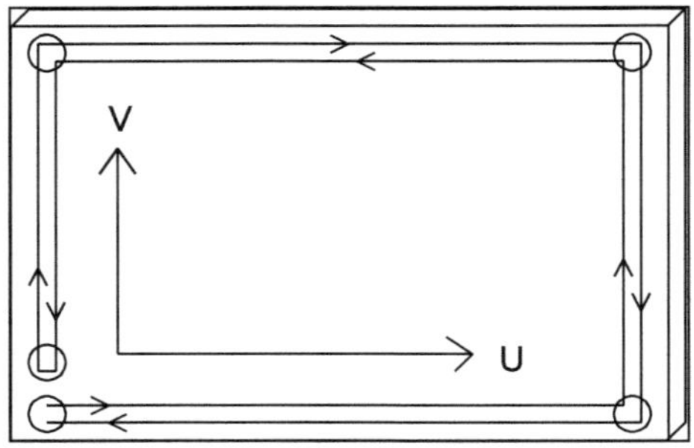

All reference elements should be measured twice (Fig. C.2) to reduce the impact of random errors.

During the measurements the temperature condition was monitored during the whole process. This simplified the localization of possible incompatibilities between repeated measurements. The number and arrangement of sensors are shown in Fig. C.3.

As a result of measurements performed with the use of the plate standard in four set positions six data files were received. These files contained coordinates of all 36 holes of plate standard centers. Comparison of the values of hole center coordinates with nominal locations, received from measurement with nominal values, was carried out in the KalKom 4.0 program, which allowed us to determine machine geometric errors and as well as calculations of CAA matrix parameters. These parameters may be used in correcting software in the form of matrix CAA—GEOComp. DEA (Figs. C.4, C.5, C.6, C.7, C.8, C.8, C.9, C.10, C.11, C.12, C.13, C.14, C.15, C.16, C.17, C.18, C.19 and C.20).

Fig. C.3 Temperature sensor arrangement. When the plate standard is used, the temperature sensor is not used due to material characteristics

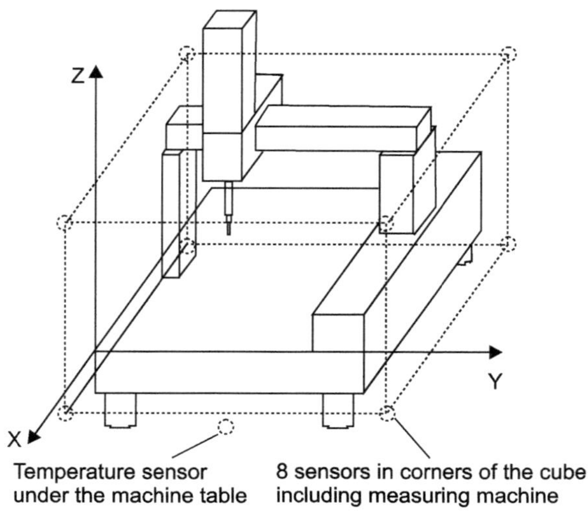

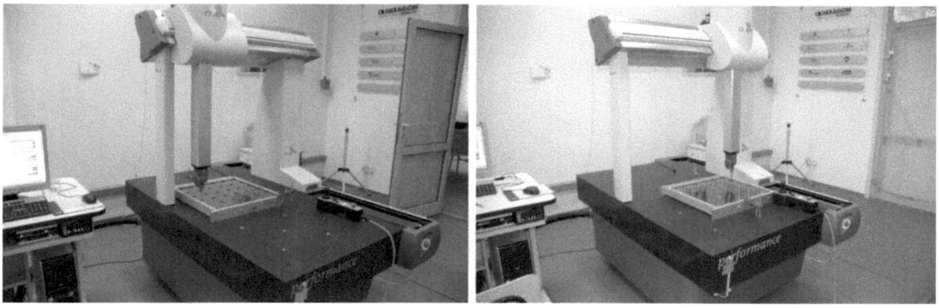

Fig. C.4 Location 111 and 211

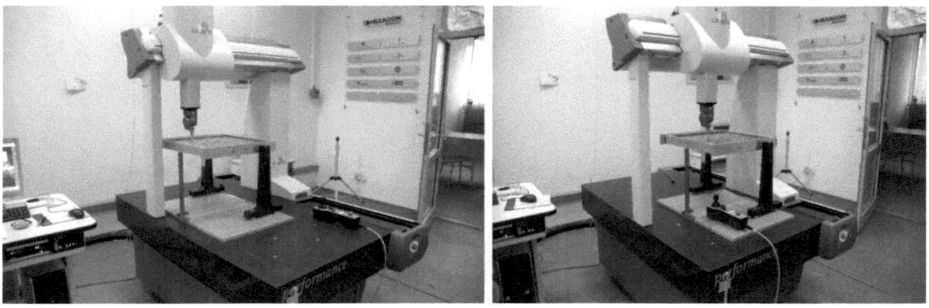

Fig. C.5 Location 112 and 212

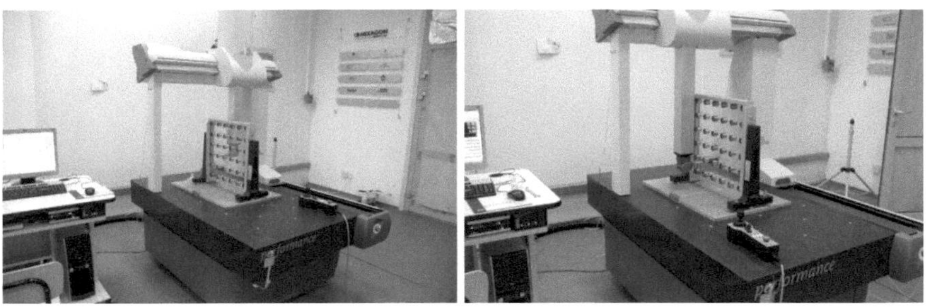

Fig. C.6 Location 121 and 122

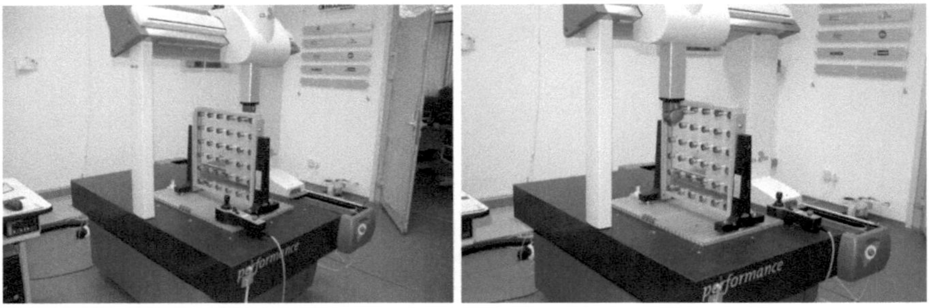

Fig. C.7 Location 221 and 222

Fig. C.8 Location 131 and 132

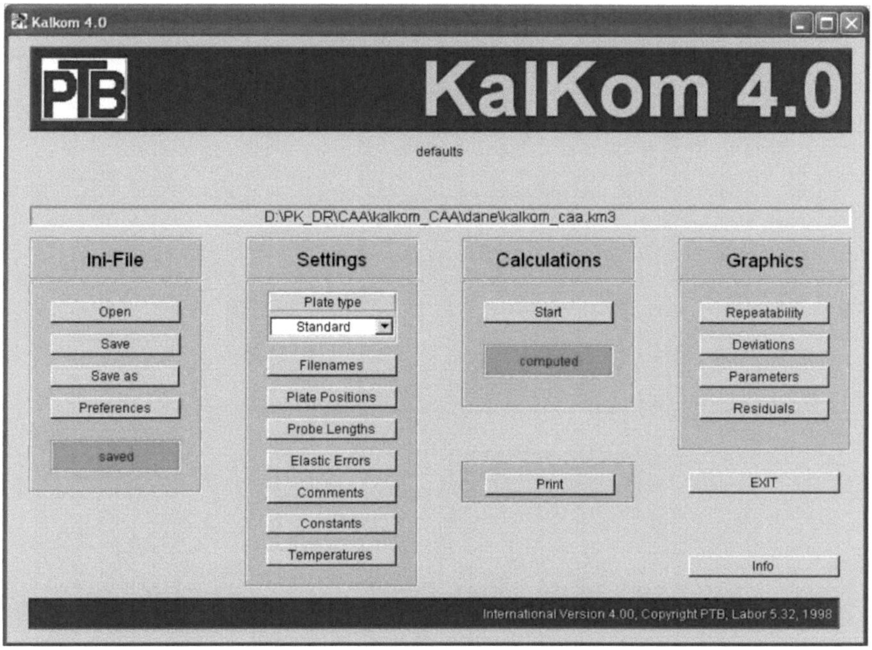

Fig. C.9 Main window of KalKom 4.0 program

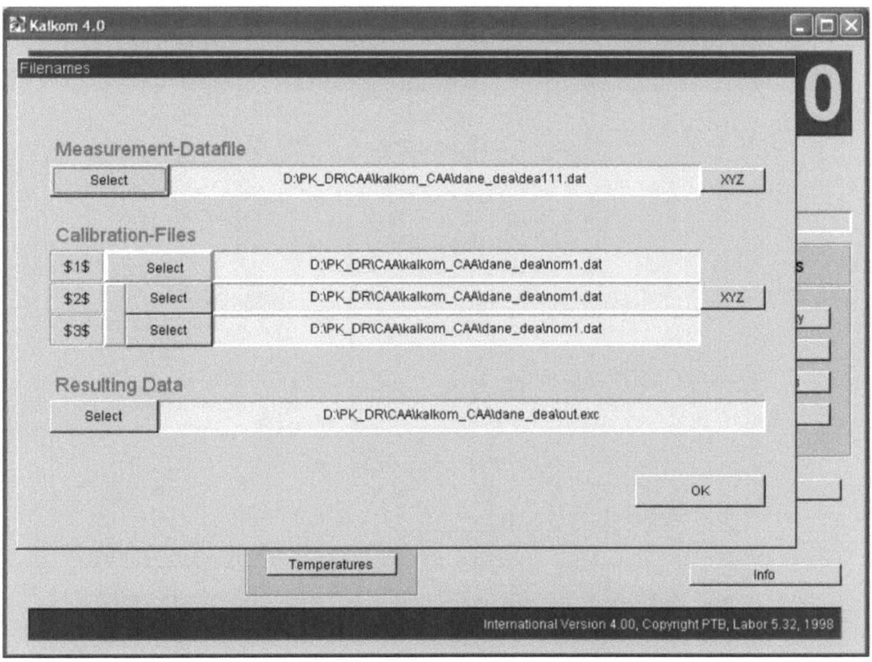

Fig. C.10 Window of KalKom 4.0 program allowing us to set measurement data

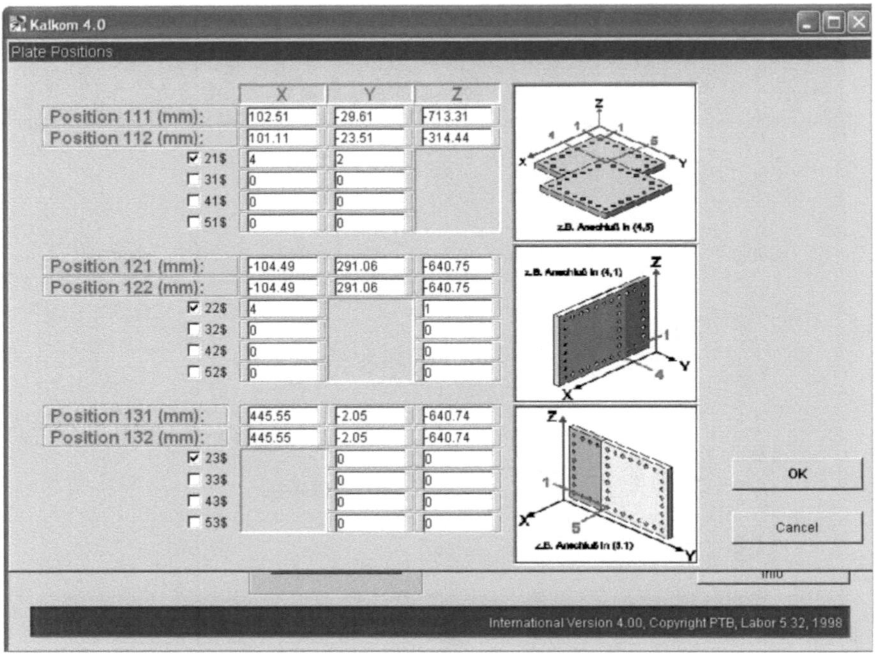

Fig. C.11 Window allowing us to set coordinates of certain locations of plate standard

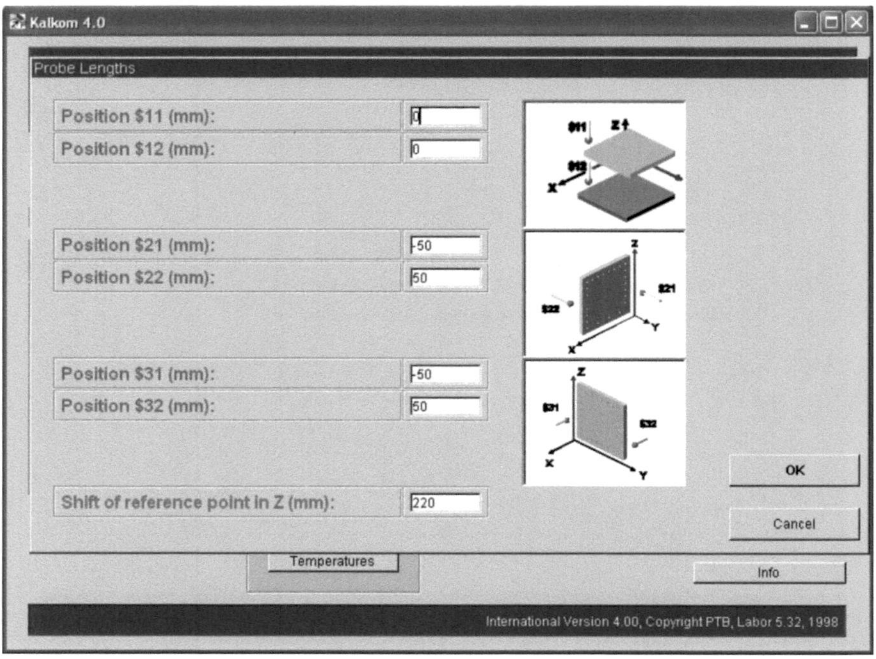

Fig. C.12 In this window we set distances of the center of measuring tip to probe reference point

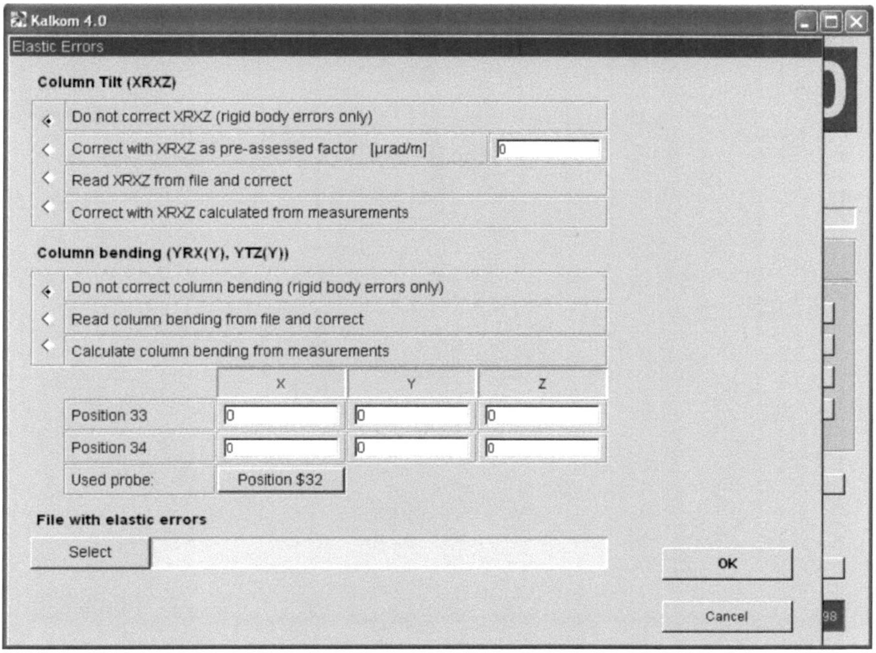

Fig. C.13 Window allowing the correction of CMM flexible errors

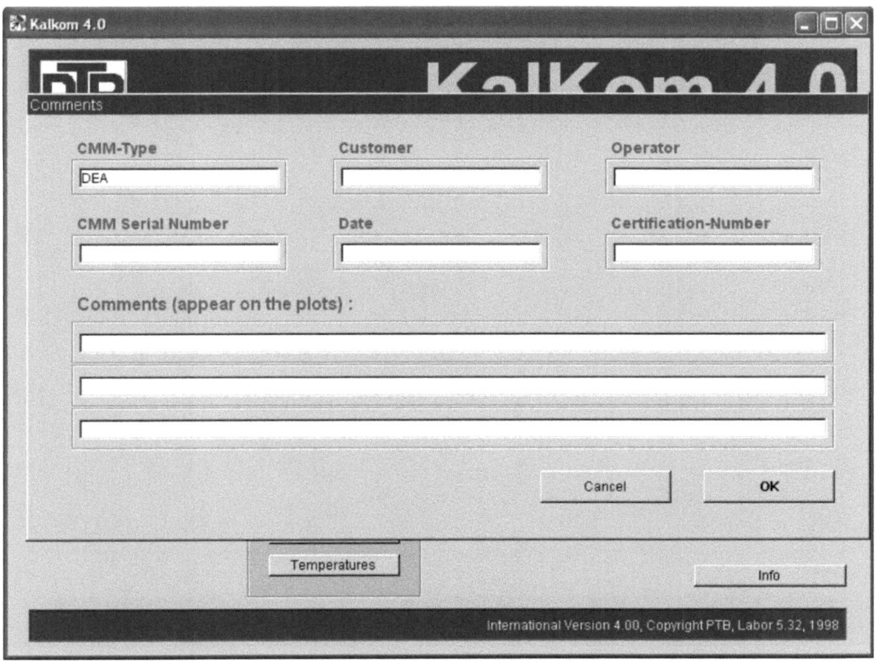

Fig. C.14 Window to comments insertion

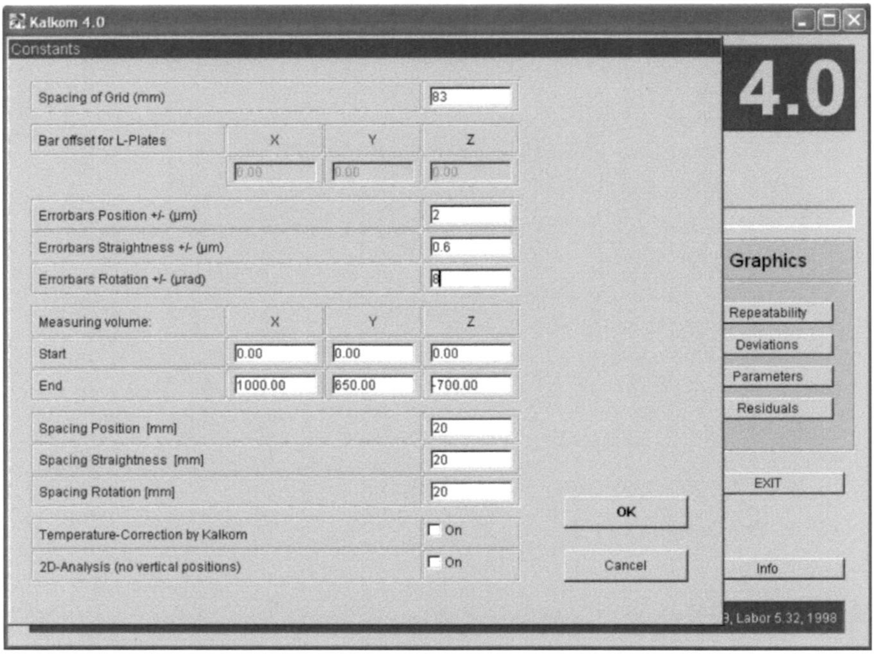

Fig. C.15 Window allowing us to set stable values needed for calculations

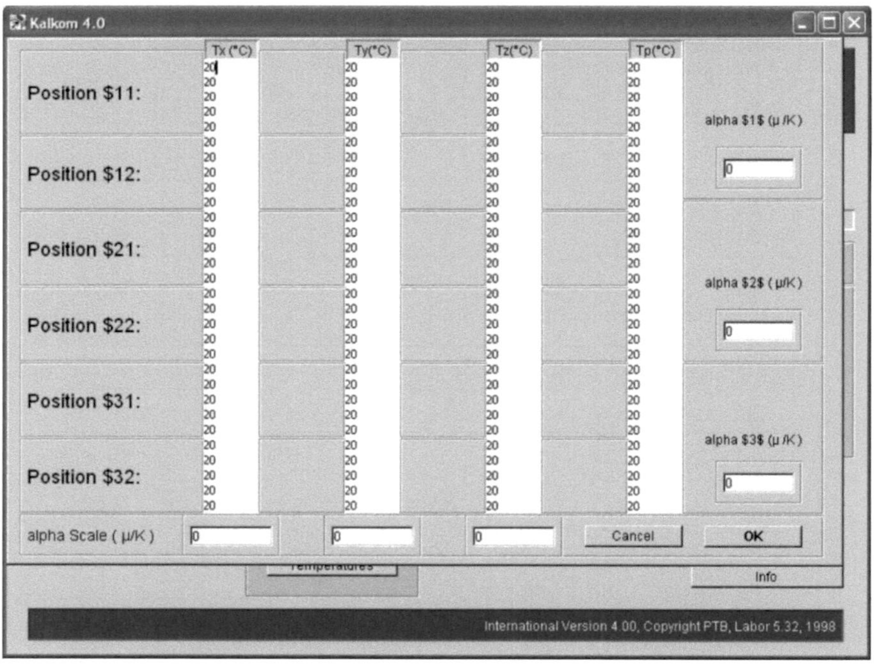

Fig. C.16 Window allowing temperature correction by KalKom 4.0 program

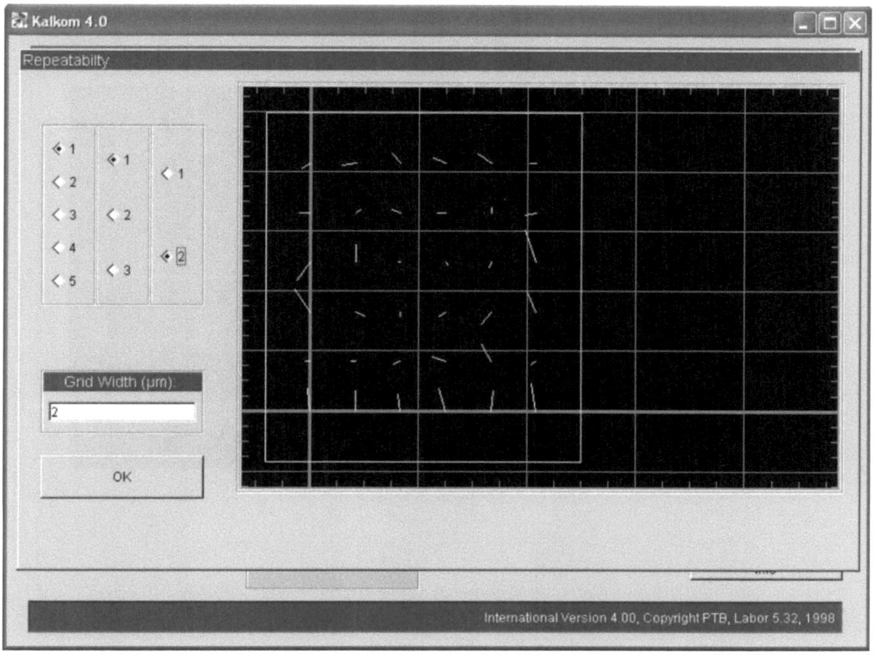

Fig. C.17 Window presenting the repeatability of plate standard measurement

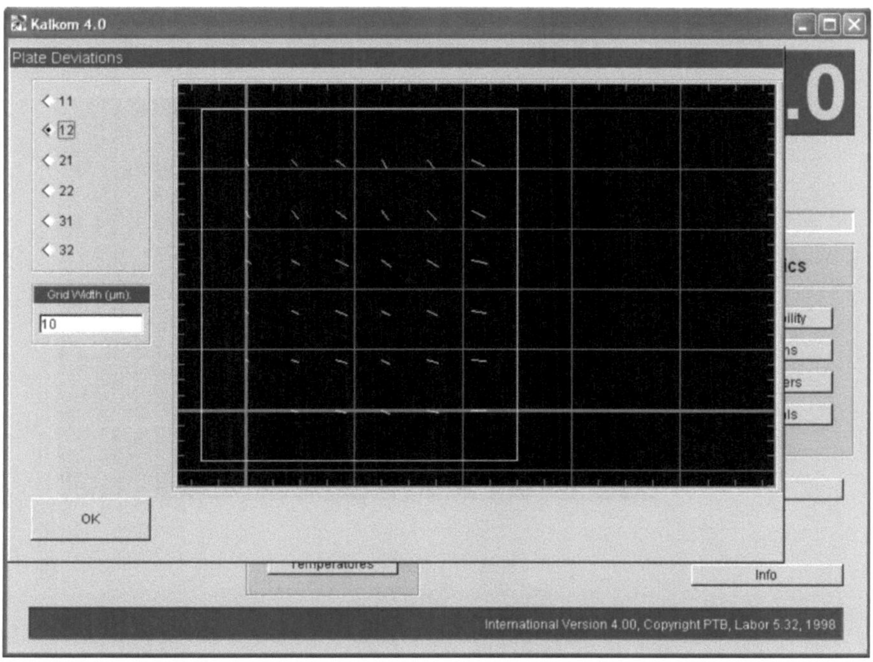

Fig. C.18 Window presenting location errors in all measured positions

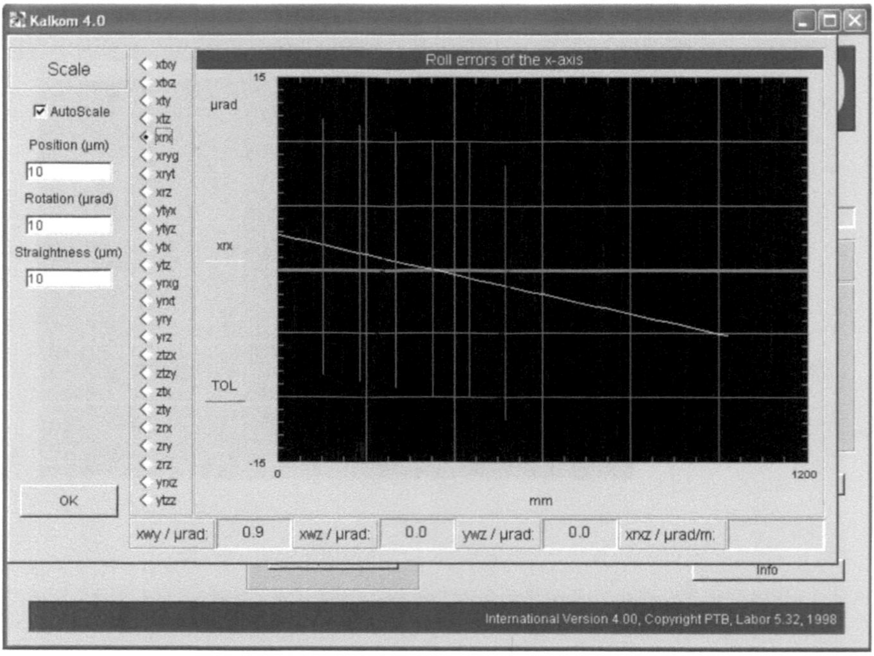

Fig. C.19 Window presenting the possibility to observe graphs of certain components of geometric errors

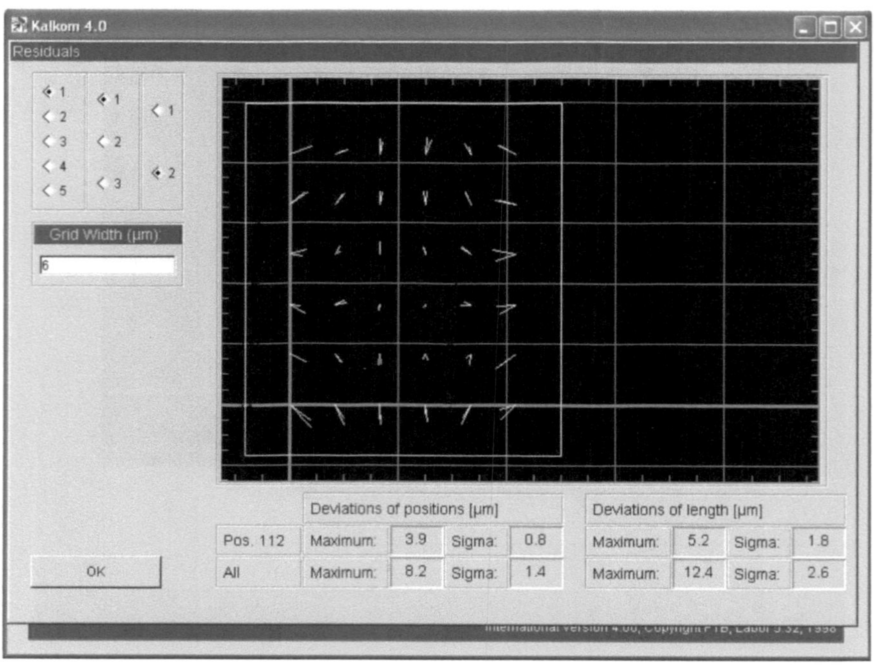

Fig. C.20 Window presenting deviations of standard sphere center positions

Table C.1 presents parameters of the CAA correction matrix containing geometric errors for the machine

Table C.1 Correction matrix containing geometric errors of: *X*-axis (**a**), *Y*-axis (**b**), *Z*-axis (**c**), and errors of mutual axes perpendicularity (**d**)

(a)

Step	*xrx* μrad	*xry* μrad	*xrz* μrad	*xtx* μm	*xty* μm	*xtz* μm
0	−22.2	2.3	−4.2	15.1	12.4	4.0
50	−18.6	2.0	−3.5	7.8	10.3	3.4
100	−14.9	1.6	−2.9	0.5	8.3	2.7
150	−11.2	1.2	−2.2	−6.9	6.2	2.0
200	−7.5	0.8	−1.5	−14.4	4.2	1.4
250	−3.8	0.4	−0.8	−21.9	2.1	0.7
300	−0.1	0.0	−0.1	−29.6	0.1	0.0
350	3.6	−0.3	0.6	−37.3	−2.0	−0.6
400	7.2	−0.7	1.3	−45.1	−4.0	−1.3
450	10.9	−1.1	2.0	−52.9	−6.1	−2.0
500	14.6	−1.5	2.7	−60.9	−8.1	−2.6
550	18.3	−1.9	3.4	−68.9	−10.2	−3.3
600	22.0	−2.3	4.1	−77.0	−12.2	−4.0
650	25.7	−2.6	4.8	−85.2	−14.3	−4.7
700	29.4	−3.0	5.5	−93.5	−16.3	−5.3

(b)

Step	*yty* μrad	*ytx* μrad	*ytz* μrad	*yty* μm	*ytx* μm	*ytz* μm
0	−12.8	−1.1	0.0	1.5	−7.3	0.0
50	−9.3	−0.8	3.0	2.6	−4.5	0.0
100	−5.8	−0.5	3.4	0.2	−2.4	0.0
150	−2.3	−0.2	2.0	−4.9	−0.8	0.0
200	1.2	0.1	−0.1	−11.5	0.7	0.0
250	4.7	0.4	−2.1	−18.6	2.1	0.0
300	8.2	0.7	−3.1	−25.4	3.8	0.0
350	11.7	1.0	−2.1	−30.6	6.0	0.0
400	15.2	1.3	1.7	−33.4	8.9	0.0
450	18.7	1.6	9.4	−32.7	12.7	0.0
500	22.2	1.9	21.7	−27.5	17.7	0.0

(c)

Step	*zrz* μrad	*zry* μrad	*zry* μrad	*ztz* μm	*ztx* μm	*zty* μm
0	−7.6	−24.7	47.6	−75.1	−9.0	−4.2
−50	−2.4	−31.9	67.6	−82.7	−11.7	−2.6
−100	4.2	−38.1	90.5	−90.3	−13.9	1.1

(continued)

Table C.1 (continued)

(c)

Step	*zrz* μrad	*zry* μrad	*zry* μrad	*ztz* μm	*ztx* μm	*zty* μm
−150	12.2	−42.7	116.3	−97.8	−15.3	7.3
−200	21.7	−45.2	145.0	−105.4	−15.8	16.4
−250	32.7	−45.0	176.6	−113.0	−15.1	29.0
−300	45.1	−41.5	211.0	−120.5	−12.9	45.4
−350	59.0	−34.4	248.4	−128.1	−9.1	66.0
−400	74.4	−23.0	288.6	−135.7	−3.5	91.5
−450	91.3	−6.7	331.8	−143.3	4.3	122.1
−500	109.8	14.9	377.8	−150.8	14.5	158.3

(d)

xwy	*xwz*	*wyz*
μrad	μrad	μrad

Geometric error graphs done in KalKom 4.0 program:

See Figs. C.21, C.22, C.23, C.24, C.25, C.26, C.27, C.28, C.29, C.30, C.31, C.32, C.33, C.34, C.35, C.36, C.37, C.38 and C.39

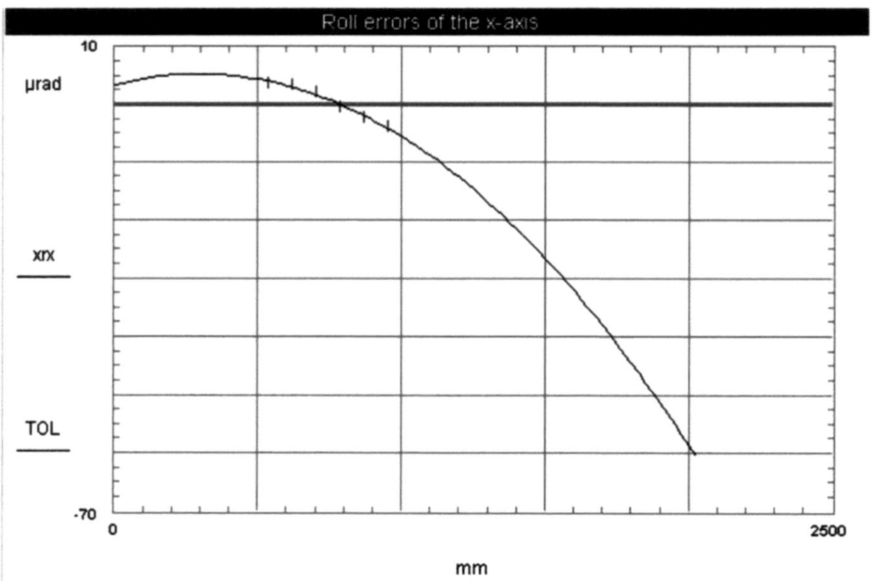

Fig. C.21 *xrx* error

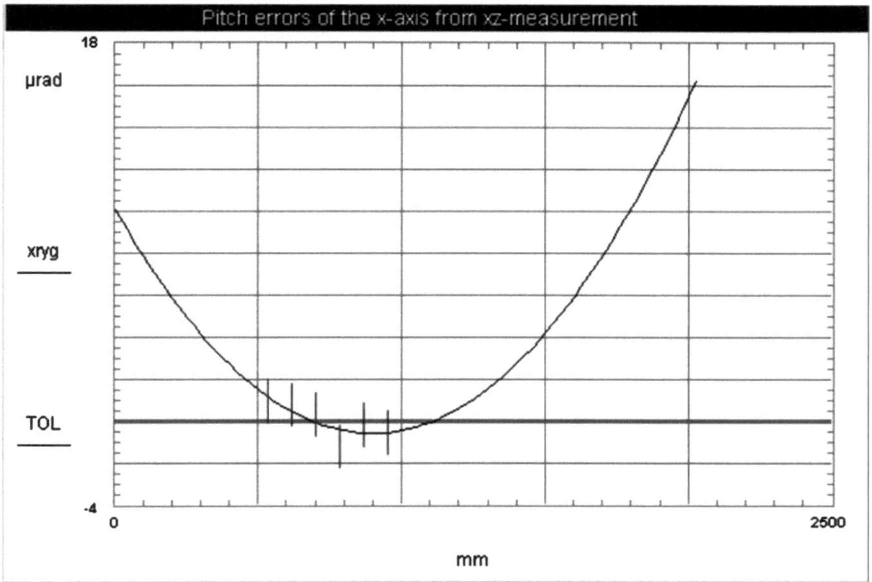

Fig. C.22 *xry* error determined in *XZ* plane

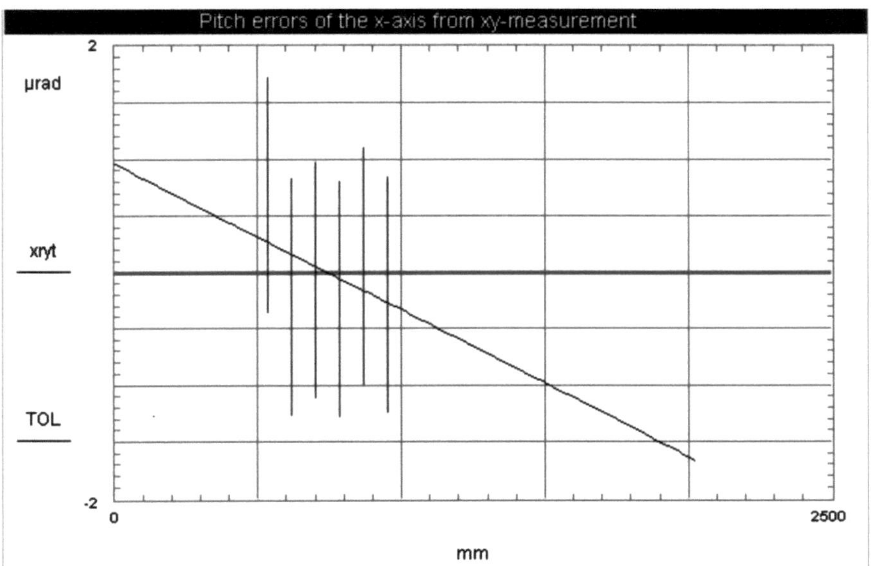

Fig. C.23 *xry* error determined in *XY* plane

Fig. C.24 *xrz* error

Fig. C.25 *xtx* error determined in *XY* plane

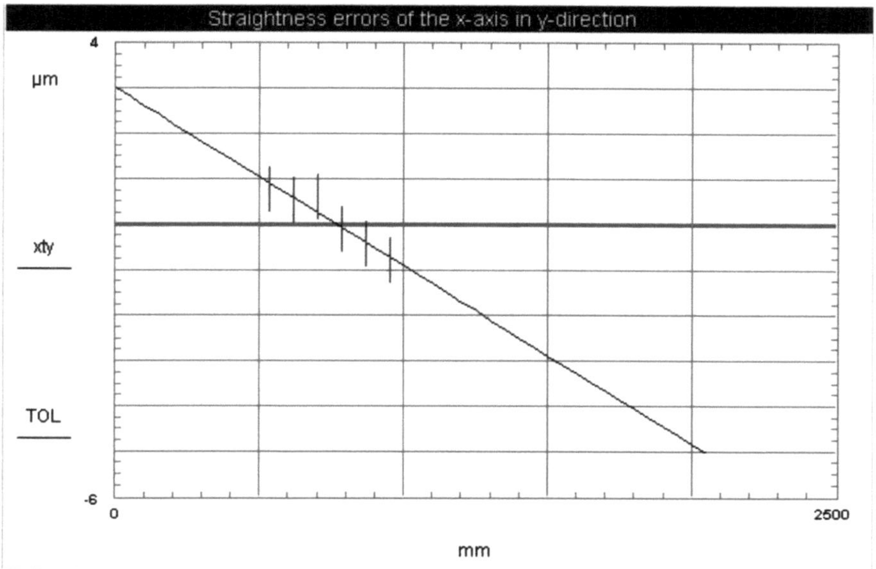

Fig. C.26 *xty* error

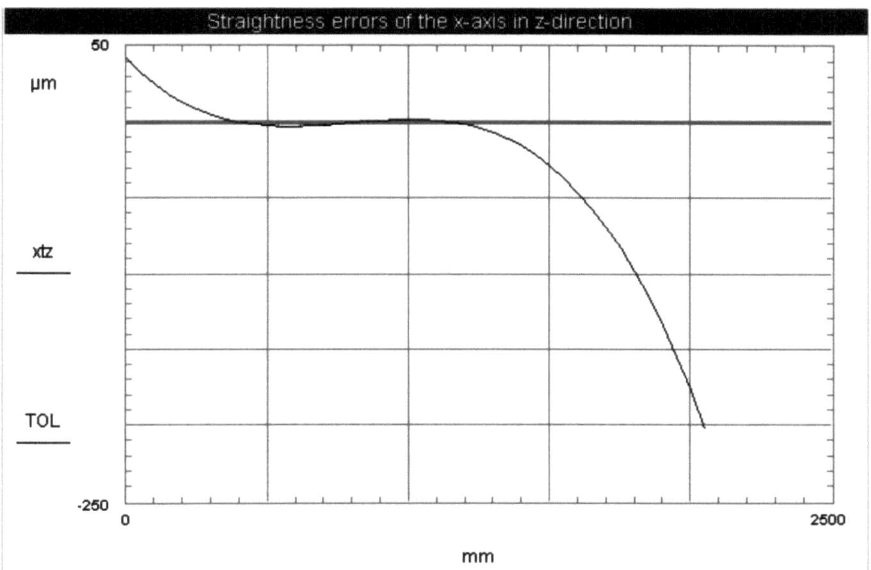

Fig. C.27 *xtz* error

Fig. C.28 *yrx* error determined in *XY* plane

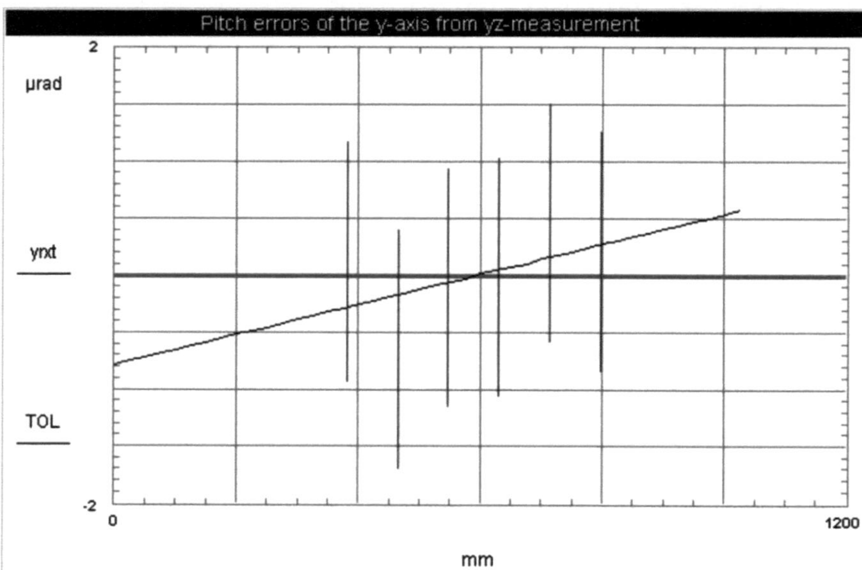

Fig. C.29 *yrx* error determined in *YZ* plane

Fig. C.30 *yry* error

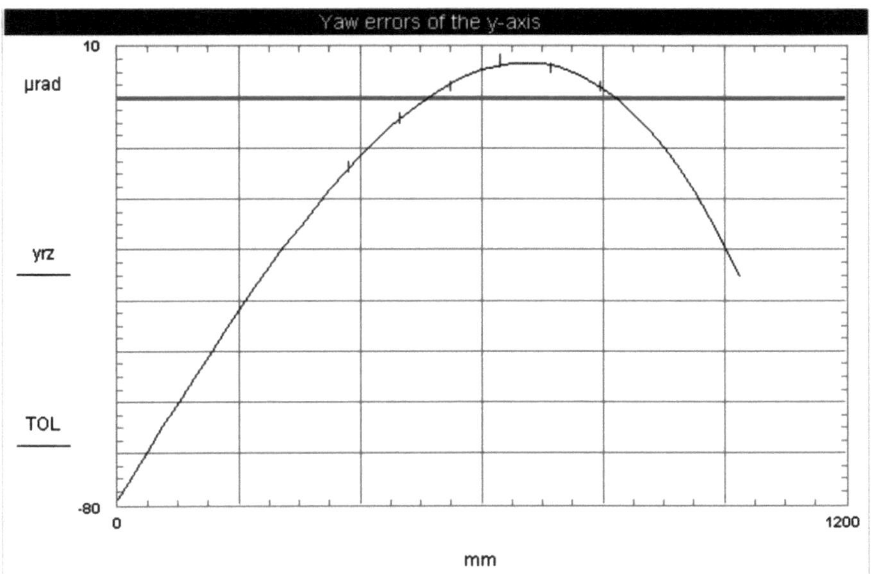

Fig. C.31 *yrz* error

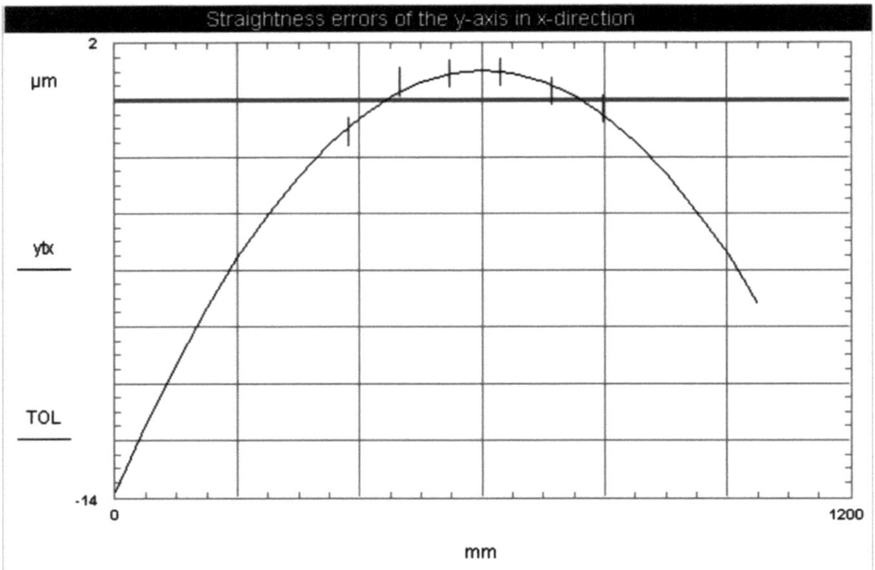

Fig. C.32 *ytx* error

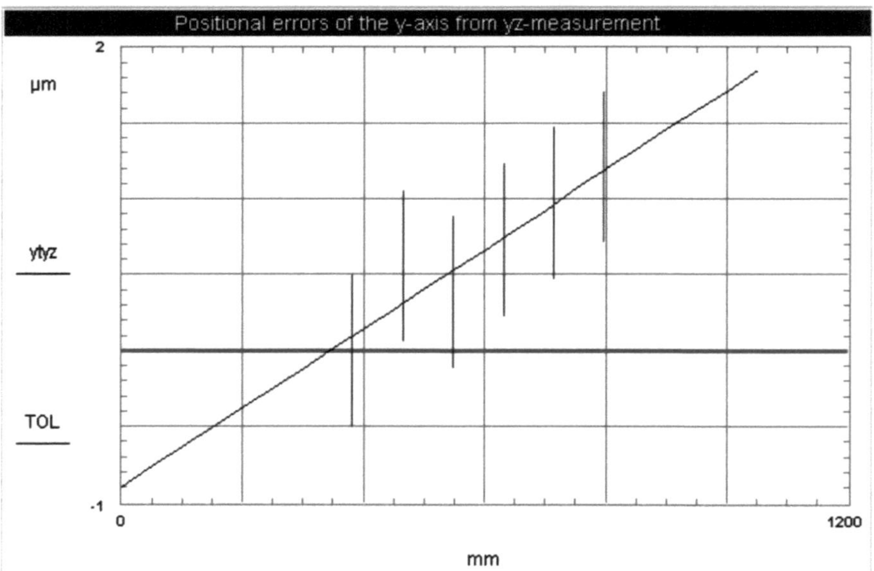

Fig. C.33 yty error determined in *YZ* plane

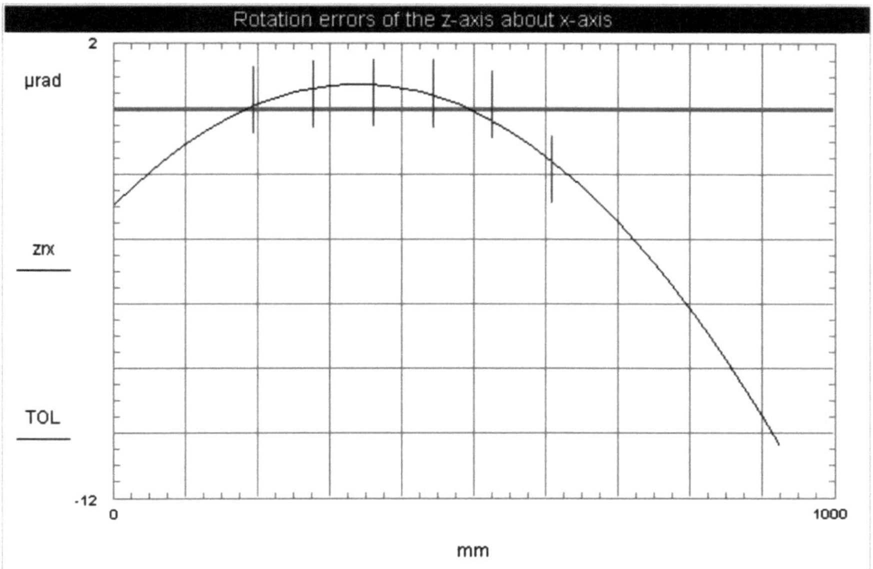

Fig. C.34 *zrx* error

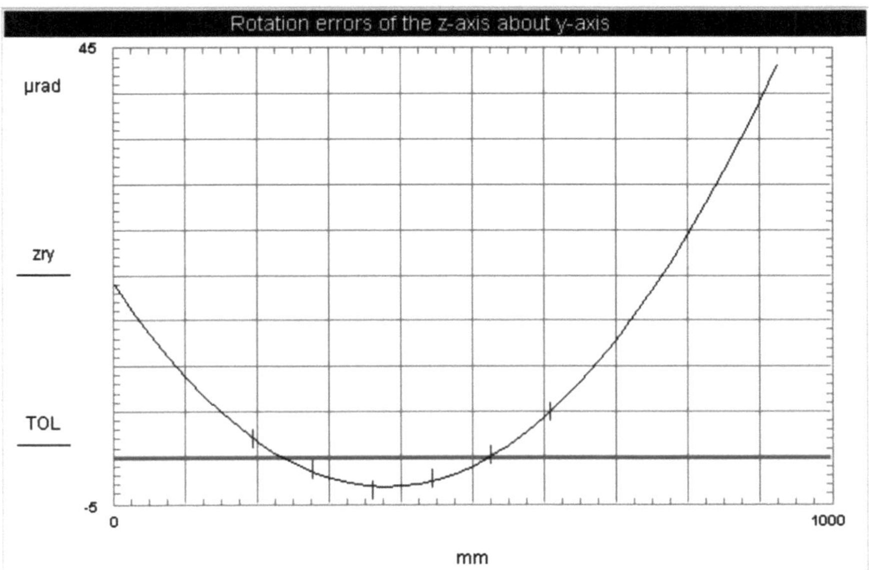

Fig. C.35 *zry* error

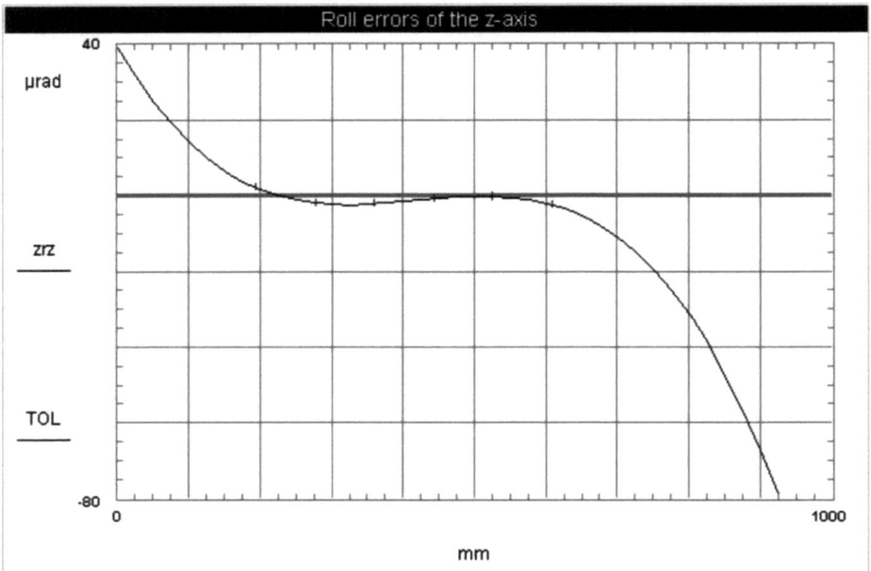

Fig. C.36 *zrz* error

Fig. C.37 *ztx* error

Fig. C.38 *zty* error

Fig. C.39 *ztz* error determined in *XZ* plane

Annex D
Example of CMM Geometric Error Identification Using Laser Tracker and Trac-Cal Software (Etalon AG)

Operation and Functions of Laser Tracker

The Leica Laser tracker LTD 840 is a system of coordinate measurement systems. It allows us to determine the location of a reflector built into a ball with a diameter of 0.5 or 1.5 in. It is equipped with laser systems for length measurements: interference interferometer one (IFM) and the absolute distance meter (ADM) and precise angle measurement systems. The first one, IFM, measures interferometrically the distance to the reflector. The accuracy of this system is ±0.5 μm/m. The ADM system also allows us to identify the distance from the reflector to the device. The indications of this system can be used by the interference system as a basis, therefore it is possible to continue the measurement after breaking the beam. The accuracy of the ADM system is 0.025 mm in whole measuring volume. The scheme of laser tracker operation is presented in Fig. D.1.

Determination of coordinate measuring machine geometry errors, with the use of a laser tracker, was carried out for DEA's Global Image machine (Fig. D.3), maximum permissible error: MPE = 0.0017 + 0.003/1000 · L mm, and measuring volume: 700 × 500 × 500 mm. This machine cooperates with PC-DMIS and Quindos software.

Identification of CMM geometric errors is realized in accordance with a measuring strategy compatible with the multilateration method. Only a distance measurement from a number of different positions is used here to determine the location position of a localized object. This method is generally used in GPS satellite navigation systems. The principle is shown in Fig. D.2. It has also been used for many years in measuring systems, so-called Internal GPS, to measure large objects, and recently is also used to correct the accuracy of measuring machines or to create coordinate measuring systems of a very large range. Knowing the distance of a localized object from one measurement station it is known only that the point is on the sphere; the center of the sphere is determined by this measurement station and when the radius of the sphere is equal to the measured distance. If the distances from two measurement stations are known, the localized object position is in the intersection of two spheres, but the figure created from the intersection of two spheres may even contain an infinite number of points. This problem can be

J.A. Sładek, *Coordinate Metrology*, Springer Tracts in Mechanical Engineering,
DOI 10.1007/978-3-662-48465-4

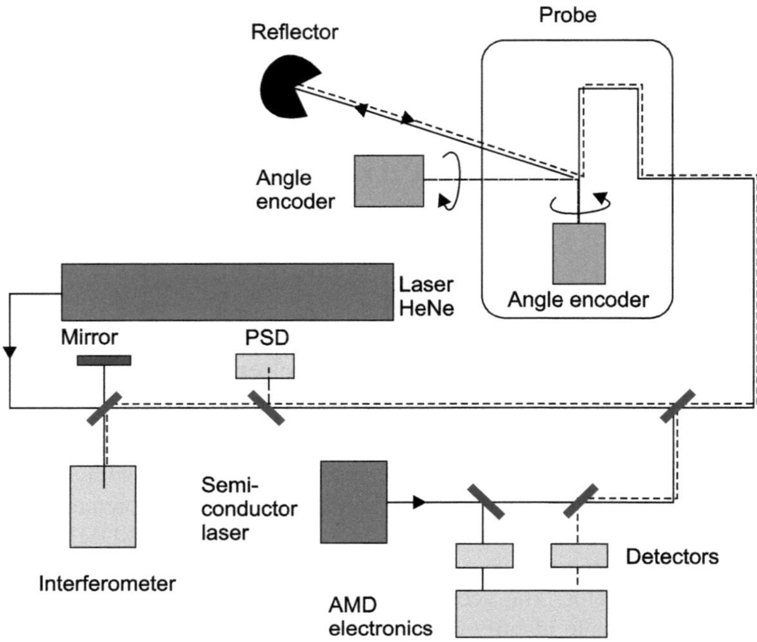

Fig. D.1 Scheme of laser tracker operation

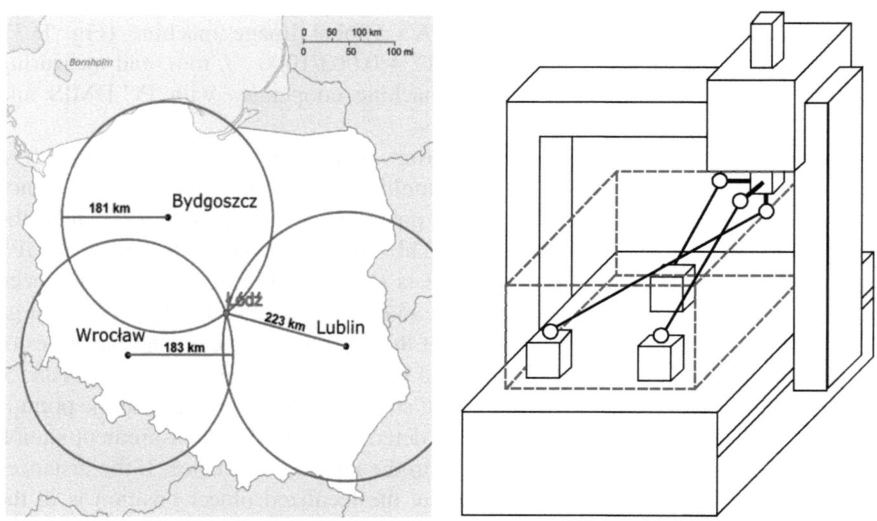

Fig. D.2 *F* principle of multilateration

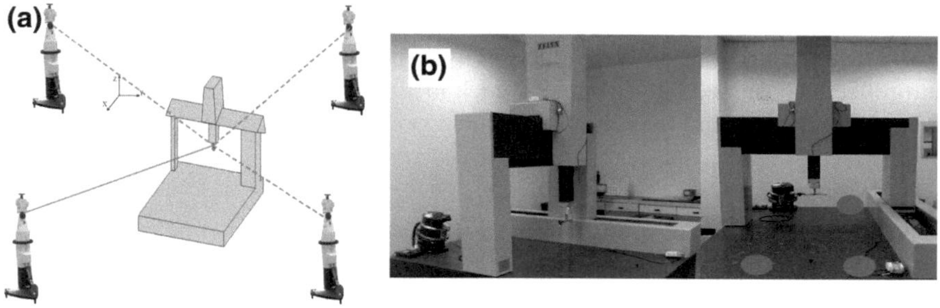

Fig. D.3 Laser tracker positions during measurements, (**a**) with the use of Leica Laser Tracker LTD 840, (**b**) with the use of Etalon AG Laser Tracker

reduced only by using a third more measurement stations (or measuring station positions). The more measurement stations or positions from which the measurement is carried out, the more accurate the determination of the location of the localized point will be.

After defining the measurement strategy the laser tracker is set in the defined position and the reflector is mounted in place of the probe. Then measurements are carried out in every single position (Fig. D.2). In geometric error tests the Leica Laser Tracker LTD 840 (Fig. D.3) and Etalon AG Laser Tracer were used

The measuring machine pinole ram moves with a mounted reflector along a specific trajectory in the machine volume. During the measurement the machine stops for a short time at each measurement point on the designated measurement path. At that time machine position, as well as the length value indicated by the laser tracker is read and saved. The measuring machine is synchronized with the laser tracker by connection realized by the Trac-Cal program. After completing the first measurement cycle (i.e., after all the defined measurement points in a given measurement path are determined) the reflector and/or laser tracker position is changed. Then another measurement path is realized (Fig. D.4).

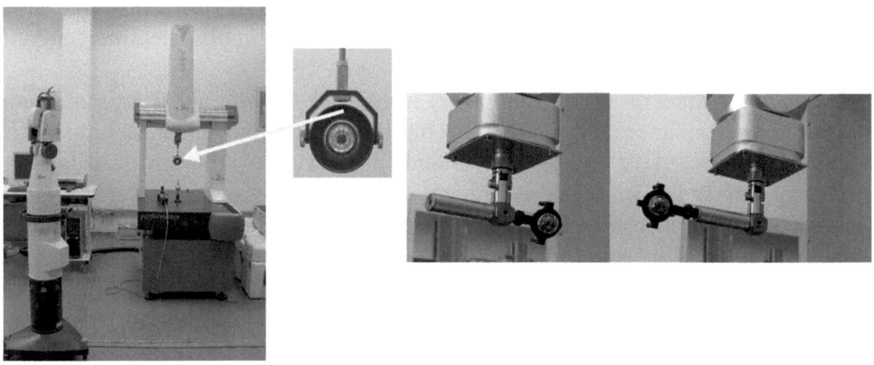

Fig. D.4 Measurements with the use of Leica laser tracker with the retroreflectors used

Measurement results and their appropriate uncertainties are calculated by the Trac-Cal program. The difference between the measured distance and the nominal retroreflector position is determined at each measurement point, and thus geometric errors of each machine axis in measurement points are determined.

Methodology of Work with the Software

This program is used generally with Etalon's laser tracer, described in Sect. 4.2.3, but it can also be used with a laser tracker system. Measurement can be done with the machine directly controlled by the Trac-Cal program or by the import of a generated CNC file into the machine controller. If the measuring machine is numerically controlled (CNC), the control code generated by Trac-Cal must be read by it. For this purpose the CNC file may be saved in different formats. Measurements are carried out independently of the program; measurement data from the calibrated machine and from the laser tracker have to be saved in a format that allows them to be read and understood by the program, which then performs calculations and analyses. On the other hand in the case of direct control of the machine by the Trac-Cal program, data from the machine are read currently and automatically, and then they are saved in the file. Measurement strategy is planned and defined in the Trac-Cal program. The number of measurement paths (trajectories) and retroreflector and laser tracker positions, as well as measurement points in the trajectory, in general mainly depend on the kinematic structure of the measuring machine, the availability, and desired accuracy.

Description of Trac-Cal 2.1 Program

The program is operated by means of using four tabs, or panels, as they are called in software.

The *Plan* panel (Figure D.5) is used to define the measurement path and laser tracker position and it also allows us to perform a simulation of a measurement and generation of CNC code.

The *Settings* function of the *Plan* panel is used for defining measuring machine parameter determination (Fig. D.6). The machine measurement range is given in columns *Start/End Axis*. In the *Step* column the distance between measurement points is given. In the *Start/End mapping* columns the machine measurement range is given, in which measurements will be carried out. The *Step (interpoll)* column defines the interval in which data in the output file will be generated. In the *Set functions is zero At* it is possible to define the beginning of the correction matrix coordinate system. By default this value is set automatically by the program.

The *configure* function is used for defining measurement strategy with a number of laser tracker positions with their corresponding measurement paths.

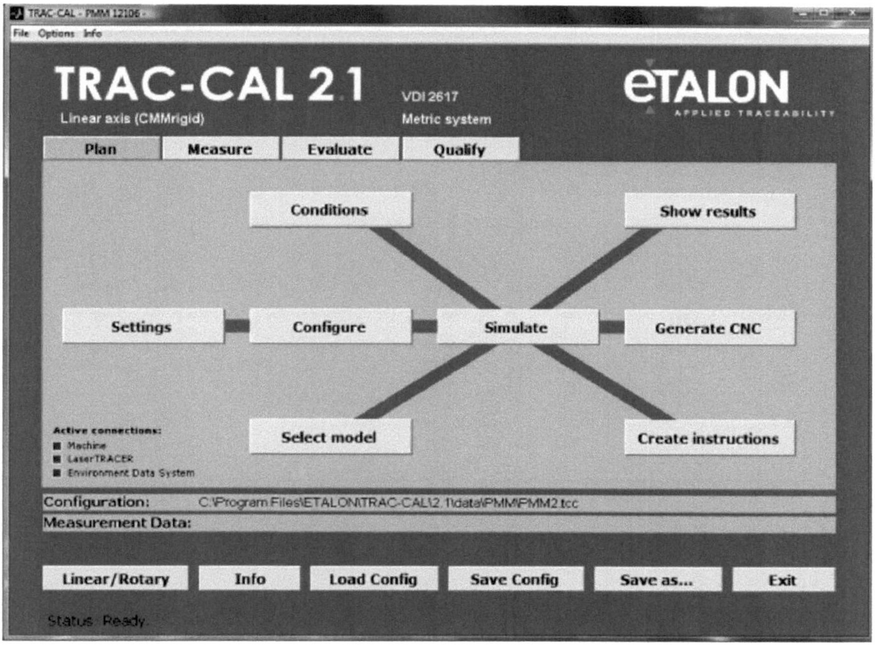

Fig. D.5 *Plan* panel window

The beginning of the matrix coordinate system is marked in the *Point pattern* window with a light green cross. The current position of the laser tracker selected in the *Setups* window is indicated by a dark green dot. Measuring paths with measuring points are marked in blue, and the active measuring path in the *Frames* window is marked in red (Fig. D.7). All data are entered in millimeters.

The *select model* function is used for selecting the kinematic model determining kinematic error components. Available models for the complete model (rigid body) allow us to determine all 21 CMM kinematic errors and errors in the cantilever construction called flexible errors as well. There are also models that are limited in comparison to the first one, for example, without flexible errors or without some rotation errors (Fig. D.8).

With the *Simulate* function it is possible to simulate the course of measurement and to determine uncertainties of geometric errors calculated with the Monte Carlo simulation method. The number of Monte Carlo simulations is put into the *Input Dialog* window which appears automatically.

The illustration *Show results* is automatically activated by the function *Simulate.*

The *measure* panel is used for performing measurements. With the *settings* function of the *measure* panel the connection of the Trac-Cal program with the measuring machine and laser tracker (Fig. D.9) is created. The interface of machine and laser tracker is chosen and the connection is confirmed by clicking the *connect* button (Fig. D.10).

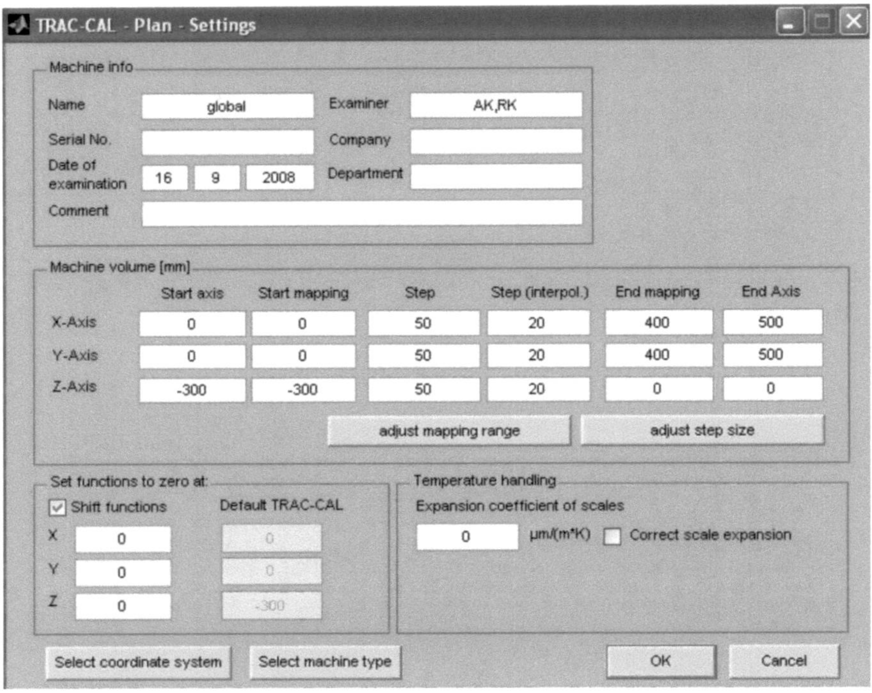

Fig. D.6 *Settings* window

Measure—Measurement

After connecting to the measuring machine and laser tracker, measurements can be started. But before that, the measurement path has to be zeroed with the *set to zero* button. Next the measurement can be started by pressing the *start* button. Errors of measured distances and coordinates are shown in graphs (Fig. D.11).

In the *evaluate* panel the calculations are performed. Here measurement data are read from the machine and laser tracker.

Read LT-Data: read load the measurement data from the laser tracker. This function allows us to read the distances measured by the laser tracker. Data from the laser tracker should be in ASCII format. The data file has to contain position, number of the measurement point, and measured distance, separated by spaces.

Posl 1 1992.686 0 0 0
Posl 2 2036.702 0 0 0
Posl 3 2080.997 0 0 0
Posl 4 2125.551 0 0 0
Posl 5 2170.349 0 0 0
Posl 6 2215.376 0 0 0

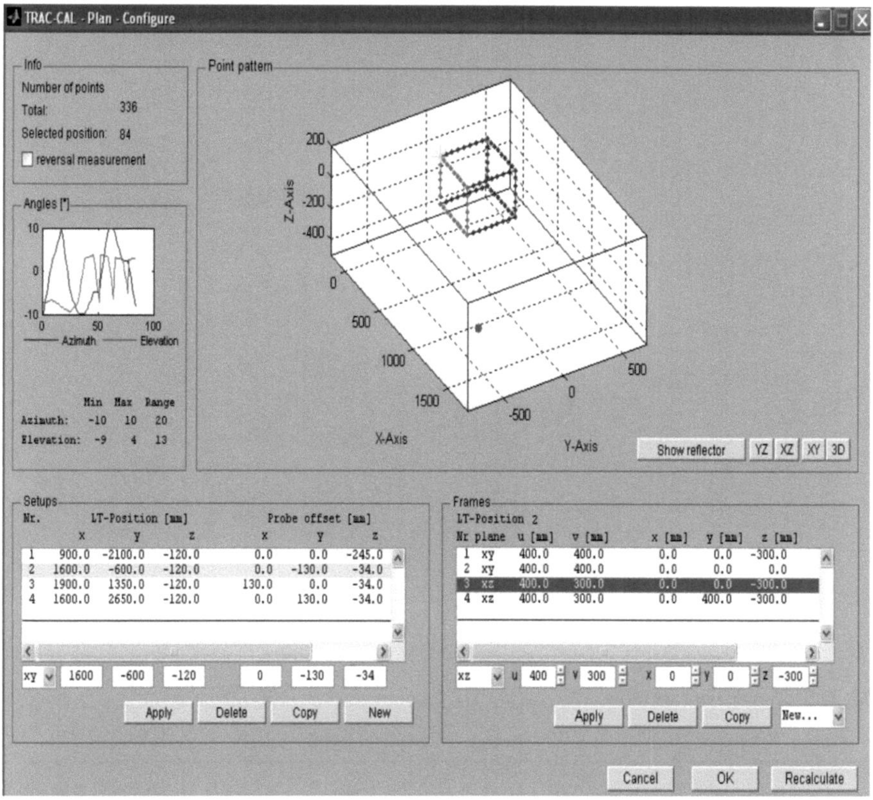

Fig. D.7 *Configure* window

Pos1 7 2260.616 0 0 0
Pos1 8 2306.058 0 0 0
Pos1 9 2351.693 0 0 0
…
Pos2 1 2072.941 0 0 0
Pos2 2 2076.925 0 0 0
Pos2 3 2082.137 0 0 0
Pos2 4 2088.565 0 0 0
Pos2 5 2096.199 0 0 0
Pos2 6 2105.027 0 0 0
Pos2 7 2115.033 0 0 0
Pos2 8 2126.201 0 0 0
…

Read CMM-Data: read load the measurement data from the machine. This function allows us to read values of measurement point coordinates measured by

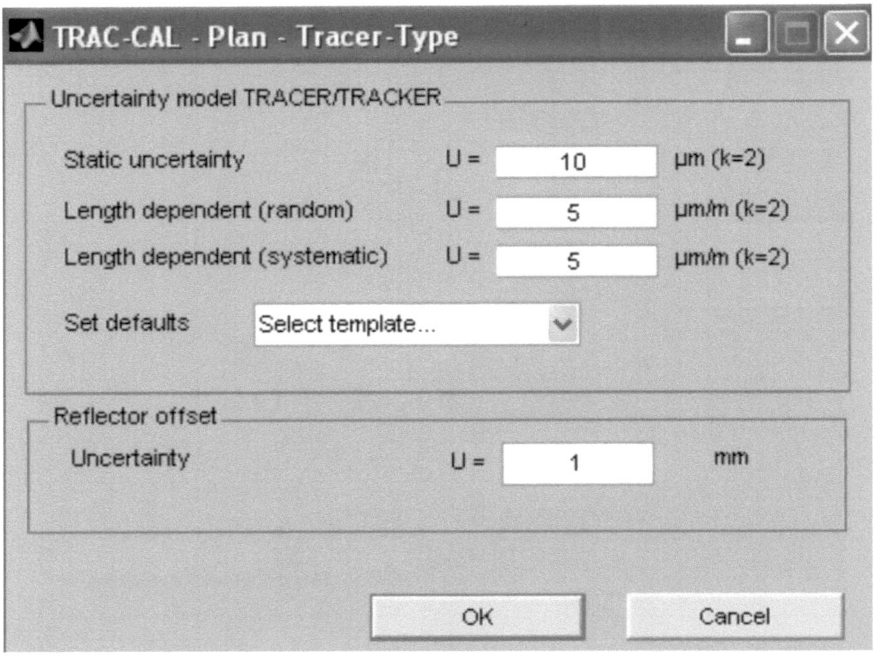

Fig. D.8 Tracer type window

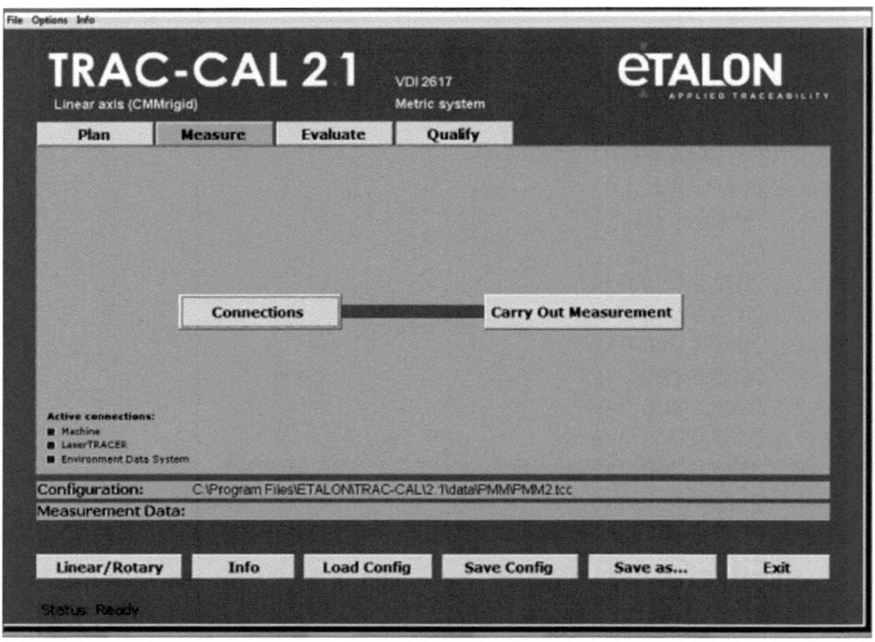

Fig. D.9 *Measure* panel window

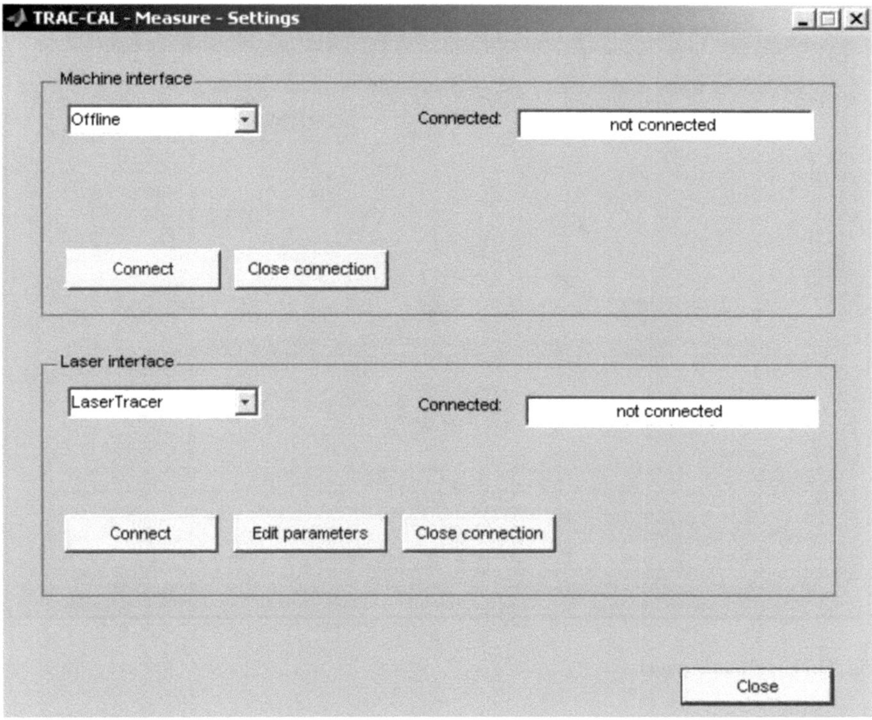

Fig. D.10 *Settings* window

the measuring machine. Data from the measuring machine should be in ASCII format. The data file has to contain position, number of the measurement point, and *X*, *Y*, *Z* coordinate values separated by spaces (Fig. D.12).

pos1 1 −0.0005 0.0003 −299.9993
pos1 2 0.0011 49.9998 −299.9994
pos1 3 0.0003 99.9995 −299.9999
pos1 4 −0.0006 149.9996 −299.9993
pos1 5 0.0003 199.9995 −300.0001
pos1 6 −0.0006 250.0005 −299.9996
pos1 7 0.0005 300.0001 −300.0002
pos1 8 −0.0002 350.0001 −299.9995
pos1 9 0.001 399.9999 −299.9996
…
pos2 1 −0.0013 -0.0005 −300.0006
pos2 2 −0.0009 49.9999 −300.0002
pos2 3 0.0008 100 −300.0003
pos2 4 0.0009 149.9995 −300.0002

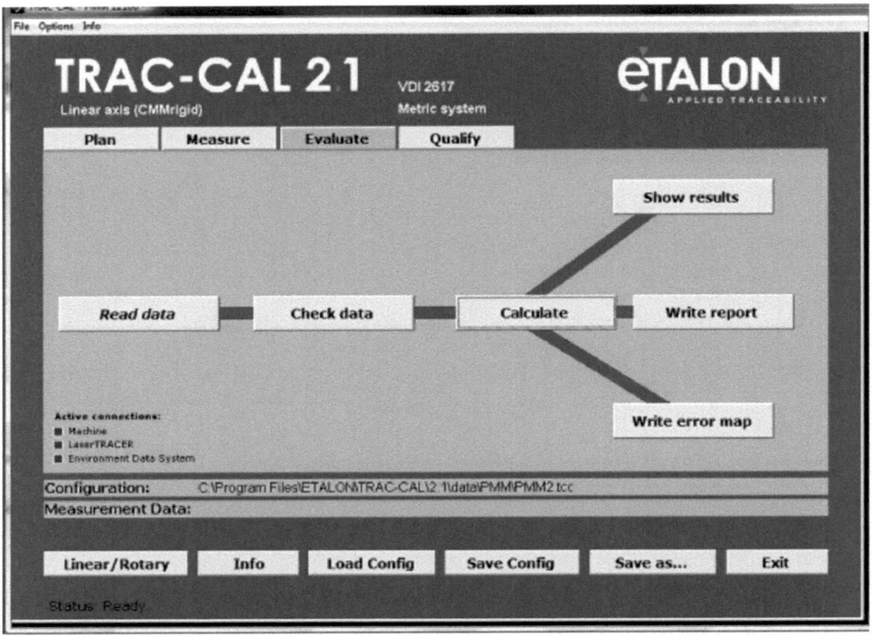

Fig. D.11 *Evaluate* panel

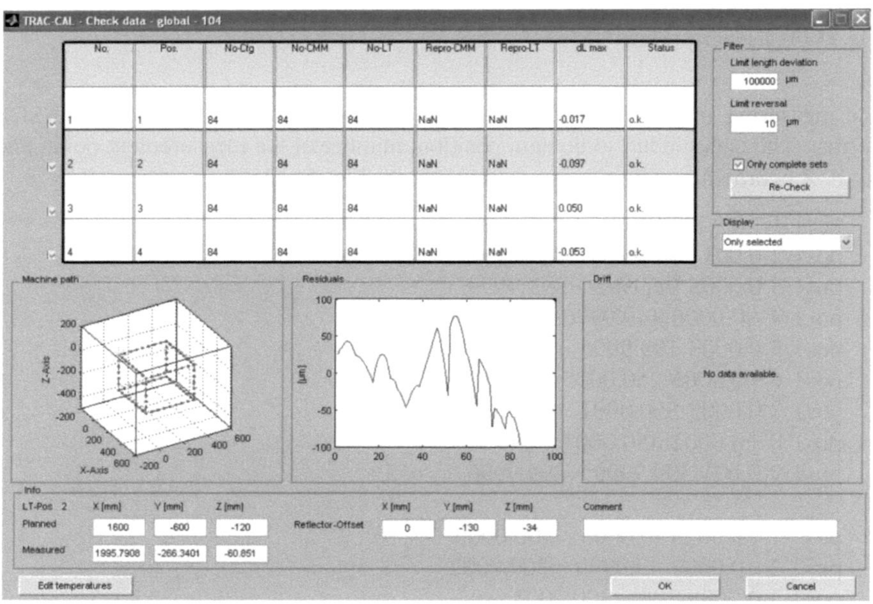

Fig. D.12 *Check data* panel window

pos2 5 −0.0003 199.9992 −299.9996
pos2 6 0.0009 250.0005 −300.0005
pos2 7 0.001 300.0005 −300.0004
pos2 *8* 0.001 349.9993 −300.0004
…

Below the table there are two separate windows. The first from the left is *Machine path* showing planned measurement points (marked with blue) and measured points (marked with red) for the currently selected position. If planned and measured measurement points are not the same, then the status of that measurement will be marked as *not o.k.* The *Residuals* window shows the course of value dL for the currently selected position.

The *calculate* function allows us to perform calculations and determine appropriate uncertainties with the Monte Carlo simulation method. For graphical interpretation of results the *show results* window is automatically activated. The *write error map* function allows us to receive all error components in the format PTB *universal* EXCHANGE (*. Exc).

The *write report* function allows us to obtain a detailed report in *doc. format. (MSWord) (Figs. D.13, D.14, D.15 and D.16):

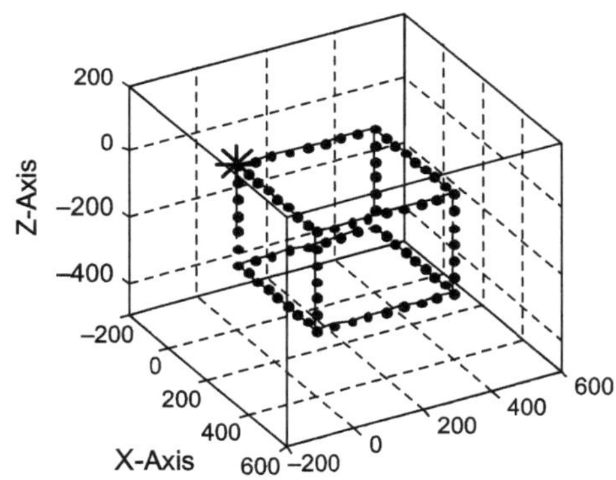

Fig. D.13 Measurement paths in during laser tracker measurement in position 1

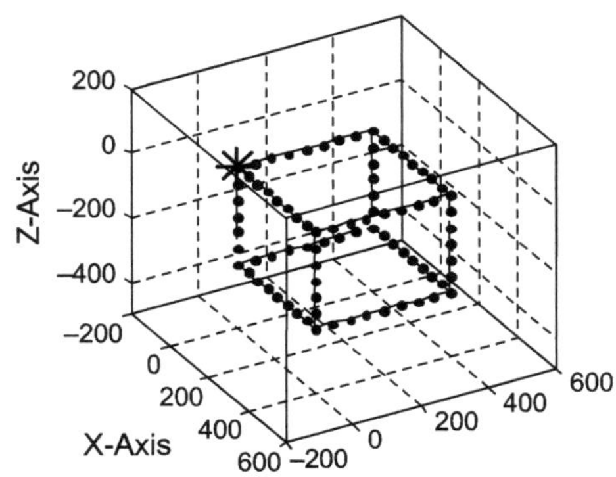

Fig. D.14 Measurement paths during laser tracker measurement in position 2

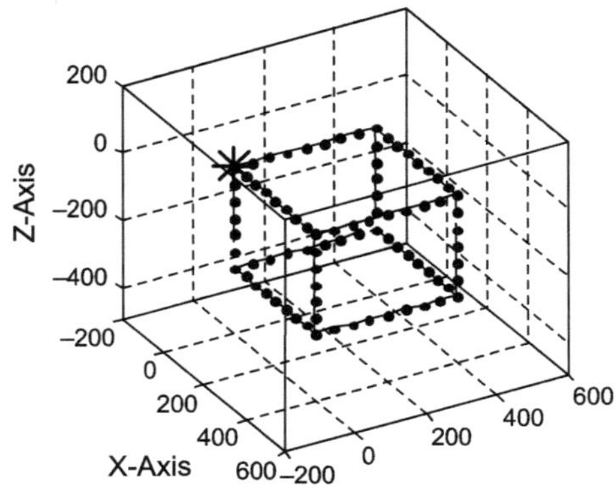

Fig. D.15 Measurement paths during laser tracker measurement in position 3

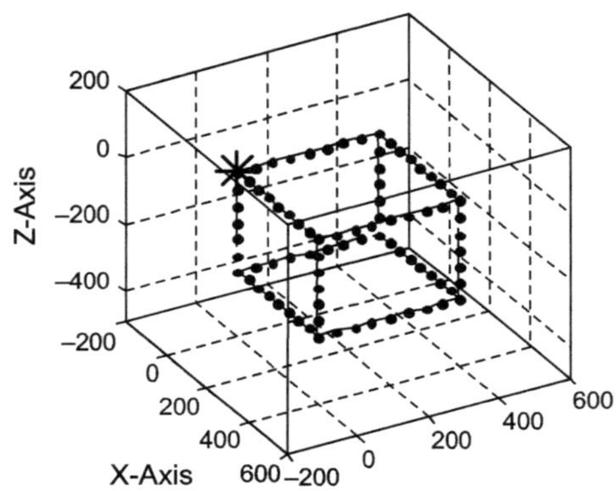

Fig. D.16 Measurement paths during laser tracker measurement in position 4

Position deviations (Figs. D.17, D.18, D.19, D.20, D.21, D.22, D.23, D.24, D.25, D.26, D.27, D.28, D.29, D.30, D.31, D.32, D.33 and D.34, Tables D.1 and D.2):

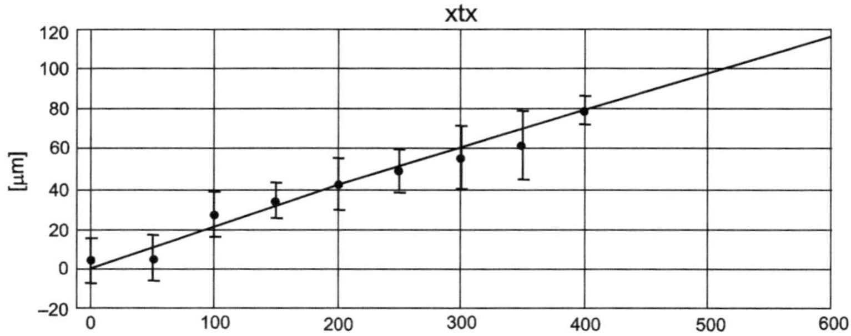

Fig. D.17 *xtx* error graph

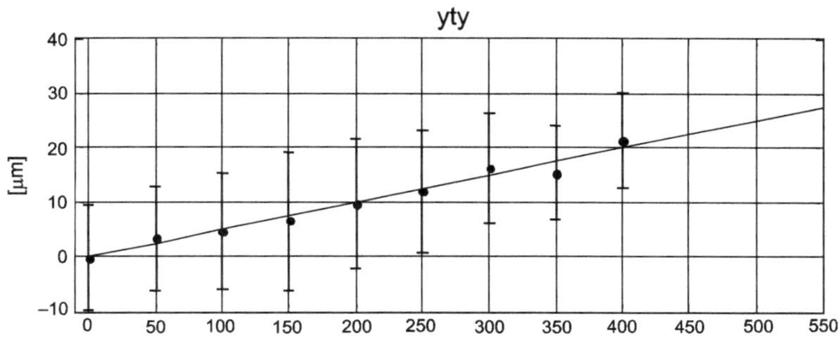

Fig. D.18 *yty* error graph

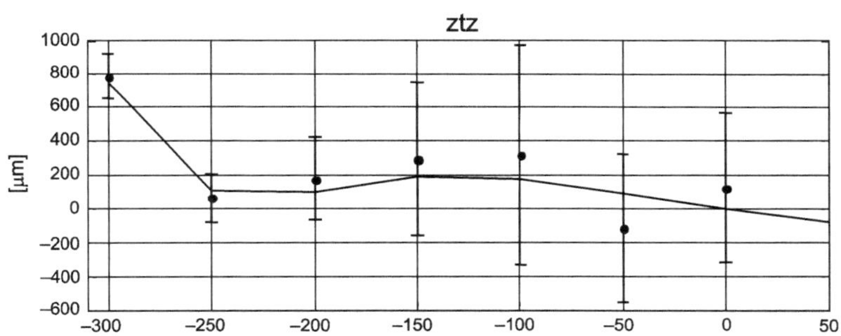

Fig. D.19 *ztz* error graph

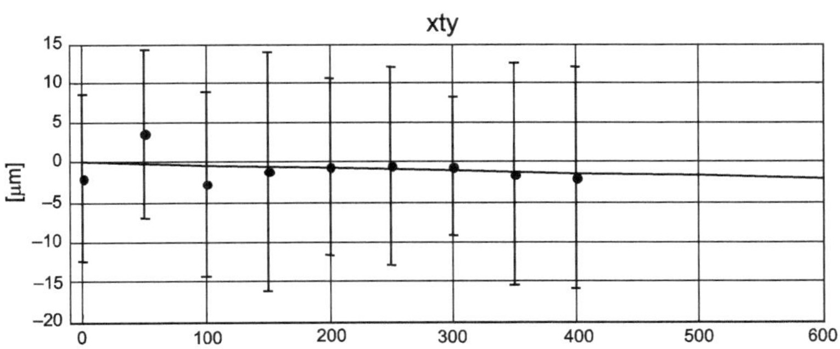

Fig. D.20 *xty* error graph

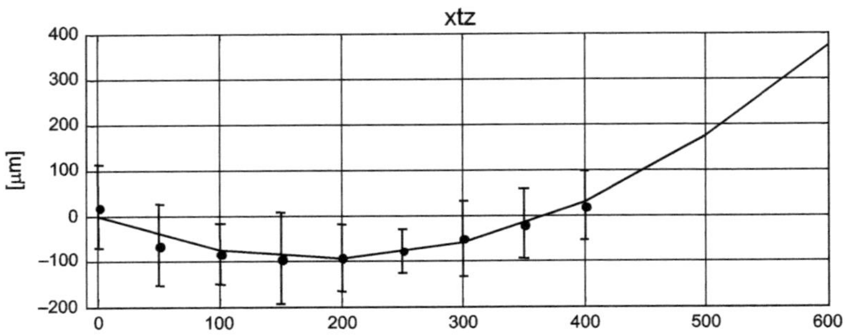

Fig. D.21 *xtz* error graph

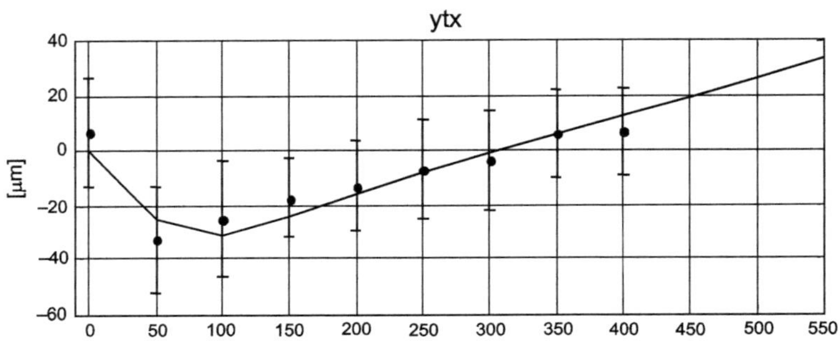

Fig. D.22 *ytx* error graph

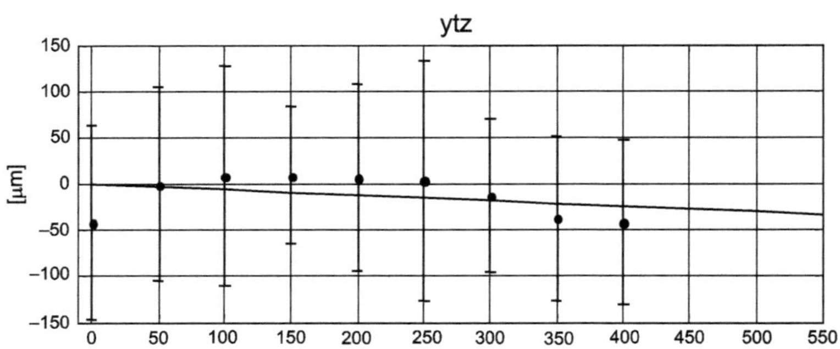

Fig. D.23 *ytz* error graph

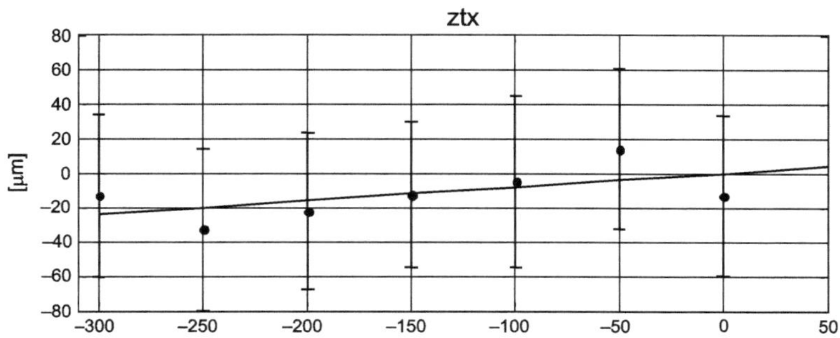

Fig. D.24 *ztx* straightness error graph

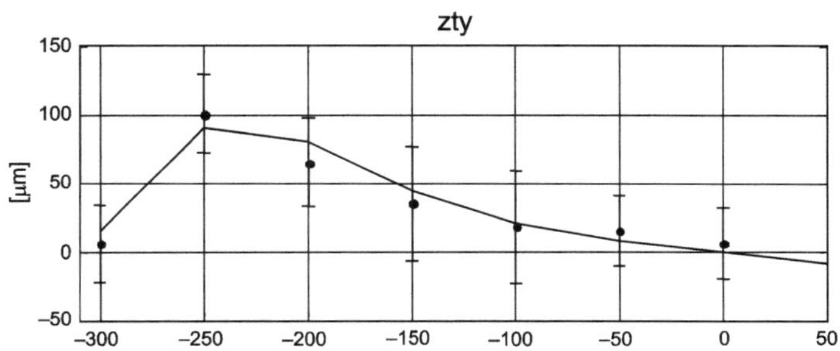

Fig. D.25 *zty* error graph

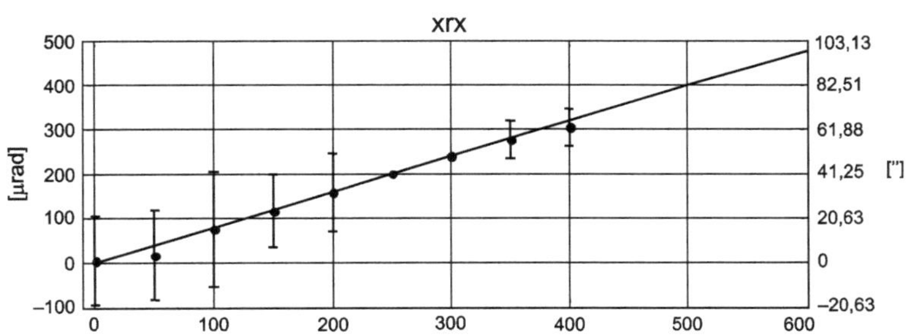

Fig. D.26 *xrx* error graph

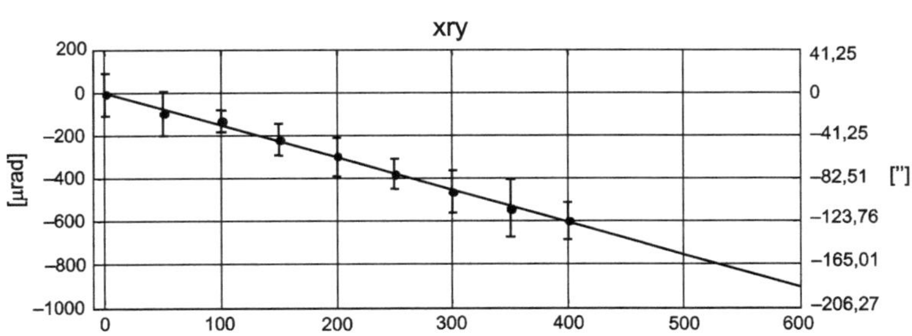

Fig. D.27 *xry* error graph

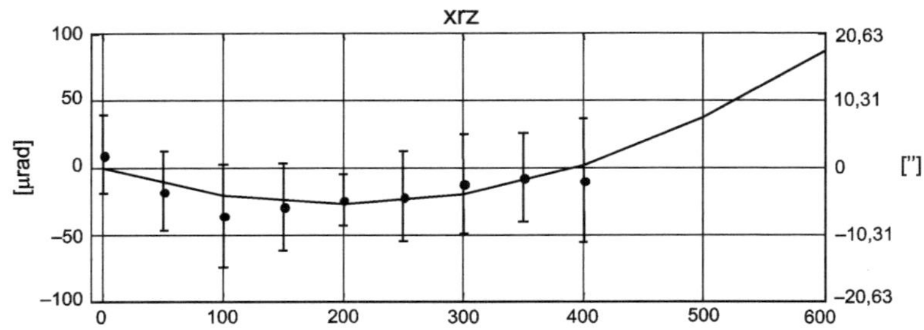

Fig. D.28 *xrz* error graph

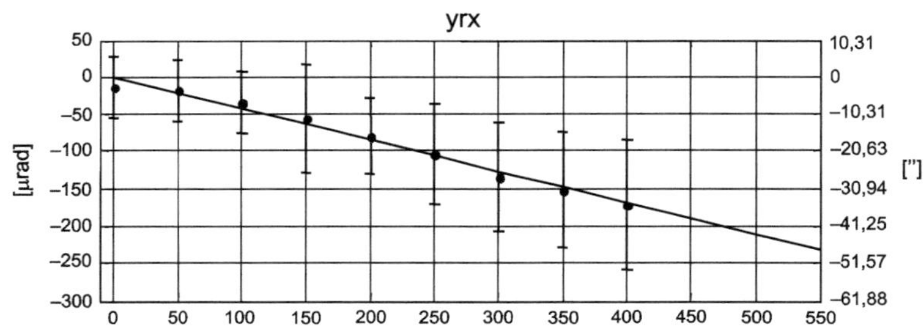

Fig. D.29 *yrx* error graph

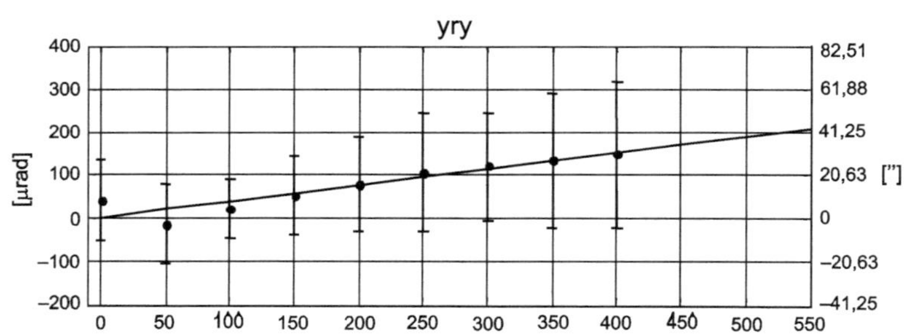

Fig. D.30 *yry* error graph

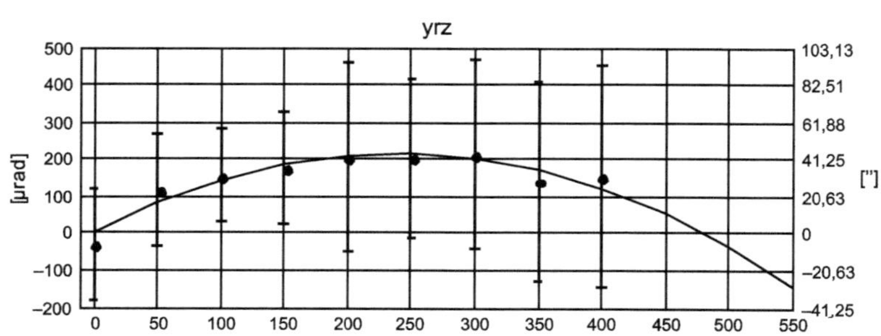

Fig. D.31 *yrz* error graph

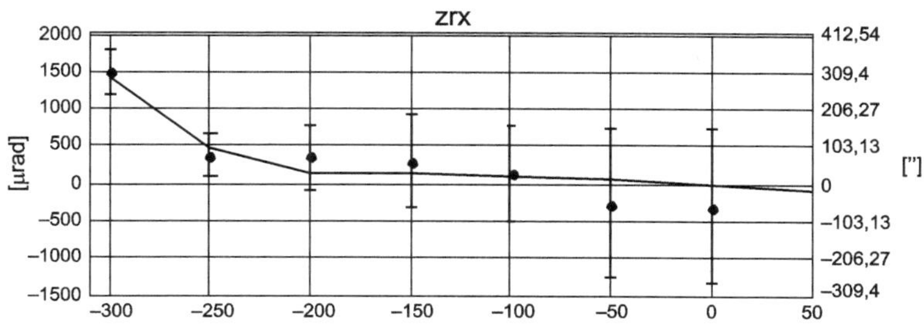

Fig. D.32 *zrx* error graph

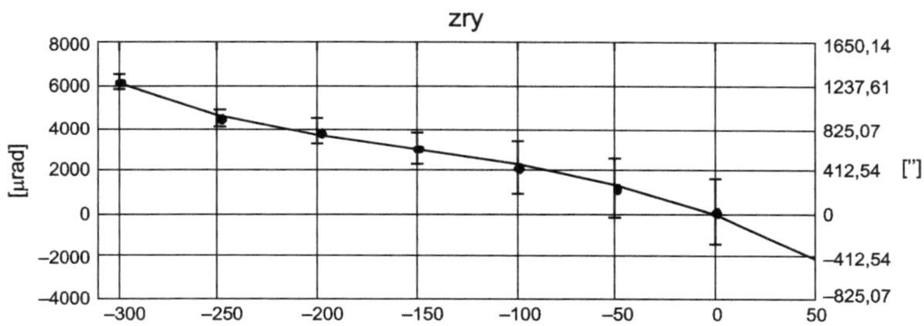

Fig. D.33 *zry* error graph

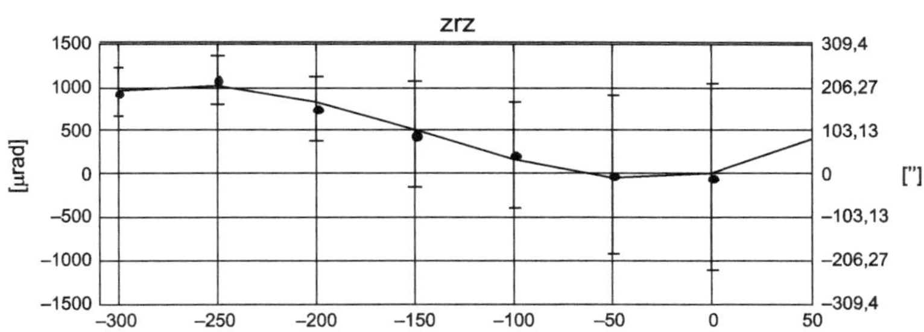

Fig. D.34 *zrz* error graph

Table D.1 Uncertainties determined with simulation method with the use of Monte Carlo method

Name	U (max)	Unit
xtx	17.8	μm
xty	15.1	μm
xtz	100.0	μm
ytx	21.3	μm
yty	12.6	μm
ytz	129.3	μm
ztx	49.8	μm
zty	42.2	μm
ztz	650.9	μm
xrx	129.3	μrad
xry	137.0	μrad
xrz	46.1	μrad
yrx	87.4	μrad
yry	170.8	μrad
yrz	299.8	μrad
zrx	1037.2	μrad
zry	1566.3	μrad
zrz	1078.5	μrad
xwy	90.5	μrad
xwz	775.2	μrad
ywz	698.4	μrad

Table D.2 Correction matrix containing geometric errors of DEA's global image machine: *X*-axis (**a**), *Y*-axis (**b**), *Z*-axis (**c**), and errors of mutual axes perpendicularity (**d**)

(a)

Step	*xrx* μrad	*xry* μrad	*xrz* μrad	*xtx* (μm)	*xty* μm	*ztz* μm
0.0	0.000	0.000	0.000	0.000	0.000	0.000
100.0	79.6	−151.1	−20.7	21.683	−0.334	−74.359
200.0	159.2	−302.3	−27.3	41.998	−0.668	−94.129
300.0	238.9	−453.5	−19.9	60.970	−1.002	−59.309
400.0	318.5	−604.6	1.5	79.515	−1.336	30.101
500.0	398.2	−755.8	37.1	98.043	−1.670	174.102

(b)

Step	*yry* μrad	*yrx* μrad	*yrz* μrad	*yty* μm	*ytx* μm	*zyt* μm
0.0	0.000	0.000	0.000	0.000	0.000	0.000
50.0	18.2	−21.1	80.6	2.506	−25.029	−3.074
100.0	37.4	−42.2	142.6	5.012	−30.985	−6.148
150.0	56.6	−63.3	185.9	7.518	−24.305	−9.222

(continued)

Table D.2 (continued)

(b)

Step	*yry* μrad	*yrx* μrad	*yrz* μrad	*yty* μm	*ytx* μm	*zyt* μm
200.0	75.8	−84.4	210.5	10.024	−15.733	−12.296
250.0	94.0	−105.5	216.4	12.530	−7.838	−15.370
300.0	112.2	−126.6	203.5	15.036	−0.603	−18.445
350.0	131.4	−147.7	172.0	17.542	6.186	−21.519
400.0	150.6	−168.8	121.8	20.048	12.747	−24.593
450.0	169.8	−189.9	52.8	22.554	19.307	−27.667
500.0	188.0	−211.0	−34.8	25.060	26.102	−30.741

(c)

Step	*zrz* μrad	*zrx* μrad	*zry* μrad	*ztz* μm	*ztx* μm	*zty* μm
−300.0	961.2	1414.9	6153.3	747.286	−23.876	14.447
−250.0	1012.4	478.3	4656.9	103.203	−19.897	91.455
−200.0	816.0	162.7	3681.1	93.346	−15.917	81.148
−150.0	494.1	135.7	2976.6	192.728	−11.938	44.536
−100.0	169.3	120.0	2294.4	175.175	−7.958	20.360
−50.0	−36.1	69.2	1385.2	87.069	−3.979	7.922
0.0	0.000	0.000	0.000	0.000	0.000	0.000

xwy μrad	*xwz* μrad	*ywz* μrad
119.768	−3800.5	−1027.4

After performing calculations by the Trac-Cal program the CAA matrix elements for medium accuracy machine were obtained.

MIX
Papier aus verantwortungsvollen Quellen
Paper from responsible sources
FSC® C105338

If you have any concerns about our products,
you can contact us on
ProductSafety@springernature.com

In case Publisher is established outside the EU,
the EU authorized representative is:
Springer Nature Customer Service Center GmbH
Europaplatz 3, 69115 Heidelberg, Germany

Printed by Libri Plureos GmbH
in Hamburg, Germany